To,

Dr John Dooley

with great admiration
and warm personal regards

M. S. Swaminathan
10 March. 2009

Life and Work of

M.S. SWAMINATHAN

Toward a Hunger-Free World

Intercultural Studies Forum, Inc. (dba: Intercultural Forum) was founded as a private, non-profit, international organization of scholars interested in intercultural studies, research and publication. Takshila Research University, a Division of Intercultural Forum, is the first step toward reviving the world's oldest research university at present-day Taxila, near Islamabad, Pakistan.

Partial List of Sponsored Publications
General Editor: Anwar Dil

1. INTERCULTURAL STUDIES SERIES (1961-)
2. ON THIS EARTH TOGETHER SERIES (1994-)

Toward Eradicating Hunger and Poverty: Life and Work of Per Pinstrup-Andersen,
ed. by Anwar Dil, 616 pages, 99 photographs, hardcover, 2003.
Published by Intercultural Forum, San Diego, California.

Science, Education, and Development: Life and Work of M. Raziuddin Siddiqi,
ed. by Anwar Dil, 560 pages, 88 photographs, hardcover, 2002.
Published by Intercultural Forum, San Diego, California.

Rice to Feed the World: Life and Work of H. M. Beachell,
ed. by Anwar Dil, 576 pages, 124 photographs, hardcover, 2001.
Published by Intercultural Forum, San Diego, California.

Hunger, Poverty, and Development: Life and Work of Sartaj Aziz,
ed. by Anwar Dil, Foreword by Norman Borlaug, Nobel Peace Laureate,
592 pages, 65 photographs, hardback, 2000.
Published by Intercultural Forum, San Diego/Ferozsons, Lahore.

The Colors of Silence: Punjabi Poems of Munir Niazi,
Rendered into English Verse by Anwar Dil,
136 pages, 64 photographs and illustrations, hardcover, 2000.
Published by Intercultural Forum, San Diego/Ferozsons, Lahore.

Bengali Language Movement to Bangladesh,
by Anwar Dil and Afia Dil, 744 pages, hardcover, 2000.
Published by Intercultural Forum, San Diego/Ferozsons, Lahore.

Norman Borlaug on World Hunger, ed. by Anwar Dil,
495 pages, 47 photographs and illustrations, hardcover, 1997.
Published by Intercultural Forum, San Diego/Bookservice, Islamabad.

On This Earth Together: Ahmed S. Bokhari at UN, 1951-1958,
ed. by Anwar Dil, Foreword by Ralph J. Bunche, Nobel Peace Laureate,
352 pages, 54 photographs and illustrations, hardcover, 1994.
Published by Intercultural Forum, San Diego/Bookservice, Islamabad.

Anwar Dil's Calligraphic Art: Series One: Six Art Prints.
San Diego: Art Vision International/Intercultural Forum, San Diego
[Numbered Limited Edition]

Humans in Universe, by Buckminster Fuller and Anwar Dil, 235 pages, 39 photographs,
hardcover, 1983. Published by Mouton Publishers, New York.

Aspects of Chinese Sociolinguistics: Essays by Yuen Ren Chao,
Selected with Introduction by Anwar Dil, xvi, 416 pages, hardcover, 1976.
Published by Stanford University Press.

Toward Developing Pakistan, ed. by Anwar Dil, Foreword by Linus Pauling,
Nobel Chemistry and Peace Laureate, 272 pages, paperback, 1970.
Published by Bookservice, Abbottabad, Pakistan.

Shahidullah Presentation Volume: Pakistani Linguistics Series VII,
ed. by Anwar Dil, 248 pages, hardcover (and paperback), 1966.
Published by Linguistic Research Group of Pakistan, Lahore.

Perspectives on Pakistan, ed. by Anwar Dil, 230 pages, hardcover, 1965.
Published by Bookservice, Abbottabad, Pakistan.

Pakistani Linguistics (1962), ed. by Anwar Dil, 210 pages, hardcover, 1963.
Published by Linguistic Research Group of Pakistan, Lahore.

Life and Work of

M.S. SWAMINATHAN

Toward a Hunger-Free World

Compiled, Edited, Introduced by

Anwar Dil

Intercultural Forum
Takshila Research University

EastWest Books (Madras) Pvt. Ltd.

• Chennai • Bangalore • Hyderabad • New Delhi

Cataloguing in Publication Data
Dil Anwar, Editor.
1928–
Toward a Hunger-Free World:
Life and Work of M.S. Swaminathan
640 Pages. Includes 81 photographs.
1. Anwar Dil, Editor, 1928- . 2. Food. 3. Agriculture. 4. Hunger. 5. Poverty.
6. India. 7.International Cooperation and Development.8. Biography.
I. M.S. Swaminathan 1925–. II. Title III. Series

ISBN: 81-88661-33-3

First Edition: October 14, 2004
United Nations World Food Day, UN/FAO International Year of Rice,
The World Food Prize 2004, Des Moines, Iowa
Intercultural Forum
Dar-ul Afia, 13756 Via Tres Vistas,
San Diego, California 92129-2755 USA
Messages and Fax: (858) 484-2228
E-Mail address: iforum@aol.com
Website: http://www.i-forum.org

Special Indian Edition: 2005 (to be sold in India only)
M.S. Swaminathan Research Foundation
Third Cross Road, Institutional Area
Taramani, Chennai 600 113, India
Telephone: +91 (44) 2254 1229/2254 1698
Fax: +91 (44) 2254 1319
Email: executivedirector@mssrf.res.in; msswami@mssrf.res.in
Website: http://www.mssrf.org

Printed at
Sri Venkatesa Printing House, Chennai 600 026.
Tel: 24838543 • e.mail : saiprints@gmail.com

Published by
EastWest Books (Madras) Pvt. Ltd.
571, Poonamalle High Road, Aminjikarai, Chennai - 600 029.
E-mail : ewb@touchtelindia.net

Dedicated
with admiration and affection to
the farm women and men
who are our hope for a
hunger-free world

Acknowledgments

It has indeed been a pleasure to work on this book of Dr. M.S. Swaminathan's life and work. I am thankful for the advice and cooperation received from Mrs. Mina Swaminathan, Professor P.C. Kesavan, Ms. Sridevi Venkatakrishnan and Mr. Nagappan Parasuraman of the M.S. Swaminathan Research Foundation (MSSRF), Chennai.

I am grateful to Ms. Gita Gopalkrishnan for assisting Dr. Swaminathan in checking the final draft and providing some useful suggestions for corrections.

A special word of thanks is due to Mr. Sartaj Aziz for contributing the Foreword.

A word of special thanks is due to the following colleagues and friends for their help in research and finalizing the camera-ready copy for printing: Bhavna Satya, Bailong Jiao, Jeri Jacquin, Zhijun Zhang, and Dr. Virginia Marable.

My wife, Dr. Afia Dil, as always, has been generous with her time and resources in making this project a reality.

All copyright materials, including photographs, have been used by permission and are appropriately acknowledged. I especially extend thanks to the following publishers and copyright holders: A.B. Volvo (Sweden); Poonam Ahluwalia; All India Radio, Delhi; *Bangkok Post; Asian Entrepreneur* (Singapore); Bharatiya Vidya Bhavan, Mumbai, India; *BioSpectrum Today* (Mumbai, India); Chinese Academy of Agricultural Sciences, Beijing; *CGIAR News* (CGIAR, The World Bank); Coromandel Fertilisers Limited; Crawford Fund (Belconnen, Australia); *Crop Science-Soil Science-Agronomy* News (American Society of Agronomy-Crop Science Society of America-Soil Society of America); *Current Science* (Bangalore, India); EastWest Books (Madras), Chennai, India; Andras Erdelyi; *Food Technology*; Gita Gopalkrishnan; *Harvard Crimson* (Cambridge, Massachusetts); *Humanscape* (Mumbai, India); *Geospatial Today* (Hyderabad, India); *Indian Journal of Genetics and Plant Breeding*; *Indiatogether*; Indian National Science Academy; International Rice Research Institute; Iowa State University Press (Ames, Iowa); R.D. Iyer; *Journal of the I.A.R.I. Postgraduate School*; Kluwer Academy Publications, The Netherlands; Konark Publishers, Delhi; Macmillan India; *Management Review* (Bangalore, India); M.S. Swaminathan Research Foundation; *Nature and Resources* (UNESCO, Paris); *New Agriculturist* (UK); New York University Press; Pakistan Agricultural Research Council, Islamabad; *Plant Genetic Resources Newsletter* (Rome); S. Ramanujam; *Science Reporter; Tertia Kiado, Budapest, Hungary;* The Ford Foundation; *The Hindu* (New Delhi); The Royal Society, London; The World Food Prize Foundation; *Time (Asia) Magazine; UNESCO Press On-line*, Paris; Venus Printers and Publishers, New Delhi; Youth Employment Summit, Washington DC; Zed Books, London.

– *Anwar Dil*

Contents

Dedication 5
Acknowledgments 6
List of Illustrations 13
M.S. Swaminathan: A Biographical Profile 15
Foreword
Sartaj Aziz
Vice-Chancellor, Beaconhouse National University, Lahore;
Formerly Pakistan's Minister of Agriculture, Finance,
And Foreign Affairs, Islamabad
Life and Work of M.S. Swaminathan: An Introductory Essay
Dr. Anwar Dil
Professor Emeritus, Alliant International University, San Diego;
Distinguished Visiting Professor, National University of
Sciences and Technology, Islamabad 29

PART I. SELECTED WRITINGS BY M.S. SWAMINATHAN 65
A. Toward a Hunger-Free India 65
1. The Origin of Early European Potato: Evidence from Indian Varieties (1958) 67
2. The Impact of Dwarfing Genes on Wheat Production (1964) 72
3. Artificial Transmutation of the Gene (1965) 79
4. J.B.S. Haldane and Indian Agriculture (1965) 86
5. Agricultural Transformation and Opportunities for a Learning Revolution (1970) 89
6. The 1971 Ramon Magsaysay Award for Community Leadership (1971) 99
7. Malthus and Mendel (1973) 103
8. Agriculture on Spaceship Earth (1973) 130
9. Our Agriculture Future (1973)
9.1 Our Agricultural Balance Sheet: Assets and Liabilities 148
9.2 The Emerging Agricultural Technology 163
9.3 Social and Educational Requirements of New Technology 178
10. Science and Integrated Rural Development (1976) 192
11. Agricultural Research in an Intellectual Property Rights Environment (1995) 210
12. Benjamin Peary Pal (1996) 218
13. Science and Sustainable Development (1997) 226

14. Indira Gandhi Prize for Peace, Disarmament and Development (2000) 234
15. Food Security and Sustainable Development (2000) 239
16. Towards a Sustainable Human Future (2002) 51
17. It's Time to Have a National Biotech Policy (2003) 262
18. Towards Hunger-Free India: Count Down from 2007 (2003) 265

B. Toward a Hunger-Free World 271

19. The Green Revolution Can Reach the Small Farmer (1975) 273
20. Science and Agricultural Progress in the Third World (1983) 275
21. Eulogy for Shouichi Yoshida (1984) 280
22. Future Horizons and Advances in Biotechnology (1984) 283
23. Commencement Address at the University of the Philippines (1984) 291
24. Rice Research: Learning from China (1987) 295
25. Our Common Agricultural Future: The First World Food Prize 1987: Award and Acceptance Speeches (1987) 299
26. Population, Environment and Food Security (1994) 311
27. Agriculture, Food Security and Employment: Changing Times, Uncommon Opportunities (1995) 320
28. Toward a Food-Secure World (1996) 338
29. Ethics and Equity in the Collection and Use of Plant Genetic Resources: Some Issues and Approaches (1997) 356
30. Science in Response to Basic Human Needs (1999) 369
31. Government-Industry-Civil Socie ty Partnerships in Integrated Gene Management (1999) 390
32. Acceptance of the Franklin Delano Roosevelt Four Freedoms Medal (2000) 403
33. A Century of Mendelian Breeding: Impact on Wheat 406
34. The Past, Present and Future Contributions of Farmers to the Conservation and Development of Genetic Diversity (2000) 422
35. Higher Education for Social and Economic Progress (2001) 434
36. Towards Freedom from Hunger: A Global Food for Sustainable Development Initiative (2002) 437
37. Acceptance Speech as President of the Pugwash Conferences on Science and World Affairs and Excerpts from "Science and Achieving the Goal of a Hunger-Free World" (2002) 450
38. Toward a Hunger-Free World: The Final Milestone (2003) 457

PART II. LIFE AND WORK OF M.S. SWAMINATHAN: CONTEMPORARY PERSPECTIVES 487

Section 1. Selected Interviews

1.1 "How to Eradicate Hunger" by Parshuram Ray, *Humanscape*, September 2000 489
1.2 "M.S. Swaminathan: A Better Harvest" by Tushaar Shah, *Management Review*, December 2000 494

1.3 "Global Agriculture - Facing the Challenges."
New Agriculturist (UK), April 2001 502
1.4 "The WTO Pact On Agriculture Will Kill Our Farmers"
by Shobha Warrier, January 25, 2001 504
1.5 "The Know-how of GIS Should be Converted to Do-how"
Geospatial Today (Hyderabad, India), February 2003 508

Section 2. Excerpts from Archives
2.1 Norman E. Borlaug, CIMMYT, Mexico, 1970 511
2.2 Felicitations on the award of the first World Food Prize, 1987 511
Ronald Reagan, President of USA 511
Dr. Joshua Lederberg, Nobel Science Laureate 511
Dr. Frank Press, President,
National Academy of Sciences, Washington DC 512
Jimmy Carter, Former President of USA 512
Javier Perez de Cueller, Secretary General, United Nations 512
2.3 Felicitations to Dr. M. S. Swaminathan on his 70th birthday, 1995 513
R. Venkataraman, Former President of India 513
K.R. Narayanan, Vice-President of India 513
P. V. Narasimha Rao, Former Prime Minister of India 513
Abus Salam, President, International Centre
for Theoretical Physics (ICTP), Trieste, Italy 513
Norman E. Borlaug, Nobel Peace Laureate 513
Jacques Diouf, Director General, FAO 514
James Gustave Speth, Administrator, UNDP 514
Elizabeth Dowdeswell, Executive Director, UNEP 515
Sir Michael Atiyah, President, The Royal Society 515
Prof. Wang Lianzheng, Vice-Chairman,
Chinese Association of Science and Technology 515
N.S. Randhawa, Vice President, National Academy
of Agricultural Sciences, India 515
Dr. R.S. Paroda, Director General, ICAR, India 516
Dr. Manju Sharma, President, The National Academy
of Sciences, India 516
2.4 Resolution by The Board of Trustees,
The Ford Foundation, 1997 518
2.5 Norman Myers, Professor, Oxford University, 1999 518
2.6 Sartaj Aziz, Pakistan Minister of Foreign Affairs,
Islamabad, 1999 524
2.7 Wakako Hironaka, Former State Minister and
Director General, Environment Agency, Tokyo, 2000 524
2.8 Ismail Serageldin, Chairman, CGIAR,
Vice-President, The World Bank, 2000 524
2.9 Her Majesty Queen Noor of Jordan, Amman, Jordan, April 2000 525
2.10 Martin Lees, Rector, University for Peace,
San Jose, Costa Rica, 2002 525

Section 3. Excerpts from Press Coverage 529
3.1 "Tribute to a Hero of the Green Revolution" by Normita Thongtham, *Bangkok Post*, April 28, 1985 529
3.2 "AWID Honors Swaminathan." *News from CGIAR*, June 1985 531
3.3 "Swaminathan: A Rare Breed of Man" by Normita Thongtham, *Bangkok Post*, July 9, 1987 532
3.4 "Indian Scientist to Receive General Foods World Food Prize." *Food Technology*, September 1987 533
3.5 "Hunger Fighter Never Ceases: Inspired by Gandhi, Indian Works to Halt Famine" by Risuko Nakamura, *Japan Times*, Tokyo, 1992 534
3.6 "Pioneer of the Green Revolution Fights Poverty with the Internet." *UNESCO Press On-line,* June 29, 1999 535
3.7 "M.S. Swaminathan" by Anthony Spaeth, *Time* Magazine, August 23, 1999 536
3.8 "*Time* (Asia) Recognizes Swaminathan Among Most Influential Asians." *Crop Science-Soil Science-Agronomy News*, October 1999 538
3.9 "The Volvo Environment Prize." AB Volvo, Sweden, October 26, 1999 539
3.10 "The Modern Gandhi's Message of Hope for Agriculture and Environment" by Heather Slater, Crawford Fund Seminar, Belconnen, Australia, November 24, 1999 541
3.11 "M.S. Swaminathan: Transforming India from Begging Bowl to Breadbasket" by Ajoy Sen, *Asian Entrepreneur*, January 2001. 543
3.12 "Genetically Modified Foods Fill Developing World Silos" by Andrew S. Holbrook, *Harvard Crimson*, March 12, 2001 544
3.13 "The Face of Food Security" by Gnana Bharathy, *Indiatogether*, May 5, 2001 545
3.14 "Swaminathan to Head Pugwash Movement." *The Hindu* (Delhi), March 18, 2002 546

Section 4. Excerpts from Selected Citations of Honorary Doctorate Degrees and Awards 547
4.1 "Award of the Silver Jubilee Commemoration Medal," The Indian National Science Academy, 1973 547
4.2 "The Borlaug Award," Coromandel Fertilizers, India, 1979 548
4.3 "Honorary Doctor of Science," University of Wisconsin, 1983 548
4.4 "Honorary Doctor of Science", University of the Philippines, 1984 549
4.5 "Honorary Doctor of Technology, " Asian Institute of Technology, Thailand, 1985 549
4.6 "Award for Outstanding Contributions to the Integration of Women in Development," Association for Women in Development, Washington DC, 1985 550

4.7 "The Tyler Prize," Los Angeles, 1991 551
4.8 "Award for International Cooperation on Environment and Development," China Council for International Cooperation on Environment and Development, Beijing, 1996 551
4.9 "Planet and Humanity Medal," The International Geographical Union, Vienna, 2000 552
4.10 "Honorary Doctor of Science," University of Massachusetts, Boston, 2001 552

Section 5. Selected Prefaces, Forewords, and Introductions to M.S. Swaminathan's Books 553
5.1 *Biotechnology in Agriculture: A Dialogue*, 1991
5.2 *From Stockholm to Rio de Janeiro: The Road to Sustainable Development*, 1991 553
5.3 *Information Technology: A Dialogue*, 1993 555
5.4 *Wheat Revolution: A Dialogue*, 1993 556
5.5 *Uncommon Opportunities*, 1994 560
5.6 *Ecotechnology and Rural Employment*, 1994 562
5.7 *Sustainable Agriculture: Towards Food Security*, 1996 565
5.8 *Sustainable Agriculture: Towards an Evergreen Revolution*, 1996 567
5.9 *Agrobiodiversity and Farmers' Rights*, 1996 568
5.10 *Gender Dimensions in Biodiversity Management*,1998 574
5.11 *A Century of Hope: Towards an Era of Harmony with Nature and Freedom from Hunger*, 1999 576
5.12 *Rio de Janeiro to Johannesburg,* 2002 579
5.13 *Groves to Beauty and Plenty: An Atlas of Major Flowering Trees in India,* 2003 584

Section 6. Selected Sections from Books on M.S. Swaminathan 587
6.1 "Editors' Preface" from *Science and Agriculture, M.S. Swaminathan and the Movements for Self-reliance,* ed. by S. Ramanujam et al, 1980 587
6.2 "Foreword" by B.P. Pal from *Science and Agriculture, M.S. Swaminathan and the Movements for Self-reliance,* ed. by S. Ramanujam et al, 1980 587
6.3 "Message" by K.R. Narayanan, President of India, from Scientist and Humanist: M.S. Swaminathan, by R.D. Iyer, 2002 589
6.4 "Foreword" by R. Venkataraman, Former President of India, from *Scientist and Humanist: M.S. Swaminathan,* by R.D. Iyer, 2002 589
6.5 "Swaminathan, The Institution Builder" from *Scientist and Humanist: M.S. Swaminathan*, by R. D. Iyer, 2002 591
6.6 "Swaminathan: The Man and His Philosophy of Life" from *Scientist and Humanist, M.S. Swaminathan,* by R.D. Iyer, 2002 606

6.7 "Foreword" by Maurizio Laccarino, Secretary General, World Conference on Science, from The *Man Who Harvests Sunshine: The Modern Gandhi: M.S. Swaminathan*, by András Erdélyi, 2002 609

6.8 "Foreword" by Poonam Ahluwalia from *M.S. Swaminathan: One Man's Quest for A Hunger-free World*, by Gita Gopalkrishnan, 2002 610

6.9 "Introducing Swaminathan" from *M.S. Swaminathan, One Man's Quest for a Hunger-free World*, by Gita Gopalkrishnan, 2002 611

6.10 "The Man" from *M.S. Swaminathan, One Man's Quest for a Hunger-free World*, by Gita Gopalkrishnan, 2002 613

Index 621

List of Illustrations

[Unless identified otherwise, all photographs are from the personal collection of M.S. Swaminathan, The M.S. Swaminathan Research Foundation, and Anwar Dil]

1. M.S. Swaminathan in the field Front Title Cover, 621
2. Portrait of M.S. Swaminathan Back Inner Flap, 624
3. Portrait of Anwar Dil Back Inner Flap, 624
4. MSS in his office, MSSRF, Chennai, 1995 26
5. Anwar Dil, Norman Borlaug and MSS, World Food Prize ceremonies, Des Moines, Iowa, October 18, 2000 56
6. MSS and Anwar Dil working on the book, 2003 56
7. Monkombu Krishna Sambasivan, M.S. Swaminathan's father 57
8. Parvathi Thangammal, M.S. Swaminathan's mother 57
9. The house Swaminathan was born in Kumbakonam, Tamil Nadu 57
10. Swaminathan's ancestral village, Monkombu 57
11. MSS with his uncle, 1941 58
12. Native High School, Kumbakonam 58
13. MSS, 1944 58
14. MSS, 1951 58
15. Marriage, April 11, 1955 59
16. Mr. & Mrs. Swaminathan, New Delhi, 1955 59
17. With wife, Mina, and daughters Nitya, Madhura, Soumya, 1968 59
18. With wife Mina in Tokyo to receive the Honda Award, 1992 59
19. With wife, daughters and sons-in-law, 1986 59
20. With children and grandchildren, 1992 59
21. In his laboratory, IARI, 1961 60
22. In the field, IARI, 1963 60
23. With Norman Borlaug, IARI, 1964 60
24. At International Rice Research Institute (IRRI), The Philippines, 1983 61
25. At the Chinese Academy, Beijing, 1986 61
26. At FAO, Rome, 1985 61
27. With Dr. Jacques Diouf, Director General of FAO, 1998 61
28. With Dr. Elizabeth Dowdeswell, Executive Director, UN Environmental Programme, receiving the UNEP Sasakawa Environment Prize, 1994 61
29. Logo of MSSRF 62
30. Digging of foundation at MSSRF, April 14, 1991 62
31. The First Team, MSSRF, 1991 62
32. Shrimati M.S. Subbulakshmi at inauguration of MSSRF, 1993 62
33. With C. Subramaniam and Norman Borlaug, at the opening of Borlaug Hall, MSSRF, 1996 62
34. A View of MSSRF 62
35. Courtyard of MSSRF 62
36. Honorary Degree, Oregon State University: MSS, Norman Borlaug, , John V. Byrne Robert MacVicar, Kenneth E. Clark, 1988 63
37. Honorary Degree, University of Bologna, Italy, 1992 63
38. Honorary Degree, University of Massachusetts, Boston, 2001 63
39. Shanti Swarup Bhatnagar Award, 1965 64
40. Ramon Magasaysay Award, 1971 64
41. The First World Food Prize, 1987 64

42. Padma Vibhushan Award by Shri R. Venkataraman, President of India, 1989 64
43. The Golden Heart Award by Mrs. Corazon Aquino, President of the Philippines, 1987 64
44. UNESCO Gandhi Award by Prof. Federico Mayor, Director General of UNESCO, 1999 64
45. MSS in the field, IARI, 1960 65
46. With Norman Borlaug, IARI, 1963 66
47. C. Subramaniam, Minister of Food and Agriculture (1964-67), India 66
48. S.P. Kohli, MSS, Norman Borlaug V.S. Mathur in the field, 1966 66
49. Shrimati Indira Gandhi releasing India Wheat Revolution stamp, 1968 66
50. At IARI with workers, 1964 66
51. MSS Launching the "Children for Happiness" Movement in India 102
52. Dr. B.P. Pal 218
53. With Prime Minister Indira Gandhi and farmers, 1968 238
54. MSS Portrait, 1999 271
55. With Syed Ahmad Marei, Norman Borlaug, A.H. Boerma at Rome World Food Forum, 1974 272
56. With Pope John Paul II for discussion on the Sahelian drought, 1985 272
57. With Nelson Mandela for discussion on eradicating hunger in Africa, 1998 272
58. With World Food Laureate He Kang for improving agricultural development in China, 1992 272
59. With C. Subramaniam, Pioneer of Green Revolution in India, 2001 272
60. At World Food Prize, Washington DC, 1987 310
61. Speaking at World Food Prize Convocation, 1987 310
62. With the Queen of Netherlands after receiving Franklin D. Roosevelt Four Freedoms Medal, 2000 405
63. MSS with the 2000 World Food Laureate Dr. Evangelina Villegas of Mexico, Des Moines, Iowa, October 18, 2000 449
64. MSS speaking at seminar on Gender, MSSRF, Chennai, 2000 456
65. Norman E. Borlaug, CIMMYT, Mexico 458
66. Dr. Borlaug and Mina Swaminathan at MSSRF, Chennai, 1996 458
67. Sartaj Aziz, Norman Borlaug, Anwar Dil, MSS at IFPRI Conference, 2002 458
68. MSS, 1999 487
69. The 70th Birthday celebration, 1995 488
70. Farewell to M.S. Swaminathan, Los Banos, 1987 488
71. With James L. Ferguson, A.L. Clausi, and Norman Borlaug at First World Food Prize 1987 528
72. Receiving the Albert Einstein World Science Award , 1986 528
73. MSS, 1974 583
74. Title cover of *Groves of Beauty and Plenty* (2003) 585
75. Title cover of *Wheat Revolution: A Dialogue* (1993) 586
76. Title cover of *Agrobiodiversity and Farmers's Rights* (1996) 586
77. Title cover of *A Century of Hope* (1999) 586
78. Title cover of *From Rio de Janeiro to Johannesburg* (2002) 586
79. Title cover of *Science and Agriculture* by S. Ramanujam et al (2002) 619
80. Title cover of biography by R. D. Iyer (2002) 619
81. Title cover of biography by Gita Gopalkrishnan (2002) 620

M.S. Swaminathan: A Biographical Profile

1925	Born on August 7, 1925 in Kumbakonam, Tamil Nadu, India. Father: Monkombu Krishna Sambasivan. Mother: Parvathi Thangammal. Elder brother: Krishnamurthi; younger brother: Ramdas; sister: Lakshmi.
1944	B. Sc. degree in Biology (Botany, Zoology), Travancore University, Trivandrum, India.
1947	B. Sc. degree in Agriculture, Coimbatore Agricultural College, Madras University, Madras, India.
1949	Research Associate (Genetics), The Indian Agricultural Research Institute (IARI), New Delhi, India.
1949-50	UNESCO Fellow (Genetics), Agricultural University, Wageningen, The Netherlands.
1952	Ph.D., School of Agriculture, University of Cambridge, UK.
1952-53	Post-Doctoral Research Associate (Genetics), University of Wisconsin, USA.
1954	Researcher at the Central Rice Research Institute, Cuttack, India.
1954-72	Researcher, Teacher and Administrator at the Indian Agricultural Research Institute (IARI), New Delhi, India. [Head of Botany Division, 1961-66; Director, 1966-72.]
1955	Married Mina Boothalingam. [Three daughters: Soumya (b. 1959-), Madhura (b. 1961-), Nitya (b. 1963-)].
1957	Fellow of the Indian Academy of Sciences.
1961	Shanti Swarup Bhatnagar Award for contributions to Biological Sciences. (Award ceremonies were held on January 4, 1965)
1962	Fellow of the Indian National Science Academy.
1965	Mendel Memorial Medal for contributions to Plant Genetics, Czechoslovak Academy of Sciences, The Prague.
1966	Birbal Sahni Medal for contributions to Applied Botany, Indian Botanical Society.
1967	Padma Shri Award by the President of India.
1970	Doctor of Science (honoris causa), Sardar Patel University, Vallabh Vidyanagar, India.

1971 Doctor of Science (honoris causa), Andhra Pradesh Agricultural University, Hyderabad, India.

Honorary Fellow, Swedish Seed Association, Sweden.

Ramon Magsaysay Award for Community Leadership, The Philippines.

Silver Jubilee Commemoration Medal, The Indian National Science Academy for contributions to genetics and agricultural research.

1971-77 Vice Chairman, Technical Advisory Committee, Consultative Group on International Agricultural Research (CGIAR), Washington, DC.

1972 Padma Bhushan Award by the President of India.

Succeeded his teacher Dr. B.P. Pal as Director General of Indian Council of Agricultural Research.

Doctor of Science (honoris causa), Andhra University, Waltair, India.

1972-77 Vice Chairman, Protein-Calorie Advisory Group, United Nations.

1972-79 Director General, Indian Council of Agricultural Research (ICAR); and Secretary to the Government of India, Department of Agricultural Research and Education (1972-79), and Agriculture and Cooperation (1979-80).

1973 Doctor of Science (honoris causa), Haryana Agricultural University, Hissar, India.

Fellow, Royal Society of London.

1974 Doctor of Science (honoris causa), Govind Ballabh Pant University of Agriculture and Technology, Pantnagar, India.

1975 Doctor of Science (honoris causa), Kumaon University, Nainital, India.

Doctor of Science (honoris causa), Marathwada Krishi Vidyapeeth, Parbhani, India.

1976 Doctor of Science (honoris causa), Burdwan University, Burdwan, India.

Doctor of Science (honoris causa), Jodhpur University, Jodhpur, India.

General President, Indian Science Congress, Waltair, India.

Honorary Fellow, National Academy of Sciences, India.

1976-83	President, International Federation of Agricultural Research Systems for Development (IFARD).
1977	Chairman, Committee of the Whole, United Nations Conference on Desertification, Nairobi, Kenya.
	Foreign Associate, National Academy of Sciences of the United States.
1977-82	Founder Member and later Chairman of the Board of Trustees of International Council for Research on Agro-forestry, Nairobi
1978	Barclay Medal of the Asiatic Society (for contributions to Genetics), Kolkata, India.
	Moudgil Prize for contributions to standardization
	Doctor of Science (honoris causa), Agra University, Agra, India.
	Doctor of Science (honoris causa), Kerala Agricultural University, Trichur, India.
	Foreign Member, All-Union Academy of Agricultural Sciences, USSR.
1978-90	President, International Bee Research Association (IBRA).
1979	Norman E. Borlaug Award.
	Doctor of Science (honoris causa), Sri Venkateshwara University, Tirupati, India.
1979-80	Principal Secretary to the Government of India, Ministry of Agriculture and Irrigation, New Delhi, India.
1980	Member for Agriculture and Rural Development and Deputy Chairman, Planning Commission, Government of India, New Delhi.
	Doctor of Science (honoris causa), University of Agricultural Sciences, Bangalore, India.
1980-82	Member, Agriculture, Rural Development, Science and Education, Planning Commission, Government of India, New Delhi.
1980-83	Chairman, U.N. Advisory Committee on Science and Technology for Development.
	Chairman, Science Advisory Committee to the Cabinet of India.
1981	Doctor of Science (honoris causa), Banaras Hindu University, Varanasi, India.

	Doctor of Science (honoris causa), Technical University of Berlin, Berlin, West Germany.
	Chairman, National Biotechnology Board of India.
	Meghnad Saha Medal of the Indian National Science Academy.
	Rabindranath Tagore Prize of Visva Bharati University, India.
	R.D. Misra Medal of the Indian Environmental Society, New Delhi.
	Book: *Building a National Food Security System.* New Delhi: Indian Environment Society.
1981-85	Independent Chairman, FAO Council, Rome.
1982	Doctor of Science (honoris causa), Mahatma Phule Agricultural University, Rahuri, India.
	Book: *Science and Integrated Rural Development.* New Delhi: Concept Publishing Company.
1982-86	Founder-Chairman, Society for the Promotion of Wasteland Development, India.
1982-88	Director General, International Rice Research Institute (IRRI), Los Banos, The Philippines.
1983	Doctor of Science (honoris causa), Chandrasekhara Azad Agricultural University, Kanpur, India.
	Doctor of Science (honoris causa), University of Wisconsin, Madison, Wisconsin, USA.
	Foreign Member, Royal Swedish Academy of Agriculture and Forestry.
	Founding Fellow, Third World Academy of Sciences, Italy.
	President, XV International Congress of Genetics, New Delhi, India.
	Book: *Science and the Conquest of Hunger.* New Delhi: Concept Publishing Company.
	Book (Ed. with P. K. Gupta, V. Sinha): *Cytogenetics of Crop Plants.* New Delhi: Macmillan India.
	Book (Ed. with E.V. Von Weizacker, A. Lemma): *New Frontiers in Technology Application*. Dublin: Tycooly International Publishing Company.
1983-85	Member, Scientific and Technical Advisory Committee, Tropical Diseases Research, World Health Organization, Rome.

1984 Bennett Commonwealth Prize (for contributions to household nutrition security), The Royal Society of Arts.

Doctor of Science (honoris causa), University of Delhi, Delhi.

Doctor of Science (honoris causa), University of the Philippines, Quezon City, Philippines.

Foreign Honorary Member, National Academy of Arts and Sciences, Massachusetts, USA.

1984-86 Chairman, Advisory Panel on Environment, Forestry and Food Security, World Commission on Environment and Development, Geneva, Switzerland.

1984-90 President, International Union for the Conservation of Nature and Natural Resources, Gland, Switzerland.

1985 Award (for serving the cause of Women in Development), Association of Women in Development, Washington, DC, USA.

Bicentenary Medal, University of Georgia, USA.

Fellow, Royal Society of Arts, London, UK.

Foreign Fellow, National Academy of Sciences, Italy.

Doctor of Technology (honoris causa), Asian Institute of Technology, Bangkok, Thailand.

Book (Ed. with S. K. Sinha): *Global Aspects of Food Production.* Dublin: Tycooly International Publishing Company.

1985-94 Trustee, World Resources Institute.

1985-87 Honorary Vice-President, World Wild Life Fund International.

1986 Albert Einstein World Science Award, World Cultural Council, London.

Doctor of Science (honoris causa), University of Mangalore, Mangalore, India.

Krishi Ratna Award (for serving the farming community), Bharat Krishak Samaj/World Agricultural Fair Memorial Trust Society, New Delhi, India.

Book (Ed.with Zhensheng Li): *Proceedings of the First International Symposium on Chromosome Engineering in Plants.* Beijing: Academia Sinica.

1986-98 Chairman, Editorial Advisory Panel for the World Resources Report.

Member, Global Board of Directors, The Hunger Project, New York, USA.

1986-92	Trustee, Better World Society, Atlanta, USA.
1987	Doctor of Science (honoris causa), University of Hyderabad, Hyderabad, India.
	Golden Heart Presidential Award by the President of The Philippines.
	Honorary Research Professor in Genetics, Chinese Academy of Sciences, Beijing, China.
	The First World Food Prize (for outstanding individual achievement in improving the world food supply).
1988	Doctor of Science (honoris causa), Agriculture University, Wageningen, The Netherlands.
	Doctor of Science (honoris causa), Assam Agricultural University, Jorhat, Assam, India.
	Doctor of Science (honoris causa), Oregon State University, Corvallis, Oregon, USA.
	Fellow, European Academy of Arts, Sciences and Humanities.
1988-90	President, National Academy of Sciences, India.
1989	Director, Center for Research on Sustainable Agricultural and Rural Development, Madras, India.
	Doctor of Science (honoris causa), Tamil Nadu Agricultural University, Coimbatore, India.
	Doctor of Science (honoris causa), Rajasthan Agricultural University, Bikaner, India.
	Doctor of Science (honoris causa), Indian Agricultural Research Institute, New Delhi.
	Doctor of Science (honoris causa), Pondicherry University, Pondicherry, India.
	Dr. J.C. Bose Medal, Bose Institute, India.
	Padma Vibhushan Award by the President of India.
	Book (Ed. with S. L. Kochhar): *Plants and Society.* London: Macmillan Publishers.
1989-95	Andrew D. White Professor-at-Large, Cornell University, USA.
1989-96	President, World Wide Fund for Nature, India.
1989-97	Trustee of the Ford Foundation, New York, USA.
1989-	Chairman, M.S. Swaminathan Research Foundation, Chennai, India.

1990 Commandeur of the Order of the Golden Ark of The Netherlands.

Doctor of Science (honoris causa), University of Tuscia, Viterbo, Italy.

Chairman, Interim Board of Trustees, Commonwealth – Government of Guyana Program for Sustainable Tropical Forestry, George Town, Guyana.

1990-93 President, International Society for Mangrove Ecosystems, Okinawa, Japan.

1991 Doctor of Science (honoris causa), North-Eastern Hill University, Shillong, India.

Honda Prize (for outstanding services to the development of eco-technologies in the field of agriculture), The Honda Foundation, Tokyo, Japan.

The Tyler Prize for Environmental Achievement, Los Angeles.

Book (Ed. with Vineeta Hoon): *Biotechnology: Reaching the Unreached - An Interdisciplinary Dialogue*. Madras: Centre for Research on Sustainable Agricultural and Rural Development.

Book (Ed. with I.P. Getubig, V. L. Chopra): *Biotechnology for Asian Agriculture: Public Policy Implications.* Kuala Lumpur: Asian and Pacific Development Center.

Book (Ed.): *Biotechnology in Agriculture: A Dialogue*. Madras: Macmillan India.

Book: *From Stockholm to Rio de Janeiro: The Road to Sustainable Development*. Madras: M.S. Swaminathan Research Foundation.

1991-94 Chairman, Governing Board, CAB International, London, UK.

1991-96 President, National Academy of Agricultural Sciences, India.

1992 Doctor of Science (honoris causa), University of Bologna, Italy.

Fellow, Bangladesh Academy of Sciences, Dhaka.

Foreign Member, Russian Academy of Agricultural Sciences, Moscow.

Jawaharlal Nehru Birth Centenary Award, The Indian Science Congress Association.

Lal Bahadur Shastri Deshgaurav Samman, India.

International Member, American Association for the Advancement of Science.

Book (Ed. with Jana S.): *Biodiversity: Implications for Global Food Security.* Madras: Macmillan India.

1992-99 Founder Chairman of the Board of Trustees of the Iwokrama International Centre for the Sustainable Management of Rainforests, Guyana.

1992-2002 Member, China International Council for Sustainable Development.

1993 Charles Darwin International Science and Environmental Medal, Trivandrum, Kerala, India.

Book (Ed. with R. Ramesh): *Sustainable Management of Coastal Ecosystems*. Madras: Chrom Grafix.

Book (Ed.): *Information Technology: A Dialogue.* Madras: Macmillan India.

Book (Ed.): *Wheat Revolution: A Dialogue.* Madras: Macmillan India.

1994 Asian Regional Award (for contributions to the improvement of agricultural productivity in Asia), Asian Productivity Organization, Tokyo, Japan.

Chairman, Governing Board of C.A.B. International, London.

Chairman, Governing Board of the Indian Institute of Technology, Madras, India.

Chairman, Board of Trustees, International Irrigation Management Institute, Sri Lanka.

Chairman, International Commission on Peace and Food, Trieste.

Chief Adviser, Chinese Academy of Sciences for the Biovillage Program, Beijing.

Corresponding Member, Academia dei Georgofili, Italy.

Doctor of Science (honoris causa), Punjab Agricultural University, Ludhiana, India.

Fellow, Linnean Society of London.

Fellow, American Association for the Advancement of Science.

Member, Board of Trustees, The World Resources Institute, Washington DC, USA.

Member, Board of Trustees, The Keystone Center, Colorado, USA.

President, National Academy of Agricultural Sciences, India.

President, World Wide Fund for Nature, India.

Special Award (for exceptional contributions to Global Civilization and Leadership in Thought that leads to Action), World Academy of Art and Science, Chicago, USA.

UNEP-Sasakawa Environment Prize (for outstanding contribution to the management and protection of the environment and to conservation and sustainable development), Tokyo, Japan.

Book (Ed.): *Uncommon Opportunities: An Agenda for Peace and Equitable Development.* Report of the International Commission on Peace and Food). London: Zed Books.

Book (Ed.): *Ecotechnology and Rural Employment: A Dialogue.* Madras: Macmillan India.

1994- UNESCO Chair in Ecotechnology.

1994-2000 Chairman of the Global Hunger Project, New York.

1994-2002 Chairman, Genetic Resources Policy Committee of the Consultative Group on International Agricultural Research, Rome.

1995 Global Environmental Leadership Award, The Climate Institute, Washington DC.

Book (Ed.): *Farmers' Rights and Plant Genetic Resources: A Dialogue.* Madras: Macmillan India.

1996 Honorary Fellow, Crop Science Society of America, and the American Society of Agronomy.

Book: *Sustainable Agriculture: Towards an Evergreen Revolution.* Delhi: Konark Publishers.

Book: *Sustainable Agriculture: Towards Food Security.* Delhi: Konark Publishers Pvt. Ltd.

Book (Ed.): *Agro-Biodiversity and Farmers' Rights.* Delhi: Konark Publishers.

1997 Doctor of Science (honoris causa), Indian Institute of Technology, Madras, India.

Doctor of Science (honoris causa), Hemvati Nandan Bahugana Garhwal University, Garhwal, Uttar Pradesh, India.

Dr. B.P. Pal Medal (for unique contributions to agricultural research and development), National Academy of Agricultural Sciences, India.

Highest Award for International Cooperation on Environment and Development, Government of China, Beijing.

Ordre du Merite Agricole (for having rendered services of the highest quality to the cause of agriculture and its development), Ministry of Agriculture, Government of France.

V. Gangadharan Award (for outstanding contributions to national development), Trivandrum, India.

1998 Doctor of Science (honoris causa), The Hebrew University of Jerusalem, Israel.

Doctor of Science (honoris causa), University of Calcutta, Calcutta.

B.P. Pal Memorial Award, Indian Science Congress Association.

Henry Shaw Medal, Board of Trustees of the Missouri Botanical Garden, USA.

Vice-President, XVIII International Congress of Genetics, Beijing, China.

Honorary Fellow, Association of Aquaculture.

Book (Ed.): *Gender Dimensions in Biodiversity Management.* Delhi: Konark Publishers.

1999 Gender Matters Award, International Development Research Center, Ottawa, Canada.

Legend in his Lifetime Award, World Wilderness Trust, Bangalore, India.

Professor P.N. Mehra Award.

Shatabdi Puraskar Award (Agricultural Sciences), Indian Science Congress Association.

The Volvo Environment Prize, Sweden/USA.

UNESCO Gandhi Gold Medal (for outstanding work in extending the benefits of biotechnology to marginalized and poverty-stricken populations in developing countries, and in securing a sound basis for sustainable agricultural, environmental and rural development), Paris, France.

Fellow, Madras Science Foundation.

Book: *I Predict: A Century of Hope. Towards an Era of Harmony with Nature and Freedom from Hunger.* Madras: EastWest Books.

1999-2000 Asutosh Mookerjee Memorial Award, Indian Science Congress Association.

2000 Doctor of Science (honoris causa), Kakatiya University, Andhra Pradesh, India.

Franklin D. Roosevelt Four Freedoms Award (for brilliant leadership that has established a goal for the new millennium – a hunger free world), The Franklin and Eleanor Roosevelt Institute, Zeeland, The Netherlands.

Indira Gandhi Prize for Peace, Disarmament and Development (for outstanding contribution in the domain of plant genetics and ensuring food security to hundreds of millions of citizens in the developing world), New Delhi.

Millennium Alumnus Award, Tamil Nadu Agricultural University, India.

Planet and Humanity Medal (for significant contribution to environmentally relevant issues), International Geographical Union, Vienna.

Fellow, Indian Society for Plantation Crops.

2000- President, Voluntary Health Services, Chennai, India.

2001 Millennium Scientist Award, Indian Science Congress Association.

Lokmanya Tilak Award, Tilak Smarak Trust, Pune, India.

Doctor of Science (honoris causa), University of Massachusetts, USA.

Doctor of Science (honoris causa), Dr. Panjabrao Deshmukh Krishi Vidyapeeth, Akola, India.

2002 The Economic Times Award for Corporate Excellence – Lifetime Achievement (for nurturing a revolution in a highly conservative sector, for laying the foundation for food security, and for using biodiversity for sustainable prosperity), Mumbai, India.

Toda Award for Peace Achievement (for contributions to wide-ranging and inter-related concerns as the eradication of poverty, preserving regional ecosystems, promoting technological innovation and research for sustainable agriculture), Japan.

Indira Gandhi Gold Plaque, Asiatic Society, Mumbai, India.

Honorary Member, World Innovation Foundation

Doctor of Science (honoris causa), Gujarat Agricultural University, Ahmedabad, India.

Doctor of Science (honoris causa), University of Bonn, Germany.

Book: *From Rio de Janeiro to Johannesburg. Action Today and Not Promises for Tomorrow.* Madras: EastWest Books.

2003 Doctor of Science (honoris causa), Soka University, Japan.

Doctor of Science (honoris causa), Bidhan Chandra Krishi Viswavidyalaya, West Bengal, India.

Biospectrum Lifetime Achievement Award.

Book (with S. L. Kochhar): *Groves of Beauty and Plenty. An Atlas of Major Flowering Trees in India*. Delhi: Macmillan India.

Member, International Advisory Council, Intercultural Forum, San Diego, California.

2004 Doctor of Science (honoris causa), Ohio State University, USA.

Chairman, National Commission on Farmers, Government of India.

Chairman, National Committees for Developing Policies for Agricultural Biotechnology and Coastal Zone Management, Government of India.

Chairman, National Task Force to Revamp and Refocus National Agricultural Research, India.

SCIENTIFIC AMERICAN 50 Award for Outstanding Technology Leadership (for promoting community-based solutions to famine in India).

2005 Dr. M.S. Swaminathan Award for Food, created by The Trust for Advancement of Agricultural Sciences (TAAS); first recipient, Nobel Laureate Dr. Normal Borlaug.

Doctor of Science (honoris causa), Sher-e-Kashmir University of Agricultural Science and Technology, Kashmir, India.

Doctor of Science (honoris causa), Iowa State University, USA.

2005- President, National Academy of Agricultural Sciences.

4. MSS in his office, MSSRF, Chennai, 1995

Foreword

Sartaj Aziz
Vice-Chancellor, Beaconhouse National University, Lahore;
Formerly Pakistan's Minister of Agriculture, Finance,
and Foreign Affairs, Islamabad

This is the fifth volume prepared by Dr. Anwar Dil to record the life and work of persons who have made an outstanding contribution to the eradication of hunger and poverty. It is a valiant attempt to capture, in a single volume, the multi-dimensional achievements of a living legend of our time – Dr. M.S. Swaminathan. It presents a wealth of information, ideas and search-lights to guide future policy makers, the academic community, and all those who wish to contribute to the well being of humanity.

The twentieth century was undoubtedly the century of science and technology, in which new discoveries, inventions and innovations transformed our lives and brought unprecedented prosperity to humankind. The two areas of science that have helped improve the livelihood of human beings most notably are genetics and agriculture. Dr. Swaminathan's distinguished scientific work in these fields has been acknowledged by leading scientists and development specialists of our time. For example, Noble Science Laureate Joshua Lederberg writes that "he cannot think of any one who has combined the insights of the cutting edges of biological sciences, with attention to the most urgent human needs, with the competence, devotion and energy that Dr. Swaminathan has given to the challenge of eradicating hunger and poverty." Nobel Laureate Norman Borlaug, Father of the Green Revolution, has acknowledged on several occasions that "a great deal of credit must go to Dr. Swaminathan for recognizing the potential value of the Mexican dwarf wheat without which it is possible that there would have been no Green Revolution in India." In the opinion of the Former Administrator of UNDP, Professor James Gustave Speth: "No figure in the second half of the 20th century has done more than M.S. Swaminathan to improve the conditions of life of the world's poor." The monumental contribution which Dr. Swaminathan and his team have made to transform Indian agriculture is clearly reflected in the sharp and sustained increase in food production. India's wheat production which was only twelve million tons in 1968 rose to twenty-five million in 1972, fifty-five millions tons in 1992 and seventy-two million tons in 2002. Similarly rice production went up from thirty-five million tons in 1972 to ninety-three million tons in 2002.

In 1982, Dr. Swaminathan moved from the national to the international stage, when he was appointed Director General of the International Rice Research Institute (IRRI) at Los Banos, The Philippines. There he played a major role over a period of six years, in spreading new high yielding rice varieties to many countries of Asia, Africa, and other areas of the world, and in establishing a network of national research institutes, particularly in South Asia. Dr. Swaminathan visited Pakistan in March 1984 to give an inaugural address at an international seminar organized by the Pakistan Agricultural Research Council at Islamabad. During his welcome visit he offered valuable suggestions for improving food production in the region.

In 1988, Swaminathan returned to India from IRRI and used the World Food Prize money to establish the M.S. Swaminathan Research Foundation, to promote "a job-led economic growth strategy in rural areas, based on pro-nature, pro-poor and pro-women orientation to technology development and dissemination". He thus created a forum to translate his vision and cumulative wisdom into a concrete program of action. The Foundation has already implemented hundreds of projects and organized conferences and seminars to disseminate the results with exemplary success.

Dr. Swaminathan is not only a scientific genius, but also a great thinker, philosopher and a remarkable human being. He has always presented his scientific knowledge and advice in the context of his universal concern for human and moral values. In recognition of his distinguished services he has received more awards, prizes and honorary degrees than any other scientist of the twentieth century. The most important of these was the first World Food Prize in 1987 for "his outstanding contribution to expanding production and availability of food throughout the World". In August 1999, *Time Magazine* (Asia) ranked Dr. Swaminathan as one of the twenty most influential Asians of the twentieth century for his outstanding service towards a hunger-free Asia.

Dr. Swaminathan has written extensively and persuasively on a wide range of topics throughout his career. Thirty-eight of his best writings are presented in Part I of this volume and represent a rich legacy of knowledge, ideas and concepts for the benefit of future generations. Part II presents selected prefaces and introductions to his books as well as important selections from his vast archives and insightful statements on his life and work from both national and international experts and publications.

All of us owe a debt of gratitude to Professor Anwar Dil for capturing in this volume Swaminathan's monumental contribution to the well being of humankind.

Lahore
September 1, 2004

Life and Work of M.S. Swaminathan: An Introductory Essay

Anwar Dil

"Concern for man himself and his fate must always form the chief interest of all technical endeavors in order that the creation of our minds shall be a blessing and not a curse."

– Albert Einstein

I first heard of Dr. M.S. Swaminathan in 1964, from Haldore Hanson, the Ford Foundation Representative in Pakistan. I wanted to know from him what kind of continuing support by the Foundation was available for our linguistics project at the West Pakistan Education Extension Centre, Lahore, where I was working following my training on a Ford Foundation fellowship at the University of Michigan. I found him more interested in acquainting me with the Foundation's high priority agricultural development program for making Pakistan self-reliant in food and agriculture, and with the pioneering work of Dr. Norman Borlaug of CIMMYT in Mexico, who had the strong support of President Ayub Khan and Pakistan's Minister of Agriculture, Malik Khuda Bakhsh Bucha. Of course, I was interested in the efforts to eradicate hunger and poverty in my homeland, so Hanson went on to explain similar work being done by Dr. Borlaug with Dr. M.S. Swaminathan, Indian Minister of Agriculture Mr. C. Subramaniam, and other scientists at the Indian Agriculture Research Institute (IARI) in Delhi. I remember the passion with which Hanson spoke about "one of the greatest revolutions of our time" taking place in Pakistan and India, following Borlaug's monumental achievement in transforming Mexico from a chronic food-deficit to a food-surplus country, and their hope that what had been done in Mexico in fourteen years would be accomplished in both Pakistan and India in five years.

Later, I saw Swaminathan in a BBC documentary, "Indian Eyes on the Future". I was visiting Michigan State University, East Lansing, where some experts associated with Dr. Akhtar Hameed Khan's Comilla Project in East Pakistan were discussing the problems of hunger and poverty in South Asia. Some prominent international observers had made highly publicized predictions that India and other nations in the region would be facing massive famines and starvation in the coming years. It was a powerful film depicting the terrible conditions of hunger in the region but what impressed me most was the closing part of the film, where an interviewer asked if Dr. Swaminathan agreed with the view that in ten years a very large number of Indian people would starve to death. Swaminathan responded with great confidence that he did not. He asked the crew to accompany him to see for themselves and take pictures of the new varieties of dwarf wheat, a result of the successful experiments with the imported Mexican wheat varieties at the IARI experimental stations, and the demonstration fields of the poorest farmers in the country, who were determined to change the entire scene: "The wheat harvest of 1968 will mark a new beginning in India's agricultural destiny." I will never forget the hopeful note on which the film ended. Not long thereafter, it was so gratifying to hear the news and read the stories in both the Indian and international media that Swaminathan's prediction had, in fact, come true. In a recent letter to me Swaminathan recalled the moment: "I made this statement because I was confident that if all the steps in the areas of technology dissemination and public policy support we had recommended to the government were taken, our farmers would be able to do the rest. This prediction came true, and Prime Minister Indira Gandhi released, in August 1968, a special commemorative stamp titled 'The Wheat Revolution' to bring to the attention of the public the impact of the synergy generated due to the introduction of a symphony approach to technology development and dissemination."

The award of the 1970 Nobel Prize for Peace to Norman Borlaug in recognition of this pioneering work toward the Green Revolution in Mexico since 1944, and in Pakistan since 1958, demonstrated how the status of agricultural development could be changed dramatically throughout the developing world. Borlaug knew that the work in India was making exemplary progress under Swaminathan's leadership, especially since the beginning of their collaboration in 1963. In fact, on the day he received the Nobel Peace Prize is Oslo (December 10, 1970), Borlaug wrote in response to Swaminathan's message of congratulations: "The Green Revolution has been a team effort and much of the credit for its spectacular development must go to the Indian officials, organizations, scientists, and farmers. However, to you, Dr. Swaminathan, a great deal of the credit must go for first recognizing the potential value of the Mexican dwarfs. Had this not occurred, it is quite possible that there would not have been a Green Revolution in Asia. ... We want all of you to know that it has been a privilege for both the entire CIMMYT staff and for me to have had an opportunity to participate in a modest way in your phenomenal increase in wheat production." In his Nobel Lecture, Borlaug paid a richly-deserved tribute to his Indian colleagues: "The All-India Coordinated Wheat Improvement Program which is largely responsible for the wheat revolution in India, has developed one of the most extensive and widely diversified wheat research programs in the world. ... The current agronomic research on wheat in India equals the best in the world. ... The breeding program is huge, diversified, and aggressive; already it has produced several varieties which surpass those originally introduced from Mexico in 1965. ... The rapidity of creation and distribution of these new varieties has already diversified the type of resistance to diseases and, therefore, minimizes the menace of destructive disease epidemics, if and when changes occur in parasitic races of the pathogens."

Dr. Swaminathan likes to sum up his years of service toward pioneering the transformation of Indian food and agriculture by identifying the following key programs in his symphonic development approach: a) initiation of effective Lab-to-Land and Land-to-Lab programs throughout the country; b) organization of

Krishi Vigyan Kendras (Farm Science Centres) to promote a techniracy (technical literacy) movement; c) organization of integrated pest management, integrated nutrient supply, and other ecotechnologies essential for sustainable agriculture toward an Ever-Green Revolution; d) training of a cadre of soil health care workers in villages; e) promotion of crop-livestock-fish integration and giving an income and employment generation orientation to farm research; f) integration of production, processing, and marketing in the form of an integrated system of agricultural prosperity; g) influencing public policy in a manner that all farmers, irrespective of the size of the holding, are able to benefit from new technologies; and h) organization of dry farming research and development in India.

Swaminathan received his first international award in 1971 for his outstanding ability to establish contact between scientific work and community life. The Ramon Magsaysay Award – established in honor of one of Asia's great leaders, who approached his mission of improving the human condition in The Philippines with exemplary selfless devotion – is given annually to recognize Asian leaders who have rendered signal service to the public good. The 1971 Award for International Understanding went to the Japanese economist Dr. Sburo Okita; for Public Service, to the Filipino educationist Dr. Pedro Orata; for Journalism and Creative Communication, to the Thai cartoonist Prayoon Chanyawongse; for Government Service, to the Governor of Djakarta, Ali Sadikin; and for Community Leadership, to Dr. Swaminathan. The editorial coverage, entitled "Asia's Pride" in *The Asia Magazine* (December 12, 1971), highlighted Swaminathan's contributions as a cytogeneticist, whose chief concern is a deficiency of proteins and calories, which causes irreversible brain damage in children; his research has led to the improvement of the diet of the Indian people and his projects have shown the poorest farmers how, by acquiring technological skills, they could produce crop yields higher than they ever thought possible. This Award was especially helpful to Swaminathan not only because the prize money of ten thousand dollars could build up his public projects, but more importantly because the farmers of India were greatly inspired by the fact that their work had been recognized internationally.

Dr. Swaminathan was awarded the 1978 Norman E. Borlaug Award for his "outstanding contributions to agricultural research, agricultural education, and agricultural development in all its aspects within the country as well as for enhancing the prestige of Indian agricultural science on the international plane." The citation also recognized him "for evolving a strategy for agriculture rooted in science but tempered by concern for ecology and human values." While presenting the cash award and the gold medal, Dr. Borlaug paid rich compliments to Dr. Swaminathan for his "scientific excellence, broad vision and ability to synthesize important aspects from many disciplines." Dr. Swaminathan's acceptance speech was a testimony to his love of both the plants and the farmers.

In 1987, Dr. Swaminathan was awarded the First World Food Prize (an equivalent of the monetary value to the Nobel Prize), which was planned by

Norman Borlaug to be awarded annually "to recognize, encourage, and reward outstanding contributions to the improvement of agriculture and food through education, research, extension, or their application toward increasing production or improving quality of food, feed, or fiber." Individuals and organizations eligible for nomination covered a number of broad fields of agricultural economics, agronomy and soils, crop sciences, development of agricultural institutions, fisheries, nutrition, food sciences and technology, forestry, plant protection, soil conservation, and government leaders who had transformed food production. In announcing that Dr. Swaminathan had been selected as the recipient of the First World Food Prize from among seventy-four outstanding representatives of eighteen disciplines in food and agriculture from twenty-two countries on five continents, Dr. Borlaug, Chairman of the Selection Committee, said: "Dr. Swaminathan has performed with consistent, energetic deftness the roles of innovator, statesman, scientist, and administrator – focusing always on the goal of improving the quantity, quality, and availability of food to national regional, and worldwide communities." The award ceremonies were held at the Smithsonian Institution in Washington, DC, on October 6, 1987, where Dr. Borlaug said that Dr. Swaminathan was being awarded "the first World Food Prize for his outstanding contributions to expanding production and availability of food throughout the world. During the 1960s and 1970s, I personally had the good fortune to see the important diverse roles he played in research, extension and agricultural policy, which ushered in the Green Revolution in India. During the 1980s, I have seen this tremendous positive impact spread to many other countries."

In his acceptance speech, Dr. Swaminathan stated, in his characteristically forthright manner: "I plead for coordinated non-governmental initiatives in: first, *generating awareness* by taking advantage of the powerful mass media tools now available; second, *analysis* of field problems and developing solutions for them jointly by farmers and scientists; and third, *action* in utilizing the growing global grain surpluses for enabling all those in need of help to earn their daily bread." He further added: "We live in an age of unparalleled opportunity for promoting sustainable nutrition security. The prospect for a world without hunger is a glorious legacy given to our contemporary world by scientists and technologists; communicators and social scientists, administrators and industrialists, and workers in the factories, fields, forests, pastures, river, and oceans. As we depart for dinner this evening, what could be a more satisfying and joyful feeling than knowing that every other member of the human family will also go to bed after a nourishing meal? Until such a wholly attainable world becomes a reality, our task remains unfinished."

The well-deserved award was hailed by many around the world. Javier Perez de Cueller, Secretary General of the United Nations, issued a widely-quoted statement: "Dr. M.S. Swaminathan ... is a living legend. His contributions to agricultural science ... have made an indelible mark on food production in India and elsewhere in the developing world. By any standard, he will go into the annals of history as a world scientist of rare distinction." Professor Joshua Lederberg, winner of the 1958 Nobel Prize in Physiology and Medicine for his work in genetics, wrote the following in his letter of congratulations (September

25, 1987): "The central focus of application of plant genetics technology has moved a long way since our association at the University of Wisconsin Genetics Department almost forty years ago. I can think of no one who has combined the insights of the cutting edges of biological science, with attention to the most urgent of human needs, with the competence, devotion and energy that you have given. The world is and will be a better place on account of your contributions to its welfare." Frank Press, President of the National Academy of Sciences, Washington, DC, sent the following message (September 29, 1987): "To make a difference in the world's capacity to feed itself is indeed an awesome accomplishment, and so I congratulate you most sincerely and warmly on the honor given you as a recipient of the 1987 World Food Prize. Your accomplishments as a scientist and a scientific leader should give you great personal satisfaction. More importantly, however, they have an impact on the lives of millions of people in Asia and other parts of the developing world." Former President Jimmy Carter, a member of the Advisory Council of The World Food Prize Foundation, wrote (October 6, 1987): "You can take pride in knowing that many generations of people will benefit from your contributions. Your accomplishments will be of great advantage to scientists and agronomists of the future who will continue your efforts to meet the needs of our expanding population." President Corazon Aquino of The Philippines, who had known Swaminathan in his position as Director General of the International Rice Research Institute (IRRI), honored him in a special ceremony on November 10, 1987, with The Golden Heart Award of The Philippines – the highest national award for a non-Filippino – "in recognition of his contribution in resolving a wide range of problems in basic and applied genetics, ... for his accomplishments in the area of agricultural science and research highly beneficial to Filipino farmers, and for having expanded considerably the International Rice Research Institute's capacity for upstream research to bring the fruits of recent advances in science and technology to Asian rice farmers."

At the international level, Dr. Swaminathan has played a key role in a number of important projects. In 1975, soon after the unification of Vietnam, he was invited to help develop rice production in the Mekong Delta. On his suggestion, a rice research institute was established to make the country self-sufficient in food. Today, Vietnam is the second largest exporter of rice in the world. Another of his admirable projects was the designing and establishing of the Iwokrama International Centre for Rainforest Conservation and Management in Guyana – the world's largest program in sustainable development of rainforests. He also served as Chairman of the Board of Trustees of the project from 1992 to 1999. In 1996, UNESCO designated him as UNESCO-Cousteau Professor in Ecotechnology for Asia.

In 1999, Dr. Swaminathan was ranked by *Time Magazine* (Asia Edition, August 23) as one of the twenty most influential Asians of the twentieth century – alongside Gandhi and Tagore from India – for his outstanding service toward a hunger-free Asia. In reporting the award, the *Crop Science-Soil Science-Agronomy News* of Salt Lake City, Utah (November 1999), in its front-page story, said that in pioneering the Green Revolution in India and transforming India from a begging bowl to a bread basket, Swaminathan had "combined all of the components of a revolution: vision, dedication, energy, and follow-through. ... The impact of Dr. Swaminathan's work has reached far beyond the borders of his homeland. His worldwide reputation has made him an easy choice to chair many prestigious international conferences. ... His foresight and inspiration have led to the establishment of renowned organizations to promote international scientific collaboration. ... Dr. Swaminathan's enthusiasm for passing on knowledge has earned him a reputation as a lucid educator and a profound humanitarian. In a career dedicated to alleviating human suffering, Dr. Swaminathan has completed the work of many lifetimes and still continues, via his leadership of the M.S. Swaminathan Research Foundation, established with his World Food Prize money." In 2002, he was elected to lead the Pugwash Conferences on Science and World Affairs, established by Albert Einstein and Bertrand Russell to bring peach to the world. Today, when the world is facing an unprecedented threat of disastrous wars that threaten the whole of humankind, Swaminathan has accepted the responsibility with humility but strong determination to do his best.

Dr. Swaminathan is perhaps the most highly decorated scientist of our time. He has been awarded nearly fifty honorary doctorate degrees and numerous international prizes. In 1999, he was awarded the UNESCO Gandhi Gold Medal "in recognition of his outstanding work in extending the benefits of biotechnology to marginalized and poverty-stricken populations in developing countries, and in securing a sound basis for sustainable agricultural, environmental, and rural development." In August 2000, he was the recipient of the Planet and Humanity Medal at the twenty-ninth International Geographical Congress held in Seoul, Korea. The citation recognized him as "a great scientist, a responsible and thoughtful caretaker of nature, and an outstanding humanist" and for his "significant contribution to environmentally relevant issues". In December of the same year, he was informed that the Franklin and Eleanor Roosevelt Institute in New York had designated him as the recipient of the Franklin D. Roosevelt Four Freedoms Medal, created to "honor individuals and institutions whose work has given special meaning to the four freedoms which President Roosevelt described in his historic speech of January 6, 1941." The Institute's President Anna Eleanor Roosevelt and Honorary Chair Arthur M. Schlesinger, Jr., in their congratulatory letter to Swaminathan wrote: "The Trustees of this award believe that you exemplify those objectives which President Roosevelt described in urging Freedom from Want as an essential condition of human liberty. Your extraordinary work as an agricultural scientist, leading the Green Revolution, and bringing hope to the peoples of the developing nations so that the age-old scourge of famine and hunger can be ended, has been particularly noted. You have taught nations how to be self-sufficient in their need for food, just as you have taught farmers how to develop and enhance the productivity of their land. Your brilliant leadership has established a goal for the new millennium – a hunger free world, an international structure of cooperation among nations, and a determination to use the miraculous technology of our times to help those in need. Your dynamism and compassion have give new meaning to Franklin Delano Roosevelt's commitment to a better world where all nations will understand and strive for Freedom from Want."

In 1983, the University of Wisconsin-Madison, awarded Swaminathan the Doctor of Science (honoris causa) degree for having been "a major force in shaping the Indian government's policies in agriculture and environmental affairs. Equipped with post-doctoral work at the University of Wisconsin, he has dedicated his career to preventing famine and ameliorating hunger in his homeland and throughout the world. ... Dr. Swaminathan has been at the center of one of the notable human achievements of our generation." The following year, the University of The Philippines gave him the Doctor of Science (honoris causa) degree "in recognition of his outstanding contributions to India and the Third World as scientist, educator, agricultural planner and administrator, community leader, international public servant, and leading exponent of modern peasant agriculture; ... and for giving others the splendid example of his life as scientist and humanist – a life of integrity and humility, of self-abnegation and total commitment to the welfare of others, especially the poor farmers of Asia, bringing to his work a missionary zeal which is infectious and a sense of dedication which is inspiring." The Asian Institute of Technology in Thailand conferred on him the Honorary Degree of Doctor of Technology, in April 1985, in recognition of his "outstanding contributions to the academic and technological development of this region of the world." On September 23, 1996, the China Council for International Cooperation, Beijing, awarded him their highest Award for International Cooperation on Environment and Development for his "outstanding contributions to the lofty cause of environmental protection and development, and for his signal accomplishments in the field of international cooperation."

In his home country, Shri R. Venkataraman, former President of India, summed up Dr. Swaminathan's distinguished career in these words: "Pleasant, soft-spoken, dignified, but unassuming, Dr. Swaminathan has a demeanor that conceals his brilliant intellect, unparalleled achievements in science and research, and countless awards, prizes, fellowships, and honors showered on him by national and international institutions. The Green Revolution, which transformed our chronic food-deficit country with three hundred sixty million people into a food self-sufficient one with nine hundred million, has earned for him an indelible place in our national history." Shri K.R. Narayanan, also a former President of India, adds the following tribute: "Acclaimed internationally for his pioneering contributions to agricultural science and practical application of his scientific genius for modernization of agriculture and enhancement of food production, Dr. Swaminathan stands out with rare distinction in the scientific community. ... In his pursuit of scientific research he has consistently emphasized sustainable agriculture and an environment-friendly approach. His scientific insights have opened new frontiers of success for our nation." And Dr. N.S. Randhawa, of the National Academy of Agricultural Sciences, voiced the high professional admiration that Dr. Swaminathan has earned throughout India: "A plant geneticist by training, he made original research contributions in the areas of plant cytogenetics through use of mutations in a number of crops like the potato, wheat, rice, etc. His work on the use of 'Norin' dwarfing genes to overcome the barriers to high yield in wheat is widely quoted. He inspired agricultural scientists, not only in professional pursuit of their choice but also in several other disciplines of agriculture, with visionary concepts and ideas, such as 'crop cafeterias', mid-season corrections in crop scheduling, risk distribution agronomy and alternative cropping strategies for different weather conditions, conservation of genetic resources, National Demonstrations, and the Lab-to-Land program, etc." Dr. R.S. Paroda, distinguished agricultural scientist and former Director General of the Indian Council of Agricultural Research, told me during my meeting with him at the 2001 Indian Science Congress in New Delhi, that he was privileged to count himself among the students of Professor Swaminathan, who is an outstanding teacher and "a human being par excellence with a most kind heart."

Professor Norman Myers of Oxford University is one of those who believe that Dr. Swaminathan, like Dr. Norman Borlaug, deserves the Nobel Prize for Peace: "Virtually alone among agricultural leaders, he has shown that the challenge of feeding developing countries is closely tied in with a number of other crucial factors, notably poverty, women's roles, and the environment. These further factors broaden agriculture to include concepts of food security, basic human needs, and social equity. It is in this larger sense that Dr. Swaminathan is far more than an agricultural pioneer. Indeed, he ranks as one of the world's finest exponents of human development in its proper broad sense." Dr. R. Martin Lees, Rector of the University for Peace in Costa Rica, extends these observations with the following assessment: "First, as a world-renowned agricultural scientist, Dr. Swaminathan is widely recognized as the father of the 'Green Revolution' which proved to be a turning point in improving the condition of hundreds of millions of human beings, not only in India but throughout the world. Abiding hunger and poverty are the breeding grounds of conflict, this transformation of agriculture has proved to be a critical contribution to world peace and progress, ... acting effectively through scientific research and at the highest levels of policy. Dr. Swaminathan has achieved extraordinary practical results in addressing poverty at the human, grassroots level. In promoting the elimination of poverty, he is also a strong advocate of responsible policies to preserve and rehabilitate the environment and to improve the condition of women. He has established a Research Center and Foundation – which has now achieved a worldwide reputation – to focus many practical applications of science and technology on eliminating poverty in the villages of India through employment, education, and empowerment. Such approaches, for example, through the establishment and support of a network of 'biovillages' in southern India, have application throughout India and, of course, in other countries. ... In confronting directly the real and urgent problems of poverty, gender, and environmental degradation afflicting hundreds of millions of people, Dr. Swaminathan has demonstrated remarkable leadership in mobilizing efforts, both national and international, to resolve many profound social injustices, such as: the condition of landless laborers, access to education, the empowerment of women, the prevention of child labor, and access to health care. His prestige is such that his leadership on these issues, both in promoting action at the grassroots level and as an advocate at the high policy level is both respected and effective." Dr. Maurizio Iaccarino, Professor of Microbiology at the University of Naples and former Assistant Director General of UNESCO, stated: "Like the late Abdus Salam, [Swaminathan] links the Western and Third World cultures." Professor James Speth of Yale University, who has known Dr. Swaminathan as his colleague in international development, adds: "No figure in the second half of the twentieth century has done more than M.S. Swaminathan to improve the condition of life of the world's poor. He has brought extraordinary humanistic vision and profound commitment to human development to all of his scientific accomplishments. During a period when science has been moving forward with unprecedented success, M.S. Swaminathan has been at the forefront of international and Indian efforts to ensure that those successes translate into better lives for the poor and marginalized." I have heard similar evaluations of Professor Swaminathan's life and work by many other distinguished scientists from all continents at the World Food Prize gatherings in Des Moines, Iowa.

Who is this man and what is his life's work that has been recognized so enthusiastically by so many in his home country and around the world?

Swaminathan was born on August 7, 1925, in a middle-class family in Kumbakonam in the state of Tamil Nadu in southern India. Kumbakonam is famous for the Mahamaham Festival, celebrated at the Mahamaham Tank for purification, symbolizing the mythological deluge. Everyone, irrespective of social status, has a dip in the tank.

He was given the name Monkombu Sambasivan Swaminathan according to the naming tradition of his people: the first name is the name of his village, the second is his father's given name, and the last is his own given name, by which he is known. Monkombu (derived from the local name for mango) was the family's ancestral village. Swaminathan remembers the delicious mangoes that grew there. The village has a number of canals running through it, like Venice, with people using boats to get around. The rice fields, coconut palms, and mango trees composed the dominant part of the landscape.

The family had a well-respected heritage. The most important ancestor was Venkatachella Iyer, an eminent scholar of the Indian classics, who had been awarded prize land by the ruler of the area. Most members of the family were farmers who grew rice, coconuts, mangoes, and pepper, and who had a successful trading network. Swaminathan's father, Monkombu Krishna Sambasivan, had studied medicine at the medical college in Madras, and after getting his degree in 1915, started as a surgeon in Kumbakonam. He also served the community as an elected Chairman of the town's municipal corporation. Swaminathan's early memories of his father are of his selfless devotion to public service, as he started a successful public campaign toward the eradication of diseases caused by mosquitoes, notably malaria, filarial, and elephantiasis. Swaminathan has never forgotten how his father mobilized the town-wide campaign of cleanliness by involving men, women, and children in filling up the pools of filthy water with sand, digging landfill dumps for garbage disposal, and disinfecting sewers and open drains to rid the town of mosquitoes. He recalls with pride the most unforgettable lessons he learned from his father, that if people work together synergetically through broad-based public education, social mobilization and community action, they can transform their society. Swaminathan has always used this "people's power" approach in all his plans and projects in mobilizing people to fight hunger and poverty at local, national, and international levels. Another lesson that he learned from his father's exemplary life was his selfless commitment as a worker for Indian independence and progress. He also likes to quote a saying of his father's that the word "impossible" exists only in our minds and in the dictionary, and that given the requisite will and effort, one can accomplish great tasks, even with limited resources.

In response to my questions about his mother, Swaminathan said: "She was a simple and spiritual person and represented all that is best in motherhood. She taught us that all rights come from a duty well done and that rights and obligations should go together. But for her strength of character and support, I do not think we would have been able to develop our full mental and spiritual potential. Till her death, she was a source of both inspiration and sane advice." Swaminathan remembers that following the sudden and premature death of his father, on October 12, 1936, at the age of thirty-six, his mother was "a pillar of strength" to her three sons and one daughter. Swaminathan was only eleven years old at the time. His elder brother was thirteen, his younger brother nine, and his sister only four. The family had to sell their car and change to very simple living. But because of the joint-family system they continued living in the same household with his uncle, Narayanaswami, as the head of the family. He grew up with his uncle's four children. This early experience ingrained in Swaminathan the value of "simple living and high thinking", sharing together as members of a human family in both good and bad times.

He was inspired by another uncle, Professor K. Swaminathan, who moved to Delhi in 1960 as editor of the *Collected Works of Mahatma Gandhi* – an awesome undertaking published in one hundred volumes with an Editor's Preface in each volume. He was a man of great learning and lived a life of synthesis between Gandhi's and Ramana's philosophies: "Gandhi exemplifies the tree's rootedness and Ramana[1] embodies its openness to light. They both knew ... that the earth that holds the tree that holds the branch that holds the leaf is, in its turn, held by the sun." Swaminathan was deeply touched by this philosophy and told me once that he owes a great debt to his uncle for inspiring him from a very early age with a sense of oneness and responsibility towards one's fellow humans. He was also inspired to read Gandhi's writings and, as evidenced in several of his writings, Swaminathan has been greatly influenced especially by Gandhi's concepts of *sarvodaya* (advance of all at all levels) and *antyodaya* (starting with the poorest person in the society). He has also been influenced by the life and

[1]Ramana was a twentieth-century Indian sage, whose Ramanasraman is about one hundred miles from Chennai.

teachings of Ramana. He has influenced Swaminathan as an exemplary model of accepting the responsibility of one's own inner light and that fulfilling the responsibility to others is the cornerstone of one's own fulfillment.

On another level, Swaminathan remembers the summer holidays, when the family used to go the ancestral village of Monkombu, and the happy times he spent with his cousins playing in the rice fields and mango groves. He relished hearing stories of Indian history and mythology from his grandmother, who was the supervisor of all day-to-day family matters. There was no gender discrimination in their household, so boys and girls all played together and grew up without any sense of gender superiority. There was also no discrimination along the caste lines, except among the very orthodox. In fact, the village temple of goddess Bhagvati was open to all people without prejudice or social distinction. Thus the temple played an important role in bringing together and unifying the village people, who, for all practical purposes, lived like one large family. Swaminathan acknowledges that he owes most of the humanistic values he has upheld all his life to his early upbringing in such an egalitarian environment.

The most shining early memory of Swaminathan's dates back to the time when he was only six years old. Mahatma Gandhi came and stayed in a shelter attached to his father's clinic: "I got three basic principles from his teachings: Truth, Non-Violence, and Self-Reliance. I have been greatly influenced by his philosophy that every Indian village should produce its own clothes and food. My father had great respect for him, and we had a *charkha* (spinning wheel) in our house; we used to spin one hour in the morning and one hour in the evening. We wore the clothes made of this home-spun cloth." However, the most lasting impact of Gandhi's life and teaching on Swaminathan has been that he came to believe very strongly that all human beings are the same before God and that caste and other social distinctions are detrimental to human society. Swaminathan's family belonged to the Brahmin caste but his father did not believe in higher or lower castes. Their home was open to people of all castes and social classes, and because of this belief and practice, he was, in fact, excommunicated from his own caste by the extreme orthodox priests of the community. Swaminathan especially remembers that when his father's mother died, the Brahmin priests refused to come to their house to perform the traditional religious ceremonies.

For his early schooling, Swaminathan was sent to the Native High School and later to the Little Flower High School (run by Catholic missionaries) in Kumbakonam. He remembers his teachers with affection and gratitude for their high educational, disciplinary, and ethical standards. A number of students from his high school went on to win scholarships and enter universities for higher education. It is worth nothing here that as one raised in a Brahmin family that was outcaste by some local orthodox Hindu priests, and educated in a Roman Catholic School, Swaminathan grew up to have respect for universal moral and spiritual values and to rise above sectarian religious dogmas and rituals. He recalls: "We weren't worshipping idols or several gods, but my mother, in particular, trained us to believe that there is somebody above you to whom you must be accountable in your life. You may call it Divine Force, Ram, Rahim, Buddha, Jesus, but the essence is that you must have a certain self-discipline, which lets you determine your actions by this force and by the sense of responsibility related to it." On a more personal level of his beliefs and behavior, Swaminathan says: "You may call me most likely a Buddhist, because complete non-violence is one of the basic principles of my philosophy." In a part of the world where religious bigotry, intolerance, and aggressiveness have played such a destructive role, Swaminathan's parents and family members were exemplary in their respect for one's fellow humans across religious, ethnic, linguistic and other variations. That is how he grew up to have loving respect for Buddha, Jesus, Muhammad and other great teachers of the world as role models of love for all that gives a person inner strength to bear the responsibility for one's upright actions under all circumstances.

Swaminathan credits his early interest in farming and agriculture to his surrounding landscape dominated by rice fields. The area of Kuttanad, where Monkombu was located, was called "the rice bowl" of Kerala. He recalls the rows of village women planting rice saplings and how even as a small child he would ask them to let him join them; but the very first day he was given the permission he found out that it was back-breaking work, like the Filipino song: "Planting rice is no fun." He has never forgotten to respect the hard work of farmers and field workers. He especially remembers the exhausting effort involved in pumping out the saline water which flooded the fields before the monsoon crop was sown and the round-the-clock vigilance against possible damage to the dikes constructed as safeguards against the flooding. He also learned the importance of good tools and machinery to relieve the back-breaking work, but also the important lesson his uncle taught him, at the time, that in an overpopulated country like India machines should not be allowed to replace people, who need work to earn their livelihood – hence the necessity of balancing the use of machinery and human labor.

As he grew up, he saw the family having to expand their agricultural activities by planting cash crops like rubber and coffee, which involved learning the complex system of marketing of these products. Swaminathan remembers how his people used to worry about the fluctuations in national and international markets affecting their day-to-day economic life, and how the monsoons and the pests and plant diseases ruined their crops, so the farmers struggled to take measures like mixed and multiple-cropping or venturing out for off-farm activities.

Swaminathan graduated from high school in 1940, and he and his elder brother went to study at the Maharaja's College at the University of Travancore (the present-day University of Kerala), in the capital city of the then princely state of Travancore. This was the time of the final stage of the Indian struggle for freedom from British rule, and Swaminathan remembers the popularity of the nationalist songs of Subrahmanya Bharati, and how passionately the people were inspired to educate themselves to replace the British rulers in all walks of life. Millions suffered unbelievable misery and death during the Bengal famine of 1942 and the epidemic that followed. This was the time that Swaminathan pondered over the question: When my country gains independence, which area of education could benefit my people the most?

By the time he earned his Bachelor's degree in Zoology, in 1944, Swaminathan started thinking seriously about studying agriculture to save his people from the kind of suffering the great famine had caused. He also had observed the hard life of the farmers and how they were at the mercy of pests, plant diseases, weather, and other natural disasters. So, he decided to study science in the hope of developing new and better seed varieties for rice, wheat, the potato, and other crops, and for improving farming methods and production. It became his dream to use science and technology to make his people self-sufficient in food as the first step toward a life of dignity. Thus, his dream of politically independent India became linked with a people self-reliant in food and other basic necessities.

So, after earning his Bachelor's degree in Zoology, he entered the Agricultural College in Coimbatore – affiliated with the University of Madras. During his stay there, he developed a special interest in working with farmers for planning programs of development by combining their practical knowledge and experience with his biological, agricultural, and technological knowledge. In 1947, he earned his second Bachelor's degree, this time in Agriculture, with distinction.

India gained independence on August 15, 1947. Three weeks later, Swaminathan left for Delhi for higher studies and research at the Indian Agricultural Research Institute (IARI) – commonly called PUSA because of the original place in the state of Bihar, from whence it was moved to Delhi after the earthquake of 1934. When his train reached Delhi, after an exhausting three-day journey because of communal rioting along the way, he had to spend the night at the railway station among hundreds of refugees. In the morning, he took a horse-drawn carriage to PUSA, but was shocked to see en route horrible scenes of streets full of dead bodies massacred during the communal riots in the city. The immensity of the human suffering caused by religious fanaticism touch him so deeply that he decided to dedicate himself to the humanistic ideals he had learned from his father and other family members – and ever since, he has been an exemplary bridge builder across communal barriers.

In his studies and research, he became fascinated by the historical development of the potato for its genetic traits and staple food value, and also because of its cultural history spanning several countries across the continents – from the Ecuador, Peru and Chile region of South America to Europe in the sixteenth century. Swaminathan was impressed by the fact that the potato plant today is very different and greatly improved from what it was at the beginning in its place of origin five centuries ago. This developmental feature attracted Swaminathan's scientific interest in a possibility of genetic research to develop new varieties of rice, wheat, and other plants. Following the pioneering work of Gregor Johann Mendel and his impact on botanical research in the first half of the twentieth century, Swaminathan devoted himself to carrying on experiments to explore new features that could be developed through cross-breeding and new methods of cultivation for improved varieties of the potato and other crops.

As luck would have it, Prime Minister Jawaharlal Nehru visited IARI at that time to inquire about the possibilities of bridging the gap between supply and demand in food and of making India self-sufficient in food production. He made a tour of different departments and experimental fields to see for himself what was being done toward a hunger-free India. Swaminathan was greatly inspired by his professors' briefings of the Prime Minister on several experiments and breakthroughs. He took detailed notes and felt that he had found the meaning for his life and work in his chosen field of combining genetics and agriculture. But he also felt frustrated, because, in spite of the important work being done by a number of scientists, their promise of making India self-sufficient in food remained a distant dream. In his struggle to find the reason, he came to the realization that there was very little coordination or synergy among the researchers, because they were working in isolation in their respective laboratories and experimental stations – and, more importantly, they had no contact with the farmers and workers in the field. Later when he came to teach at IARI, he developed his famous Lab-to-Land and Land-to-Lab Program to bridge the gap between the researchers and the farmers in the field.

In 1949, when he earned his post-graduate diploma from IARI (equivalent to a Master's degree), with distinction in cytogenetics, his family wanted him to take the Indian Administrative Service examination. He did not want to leave his chosen field of study and research, but out of respect for the family's wishes, he competed in the All-India competitive examination and was selected for the Indian Police Service. At about the same time, he received the news that he had been awarded a UNESCO fellowship for higher studies and research on the potato for one academic year in the Department of Genetics at the Wageningen Agricultural University in The Netherlands. He told me that he was inspired by its mission statement: "Wageningen University wishes to develop and disseminate the scientific knowledge needed to sustainably supply society's demand for sufficient, healthy food and a good environment for humans, animals, and plants." He accepted the fellowship and went to The Netherlands.

His average working day at the University was from 7:30 a.m. to 9:30 p.m. on "crossing frost-resistant potato species from areas which were far apart from one another, with focus on a parasitic worm problem called the golden nematode, and the problem of frost which damaged potato crops especially in areas reclaimed from the sea." He recalls being inspired in particular by a paper by Dr. Gerhard Stelzner at the Max Planck Institute, published in a German journal, on standardizing procedures for transferring genes from a wide range of wild species of *Solanum.* In June 1950, his professors helped him make a visit to Frankfurt, Germany, to meet Professor Stelzner at the Max Planck Institute for Plant Breeding. When he reached the war-torn city, he learned from Mrs. Stelzner that the Professor had gone to Russia, and nobody knew what had happened to him. In any event, Swaminathan was able to bring back some of his papers and notes, which proved invaluable in his cross-breeding of potatoes. In 1950, he had his first major scientific publication in *Der Zuchter*, the internationally renowned scientific journal of Germany, bringing him to the attention of eminent scholars in the field. He was encouraged to go to the Plant Breeding Institute of the School of Agriculture at Cambridge University, the best place to carry on advanced research in his field of specialization at that time.

At Cambridge, he was admitted to the Fitzwilliam House, where there was a rich collection for studying distant species from Peruvian, Bolivian, and Chilean regions, which he crossed with European species. His research guide was Professor H.W. Howard, but Swaminathan also worked closely with a number of other distinguished scientists, notably G.D.H. Bell, world famous wheat breeder; Sir Roland Fisher, eminent geneticist-statistician; and Sir Joseph Hutchinson. He also came to know and admire Joseph Needlam whose monumental work on Chinese sciences inspired him to develop a special interest in the Chinese scientific tradition, Swaminathan's Ph.D. dissertation on "Species Differentiation and the Nature of Polyploidy in Certain Species of the Genus *Solanum-Section Tuberarium*" (1952) presented a new concept of the relationships within the tuber-bearing *Solanums* – an important contribution in both basic and applied science. An especially notable event during his Cambridge days was his visit as the houseguest of F.L. Brayne, a retired Indian Civil Service officer, who had served in India and had written famous books like *Socrates in an Indian Village,* on rural development and organic farming in the Punjab province. Swaminathan's stay in his home for the Christmas holidays became a priceless experience, as Brayne enriched him with his vast knowledge and experience of rural India, land consolidation, organic recycling, and other critically important matters. Swaminathan says he owes a great debt to Brayne for several of the ideas in rural development, which he later applied in India.

Swaminathan's research at Cambridge was published in a series of articles. His article on the potato, published in the *American Potato Journal,* led to an offer of a position as a post-doctoral research associate in the Department of Genetics at the University of Wisconsin at Madison, with the assignment to establish an Inter-regional Potato Center at Sturgeon Bay in Lake Michigan – a project of the US Department of Agriculture. This helped him extend his research in the United States for about a year and a half and enabled him to publish several research papers on the *Solanum* species in American and European scientific journals. He was lucky to meet the brilliant geneticist Joshua Lederberg, who was working at the time on heredity and cross-breeding of bacteria and was developing his theory of transduction. Swaminathan was greatly inspired by the high level of research in genetics that was being carried on at Wisconsin and other centers of research in the United States. Another especially noteworthy memory of his Wisconsin days was his sharing of an office with Professor James Crow, a leading geneticist who was a human rights activist. Before the completion of his post-doctoral work in December 1953, Swaminathan was offered a position by the President of the University as a regular member of the faculty in a vacancy caused by the death of Leonard Huskins, a well-known cytologist. Also there was a very attractive opening at a potato chip factory, Red Dots, in Maine with a beginning annual salary of $14,000; but he was more interested in returning to India and serving his people, who needed the benefits of the new scientific approaches and findings applied to agricultural development.

Swaminathan returned to India in early 1954. But because genetic research was such a new field, he could not find a suitable job, so he had to wait for three months at home. The University of Wisconsin renewed their offer of a faculty position in March, but Swaminathan was determined to find work in India. By chance he met his former professor, N. Parthasarathy, who was working at the Central Rice Research Institute at Cuttack in Orissa. He offered Swaminathan the temporary job of Assistant Botanist, to work on transferring genes for fertilizer response from *Japonica* varieties of rice grown in Japan to the *Indica* varieties grown in India. Swaminathan had hoped to continue his research on the potato, but he found research on rice equally interesting and even more important because rice was the staple food of a very large number of Indian people. Erstwhile, he had primarily researched diseases and methods of their prevention for the potato, but for rice he was challenged to work on developing new varieties for growing more and better quality rice. Within a few weeks, the results of his experiments started showing hopeful promise and his vision was raised higher than ever before toward making a major difference in solving the hunger problem of his people.

In October 1954, as if destiny were charting the course of his life's work, Swaminathan was selected as Assistant Cytogeneticist at the Botany Division of the Indian Agricultural Research Institute (IARI) in New Delhi. India was threatened at the time by a series of famines, and some prominent international economists had predicted severe famines for the country. IARI had already built a reputation in the field of wheat research on advancing the work of the nineteenth century hybridization researchers, who had produced improved varieties of wheat. So, his first assignment at IARI was research on wheat. He soon realized the tremendous possibilities of Mendelian genetics for a breakthrough in wheat production in India, to change the 'ship-to-mouth' situation of dependence on foreign food aid. He also looked upon the foreign aid as not only a humiliation but also a factor in causing more unemployment among Indians. He believed that if dwarf wheat varieties were introduced in India, the situation could be reversed because the use of fertilizers mainly increased the height of the wheat plants and more pods broke down the stalks. In contrast, the dwarf varieties held their leaves more vertically, allowing more rows to be planted closer to each other, thereby making maximum use of the fields. Furthermore, the dwarf varieties lent themselves to be planted to a variety of changing day lengths and, thus, could be planted across different latitudes, even more than once a year. He had already seen the breakthrough results of developing non-lodging and fertilizer responsive varieties of the Indica-Japonica hybridization of rice at IRRI and believed that similar work on wheat was destined to yield revolutionary results. So Swaminathan began his researches on wheat production at IARI in this direction.

In 1958, he read a paper by Orville Vogel of Washington State University, who was experimenting on Norin-10 dwarfing genes. S.D. Salmon, who had been with General McArthur in Japan, had brought the variety to the United

States, passing it on to Vogel, who then developed the semi-dwarf variety of Gaines, which raised the production to nearly fifteen tons per hectare – a world record at the time in wheat production. Swaminathan wrote a letter to Vogel, who sent him a set of Gaines seeds. But Gaines was a winter wheat unsuitable for India, where wheat is grown in spring. Vogel advised Swaminathan to get in touch with Norman Borlaug at CIMMYT in Mexico, who was working with the same material but had developed a spring wheat variety. Swaminathan had met Borlaug in 1953 at the American Institute of Biological Sciences in Wisconsin, where he had heard his paper on multilane breeding for resistance. He wrote to Dr. Borlaug asking for his advice and help.

In March 1963, Borlaug's visit to India was sponsored by FAO and, accompanied by Swaminathan, he traveled in selected areas of the country to assess the soil, speak to the farmers, and examine the proposed experimental sites. As a follow-up of Borlaug's recommendations, his experimental seeds were sown in a plot of two hectares (about five acres). The plants responded well to the high doses of fertilizer and irrigation, and the yield was more than five metric tons per hectare – about three times that of the traditional varieties. It was a major breakthrough. The harvest was kept for sowing, and more testing was carried on in different locations under a variety of land and weather conditions with equally favorable yields.

In 1964, Swaminathan carried on extensive consultations with the Ministry of Agriculture for arranging demonstration programs in different parts of the country on one-hectare plots in the fields of small farmers. He also succeeded in getting the government to agree to import 18,000 metric tons of Mexican seeds. Shri C. Subramaniam, the visionary Minister of Agriculture of India, was convinced that the small cost of funding one hundred and fifty demonstration plots was nothing compared with the possibilities of promising breakthrough results. In Swaminathan's own words: "[The five hundred rupees per demonstration] covered the cost of seed and fertilizer. The farmers did all the rest." Swaminathan and Subramaniam were proved right beyond the wildest hopes of all concerned, and India was on its way to self-sufficiency in food.

But there were several problems that Swaminathan and his colleagues had to overcome. For example, the Mexican wheat was red in color, while the Indians were used to an amber-colored wheat bread. So, experiments were conducted, under Swaminathan's supervision at IARI, using gamma rays and ultraviolet light to change the red colour to amber. Pusa Lerma and Sharbati Sonora, the two gene dwarfs, were developed from Lerma Rojo and Sonora 64 by mutation breeding. The new hybrids proved to be better than the Mexican varieties. One special effort for which Swaminathan deserves special credit is that he ensured that the new varieties would be cultivated in the fields of the poorest farmers. Two varieties, Lerma Rojo 64A and Sonora 64 were approved by the Government of India for cultivation in irrigated wheat districts. About two hundred and fifty tons of these two varieties were planted in 1965 and about 18,000 tons in 1966, and by 1968, over 200,000 hectares were under cultivation. Swaminathan recalls those exciting days: "There are experts who put down the people from the Third World saying that their economic backwardness is the result of their stupidity and lack of ability to learn new ideas and technology.

Jared Diamond, an American scientist, investigated the last 13,000 years of human history and proved that people who until recently were technologically primitive, such as aboriginal Australians, speedily master industrial technologies when given the opportunity to do so. I have had the same experience with the Green Revolution in India. Barely literate farmers have been acquiring sophisticated skills." Borlaug also has a similar observation: "There were many people in the developed countries who said that with their crude planting equipment Indian farmers would not be able to control the depths of sowing. But we found that as soon as the farmers saw that this was important, they learned to adjust the depth. There was no problem." Prime Minister Jawaharlal Nehru wrote in his letter of April 14, 1964, awarding Dr. Swaminathan the prestigious Shanti Swarup Bhatnagar Memorial Award: "I have been watching with some pride and satisfaction the wonderful work which is being done by some of our scientists. I am happy to find that your work has been recognized in this befitting fashion. I am sure you will continue in your endeavors and also inculcate amongst your junior colleagues the same sense of devotion and dedication as have led to your present recognition."

In 1966, Swaminathan was appointed Director General of IARI. Dr. B.P. Pal welcomed his successor with the following words: "I have watched with great interest and intense pleasure the unfolding of your genius, and I feel very happy that you are now heading the great Institute which I also was given the privilege of serving. As an older person, allow me to invoke blessings so that your mission for the uplift of Indian agriculture may be amply fulfilled." Swaminathan's mentor lived long enough to see that, under his visionary leadership, the IARI greatly developed its teaching, research, and laboratory facilities. Swaminathan initiated a number of innovations in multi-cropping and crop rotation on scientific foundations because he strongly believed: "It is still not fully realized that the serious poverty, unemployment, and underemployment problems facing India can be overcome only through the scientific exploitation of the plant and animal wealth of the country, and that agricultural development is not merely a tool for achieving food self-sufficiency but is the most feasible and speedy method of economic growth." He also propagated the idea that "agriculture is the most import solar energy harvesting enterprise in the world, and India is blessed with abundant sunshine", so it is necessary to de-link agricultural fortunes from the monsoons. He wrote inspiringly, upholding the idea that "no other country the size of India has so much irrigation potential available for raising crops." Of course, there are dry-farming areas lacking that potential, but even there IARI experiments were carried on with the possibility of planting a second crop by developing short-grown-duration wheat and using new scientific approaches. On his initiative in 1967, IARI started a television program in collaboration with All India Radio that reached seventy-two villages in Delhi State. Drawings and instructions were supplied free to anyone wanting to start a small business of manufacturing them. He also took steps toward increasing the IARI collection of specimens of plants, insects, and fungi, and devoted special attention to building up a first-rate library of books, journals, and audio-visual materials. Swaminathan, thus, gained reputation as a legendary figure for his scientific talents, his administrative abilities, his extraordinary skills as a communicator with political leaders, and as a powerful writer, speaker, and teacher.

By 1968, Indian wheat production had gone up from twelve million to seventeen million metric tons – a spectacular jump attracting world-wide admiration. At Swaminathan's suggestion, Prime Minister Indira Gandhi ordered the issuing of a commemorative postage stamp to celebrate the Wheat Revolution in India. The stamp carried a view of the IARI Library as acknowledgment of the Institute's contribution to the wheat revolution. In two decades, the life expectancy of the Indian people increased by ten years. Swaminathan never fails to give credit for the revolution to Norman Borlaug's pioneering work at CIMMYT and to the Indian farmers and the cooperative experiments established by ICAR through the All India Coordinated Projects. But above all, he says that it was the coordinated collective effort of the farmers that made it possible. In 1971, India was officially declared self-sufficient in food production. The success with wheat was carried over to other crops, especially rice, which, in fact, is the principal diet of the Indians – twice as much as the consumption of wheat.

In 1972, Swaminathan was appointed Director General of the Indian Council of Agricultural Research (ICAR). One of his outstanding innovations in his new position was to introduce the concept of "Techniracy" (Technical Literacy) by providing training to workers in the latest technical skills through actual work experience. Swaminathan has always believed that in our technological age today, a person can make a jump from the pre-literacy stage to the post-literacy stage without having to go through all the intermediate steps, and that countries like India should take this leap because they do not have enough time or resources to go through the so-called logical steps. His pro-poor techniracy programs played a major role in providing jobs to thousands of disadvantaged members of the society. He also helped establish a chain of Farm Science Centres and Forestry Science Centres throughout the country, as well as a National Bureau of Animal Genetic Resources.

A lesson from his early childhood that has inspired Swaminathan in all his development plans is that humans are only one component of the great web of life in the cosmos, and that every action leads to a chain of reactions in the environment – some expected, others unexpected; some positive, others negative. He likes to cite the examples of the pesticide DDT and the fertilizers causing so much environmental harm. Swaminathan believes that we should develop a mind-set whereby we should carry out a comprehensive benefit-risk analysis to guard against the overuse of pesticides and fertilizers and against the coverage of large areas with the same genotypes, because a new pathogen could cause tremendous devastation. His balanced, far-sighted approach has always provided for precautions against pollution, diseases, and other problems for ensuring positive results.

In 1972, wheat production in India had surpassed twenty-six million tons. The farmers had proved that they could learn the latest technology and use it in creative ways to produce results. In his Rajendra Prasad Memorial Lecture that year at the Indian Society of Agricultural Statistics, Swaminathan raised the question: Can we face widespread drought again without food imports? His lecture elaborated why his answer was: "Yes, but only if the country takes the path of building grain reserves, introducing mid-season corrections in cropping patterns, building seed reserves, promoting crop life-saving techniques and a risk-distribution agronomy." He was proved right during the severe drought of 1979.

In 1980, Prime Minister Indira Gandhi appointed Swaminathan as a member of the Planning Commission of India. During the two years he served in that position he was able to introduce two new chapters for the Commission's Plan on "Women in Development" and "Environment and Development"; in addition, he introduced a new section on a "New Deal for the Self-Employed". He also served as Chairman of the Scientific Advisory Committee of the Cabinet of India, and in this capacity helped to set up a National Biotechnology Board, of which he was also the first Chairman, and a National Science and Technology Entrepreneurship Development Board, of which also he was the first Chairman. He also helped to reorganize the Indian Meterological Department to make it a powerful instrument for short-, medium-, and long-term weather forecasting.

In 1982, Swaminathan was approached to succeed R.F. Chandler, the Founding Director General of the International Rice Research Institute (IRRI) at Los Banos in The Philippines. Prime Minister Indira Gandhi did not want Swaminathan to leave his position because, in her words, he was "needed in India". Swaminathan chuckles when he narrates his rejoinder to her remark: "That is exactly when one should leave rather than having to leave when one is not needed!" Indira Gandhi reflected for a moment before she smiled, and agreed to let him go for a term or two. In fact, on the eve of his departure (March 30, 1982), she wrote to him: "As you prepare to leave for Manila, I should like to express my appreciation of your work in India and also to tell you how much we shall miss you. You have made significant contribution to our scientific and economic policy making. ... To your tasks in the Planning Commission, you brought deep concern for the nation's realities, besides the intellectual discipline of a scientist. My good wishes to you in the new assignment you have taken up."

Swaminathan accepted the challenging responsibilities at IRRI in developing rice production for helping not only India, but also China and other neighboring nations in Asia, where rice has been for centuries the staple food of the people. He started making some substantive changes, as well as some symbolic alterations reflecting his Ecology-Ecosystem Approach: for example, he changed the name of one of the big IRRI projects from "International Network on Fertilizer Response" to "International Network on Sustainable Rice Farming", to show that there was a different approach at work. One of the first projects he initiated was to map the countries in which rice research did not exist or was at a very low level. The first year he was in office, he visited a number of these countries to see for himself and understand the state of affairs. His first visit was to China, where he helped to select the site and discuss the priorities for the Chinese National Rice Research Center, and on his return, he prepared the project and got necessary financial support from UNDP, the World Bank, and the Rockefeller Foundation.

During his six years as head of IRRI, he made important contributions, not only in upgrading research, training, and technology transfer programs at IRRI, but also in establishing a network of national rice institutes in several countries, notably Burma (Myanmar), Cambodia, China, Egypt, Iran, Korea, Madagascar, Nigeria, Pakistan, The Philippines, Sri Lanka, Tanzania, Thailand, and Vietnam. His experience during these surveys increasingly convinced him of the central role women play in the production, marketing and consumption of rice; so he designed a project called Women in Rice Farming Systems (WRFS). Another notable program that Swaminathan developed at IRRI was Paddy and Prosperity, for organizing training workshops to help rice farming families produce value-added products from rice straw, husk, and bran as well as the use of rice biomass as a source for edible oil, animal feed, soap manufacture, biogas generation, and other useful purposes. Several new uses of the rice biomass were researched, for example, developing husk as a source of energy for grain drying and charcoal production, and extraction of solar grade silicon for use in photovoltaic cells. These and other projects ensured that the link between paddy and poverty would be broken and a new link between paddy and prosperity established.

When the Indonesian rice crop was struck by the brown plant hopper, President Suharto appealed to IRRI for help. Dr. Swaminathan and Dr. Peter Kenmore of FAO advised banning about fifty large pesticides that were killing the natural enemies of pests in Indonesia. In light of his experience in India, Swaminathan was able to convince the policy makers, and President Suharto made Integrated Pest Management compulsory throught the country. After two years, the suggested policy paid off: the pesticide consumption came down by sixty-five percent while the production of rice increased by eighteen percent, greatly reducing the cost of cultivation and simultaneously resulting in a cleaner environment. During this period he also served in several key assignments internationally, including Chairmanship of the International Council for Research on Agro-Forestry and of the Society for the Promotion of Wasteland Development. In 1984, he was elected President of the International Union for the Conservation of Nature (IUCN). Under his leadership, IUCN prepared a draft Biodiversity Plan that was discussed at the IUCN General Assembly at San Jose, Costa Rica. He served as Vice-President of the World Wildlife Fund and actively worked for the Draft Convention on Biological Diversity. He also served as Chairman of the Advisory Panel on Food Security, Agriculture, Forestry, and Environment, of the World Commission on Environment and Development. He produced the landmark report, *Food 2000: Global Policies for Sustainable Agriculture* (1987). Among the numerous awards and honors he received during his IRRI years, the most notable was the First World Food Prize in 1987.

On completion of his contract with IRRI in 1988, Dr. Swaminathan decided to return to India and use his World Food Prize money to establish the M.S. Swaminathan Research Foundation. It would be of interest to note here that in the

1960s, Sir C.V. Raman, the Nobel Laureate physicist, had approached Swaminathan to join his Indian Institute of Science, which he had founded in 1934 in Bangalore, South India. The Institute was doing outstanding research, but Raman was growing old and needed someone to carry on his vision. He knew the scientific outreach and visionary leadership abilities of Swaminathan and wanted him to devote his extraordinary talents as a scientist-statesman toward building up the Institute by adding divisions of biology and genetics research and other emerging frontier sciences. But Swaminathan had his own life's work focused on agricultural research and food production, so he could not accept the offer. In 1970, Dr. Raman came again to stay with him for a couple of days, and before leaving got Swaminathan to promise that when he retired from government service, he would start his own research institute.

So, within days of his arrival in India, Swaminathan started working toward the establishment of the M.S. Swaminathan Research Foundation (MSSRF). It was registered in New Delhi with the initial funds of the World Food Prize of two hundred thousand dollars. The stated mission of MSSRF was "to organize research and training designed to promote a job-led economic growth strategy in rural areas, based on a pro-nature, pro-poor, and pro-women orientation to technology development and dissemination." Two years later, he decided to shift the headquarters to Madras (now Chennai), because of his special interest in the region, to initiate projects like the coastal ecosystems.

The logo he designed for MSSRF reflects "the open-ended, many-sided, and continuous evolution of humankind-nature interaction". All the plans for various departments and buildings were to reflect this spirit and all the buildings were to have a message of "Harmony with Nature". He worked with a cousin who was an architect to ensure that the building would have provision especially for harvesting both sunshine and rain. The garden in the main building was planned to be based on the Sangam literature, which divided Tamil Nadu into five agro-ecological zones: *Kurinji* (hill area), *Mullai* (forest), *Marutham* (farmland), *Neithal* (coastal area), and *Palai* (desert). (It is interesting to note here that when this classification was made over two thousand years ago, there was no *Palai* in Tamil Nadu; however, Tolkappiyar, who made this classification, had felt that human beings would, in course of time, convert good land into desert, so made provision for it.)

The keynote of MSSRF was to be "partnership" as an "institution without walls", involving everyone without any barriers. Swaminathan made a special effort to look for young men and women who were interested in dedicating their lives to serving the people by working with farming families on their fields. The Foundation was to function as a teaching, training, research, and archival institution by involving the workers to gain new knowledge and skills to use in the field, and to bring back their experience to help others both at the grassroots level and for advanced research in frontier sciences in line with Swaminathan's definition of sustainable development as producing high yields without any harm to the environment.

The programs of the Foundation were planned to combine the best of traditional and new frontier knowledge and technologies needed for transforming agriculture and other key areas of socio-cultural development. The following five major areas were chosen to form the core of the research activities: 1) Coastal Systems Research, 2) Biodiversity and Biotechnology, 3) Ecotechnology and Food Security, 4) Gender and Development, and 5) Informatics. All these areas were to be coordinated to serve as a symphonic and synergetic unity for social development. Among the many excellent projects and programs, the following deserve special mention.

The MSSRF Gene Bank is a walk-in cold storage where seed materials of cereals, pulses, and medicinal and endangered plants are preserved to represent the traditional biodiversity of the region as well as new materials accessible to all the farmers. The MSSRF Community Agrobiodiversity Centre at Wayanad in Kerala, an extension of the Gene Bank, has the mission to strengthen community biodiversity conservation systems by collecting and categorizing rare and endangered plants in the region. Swaminathan played an important role in the adoption of the Convention on Biological Diversity at Rio de Janeiro, in 1992, which defined biological diversity as "variability among living organisms from all sources, including, among others, terrestrial, marine, and other aquatic ecosystems and the ecological complexes of which they are part; this includes diversity within species, between species, and of ecosystems." According to a recent global biodiversity assessment report published by the United Nations Environment Program, there are approximately 13.6 million species on our planet of which only about two million are known. Swaminathan is greatly devoted to doing his best to learn and preserve the uses of as many of these as possible, because of their great value to the environment and the future development of humanity.

In 1989, Swaminathan proposed, at the World Climate Conference in Geneva, to initiate an anticipatory research program to meet the challenge of the adverse effects of global warming, especially the rising of sea levels. The following year he started a Genetic Resource Centre at MSSRF for adaptation to sea level rise for carrying on a scientific program designed to conserve mangrove wetlands in the coastal areas, and to transfer genes for sea water tolerance to rice and other crops. The program has been notably successful, especially with rice plants with genes for salt tolerance tranaferred from mangrove trees in South India.

In 1991, he launched a Biovillage Program in Pondicherry, with nineteen villages, designated as "Biovillages", where the rural poor are introduced to starting small enterprises like raising vegetables, mushrooms, floriculture, aquaculture, and poultry; for designing and carrying on these projects on the principle of self-help, free from outside aid and control; and for training to help them acquire the necessary agricultural and administrative knowledge and skills to become self-reliant. This program has been an exemplary model throughout the country for bringing semi-literate young men and women to the forefront of new and dynamic leadership roles in community development. By encouraging the emergence of a new generation of leaders, Swaminathan has demonstrated that poverty and lack of formal education are no barriers to acquiring new scientific knowledge and skills, when their application and benefits are demonstrated to the people and the programs are participatory and environmentally sound.

The MSSRF Technical Resource Centre, the first of its kind, was established: for implementing the recommendations of the 1992 Rio de Janeiro Convention on Biological Diversity, for collecting primary data from all the states of India, for recording the contributions of rural and tribal families to the conservation and enhancement of agrobiodiversity for training young tribal and rural families to compile local biodiversity inventories, for revitalizing their traditional practices, for establishing multimedia databases and reference centers for sharing information and knowledge, and for developing new community projects.

In 1998, a new dimension, "Village Knowledge Centres", was added to the biovillage program to connect villages through a two-way wireless system for sharing simultaneous information, data, and needed experience. These Centres are also simultaneously connected with MSSRF through telephones and computers twenty-four hours a day. Each community provides the necessary volunteers, space, equipment, and funds to ensure the continuity of the programs – focused on health, education, markets, weather forecasts, government schemes, bank loans – through information technology in societies where print literacy has held back millions for centuries. Many of the beneficiaries of these programs are making important contributions, for example, through their mutual sharing of first-hand knowledge of hundreds of herbs that have been traditionally grown in their respective areas.

Over the years, MSSRF has organized several hundred conferences, seminars, and consultation programs to explore the ways and means of stimulating food productivity, of eradicating hunger and poverty, and of creating employment opportunities, and sustainable development. Training programs in food processing and water and sun harvesting and management help to snowball such activities. Community health care and education are at the root of all such activities. Partnership and alliance are the key principles of people empowerment programs for mobilizing their initiative and creativity for employment generation. Small-scale credit facilities and community banking systems support such enterprises.

Swaminathan has also worked toward establishing a global network of Community Food Banks (CFBs), especially at strategic spots in Asia and Africa, to store quantities of grain staples – averaging about one ton for every four persons, to meet the needs of the people for at least a year. These banks also aim at serving the bigger purpose of providing nutritional needs of pregnant and nursing mothers, infants, pre-school children, and the elderly, and to ward off the catastrophic effects of natural disasters. For India to become hunger-free by August 15, 2007 (the sixtieth anniversary of India's independence), Swaminathan has launched a movement to establish one hundred thousand CFBs. For a Hunger-Free World, he has proposed the idea of establishing an International Bank for Nutrition for All, to strengthen the work of the United Nations World Food Program.

Many of Swaminathan's ideas and projects have attracted world attention, but perhaps none like his ideal of the Ever-Green Revolution. For example, E.O. Wilson writes in his book, *The Future of Life* (2002): "The aim of this new thrust [the Ever-Green Revolution) is to lift food production well above the level attained by the Green Revolution of the 1960s, using technology and regulatory policy more advanced and even safer than those now in existence." Swaminathan believes that such a revolution can only come with the help of small-scale farmers by combining their traditional approaches with breakthroughs in modern science and technology. He also lays special emphasis on Ecotechnology, defined as "the blending of the ecological prudence and technologies of the past with the best in frontier technologies, particularly biotechnology, information and digital technology, space technology, nuclear technology, and management technology." In this process, he lays great emphasis on the importance of ethics and integrity as the foundation of progress.

As I see it, Professor Swaminathan's excellence, as both a person and a scientist-reformer, is rooted in his love of Life – not only of human beings, but of every life form in the Universe, especially flowers, plants, and trees. We all know how often he likes to quote these lines from Joyce Kilmore's poem:

I think I shall never see
A poem as lovely as a tree.

Another favorite quote is a folk saying he heard in Guyana: "The sky is held up by trees. If the forest disappears, the sky – the roof of the world – collapses and Nature and humans perish together." He once told me that when I visit him in Chennai, he will take me to see the famous banyan tree said to be more than five hundred years old, which now covers over 40,000 square feet with more than five hundred aerial roots, which are home to an amazing variety of birds, rodents, reptiles and insects – a wonderful ecosystem. One of his well-known sayings is: "At the heart of agriculture, there is a simple truth that in the growing of green plants, we can capture the energy of the sun, our most precious asset, and turn it into food, fuel, fodder, and other products. Hence, wherever man plants trees, grows vegetables in small pots, covers a tiny patch with grass or flowers, or covers the roofs of mud huts or brick buildings with vines, he is playing a role in our agricultural development." Another of his oft-quoted sayings is: "Agriculture is what you call applied ecology. Ecology means remaining in harmony with nature, and agriculture means using natural resources in a sound and scientific way. Therefore, anybody who is interested in agriculture, not only today but also for the future, has to think of the conservation of nature and natural resources."

Dr. Swaminathan's great love of the environment is at the core of all his work. I find his concept of environment very similar to Buckminster Fuller's in his little poem "Cosmic Plurality", composed as a gift for Indira Gandhi (Buckminster Fuller and Anwar Dil, *Humans in Universe,* New York: Mouton Publishers, 1983, p. 37):

Environment to each must be
All there is, that isn't me.
Universe in turn must be
All that isn't me AND ME.
Since I only see inside of me

What brain imagines outside me,
It seems to be, you may be me.
If that is so, there's only we.

In my estimation, Swaminathan is a beautiful exemplar of Buckminster Fuller's idea of a good human in the universe. He has devoted himself with exemplary missionary zeal to protecting and improving human environment, as evidenced, for example, in his protection of the mangrove forests. From 1990 to 1993, he served as Chairman of the International Society of Mangrove Ecosystems. In appreciation of his dedication to the cause, the Swaminathan Foundation has been designated as the Indian Ocean Regional center under the Global Mangrove Information Service.

Another most commendable aspect of his life's work in his noble struggle for women's rightful position in society. It has been acknowledged by many that Swaminathan, more than any other leader of India, has helped improve the status of Indian women. He believes that all women, irrespective of their origin and social position, are very hardworking and aspire for human dignity and full participation in all walks of life. He has written and spoken on integrated programs for achieving the empowerment of rural women in agriculture. All MSSRF programs give great prominence to providing new skills and training in new technology to help women.

Swaminathan has always believed that because women are closer to the soil and do most of the work of sowing, maintaining and harvesting crops, their knowledge should be gathered and shared for the overall improvement of food and agriculture. He refers to the experience in one village in Thailand where by pooling together the knowledge of rural women they were able to identify two hundred and thirty different plant species – many of which were rescued from extinction by women from a nearby forest as it was being cleared. He believes that women are the best resource for preserving the biodiversity of nature. MSSRF has set up a Women's Biotech Park near Chennai, where women professionals are provided opportunities for innovative projects, for example, the manufacture of herbal cosmetics.

Swaminathan's great love of children and his devotion to improving their nutrition, health, and education is another praiseworthy dimension of his personal and professional life. I have heard him quote from memory some of the most beautiful sayings about our duty towards children, especially before they are born and during their early years. He speaks with special concern about low birth weights that are the results of poor nourishment of mothers and that cause severe damage to the physical and mental health of children, especially in the developing countries. For example, in his keynote address at the World Conference of Science in Budapest, in June 1999 (Chapter 30 in this volume), he called upon all United Nations agencies, in collaboration with all national governments, to address the problem by launching a high-priority world program focused on fighting maternal and fetal under-nutrition. He called this initiative the "Global Movement for Children for Happiness". He believes that the content of education at the elementary and secondary levels must be changed drastically to include topics that are relevant to the children's real needs, for example, how to grow up healthy and strong.

Professor Swaminathan is a most inspiring teacher, Dr. R.D. Iyer,[2] his biographer and devoted student-friend from his IARI days, recalls: "Those were exciting days in the IARI experimental fields, when Swaminathan would come on his 'Robinhood' bicycle with a cane basket hanging in front of the handlebar, weighed down with books and notes meant for the morning 8:15 class. After the field visit, he would be in the classroom by 7:45 a.m., in rain, shine or winter of Delhi, and would fill up all available space on the blackboard with tables, charts, references, and figures meant for the morning lecture on the Cytogenetics-I Core Course." One of his inspirational sayings for his students and co-workers, many of whom have carried it around the world, was: "Choose a problem that is relevant to the needs of the people and put in excellent and thorough efforts to get meaningful results."

In May 2003, Professor P.C. Kesavan, the Executive Director of MSSRF, who also holds the Homi Bhaba Chair for Nuclear Sciences and Rural Development, told me, during our meetings at MSSRF, that he is very proud to have been one of the scores of Dr. Swaminathan's doctoral research students at IARI: "In the 1960s, he was already an internationally acclaimed cytogeneticist, a radiation geneticist, and a mutation breeder of crop plants. He had developed Cobalt-60 gamma garden and used ionizing radiation to induce beneficial mutations in crop plants, to overcome cross-incompatibility barriers in plants, and to elucidate 'genetic buffering' effects in the irradiated polyploid wheats. His work in the area of polyploidy and radiosensitivity is as epoch-making as his treatise on cytogenetics of potatoes. One fine morning in November 1963, Professor Swaminathan walked into the lab and handed me a piece of paper with two words on it – 'Oxygen Effect' – and said: 'Oxygen enhancement of radiobiological effect needs greater investigation in radiation killing (or not efficiently killing the less oxygenated cancer cells!) and radiation mutagenesis.' Those two words determined the course of my research as well as that of more than thirty of his students over the next two decades." Dr. Kesavan added that, by some divine design, he is back with Professor Swaminathan after some thirty years of parting and enjoys being his student all over again. They only change is the area of his activity, from the laboratory-oriented basic research to more field-based approaches, involving the use of science and technology to alleviate rural poverty.

Swaminathan is very lucky to have married Mina Bhoothalingam, who was been a major inspirational influence in his life since 1955. Her father was an eminent economist and civil servant, and her mother is a well-known writer and lecturer. Mina has a Masters degree from the University of Cambridge, and

[2]Dr. R.D. Iyer's *Scientist and Humanist: M.S. Swaminathan* (Mumbai, India: Bharatiya Vidya Bhavan 2002) is an excellent book on Swaminathan's life and work.

worked in the Planning Commission in Delhi before they were married. Since her marriage, she has carried on her independent career as a teacher and social worker for women's and children's rights, while supporting her husband in his work. Swaminathan acknowledges: "My wife has been my principal guide and source of inspiration. She has encouraged my work, at considerable sacrifice to her own professional and personal life. Mina is a person with a unique combination of qualities. Her sense of values and her conviction that the future of India depends upon the education of children have provided much of the stimulus for my work." They have three daughters – Soumya, Madhura, and Nitya – who were provided excellent opportunities for education and learning and who all have careers of their own. Soumya is a pediatrician, Madhura is an economist, and Nitya works in the field of gender and rural development. Swaminathan says: "They have provided me windows into the thought processes and dispositions of the post-Independence generations. My daughters have helped me remain young in my thinking and have helped me bridge the generation gap in terms of values and aspirations. My three sons-in-law and five grand-children make me a very happy and proud parent and grandfather."

Dr. Swaminathan is both a good human being and an important transformational leader of our time – a sage in the classical tradition of Indian knowledge and wisdom, who is shaping the course of humanity for a better future for all. Borrowing the words of the German poet Rainer Maria Rilke, I should say that Dr. M.S. Swaminathan is one of the "special people" who "move to broader laws" and change the world for a more hopeful tomorrow.

It has, indeed, been a most enriching experience for me to put together this book in the hope that this truly great man's life and work, as it unfolds here mainly through his own writings and some by his close associates, will inspire readers around the world – especially our young men and women – to understand what one man can do for the good of millions for generations to come.

5. Anwar Dil, Norman Borlaug and MSS, World Food Prize ceremonies, Des Moines, Iowa, October 18, 2000

6. MSS and Anwar Dil working on the book, 2003

7. Monkombu Krishna Sambasivan
M.S. Swaminathan's father

8. Parvathi Thangammal.
M.S. Swaminathan's mother

9. The house Swaminathan was born in
Kumbakonam, Tamil Nadu

10. Swaminathan's ancestral village, Monkombu

Growing Up in Tamil Nadu

11. MSS with his uncle, 1941

12. Native High School, Kumbakonam

13. MSS, 1944

14. MSS, 1951

Family

15. Marriage, April 11, 1955

16. Mr. & Mrs. Swaminathan
New Delhi, 1955

17. With wife, Mina, and daughters, (L to R) Nitya, Madhura, Soumya, 1968

18. With wife Mina in Tokyo to receive the Honda Award, 1992

19. With wife, daughters and sons-in-law, 1986

20. With children and grandchildren, 1992

At IARI (Indian Institute of Agricultural Research, New Delhi)

21. In his laboratory, IARI, 1961

22. In the field, IARI, 1963

23. With Norman Borlaug, IARI, 1964

International Service

24. At International Rice Research Institute (IRRI), The Philippines, 1983

25. At the Chinese Academy, Beijing, 1986

26. At FAO, Rome, 1985

27. With Dr. Jacques Diouf, Director General of FAO, 1998

28. With Dr. Elizabeth Dowdeswell, Executive Director, United Nations Environmental Programme, receiving the UNEP Sasakawa Environment Prize, 1994

M.S. Swaminathan Research Foundation (MSSRF)

29. Logo of MSSRF

30. Digging of foundation at MSSRF, April 14, 1991

31. The first team, MSSRF, 1991

32. Shrimati M.S. Subbulakshmi at inauguration of MSSRF, 1993

33. With C. Subramaniam, Norman Borlaug, at the opening of Borlaug Hall, 1996

34. A view of MSSRF

35.Courtyard of MSSRF

Honorary Degrees

36. Oregon State University: MSS, Norman Borlaug, John V. Byrne, Robert MacVicar, Kenneth E. Clark, 1988

37. University of Bologna, Italy, 1992

38. University of Massachusetts, 2001

Awards

39. Shanti Swarup Bhatnagar Award, 1965

40. Ramon Magsaysay Award,
The Philippines, 1971

41. The First World Food Prize, 1987

42. Padma Vibhushan Award by
Shri R. Venkataraman,
President of India, 1989

43. The Golden Heart Award by
Mrs. Corazon Aquino,
President of the Philippines, 1987

44. UNESCO Gandhi Award by
Prof. Federico Mayor,
Director General of UNESCO, 1999

PART I: SELECTED WRITINGS AND ADDRESSES BY M.S. SWAMINATHAN

I.A
Toward a Hunger-Free India

45. MSS in the field, Indian Agricultural Research Institute (IARI), 1960

46. With Norman Borlaug, IARI, 1963

47. C. Subramaniam, Minister of Food and Agriculture, India (1964-67)

48. S.P. Kohli, MSS, Norman Borlaug, V.S. Mathur in the field, 1966

49. Prime Minister Indira Gandhi releasing India Wheat Revolution stamp, 1968

50. At IARI with workers, 1964

1

The Origin of the Early European Potato: Evidence from Indian Varieties

There are two centers from which potatoes could have been introduced into Europe. These are first, the Chiloe region of South Chile (40° to 45° latitude) and second, the Andes at tropical latitudes. Bukasov (1930) and his co-workers suggested that the Chiloe region was the place from which the original introductions were made since the Chiloe potatoes were morphologically more similar to European potato varieties (*Solanum tuberosum sensu stricto*) and also were adapted to long days in contrast to the potatoes from the region of the Andes, which were typical short-day types doing best in a 12-hour-day. On the strength of this reasoning, Juzepczuk and Bukasov grouped the cultivated Andean tetraploid varieties (2n=48) into a new species *S. andigenum* and considered that they had an independent origin from that of *S. tuberosum* (this name was given to the European potato by Gaspard Bauhin in 1596 and was retained by Linnaeus).

Salaman (1937, 1946, 1954) and Salaman and Hawkes (1949) have disagreed with the view of the Russian workers and suggested that the original introductions to Europe came from Colombia, Ecuador or Bolivia (the Andean region). Van der Plank supported Salaman's views from the results of his study of the potato varieties growing in Basutoland in South Africa, where potato had been introduced from Europe in 1833, i.e., before the blight epidemics of 1845 had destroyed most European varieties and led to the extensive breeding of new varieties. These Basutoland varieties, like the old Andean potatoes, were typical short-day types.

Though little is known about the history of the introduction and spread of the potato in India, it is likely that the potato was brought into India either by the Portuguese or later by the British. Watt has stated that the first mention of the potato in India occurs in Terry's account of the banquet given by Asaph Chan to Sir Thomas Roe at Ajmer in 1615. Pal and Pushkarnath (1951a) whose detailed study has cleared the confusion that prevailed in the nomenclature of the Indian potato varieties, have recognized sixteen distinct *desi* (native) varieties, besides many foreign named varieties in India. It is probable that the *desi* varieties are relics of some of the early introductions into India from Europe. This view receives support from the fact that, of the Indian potato varieties sent by Pal and

---

First published in *Indian Journal of Genetics & Plant Breeding* 18(1): 8-15 (1958).

Pushkarnath (1951a) to the United Kingdom for identification, a large proportion proved to be very old varieties which have gone out of cultivation in that country. Practically no conscious selection or hybridization had been practiced in the *desi* varieties until the year 1935 when Dr. B.P. Pal initiated the potato improvement work at the Indian Agricultural Research Institute. Also, there is little scope for natural hybridization to have taken place among the *desi* varieties since, with a few exceptions, potato varieties rarely flower or form berries in the plains of India. Thus, while some of the *desi* varieties may be the result of chance selection of naturally occurring seedlings in varieties like Phulwa which set berries and seeds, the bulk of the genetic changes in the introduced potato material would have taken place only through somatic mutations. Such changes do not seem to be many at least with reference to commercial characteristics like plant habit, stolon length, depth of eyes, etc., since the *desi* varieties largely resemble the unselected *andigenum* types growing in South America in this respect. A study of these varieties is, therefore, likely to yield some information on the characteristics of the potato varieties grown in Europe in the late sixteenth and the early seventeenth centuries, i.e., soon after the introduction of the potato into Europe from the New World. The present study of some important *desi* potato varieties was hence started.

Materials and Methods

The *desi* varieties studied were Phulwa, Gola and Darjeeling Red Round. These had been grown in the fields of the Botany Division. Phulwa was the only variety that flowered at Delhi and cytological studies were consequently confined to that variety. Phulwa plants grown in some farmers' fields at Delhi and at Mohidinpur near Meerut were also used in the study. The morphological characters of the variety Up-to-date grown at Delhi were studied and compared with those of the *desi* varieties. Some plants – believed to be long-day adapted mutants of the variety Phulwa – which occurred in a farmer's field at Gola Gokarnnath (U.P.) were studied for their morphological characters and cytological behavior during the years 1956 and 1957.

For study of meiosis, the propino-carmine PMC schedule outlined by Swaminathan et al (1954) were used. Leaf indices were calculated in the manner mentioned by Pal and Pushkarnath (1951a).

Discussion

The fact that the present-day commercial potato varieties of Europe grow and tuberize best during the long summer days of temperate regions induced Bukasov and the Russian workers to postulate that the short-day types of the Andean region could not have been the progenitors of the European potato. As against this view, a formidable array of evidence has been gathered by the late Dr. R. N. Salaman, the chief among which are listed here: a) At the time when it is known that the potato introductions took place into Europe, the Spanish hold on Chile was tenuous, but they had conquered the Andean region some fifty years earlier. b) Magellan had not then investigated the straits bearing his name and it seems unlikely that the potatoes from Chile could have survived the long overland journey back to Europe. c) The study of a series of some twenty herbarium specimens of the potato introduced or cultivated in England and Western Europe during the seventeenth and eighteenth centuries, starting with a preparation left by Gaspard Bauhin himself, reveals the gradual evolution of the European commercial potato from a typical *S. andigenum* type. d) The woodcut of the potato found in the Plantin Museum at Antwerp, which depicts the type of potato plant received by Clusius and distributed by him throughout Western and Central Europe, conforms to the *S. andigenum* type. e) A study of the leaf index and other characters of the native present-day potato varieties from the Chiloe province of Chile shows that these varieties are straightforward examples of the species *S. andigenum*.

On the basis of the above evidence, Salaman (1954) has concluded that the domestic potato of Europe, though given the specific name of *S. tubersosum*, is in a variety of the species *S. andigenum* and has acquired the present range of characteristic forms of leaf and habit as a result of prolonged selection for higher cropping on the one hand and early or late maturity on the other.

Hawkes (1944) has supported the view of Salaman and in addition has suggested that the first European potatoes may have been long-day or day-neutral types since such types do occur occasionally among Andean varieties, and that, even if they were short-day types, they may have grown fairly well in the low latitudes of Southern Europe to which they were first introduced. The short-day character might have been subsequently lost when selection for high yield took place in progenies grown from true seed. Hawkes has also pointed out that *S. andigenum* and *S. tuberosum* differ by a number of minute features and not by one sharp, clear-cut character. Swaminathan has reported that quadrivalents are formed at first metaphase of meiosis in varieties of *S. andigenum* as in varieties of *S. tuberosum*. The two species cross readily and the hybrids are fertile. There is hence no cytogenetic basis for treating these two species as distinct.

There is also evidence to suggest that the genes governing photoperiodic reaction in *S. tuberosum* are relatively unstable. In many cultivated varieties, a frequency of "bolter" plants (plants with greater height, stiffer foliage, smaller leaflets, later maturity, greater capacity for flower-bearing, higher pigmentation and longer stolons and coarse tubers) occur. Hawkes (1947) showed that bolters, when grown under short days, are indistinguishable in most of their features from normal plants and thus confirmed the suggestion of Carson and Howard (1944) that the change from normal to bolter is due to a mutation of one or more unstable genes controlling response to day-length. Hawkes also suggested that the bolter is a reversion towards the ancestral type of short-day adapted potato found in the Andean region. The fact that the odd plants reported in this paper are most probably long-day adapted mutants of a *desi* variety strongly suggests that the reverse mutation i.e., from a short-day or day-neutral type to a long-day adapted condition, also occurs in the potato. The conversion of *andigenum* strains into long-day adapted *tuberosum (sensu stricto)* varieties could have taken place by some such process.

The *desi* potato varieties of India have many of the characters of the *andigenum* type of potato. Due to the fact that only Phulwa flowers at Delhi, cytological studies could be carried out only in that variety. At first metaphase of meiosis in Phulwa, a mean frequency of four quadrivalents per cell occurred. In most of the European potato varieties, two to three quadrivalents per cell are found at first metaphase. The writer has also pointed out that *S. tuberosum* originally arose as an autotetraploid and that in the current commercial varieties the frequency of quadrivalent formation is to some extent correlated with the degree of hybridization involved in the evolution of those varieties. Seen in this context, the higher frequency of quadrivalent formation in Phulwa may be due to the fairly unaltered conditions in this variety with reference to the homology among the four sets of chromosomes. This may be due to the absence of extensive hybridization and sexual reproduction and also to a low selection pressure. Thus, if Phulwa is a derivative from one of the early introductions into India from Europe, its characters may be comparable to those possessed by the early European potato. The fact that it has, like *S. andigenum*, an open leaf habit, a low leaflet index, long stolons, deep eyes in tuber and a day-neutral or short-day photoperiodic behavior would further support the view that the early European potato possessed such characters.

Besides being of historic and academic interest, the study of the relationship between *S. tuberosum sensu stricto* and *S. andigenum* and of the origin of the early European potato is of particular interest in relation to the breeding of higher yielding varieties suitable for cultivation under short-day conditions in the plains of India as well as in other sub-tropic regions. The varieties of *S. andigenum* yield much less even in their native habitats than the European potato varieties. Selection for yield and culinary characteristics and hybridization to combine desirable characters during the last two centuries has led to such an accumulation of the polygenes controlling these characters in the European varieties that the transformation of *S. andigenum* of South America into the *S. tuberosum* of Europe has been compared by the late Dr. R.N. Salaman to the evolution of the domesticated horse from the shaggy, sturdy pony of paleolithic times.

From the point of view of breeding suitable varieties for cultivation in the sub-tropics and tropics, it will probably be more rewarding and less time-consuming if emphasis is laid upon incorporating the day-neutral or short-day photoperiodic behavior in the most high yielding and desirable *tuberosum* varieties instead of trying to improve the yield potential of the *andigenum* types. So far, attempts have been made to evolve suitable varieties for the plains by crossing the *desi* varieties with the best commercial potatoes of Europe and North America. Pal and Pushkarnath (1951b) have pointed out that in such crosses the semi-wild characteristics of the *desi* types are dominant and that the scope for selection of desirable characteristics is limited. It appears to the writer that a fruitful approach will be to try to induce mutations of the genes controlling photoperiodic reaction, which may not be difficult in view of the rather unstable nature of these genes as suggested by the high spontaneous mutation rate for this character. As a preliminary to such studies, tubers from bolter plants of improved European varieties should be grown in the plains of India and their growth and yield characteristics compared with the plants raised from the normal tubers of the same varieties. If bolters represent only photoperiodic mutants, their yield potential may not be greatly affected in comparison with that of the parental non-bolter plants. This will however hold good only if there is no positive correlation between long-day length response and high tuber yield. Work on these lines is in progress at the Indian Agricultural Research Institute and the data already obtained suggests that this approach may be a fruitful one.

Summary

The *desi* potato varieties of India appear to be relics of strains introduced into Europe during the late 16th century. Since the *desi* varieties possess all the characters attributed normally to the *andigenum* type of potato, it is likely that the potatoes first introduced into Europe were, in accordance with the views expressed by the late Dr. R.N. Salaman, from the Andean region. Many of the long-day adapted European commercial potato varieties appear to have been evolved from the original Andean introductions. Evidence from the study of bolter plants of European and Indian varieties suggests that the genes controlling photoperiodic response in the potato are rather unstable in nature. From the point of view of breeding potato varieties suitable for cultivation in the sub-tropics and tropics, it would therefore be desirable to try to induce day-neutral or short-day adapted mutants in the most high-yielding and desirable *tuberosum* varieties, instead of merely trying to improve the yield potential of the *andigenum* types by selection among hybrids between them and the European and North American commercial varieties. The latter course is subject to the limitation, as pointed out by Pal and Pushkarnath (1951b) that in such crosses the semi-wild characters of the *desi* types are dominant and that the scope for selection of desirable characteristics is poor. Initiation of mutation research in the potato with particular reference to the induction and isolation of photoperiodic mutants is hence desirable.

References

Bukasov, S.M. 1930. The cultivated plants of Mexico, Guatemala and Colombia. *Bulletin of Applied Botany, Leningrad Supplement* 47:191-226.

Carson, G.P. and Howard H.W. 1944. Inheritance of the "bolter" condition in the potato. *Nature* 154: 829.

Hawkes, J.G. 1944. Potato collecting expeditions in Mexico and South America II: Systematic classification of the collections. *IAB. Bulletin*, p. 142.

-----. 1947. The photoperiodic reactions of potato bolters. *Emp. Journal of Experimental Agriculture 15:216-26.*

Pal, B.P. and Pushkarnath. 1951a. Indian potato varieties. *ICAR Bulletin* 62: 63.

-----. 1951b. Potato breeding investigations in India. *Emp. Journal of Experimental Agriculture* 19:87-103.

Salaman, R.N. 1937. The potato in its early home and is introduction into Europe. *Journal of R. Horticulture Society* 62:61-77, 112-23, 153-62, 253-66.

-----. 1946. The origin of the early European potato. *Journal of Linn. Society (Botany)* 53:1-27.

-----. 1954. The origin of the early European potato. *Journal of Linn. Society (Botany)* 55:185-90.

-----. and J.G. Hawkes. 1949. The character of the early European potato. *Proceedings of Linn. Society of London* 161:71-84

2

The Impact of Dwarfing Genes on Wheat Production

Introduction

During 1950s several dwarf wheat types were isolated in Japan from the cross Turkey Red Fulz-Daruma leading to the origin of Norin 10, Norin 14, and other Norin varieties. Unlike the earlier dwarfs known in wheat, like compactum, sphaerococcum and forms with club and dense ears, the Norin dwarfs revealed no association between plant height and ear density. Thus, long and lax ears could occur in plants not exceeding one to two feet in height. The dwarf genes were transferred by wheat breeders in the United States and Mexico to commercially acceptable wheat varieties. The "Gaines" wheat released in the United States was one of the first dwarf varieties to be evolved by making use of the Norin dwarf genes. This variety of winter wheat, because of its short stature, can withstand fertilizer applications of the order of 100 lbs nitrogen per acre and can give yields as high as seven to eight thousand kg/hectare. It created a new record for wheat yield in the world. The dwarf genes were also transferred quickly into spring varieties at the International Center for Maize and Wheat Improvement (CIMMYT) in Mexico, which is a cooperative program between the Mexican Ministry of Agriculture and the Rockefeller Foundation. Varieties like Sonora-63, Sonora-64 and Lerma Rojo-64A were thus developed and released for cultivation in Mexico in 1963. These varieties, when given fertilizer dosages of 100-150 lbs nitrogen and 75 lbs each of P and K along with adequate water, gave in Mexico 5,000 to 7,000 kg/hectare. From the studies carried out in Mexico by Dr. N.E. Borlaug and his colleagues, it is clear that in the original dwarf material obtained from Japan, there are three major genes for dwarfing. When all the three genes are present, the plants are very short and when only one of them is present, they grow to about three feet in height. The availability of these genes has hence made the breeding of wheat varieties of different heights possible. A wide range of segregation for height occurs when a triple dwarf is crossed with a tall strain. Another interesting feature of these dwarfing genes is their insensitivity to exogenous applications of gibberellins.

It is now known that if lodging can be prevented, the wheat plant can make use of much higher levels of soil nutrition and give much higher quantities of grain. If, therefore, the yield of wheat is to be increased considerably in irrigated areas through suitable fertilizer and water management practices, the first requisite is a morphological frame which is conducive to the efficient utilization of chemical energy in the form of applied fertilizers and solar energy through

---

Presented at the All India Wheat Research Workers' Seminar, New Delhi, August 1964.

Table 1. Material received from Mexico

S. No.	Material	No. of lines	Places where sown during rabi 1963
1.	Released varieties	4	Delhi, Ludhiana, Kanpur, Pusa, Pant Nagar
2.	Near East American Spring Wheats	25	Delhi, Ludhiana, Pusa
3.	Highest yielding lines from Sonora	29	Delhi, Ludhiana, Pusa, Wellington, Bhowali
4.	Best F_7 lines from Sonora	67	–do–
5.	Best F_5 and F_6 lines from Sonora	42	–do–
6.	F_4 progenies from Sonora	30	Delhi, Ludhiana
7.	F_3 progenies from Sonora	211	Delhi, Ludhiana, Pusa
8.	F_2 progenies from Sonora	209	–do–

increased photosynthetic activity. From the work done in Mexico, reduction in plant height does not appear to involve a coincident loss in straw yield because of the high tillering ability of the dwarf strains acting as a compensatory factor.

Wheat Material Obtained from Mexico

Realizing the importance of breeding as quickly as possible short and stiff strawed varieties for being grown in the Intensive Agriculture Development Programs Districts, a request was made on my suggestion by the Indian Agricultural Research Institute (IARI), through the Ministry of Food and Agriculture, to the Rockefeller Foundation for the services of Dr. Norman E. Borlaug in 1962. Dr. Borlaug visited India in March 1963 and paid brief visits to the important wheat breeding centers in the country. On the basis of his observations, Dr. Borlaug assembled a large assortment of wheat material and sent them to the IARI in November 1963 for being tried under our conditions. The types of material received and the places where they were grown are shown in Table1.

Observations were recorded in this material by Dr. S.P. Kohli, Ms. P. Reddy and Mr. George Verghese at Delhi, Dr. D.S. Athwal, and his colleagues at Ludhiana, Dr. Y.R. Mehta and his colleagues at Kanpur, Dr. S.V.S. Sastry and Mr. P.S.L. Srivastava at Pusa, Mr. R.K. Agarwal and Mr. K.L. Sethi and Shri Butchi Raju at Wellington. Dr. Borlaug came to India on a short visit in March 1964 and saw the material at Delhi, Ludhiana and Pusa.

Dr. I. Narvaez of Mexico and Dr. R.G. Anderson of Canada also went through the material at Delhi and Ludhiana and gave us the benefit of their advice with regard to the handling of the segregating generations, besides helping in making selections.

Performance of the Four Released Varieties

Sonora 64, Mayo 64, Sonora 63 and Lerma Rojo 64 had been received in bulk quantities and were hence included in major yield trials along with local checks at Delhi, Kanpur, Pant Nagar and Ludhiana. The yield data are summarized in Table 2.

Table 2. Yields recorded in trials conducted in 1963-64 with Mexican and some Indian wheats

S. No.	Variety	Delhi				PantU.P. Nagar	Ludhiana	Kanpur
		Trial I	Trial II	Trial III	Multiplication			
1.	Sonora 63	3833	3756	4559	4495	2496	3836	3433
2.	Sonora 64	4287	–	–	4084	2947	3954	4015
3.	Lerma Rojo 64A	2950	3790	–	3667	3759	4598	3243
4.	Mayo-64A	3558	–	3650	–	2947	3937	2675
5.	N.P. 824	–	–	2800	3737	2315	–	2246
6.	N.P. 876	3796	–	–	2928	2672	–	2740
7.	N.P. 852	3684	–	–	–	–	–	–
8.	C. 273	–	–	–	–	–	3584	–
9.	C. 302	2588	–	–	–	–	–	–
10.	C. 306	–	2747	–	–	–	3596	–
11.	C. 313	1587	–	–	–	–	–	–
12.	Rs. 31-1	2153	–	–	–	–	–	–
13.	MP./Hyd. 65	2615	–	–	–	–	–	–
14.	K. 64	–	–	–	–	–	–	2186
15.	K. 68	–	–	–	–	–	–	2186

It will be seen from the data that Sonora 63, Sonora 64 and Lerma Rojo 64 on the whole gave good performance. At Kanpur, Sonora 64 was the best and was followed by Sonora 63 and Lerma Rojo in that order. At Ludhiana, Lerma Rojo showed the best performance and was followed by Sonora 64 and Sonora 63. At Pant Nagar, Lerma Rojo was again the best performer followed by Mayo 64, Sonora 64, N.P. 876, and N.P. 824. Mayo 64 generally yielded less in comparison to the other Mexican wheats.

Sonora 63 and Sonora 64 showed a high degree of resistance to black and brown rusts but developed yellow rust at Ludhiana. This was the reason for their poor performance at that place. Lerma Rojo, on the other hand, was quite resistant to all the three rusts. Sonora 63 and 64 may be suitable for the Central Indian conditions where black and brown rusts are the predominant diseases. Sonora 64 is about fourteen days earlier than Sonora 63. The former may be better for shorter growing conditions and for late sowing after paddy or cotton. Mexican wheats withstood lodging much better than all the Indian varieties. Of the latter, N.P. 876 and N.P. 852 were the best.

Performance of the Highest Yielding Lines

The twenty-nine highest yielding lines received from Mexico were grown along with N.P. 824, N.P. 876, C. 281, Sonora 63, Sonora 64, Mayo 64 and Lerma Rojo in a trial at Delhi. The design adopted was randomized block with three replications. The data are given in Table 3. None of the selections was significantly superior to the four already released Mexican varieties. However, one selection (S. 227, same as V. 17 in the American Spring Wheat Nursery) was dwarf with heavy tillering, fully awned ears and bold, amber grains. This selection also did well at the other centers and was the best performer at Bhowali. This line segregated for resistance to brown rust and the rust-free plants were separately harvested. This selection offers possibilities for direct introduction.

Discussion

The material from Mexico was received only during the middle of November 1963. Consequently they were sown mostly during the third week of November, and the yields obtained from these varieties during 1963-64 might not reflect their full yield potentialities. Due to the prolonged cold spell in January 1964, the rate of growth was slow. The plants were dark green until the third week of March 1964, and then they suddenly turned yellow. The transition from active growth to senescence was, thus, a sudden one and this seems to have affected the filling of grains. In this connection, Dr. Borlaug observed in a personal message: "I think that it is now very evident from the past year's results that a number of Mexican wheat varieties are very well adapted to Indian conditions. The two Sonora varieties and perhaps also Mayo 64 will be suitable for semi-commercial evaluation in areas where stripe rust (*Puccinia glumarum*) is not a problem. This means that none of these three varieties should be used in demonstrations or multiplication plots where this disease is likely to be serious. On the other hand, Lerma Rojo 64A is adequately resistant to stripe rust and can be grown successfully in such areas as the Punjab. After summarizing the data from the demonstration plots and the maximum yield potential plots at Delhi, as well as

Table 3. Yields obtained with the highest yielding lines from Sonora

Sample No.	Origin	Pedigree	Mean yield per plot in oz.
		Sonora 63	43.67
206	V163	Sonora 64A Reselection II8469A–2y–6c–6y–4c–10–1y–1c	35.00
226	V1034	$Yt54^4$×N10–B 9139–6y–2c–2y–2c	35.00
205	V170	Lerma Rojo 64	34.33
224	V1013	(Yt54×N10–B) IR^4 II9369A–12M–4R–1M	33.33
227	V17	Penjamo 62×Gabo 55 II8156–1M–2R–4M	33.00
212	V230	(21931–Ch53×AN)Y50 11893–4y–2c	32.67
		Sonora 64	32.00
213	V228	[($Cj54^2$×36896) Gb56] Y50 II 11881–17y–1c	31.67
		Mayo 64	29.00
228	V1091	Sonora 64	28.67
214	V231	(21931–Ch53×AN) Y50 II 11893–4y–3c	28.00
		Lerma Rojo 64	28.00
217	V1272	(21931–Ch53×An) Gb56 II 12300A–1y–1c–2y	27.67
222	V1309	(Lee^2 –KF×RL2944) (Sel.5×N10–B)$G60^2$ II 13139–11y–4c–3y	26.67
211	V216	(Kt54×N10–B) Cj60 II 13154–3y–7c	26.67
216	V267	($Cj54^2$ ×36896) $Gb56^2$ II 9445A–14c–3y–2c	26.33
209	V179	[(2193–Chapingo 53) AN] Gb56 II 12300A–Sy–14c	26.33
225	V1048	33896–Ch53×LR^3 II 9648–2R–2M–3R	26.33
208	V161	($Cj54^2$ ×36896) $Gb56^2$ II 9445A–14c–3y–2c–2y	26.33
223	V1326	[(Lee–Fn) (Yt54×N10–B)] II 53–526 II 1345–1y–2c–2y	26.00
201	V15	Andes 64A 8739–4R–1M–1R	26.00
210	V183	[($Cj54^2$ ×36896) Gb56] Y50 II 11881–18y–1c	25.33
221	V1308	{[(Lee^2 –KF×RL2944) (Sel.5×N10–B)] $Cj60^2$ } II 1181–18y–1c	25.33
207	V173	(36896×$Cj54^2$) Yt54A II 9434–4c–4y–3c–3y	25.33
220	V1313	(Kt54×N10–B) Cj60 II 13154–3y–4c–2y	24.67
215	V274	($Cj54^2$ ×36896) $Gb56^2$ II 9445A–14c–3y–2c–2y	24.67
202	V1071	(My54×N10–Brevor) Y54 II 7136–6R–5N–1R	24.00
218	V1236	(21931–Ch53×AN) Gb 56 II 12300A–19y–3c–3y	24.00
		N.P. 876	23.00
229	V1119	(1144–20×Lee) Sel.5×N10–B II 11041–1M–1R–1M	21.00
		N.P. 824	19.67
219	V1285	[(Fn×K58–N) N10–B] Y54 II 12963–5y–1c–1y	19.33
		C.281	18.67
		C.D. at 5% level = 7.72	

the data received from Dr. Athwal at Ludhiana, I am convinced that another factor, namely inadequate irrigation, limited the yield of all varieties. Sonora 64 and Sonora 63, the two earliest maturing varieties with the maximum yield potential trials were the highest yielding giving over 4,000 kg/ha. These yields are, however, not nearly as high as they should be for these varieties under conditions such as we have in New Delhi. We can expect them to yield in experimental plantings, when properly fertilized and adequately irrigated, between 5.5 and 6 tons per hectare. Why were these yields much lower than anticipated? The intervals between irrigations and perhaps also the lack of penetration of irrigation water during each irrigation were limiting factors. It may also be that the experiment was planted too late for maximum yield of these varieties. Another reason for water, or lack of water, being the limiting factor for yield, is the relative rank in yield of Lerma Rojo 64, which averaged only 2,992 kilos in one trial. This wheat, in last year's Near East American Yield Nursery was widely adapted everywhere. It was tested and was high yielding. It appears, therefore, that this variety suffered severely from lack of adequate irrigation."

According to Dr. Borlaug, the yields of Lerma Rojo 64, Sonora 64, Sonora 63, Variety V 17 (cross 8156 IM-2R-4M, amber grained) should be at least six tons per hectare. Until this is attained it should be recognized that there are agronomic factors which are limiting their yield potential and studies should be undertaken of all the different factors governing yield and attention paid to those which are limiting. On the factors affecting the yield of paddy, Chandler (1963) observed: "It is disturbing to read paper after paper, from various research and educational organizations, experimenting with rice, in which yield data ranging from 1,500 to 3,000 kilograms per hectare are reported and no reasons are given for the low yields. The production of grain is obviously the objective of rice growing. Certainly, therefore, all field research should be concerned, in part with production. Though unquestionably the differences in yield between treatments are highly important, the actual level of production is also extremely relevant."

The Mexican material was grown at Delhi as well as at other places at a spacing of 1½' between rows and 4'-6" between plants. Over sixty kg of K had been given to the soil. Four to five irrigations were given. This method of growing the material has many advantages: a) plants which are susceptible to rusts will immediately reveal this weakness; b) plants which have weak straw will lodge heavily; c) plants which have poor tillering ability can be identified without difficulty; d) plants which have poor yield components such as grain number per year and grain weight will also be obvious. Thus, selection can be done for a number of important characters at the same time, thereby rendering the screening of much larger populations possible in a given time.

One striking factor which has come to light from the maximum yield potential trial carried out at Delhi is that the varieties which have done well at 120 lbs of nitrogen have also been the top yielders at the lower doses of fertilizer application. It, therefore, appears that when a variety is evolved under high fertility conditions, it can also yield well when grown under moderate fertility conditions. It is, however, not possible to achieve the reverse results since all the weaknesses of a strain will get accentuated when it is grown under fertility. It will be useful to keep this information in view while formulating the sowing plans for segregating generations.

The breeding material obtained from Mexico has opened up new possibilities in re-patterning the plant type to suit the needs of intensive agriculture. Some of the material like V.17 may be of direct introduction value. The red grain of the Sonora varieties is now being changed into the amber type, both by mutation and conventional breeding methods.

The genes conferring a short plant stature have also had a profound impact in increasing the yield potential of rice. Many of the indica varieties developed in Taiwan and many currently under large-scale testing at the International Rice Research Institute, Philippines, have a short and non-lodging habit. The variety Taichung Native No. 1 which has yielded more than 6,000 kg per hectare in some parts of India has in its pedigree the recessive dwarfing gene which occurred as a spontaneous mutation in the variety Chou-Wu-gin. This dwarf (80 cm in height but with good tillering and high yielding capacity) was crossed with Tsai-Yuen-Chung (130 cm in height) and Taichung Native No. 1 was developed by selection in the segregating population. Genes for dwarfing are also being incorporated in India in hybrids of jowar and bajra. They have thus added a new dimension to breeding for yield. ▯

Reference

Chandler, R.F. Jr. 1963. An analysis of factors affecting rice yield. *I.R.C. Newsletter* 12: 1-17.

3

Artificial Transmutation of the Gene

When I was informed that I had been chosen for the Bhatnagar Memorial Award for Biological Sciences (for the year 1961), I took this as a recognition, not so much of my own achievements, but of the science of Genetics, of which I have the good fortune to be a student. I was happy that at long last Genetics is coming of age in India. It is one hundred years since Gregor Mendel presented his paper on "Experiments in Plant Hybridization" in which he propounded the laws of heredity. The science of genetics was, however, born only in this century largely because of the failure of earlier workers to separate continuous from discontinuous variation in their thought and work. Though young in years, genetics has grown with remarkable vigor and occupies today a core position in the conceptual structure of biology. The laws of genetics have helped to solve the centuries-old puzzle of the mechanisms underlying the basic unity as well as the breadths of diversity and depths of differentiation of living organisms. More important, a new biological era has been opened up where man can exert either a purposeful or unfortunately a destructive influence on evolution.

As a science, genetics has demonstrated the value of "hybrid vigor" not only in thwarting Malthusian predictions on the relationship between population growth and food production, but also in fostering the growth of a scientific discipline itself. The "narrow domestic walls" which Tagore deprecated in a human society are even more harmful in a scientific discipline, since by causing inbreeding depression, they promote conclusions based on hallucinations. By integrating the principles and tools of mathematics, chemistry and physics with those of conventional biology, genetics has exploited hybrid vigor in its own growth. Genetic determinants occupy a dominant place today in the scientific work reported at all International Congresses dealing with the life sciences – whether it be a congress of biochemistry or biophysics or botany. It is unfortunate that much of the work on genetics in India is still confined only to Plant Breeding Institutes and the immense implications of this science for medicine and animal breeding have remained practically ignored. No university in our country yet has a department or professorship in Genetics and the solitary question on Mendel's work which usually finds a place in B.Sc. papers in Botany and Zoology is the hot favorite for omission under the democratic right of "choice" vested in the students. This is particularly unfortunate since our country, as in many other fields had in ancient times developed the most advanced ideas on evolution as evidenced by the concept that God expressed himself through different life-forms, first as fish, and later as mammal. The ancients had also

---

Presented on the occasion of receiving the 1961 Shanti Swarup Bhatnagar Memorial Award for Biological Sciences, Delhi, January 14, 1965. First published in *Journal of the I.A.R.I. Postgraduate School*, Vol. 3, pp. 83-88, 1965.

recognized the importance of eugenics as is shown by the postulates for the choice of bride and bridegroom in the *Grihya Sutras*. In the first century A.D., Charaka and Susruta discussed the principles governing the transmission of parental traits through the seed and went as far as to assume that the sperm cells of the parents contain minute elements derived from each of the organs and tissues of the parent, thus anticipating Darwin's "gemmules" and Spencer's "ids". I do hope that the choice of a geneticist for the award of this prize endowed in honor of one who contributed so much to the growth of Indian science, is a forerunner to greater support and interest in the spread of genetical knowledge and its exploitation in solving the problem of growing more food in progressively diminishing areas of land.

In this talk, I wish to deal briefly with the artificial transmutation of the gene, the aspect of experimental genetics in which I have spent most of my time in recent years. Every educated individual recognizes the fact that the history of the earth is written in its layers. Few, however, are aware that the history of a living organism is inscribed in its chromosomes. It was recognized towards the end of the last century that chromosomes have the faculty of replication; in other words, the capacity for continuous expansion. Early in this century the Dutch scientist. Hugo de Vries postulated that the genetic determinants located on chromosomes occasionally undergo mutations, or permanent changes, which are themselves replicable. Thus, an unlimited degree of elaboration of hereditary characteristics can be attained. De Vries saw in this process Nature's most potent evolutionary tool. As a broad generalization, evolution can be pictured as a process of the "muddling through" of microbes into men as a result of recombination and selection acting upon blind mutations. Simple organisms such as bacteria or viruses, with a new generation every ten or twenty minutes and with enormous populations consisting of millions and billions of individuals could well adjust to the vagaries of the environment by mutation alone. Due to the low frequency of spontaneous mutations, evolutionary progress in more complex organisms would result in a blind alley if it were entirely dependent on the occurrence by chance of mutations of the right type. The higher organisms have hence relied heavily on the exchange of genetic information with each other, a process popularly known as sex. No wonder then that De Vries wished very much that man could find a way of artificially altering the gene, thereby gaining control over the most powerful of Nature's evolutionary mechanisms.

This dream was realized when Prof. H. J. Muller of the. United States showed in 1927 that the sluggish natural mutation rate which forms the chief basis of organic evolution, can be artificially accelerated by exposing living organisms to X-rays. A new hope was thus raised for the practical breeder who had until then been compelled (to quote Muller's words) "to remain content with the mere making of recombinations of the material already at hand, providentially supplemented, on rare and isolated occasions, by an unexpected mutational windfall". The discovery of methods of causing the artificial transmutation of the gene opened up new vistas in genetic research. The possibility of obtaining mutants in which the synthesis of specific metabolites was blocked ushered in the science of biochemical genetics and the elevation of viruses and bacteria to the genetic throne, in the place of the fruit fly, *Drosophila melanogaster,* which had reigned as the "Queen of Genetics" for over four decades! The knowledge gathered in recent years on the structure, function and mode of organization and replication of the genetic determinants has already been recognized in the form of several Nobel awards. We now know that except in a few viruses, deoxyribose nucleic acid (DNA) is the carrier of the hereditary codescript of the cell. The information needed to specify all the species and individual characteristics of every organism is recorded in its DNA in terms of non-overlapping triplet code of bases. During gene replication this information is reproduced by the unidirectional assembly of new DNA strands, each complementary in base sequence to the single strand serving as template, the result being the formation of two daughter molecules of DNA, which are exact replicas of the parent molecule. The translation of genetic information into protein structure is achieved by the transcription within the nucleus of a specific DNA code sequence into the complementary sequence of bases in single stranded RNA molecule. This messenger RNA leaves the nucleus and serves as the template for the assembly of amino acids into a polypeptide chain, the ribosomes in the cytoplasm serving as the sites for this reaction. Each specific DNA base triplet would specify a particular amino acid and the sequence of base triplets determines the sequence of amino acids in the final polypeptide product. Thus, we now know how genes beget proteins and this knowledge has helped in designing experiments for attacking the gene at the molecular level and also for ascertaining the effect of a gene mutation on the protein. An early study of this type is that of Ingram of UK, who discovered that in the abnormal hemoglobin causing the disease known as "sickle-cell anemia", a glutamic acid residue is replaced by valine in the protein chain.

Prof. Melvin Calvin remarked in a recent symposium that the nature of matter can be conveniently considered at four different levels – axioms, aons (a new term for the leptons and baryons), atoms and aggregates of atoms. In a similar way, the possibilities for experimentally manipulating the gene can be considered at four different levels – a base pair in DNA (the purines, adenine and guanine and the pyrimidines, cytosine and thymine), a nucleotide (nitrogenous base-sugar phosphoric acid combination), a cistron (a section of a chromosome that has a unitary function and may consist of several thousand nucleotides) and a chromosome (the gene strings of one human chromosome have been estimated to contain some 4×10^9 nucleotide pairs). Elegant experiments demonstrating alterations, both structural and numerical, at the level of the individual base pairs have been performed in micro-organisms. In higher organisms like the crop plants with which I work, experiments lending themselves to a precise interpretation of the causal mechanism involved in an induced mutation cannot be performed. Only two major categories of mutations can be usually recognized – those associated with visible chromosome structural changes and those with no such detectable change. The new knowledge from molecular biology has, however, provided many new classes of mutagens and attempts can be made to relate an observed effect to a change at a molecular level, although these cannot be definitely proved. The successful survival of a higher organism depends upon a high degree of integration and coordination of function at the level of the individual cell and of the organ. Because of their delicately balanced co-adapted genotype, such organisms are more often than not affected unfavorably when a mutation arises. It is hence that geneticists have been very concerned not only about the damage that will be done to later generations by the fallout from the nuclear test explosions but also by the increasing use of radiations and radiomimetic chemicals in medical diagnoses and cure. In plants, however, a significant advance in crop production can be made even if one out of 1,000 mutants is of a progressive type. Mutation breeding has, therefore, become a powerful instrument in modern plant breeding and I shall briefly summarize some results of applied value in a few crop plants and some of wider biological significance from the work carried out during the last few years by my students and myself at the Indian Agricultural Research Institute, New Delhi. Results of considerable economic value could be obtained only because of the effort expended in standardizing methods for raising the frequency of induced mutations to very high levels (every treated seed in barley and wheat can now be made to yield one or more mutations) and for altering the proportion of mutations associated with invisible in contrast to visible chromosome changes.

The mutations induced can be grouped into the following four classes according to the number of genes involved and the phenotype produced.

1) Micro-mutation: These involve changes in quantitative characters and are hence of the greatest value to the breeder. Such mutations affecting several genes can be detected only through the use of suitable biometrical procedures. Our work has shown that there is immense scope for increasing the number of ear-bearing tillers, the number of grains per ear, the weight of grains and the protein content in the grains in wheat. In the oilseed plant, *Brassica campestris* var. *toria*, in which the chromosome number had been doubled two decades ago through the use of another method of chromosome engineering, progress in the development of superior tetraploid lines had come to a standstill, owing to the lack of variability in the population for the number of secondary and tertiary branches. By subjecting the tetraploid line to treatment with gamma rays, this deficiency was overcome and the way paved for attempting a major break-through in the yield of this plant.

2) Mutations affecting single character: One of the principal attractions of the mutation breeding technique is the possibility of rectifying solitary defects in a variety without disturbing its good qualities and adaptive characteristics. Two good examples of this category of mutation from our work are: a) the development of the awned wheat variety N. P. 836 from the awnless strain N. P. 799; and, b) the introduction of dense hairiness and consequently of resistance to jassids in the long-staple cotton variety *Mescilla acala.*

3) Macro-mutations: Such mutations involve changes in a whole constellation of characters and have hence a striking phenotypic effect. The stiff strawed and non-lodging mutants isolated by us in wheat, barley and rice are examples of this type. These mutants have not only a shorter and thicker culm but invariably also a more dense ear and more stiff leaves. Such mutants are of

great practical interest since they confer upon the plant a morphological frame vital for it to be grown under conditions of high soil fertility and abundant water supply.

4) Systemic mutations: Mutations which either simulate an already existing taxon or necessitate the creation of a new systematic unit fall under this category. Such mutations are of great value in tracing phylogenetic trends and demonstrate that the evolutionary scale can sometimes be traversed by jumps of larger quanta. They raise several questions in our current ideas in the field of plant geography. We have been able to create from one variety of bread wheat (*Triticum aestivum*) all the known sub-species of this species, namely spelta, macha, vavilovi, compactum, and sphaerococcum. Also, from a variety of the tetraploid wheat species, *Triticum turgidum,* mutants having the key characteristics of *T. durum* and *T. pyramidale*, have been isolated. Thus, the artificial transformation of one species into another, which is one of the two major goals set for 20th century biology by Jacques Loeb in his book, *The Dynamics of Living Matter* published in 1904, may be said to have been attained (the other goal is experimental abiogenesis, a field in which also great progress has been made).

While we can now artificially generate in a single variety of wheat or barley almost the entire range of variability found in a world collection of germplasm, it is still rather difficult to obtain mutations giving rise to characteristics seldom observed naturally. This is because of the fact that all non-lethal mutations, which can pass through the somatic and gametic sieve, tend to survive in populations unless they arc of negative selection value. One may ask whether mutation research will represent a major advance from the standpoint of plant breeding, if the mutations which can be induced artificially are the same as are found in nature. Our work has shown that for the following reasons the answer to this question is in the affirmative. First, the ability to create in a single well-adapted strain the variability found in a world collection is definitely an advantage, as has been clearly shown by the new varieties developed in a short period of time in wheat, barley, rice and cotton. Secondly, new characters which are not found within the limits of a species are observed in material treated with mutagens. Thus, we were able to isolate a mutant in the variety N. P. 797 of *Triticum aestivum* with adventitious branching in the ear, a characteristic recorded in nature only in *T. turgidum* var. *mirabile* (popularly known as "Miracle" or "Mummy" wheat). The incorporation of this branching character in commercial wheat varieties could lead to substantial gains in yield. Thirdly, new alleles of already known loci occur in mutation experiments. Fourthly, more subtle variations in phenotype arising from new recombinations of genes can be created. Finally, mutations for polygenically determined characters and the release of hidden genetic variability due to crossing-over in the regions adjacent to the centromere (where normally crossing over is rare) are of definite value in breeding work. Mutation research should, however, be regarded only as a valuable supplement and not as a substitute to the conventional methods of plant breeding. Mutation breeding is unique in one respect. While with some of the great practical achievements in plant breeding like the exploitation of hybrid vigor in maize, a theoretical understanding of the reasons for the enhanced vigor of the hybrid has not been an essential pre-requisite for economic gains, the reverse is true of mutation breeding. The more one knows about the genetic architecture of the material and the mechanism of action of the mutagen, the greater will be the chances of getting worthwhile results. It is hence that I believe that the theoretical understanding currently being built up at the Indian Agricultural Research Institute of the various processes connected with induced mutagenesis would yield rich dividends in the near future.

I mentioned at the outset that genetics provides the key to the understanding of the basic unity of the living world. It follows, therefore, that the genetical knowledge gained from one test organism may lend itself for extrapolation to totally different organisms. It also emphasizes the importance of avoiding the "single track" approach to taking decisions of biological significance. I would briefly cite one example from our work to illustrate each of these two points.

In 1956, we observed that in plant species in which multiples of a basic chromosome number exist (technically known as polyploids), fast neutrons produce very many more chromosome breaks in the polyploid than the diploid. Thus, in species of wheat (*Triticum* sp.) fast neutrons produced ten times as many breaks in the hexaploid species with forty-two chromosomes as in the diploid or the tetraploid species with fourteen and twenty-eight chromosomes respectively. In contrast, X- and gamma-radiation did not reveal such a property. Such an enhanced ability for chromosome breakage in polyploid cells together with the relative independence of neutron action of cellular oxygen levels, suggested that neutrons may be of great value in the treatment of advanced malignancies characterized by a high degree of endopolyploidy in the constituent cells.

The action of radiations on cells can be classified into two major groups – direct and indirect. These two groups, however, are not mutually exclusive. In both, one of the main initiating events seems to be the formation of a free radical. In the direct action, this radical is formed in a genetically important molecule and may destroy that molecule by chemical rearrangements, while in the indirect action the radical migrates from an unimportant molecule to an important one. We have carried out several studies to assess the radiomimetic potency of the indirect component of radiation action. Our data reveals that: a) chromosome aberrations occur in the root meristems of barley, onion and broad bean when they are cultured on irradiated substrates; b) potatoes exposed to 20,000 rads of gamma rays for inhibiting sprouting as well as orange and apple juices pasteurized with 200,000 rads of gamma rays cause a low frequency of chromosome breaks in barley root meristems; c) there is a two- to four-fold increase in the spontaneous mutation frequency in the fruit fly (*Drosophila melanogaster*), when the flies are reared on a medium pasteurized with 150,000 rads of gamma rays. The extensive research on the wholesomeness of irradiated food carried out in the USA and Europe prior to our work, had all been designed to detect possible somatic effects but we had not at all planned for identifying probable genetic harm. The recent tragic experience with *a*-(N-phthalimido) glutarimide (commercially known as thalidomide and by several other names), which induces chromosome breaks in onion root tips, has stressed the necessity of remembering the generation as yet unborn in all our attempts to make life easy for ourselves. Chemicals with carcinogenic and mutagenic properties are finding ever increasing applications in various spheres of life and unless serious biological thinking is directed towards such problems, on the lines so convincingly pleaded by Rachel Carson in her book *The Silent Spring*, we may cause the gradual deterioration of the genetic endowment of man.

One of the most significant discoveries of this century is the unraveling of the biochemical constitution and mode of functioning of the gene. Much of this knowledge has come through our ability to cause experimentally the transmutation of the gene and observe its consequences in the proteins produced. Prof. J. Lederberg has drawn attention under the term "euphenics" to the enormous further possibilities for rectifying sporadic infirmities in man, that will be opened up by modern genetics in its attack on problems of development and of gene mutation. To a country like ours faced with the urgent task of enhancing food production, the significance of modern genetics for the production of new varieties of plants and new breeds of cattle is even greater.

Since we are still close to the season of "Peace on earth and goodwill to all", I would like to end this talk with a few lines from a poem by George Herbert, on the true source of peace – with the added suggestion that the stalks of wheat which sprang out of the grave represent high yielding mutants!

Sweet Peace, where dost thou dwell?
    I humbly crave,
    Let me once know.
I sought thee in a secret cave,
And ask'd, if Peace was there,
A hollow wind did seem to answer, No:
    Go seek elsewhere.
At length I met a rev'rend good old man;
Whom when for Peace
I did demand, he thus began: ...
"Take of this grain, which in my garden grows.
And grows, for you;
Make bread of it: and that repose
And peace, which ev'ry where
With so much earnestness you do pursue,
    Is only there." ▮

4

J.B.S. Haldane and Indian Agriculture

Haldane was one of the most gifted of the fast vanishing species biologist. His interests and study encompassed all living organisms ranging from viruses to man. Since genetics provides the major key to the understanding of the mechanisms underlying the unity as well as diversity of life, he naturally became, in addition to a researcher, a teacher and popularizer of this science. He was the first to draw the attention of the western world to the fact that the earliest attempts to apply genetical ideas to human societies are found in ancient Indian writings. Because of the breadth and depth of his vision, he could look far ahead and indicate what human and natural selection might do to evolution.

The following example from a lecture he derived at the XI International Congress of Genetics held at The Hague in 1963 would be sufficient to illustrate the uniqueness of his thought process. "It is rash to speak of the biological function of a gene. What is the function of the gene for haemophilia? Today we say that it fails to perform a normal and necessary function. A thousand years hence our descendants may say, 'The function of this gene is to prevent the blood from clotting rapidly.' This causes some inconvenience in the first century of a human life, but is most valuable as soon as one is fitted with an artificial heart, as most people are before the age of a hundred and fifty."

In the few years India had the privilege of having Prof. Haldane as an honored citizen, he endeavored to introduce a sense of purpose and originality into the thinking and experimental approach of Indian biologists. I shall illustrate with two examples the type of experiments he stimulated and which are of very great significance to Indian agriculture.

Cooperation and Competition Among Varieties

A simple experiment performed by Dr. Subodh Kumar Roy at the suggestion of Prof. Haldane revealed that when two varieties of rice are planted together, either fully mixed, or in alternate rows or even in separate halves of the same plot, each may influence the yield of the other. This interaction may have either favorable or unfavorable consequences upon yield, depending on the particular combination of varieties used. When the yield is enhanced, the varieties are said to cooperate, while a depression in yield is assigned to competition between the varieties. The important finding that came out of this study was that certain varieties like the winter (aman) strains, 36 BK BR7 and 498-2A BR 8 may yield 26 percent more when they are grown in alternate rows than when they are grown separately. In Kerala, the rice varieties of P.T.B. 2 and P.T. B.5 cooperated in a similar way. The following statement of Haldane brings out clearly the importance of such studies."One of the greatest differences between a tropical

---

First published in *Science Reporter* 2:510-512 (November 1965).

flora and a temperate one is the far greater number of species in the former (apart from very arid regions). Thus, India has over 20,000 species of flowering plants, a number far surpassing states of much larger area such as the USSR, and only surpassed by Brazil. One would expect symbiotic relations between flowering plants to be much commoner in regions with a diverse flora than in those where a natural plant community often consists of a few members. Further, where harvesting is not mechanized, there is often no difficulty in growing mixed crops and they are in fact grown largely in India. While the most obvious symbiosis to look for and to exploit, if discovered, to increase agricultural production, is between different species, particularly cereals and legumes; nevertheless if symbiotic relations are common, they should be looked for between different genotypes of the same species. It is possible that some of the advantage of outbreeding in such a crop as cotton may be due to the compresence in a field of diverse genotypes, that is to say to heterogeneity between plants as well as within them." The implications to agricultural production of the research carried out on these lines are obvious.

Coconut Improvement

Following a brief visit to the Coconut Research Centers at Kayangulam and Kasaragod in Kerala, Prof. Haldane wrote a most thought provoking article on coconut improvement. Considering the importance of this crop in the trade and economy of South India, it is worthwhile reiterating its significance. The first suggestion relates to the study of xenia, i.e., the influence of the pollen and parent on the characteristics of the endosperm. In coconut the endosperm is the most important tissue, since it provides the copra of commerce. Although the nuclei of the endosperm contain two maternal sets of chromosomes and only one paternal set, the latter could influence the carotene content, the type of carbohydrate, the content of several vitamins and the anthocyanin coloration. A comparative study of the size, oil content and other characters of the nuts borne on the same tree following crosses with several distinct pollen parents would help to assess the extent to which xenia effects are manifested.

The second important suggestion relates to the need for gathering statistical data on the yield of copra by each of several hundred trees over a period of ten years. If such data are available it would be possible to estimate: a) the number of years needed to assess the tree's performance and b) the possibility of predicting the yield of a tree in 1965 from a knowledge of its yields during 1958 to 1963. The use of quantitative methods in studying the value of an individual tree is particularly important in a perennial and obligatorily cross-pollinated plant like coconut, since these alone can provide an insight into problems such as the genetic potential for yield in different palms, the rates of ageing, and the existence of correlations between characters like copra yield and oil content.

In coconut, various types of freak palms are often described. Haldane emphasized that what may be a freak at one place, may be a valuable character in another environment. He illustrated this by citing the following two examples. Small-combed poultry are bred in South India to lessen the danger of bleeding during cock fights. The genes which reduce the comb size in these breeds have been transferred to high laying breeds to avoid injury to combs by frost, both in the USSR and Canada. Genes, which lessen the feathering, characterize "fancy" breeds in temperate climates but these are found in tropical breeds which are kept entirely for egg and meat production.

The coconut palm in many parts of India appears to suffer from various degrees of inbreeding depression. This is because of the fact that cross-pollination per se does not ensure genetic heterozygosity. Unless there is variability in the material, cross-pollination and self-pollination will have the same genetic consequences. It is important to recognize this and make use of this phenomenon to induce some heterotic vigour in the population as well as to study the genetics of some of the economically important traits like dwarfing. Crosses between dwarf and tall palms are known to produce vigorous progeny and there is immense scope for improving the productivity as well as the quality and yield of oil of the coconut palm by exploiting genetical and biometrical procedures. It fell to the lot of Prof. Haldane to draw the attention of those in charge of coconut improvement to this simple yet very important fact.

A co-worker of Prof. Haldane, Prof. T.A. Davis observed that the leaves of coconut palms are arranged in five right-handed or left-handed spirals and that the two types of trees are almost equally common. What made this observation very important was the coincident finding that trees with a left-handed foliar spiral yield, on an average, more nuts per year than those with a right-handed spiral, the increase being as much as twenty-one per cent. Similar studies on the frequency distribution of foliar asymmetry in populations and the correlation of this phenomenon with yield have now been initiated in several other crops. When clearly understood, this phenomenon may open up yet another avenue of manipulating yield.

Consciously or unconsciously, Prof. Haldane initiated a new era in applied genetic research in India. By emphasizing the need for studying variation in relation to selection, either natural or human, Haldane introduced an interest in the understanding of the physiological and functional significance of morphological characters. Since modern plant breeding is largely applied gene ecology and physiology, an interest in the basic processes underlying gene function expression and adaptation would help to place Indian plant breeding on a rational and effective base.

5

Agricultural Transformation and Opportunities for a Learning Revolution

I feel privileged to have this opportunity to pay homage to the memory of one of our greatest educationists during 1970 which has been designated by the United Nations as the International Education Year. This title for the year, in spite of the weakness of its ritualistic approach, has helped to stimulate fresh thinking on the role and relevance of education in achieving the political goal of rapid economic growth coupled with social justice. The most outstanding characteristics of the decade that has just been completed with reference to our country, are the beginning of a process of technological change in agriculture on the one hand and the beginning of a mass unemployment problem on the other.

The number of educated persons registered on the live registers of the employment exchanges as job-seekers generally exceeds a million. The vacancies reported in the *Quarterly Bulletin on Job Opportunities* issued by the Ministry of Labor, Employment and Rehabilitation usually number about 15,000. These include the vacancies notified by Central and State Government as well as several public and private sector companies. Research theses by economists and educationists all reveal that education is basic to the modernization of agriculture and industry and the transmutation of our natural endowments into wealth meaningful to our present-day needs. In contrast to these scholarly conclusions, we see in newspapers and in our daily life many highly educated doctors, engineers and technologists complaining about a lack of job opportunities. Also, a large proportion of those who get salaries complain about "mis-employment", such as science and agricultural graduates working as clerks.

Those in authority rightly ask, "Does not a medical or agricultural or engineering degree equip a person to create employment for himself?" Studies at the Delhi School of Economics and elsewhere have clearly shown that educational differences do affect agricultural productivity and schooling should be an essential constituent of agricultural progress (Amartya Sen, 1970). Other surveys, however, show that the rate of adoption of high yielding varieties of crops has been the lowest in Kerala and highest in the Punjab, an enigma since Kerala has the highest percentage of literacy in our country. Though production is a multichannel end-point, the relevance and quality of education would obviously influence such correlations. One part of the Scientific Policy Resolution adopted by Parliament in 1958 states: "The wealth and prosperity of a nation depend on the effective utilization of its human and material resources through industrialization. The use of human material for industrialization

---

Presented as the First Zakir Hussain Memorial Lecture, University Grants Commission, New Delhi, 1970.

Table 1. University enrollment by faculty (thousands)

	1958-59	1967-68
Arts	438.5	918.3
Science	271.9	737.5
Commerce	85.9	219.8
English	34.0	104.3
Education	14.8	44.1
Medicine	30.7	83.4
Agriculture	15.2	51.6
Veterinary Science	4.5	6.6
Law	24.0	44.0
Others	1.1	9.5
Total	917.8	2,217.0

Source: University Grants Commission, India.

demands its education in science and training in technical skill. Industry opens up possibilities of greater fulfillment for the individual." Twelve years later, we find that there has been a great upsurge in educational activity (Table 1). In spite of the very impressive advances in the out-turn of technical personnel, it is widely recognized that the efficiency of our production processes as reflected in both the cost and quality of the product is often low and that the performance of our industries, both in the public and private sectors, is not as impressive as it ought to have been in the light of the output of University trained technical personnel.

A consequence of economic advance not having kept pace with the out-turn of technical personnel is the lack of job opportunities for educated youth. In my field of agriculture, 8,950 agricultural graduates and 705 agricultural engineers would become "surplus" to our requirements by 1973, according to a calculation made by the Manpower Planning Unit of the Union Ministry of Food and Agriculture. Why are we witnessing this coincident growth of education and unemployment? The slogan "Every house a factory and every acre a pasture" obviously prescribes a very high degree of self-employment and the educational path often indicated to achieving this goal is alteration of the curriculum at primary, secondary and university levels in a way that would confer on it a vocational bias. Thus, one could argue that if farming were effectively taught in schools, particularly in rural areas, then school-leavers would become farmers and not drift to the cities and towns. There are, however, others who believe that primary education cannot be narrowly vocational. Students should be able to read and write, to do a certain amount of arithmetic, to understand enough science and history to interpret the world around them and to learn enough civics to be aware of their rights and responsibilities as citizens of their nation. Education, according to this concept, is not meant merely to adapt pupils to their society but also to equip them to alter it.

It is a tribute to the genius and vision of Mahatma Gandhi that he identified and discussed at length over forty years ago these very problems. Gandhiji's contribution to educational thought was unique and typical of him. The wonder of his approach is that, merely by applying himself to the problems of Indian education as it was and as it should be, and out of his intuitive understanding and judgment, he arrived at the same conclusions as those which the best educational thinkers had evolved during the last century. I wonder whether he was acquainted, except in a casual manner, with the writings of Rousseau, Froebel, Pestalozzi, Dewey or Maria Montessori? Yet the principle behind what later came to be known as Basic Education is identical to the one stressed by all these educationists – the principle of activity as the basis of learning. Gandhiji's concept of Basic Education was essentially that of activity education – it was both craft-centered and child-centered and was thus diametrically opposed to the academic and book-centered system.

Dr. Zakir Hussain, who was in many ways the architect of the Basic Education system, was able to provide an operational foundation to Gandhiji's ideas. Even at the Educational Conference held at Wardha on October 22, 1937, where Gandhiji pleaded for a radical transformation of the Indian educational approach, Dr. Hussain warned, "Mahatmaji thinks that the scheme which he has placed before you is absolutely original and that it can be accepted only by those who believe in non-violence and in rural civilization. But those working in the educational field will not find Mahatmaji's scheme very new. They know that true learning can be imparted only through doing. This method is called the Project Method in America and the Complex Method in Russia. We can certainly impart education to our children through the *takli* and the *charka* and some other suitable handicrafts. But the greatest difficulty in carrying out this scheme will be the paucity of trained teachers. Prof. Dewey in America had a similar plan which was welcomed enthusiastically but he had to close down his school in three years because he had no men to run it."

Having identified the key input for the successful launching of this new movement in education, Dr. Zakir Hussain tried to build through the Jamia Millia Islamia and associated institutions the infrastructure for basic education. Two decades later, in the Sardar Patel Memorial Lectures delivered in 1958, Dr. Zakir Hussain concluded, with what pain we can only imagine from one who was the father of this educational movement, that Basic Education in our country had been a failure, not because it was educationally unsound but because it was never given a fair trial. I quote his words: "We have turned the so-called intellectual book-school into a mechanical memory training school and succeeded in making our Work School, the Basic School, a place of mechanical work. The work is extraneously and uniformly prescribed; there is no semblance of a spontaneous motivation in the child and he is supremely ignorant of any personal or social purpose behind his work. Work which is mechanical, work in which no mental exertion is involved, work in which one is satisfied with just any result and there is no constantly prodding urge to aim at its possible perfection, work in which there is no self-criticism and so no real progress, is in no sense educative. Schools that have such work are not Work Schools in any sense." Since the concept of work experience as the medium and means of education had been given practically a burial because of our tendency to discuss ideas out of existence without ever seriously giving them a try, Dr. Zakir Hussain could only lament with gentle wit, "I cannot very well deny the wisdom of learning by experience. But I cannot easily concede the claim to be wise without any real experience."

The Education Commission has offered several practical suggestions to make University students "employable" or capable of self-employment. In agriculture, the Commission recommended the establishment of Polytechnics where students could acquire the necessary technical skills so as to contribute towards the growth of scientific agriculture. The Commission has also suggested ways of achieving a qualitative change in education and of linking admissions to the country's ability to absorb University trained personnel. Many of the Commission's suggestions are yet to receive serious consideration from the viewpoint of implementation. Meanwhile, the problem of jobs and student unrest is growing to explosive proportions.

Even more serious than the inability of our University trained personnel to get or create jobs, is their total insensitivity to the realities of our economic situation. While every educated Indian ought to know that our average per capita national income is only about Rs. 300 per year, it would provide an interesting, though sad, insight into the psychology and outlook of the educated-employed, if the Third Pay Commission would publish, in addition to their recommendations, all the memoranda received by them, reflecting the beliefs of the educated on the incomes they should have. Our education, while failing to increase our wealth-creating capacity, has enhanced our wealth-absorbing desire. The millions in Calcutta and elsewhere who live on the pavements do not stir the educated heart into action. Mahatma Gandhi's oft-quoted definition of God, "To those who have nothing to eat, God is bread; to those who have no home, God is a roof over their heads; to those who have no clothes, God is a piece of cloth" – has come to occupy the same place as far as practice is concerned, as the three monkeys of ivory found in the drawing rooms of the educated affluent. This act of pious regard for the symbols of the wisdom of the great, makes an ironic contrast to the simultaneous neglect of wisdom itself.

Significance of Recent Trends in World Youth

Having failed to see light within our own educational effort, I would like to take a peep into the educational arena of the developed nations of Europe and North America. It seems to me that the most outstanding characteristic of the decade that has just been completed, in the West, is the revolt of youth. Much has been written about the various forms this revolt has taken and its causes, and the nature of the remedies that society must begin to apply. Here I would like only to dwell on the significance for us of some of these characteristics. Sometimes violently, sometimes through escapism and turning away, and sometimes through political expression, youth is proclaiming its rejection of society, turning its back on two facets of Western society as they know it – affluence and the impersonal organization of mass industrialized society. The turning away from affluence and material values is shown in many ways – partly in the desire of youth to learn more about life in countries less fortunate in material blessings, to serve and respond to human need, and to make a meaningful life in terms of service to fellow human beings. A remarkable characteristic of such young people who are moved by the desire to work and serve in less affluent countries, and they are not few in number, is their acceptance of voluntary poverty as an ideal, and as a necessary condition of service. I need hardly point out that this is a return to one of the most ancient and constantly reiterated ideals of all world religions. The acceptance of voluntary poverty has been a condition for all the great monastic service orders of the past, and in recent times, was given meaningful expression in our own country by Gandhiji. The numbers of such young people and the urge which drives them away from all the wealth and convenience which they can command in their own homes is of no small significance. While we have had an outflow of talented young men and women from our own country, and other developing nations to the West, in search of the material and other satisfactions that life in an affluent country can bring, we have not taken note of the great return influx of young people from the West accepting poverty in the search for meaning, for challenge and for service.

The second important feature that I would like to comment upon is the expression by youth of the longing for community. In the impersonal, highly organized mass society that is typical of both the West and the communist world, Man fails to find the satisfaction that comes only from close integration with and belongingness to a community that can be seen and grasped in its entirety. Man, as a primate, is essentially a social being. The findings reported by biologists working in the field of animal societies are of interest. Desmond Morris notes that the average size of the group with which a human being comes into close contact at any time in his life, including his working colleagues, neighbors and relations is roughly one hundred, which is the same as the size of a small tribe, or a typical baboon or primate group. This need to be a member of a smaller, recognizable and well understood group appears to be an integral part of man's innate equipment. It is this longing for a meaningful community which has led to various communitarian and educational experiments in the West in the last few years.

The other side of this coin is the recognition and acceptance of the worth of the individual human being. It is this recognition which is leading young people of the West away from the so-called hard sciences and towards the life sciences, away from manipulative strategies and organization and ways of thinking and towards a more humanistic approach. This refusal of the young either to manipulate or to be manipulated by others is very important in relation to their attitude towards development. This same feeling of the oneness of humanity, the need for relation with others, and the uniqueness of the individual is also expressed in the new awareness of the natural environment, or the emphasis that is being given to the science of ecology. The realization that man is part, not only of a human, but of a natural and biological community, and that he has important responsibilities in this respect is a part of this newborn feeling.

You may wonder how this analysis of the attitude of the foreign young is relevant to our own situation, which is characterized by poverty, unemployment, frustration, a great longing for material satisfactions on the part of those who already have a few, and the rise of a highly competitive and egoistical society in our day. It is precisely this contrast which should give us food for reflection. We have no need to look for challenges and opportunities for service; these exist all around us. We are also fortunate that large parts of our country are still a part of an older social organization in important part. Unfortunately, it is no longer the ideal towards which we strive, though meaningful tasks can only be performed on a small scale and in relation to the face-to-face community.

Our village communities exist, but we have not seriously used them for developmental purposes yet. Caste, communalism and other disintegrating forces are leading towards the break up of the once integrated communities. Youth is particularly alienated from this kind of thinking and living. In our case, the reason for alienation is not industrialization or the creation of a mass consumer society, as it is in the West. The reason can very largely be attributed to our educational system, which is academic and book-centered, and concentrates on non-relatedness to the environment of the person at every stage in his education. It is the divorce of education from the real tasks before us that I would like to take up next.

It is eighteen years since I started doing some teaching, mostly to post-graduate students. Looking back, I can see clearly one trend – the students whom I have "helped" most by writing part of their theses have been the most resentful in later years of the help so rendered! I guess this is due to the mental burden of gratitude that the student is made to carry when the teacher starts working for him. The same reaction is exhibited at all levels of society. Sovereign nations, for example, are most resentful of those countries which dole out the greatest quantum of aid to them, again the by-product of counter currents of expectation of gratitude on the one side and resentment of having to be obliged, on the other. The concept of hierarchical needs in man, developed by psychologists in recent years, suggests that within each one of us, child or adult, there dwells a captive spirit struggling to find fulfillment. The following extract from a poem entitled "Prayer Before Birth" by Louis Macniece expresses this idea beautifully:

I am not yet born; O hear me,
Let not the man who is beast or who thinks he is God
Come near me.
I am not yet born; O fill me
With strength against those who would freeze my
humanity, would dragoon me into a lethal automation,
would make me a cog in a machine, a thing with
one face, a thing, and against all those
who would dissipate my entirety, would
blow me like thistledown hither and
thither or hither and thither
like water held in the
hands would spill me.

Let them not make me a stone and let them not spill me,
Otherwise kill me.

There is little doubt that education, however imperfect it may be, kindles this captive spirit. If we wish to take cognizance of this fact, we have to review afresh both our educational methods and our schemes for mobilizing students for social service. In my view the era of working for others has to give way as soon as possible to the era of working with others. The concept of social service as practiced generally in our country confers a spurious psychic satisfaction and thereby a sense of complacency to the one who "serves" and produces a feeling of distrust and derision in the minds of the one who is "served".

Education and Agricultural Advance

We are thus faced with several exacting requirements in our educational endeavor. On the one hand, we have to introduce work experience as an integral part of education both for equipping the student for self-employment and for harmonizing his personality and outlook with the socio-economic needs and conditions of his environment. On the other hand, opportunities for work experience have to be provided in such a way that the evils of mechanical work or slave-driving to which Dr. Zakir Hussain had drawn our attention, as well as of the psychic perils which I mentioned earlier are both avoided. Thirty years ago, addressing young educated volunteers who were to go to villages for rural uplift work, Gandhiji said, "The fact is the villagers have lost all hope. They suspect that every stranger's hand is at their throats and that he goes to them only to exploit them. The divorce between intellect and labor has paralyzed our rural society. The worker should enter villages full of love and hope, feeling sure that where men and women labor unintelligently and remain un-employed half the year round, he, working all the year round and combining labor with intelligence, cannot fail to win the confidence of the villagers." Gandhiji rightly diagnosed that the marriage of intellect with labor would provide the most powerful instrument of achieving agrarian advance and rural prosperity.

Agriculture not only provides more than half of our national income and export earnings but also the means of employment and livelihood to over seventy percent of our population. It has been calculated that even by 1981, agriculture will have to provide employment to the large majority of our work force (Table 2). Agricultural work-experience should, therefore, occupy a dominant place in our educational set-up, if it is decided that we must make education more meaningful and relevant to our present-day needs. Unfortunately, during

Table 2. Selected statistics on the work force of India 1951-81

(Figures in millions)

Year	Population	Working force	Agricultural workers	Non-agricultural Workers	Percentage agricultural to total workers
1951	356.88	139.52	97.30	42.22	69.74
1961	439.23	188.78	131.22	57.56	69.51
1971	559.62	240.52	167.19	73.33	69.51
1981	694.90	298.67	207.61	91.06	69.51

Population estimates for 1971 and 1981 are those recommended by the Expert Committee on Population Projections in 1964 for use in the formulation of the Fourth Five Year Plan.

Gandhiji's time, the scope for introducing intellect into farming was relatively limited, since our extension agencies then had little to extend by way of new knowledge. The situation today is entirely different and it is hence that I have linked up in the title of my talk agricultural transformation with opportunities for a learning revolution.

New Vistas in Agriculture and Animal Husbandry

What are the major differences in our agricultural technology between the days when Gandhiji tried to combine knowledge with labor and now? With regard to the general features of our agrarian economy, the following characteristics have remained either the same or turned more unfavorable due to the population explosion: a) small land holdings, which are often spatially fragmented and poor income with consequent low purchasing power; b) large farm families with consequent under-employment and unemployment; c) wide seasonal fluctuations in yield and income, due to only a small proportion of the cultivated area having assured irrigation facilities; d) widespread under-nourishment and malnutrition in animal as well as human populations with attendant low productivity.

In the past, we had four major handicaps in raising the output per unit of land and per unit of water. First, the varieties of crop plants then cultivated were mostly season-bound and had been selected for survival under adverse conditions rather than for performance under good management. Secondly, the methods of feeding the plant with nutrients were based solely upon the addition of organic manure either in the form of cow dung or green manure or upon a system of shifting cultivation. Thirdly, water management was poor with the result that the output of food per unit of water was low. For example, we use about 1,800 gallons of water to produce a pound of rice, while only half this quantity is used in Japan. Finally, pest control methods were poor both before and after harvest and the gross national waste of food was high.

Agricultural science has taken giant strides, particularly since World War II, although it must be stressed that had steps been taken even prior to the first Five Year Plan period to introduce science into farming, we would never have had to import food. For example, the theory of diminishing returns of the soil was put to rest by Karl Marx as early as 1866, when in a letter to Friedrich Engels explaining the significance of the discovery of fertilizer in maintaining and enhancing soil productivity, he wrote: "The new agricultural chemists in Germany, especially Liebig and Schonbein, are more important for this matter than all the economists put together." Fertilizers freed us from one natural limitation of crop production – the supply of nutrients from soil.

The next major event was the rediscovery of Mendel's laws of inheritance in 1900, leading to the birth of the science of genetics. Genetic principles made it possible to tailor crop varieties to meet specific needs. The exploitation of hybrid vigor in maize was the first major advance in breaking traditional yield ceilings. It also became possible to build into plants a high degree of resistance to pests and diseases, so that the chances for widespread devastation of crops by pathogens, such as those which preceded the Irish famine of the last century and the Bengal Famine of World War II, were greatly diminished.

During and after World War II, came a succession of significant findings such as the discovery of potent pesticides, new plant types which could use both nutrients and solar energy better, new concepts in agronomy such as minimum tillage, chemical control of weeds and foliar feeding, coordinated strategy of pest and disease control, better techniques of moisture conservation and water use leading to a greater return per unit of water and the emphasis on population performance rather than on individual plant characteristics. When these new findings were put together in a synergetic package, as was done under the High-yielding Varieties Program in 1966, rapid progress took place in increasing productivity and in transforming the outlook and receptivity to new practices among the farming community.

Considering the fact that over ten million children are born in our country every year and that ninety percent of them may belong to rural and low-income families, it would be obvious that programs like fortification, free milk supply and feeding of food mixtures can only touch the fringe of the problem. Even these limited programs seem to be facing administrative difficulties as exemplified by the recent announcement of the Bihar Government that it is closing down the CARE program. It is in this context that the new movement in plant breeding of genetically upgrading the nutritive properties of the basic staples is particularly relevant to our needs.

As a result of intensive research carried out at IARI and some agricultural universities, we are now in a position to popularize high-yielding-cum-good quality varieties in many cereals, pulses, vegetables and fodder crops. Since protein deficiency in our country is preponderantly an indirect result of energy deficiency and hence cannot be corrected by the addition of extra protein alone, the dual combination of yield and quality is essential. The great merit of this approach is the possibility of enriching our food without added expense and without the primary product having to go through a processing chain. I believe that the distribution of nutritional seed kits comprising seeds of suitable strains of wheat, rice, maize, nutritious millets, high and quick yielding pulses, good quality vegetables and fruits would help in the rapid eradication of protein-calorie malnutrition.

If high quality fodder crops are grown, milk yield can be increased without having to supply more concentrates. The cultivation of better fodder crops coupled with the rapid genetic upgrading of the nondescript cattle population through artificial insemination would lead to a great rise in milk production.

From what I have said so far, it would be obvious that we are on the threshold of a major technological change in our agriculture. Old varieties will give place to new ones characterized by a great capacity for yield per day and per unit of water, new methods of breeding for high productivity will be introduced, sunlight will be tapped to the maximum both by new plant types and multiple cropping, plant protein will become similar to animal protein in quality, ploughing will be reduced to the minimum, factors causing instability in production will be scientifically destroyed, new procedures of pest control will be introduced, credit and marketing will be better organized, deserts will be made to bloom, dry farming will become profitable and animal productivity will go up, if we take advantage of recent progress in science and technology. To achieve all this in a short period of time requires a gigantic educational effort and mass participation. This new movement in agriculture thus provides unique opportunities for both learning and service.

Both the needs and the opportunities for a learning revolution have never been so great as they are now. What is needed is the will to act and not mere discussion and analysis. "For sheer size, the tasks ahead of us are so demanding that no one can afford to sit back and just watch or let frustration become endemic in our country. The situation demands of us work and more work, silent and sincere work, solid and steady reconstruction of the whole material and cultural life of our people" – these words are even more relevant today than when Dr. Zakir Hussain spoke them while assuming the office of the President of India on May 13, 1967. ◼

Reference

Sen, Amartya. 1970. *The Crisis in Indian Education*, Delhi: Delhi School of Economics.

6

The 1971 Ramon Magsaysay Award for Community Leadership

Citation read by The Hon'ble Dr. Jose V. Abueva,
Member Board of Trustees,
Ramon Magsaysay Award Foundation, The Philippines,
August 31, 1971

Whether India, with nearly one-sixth of the human race, can provide sufficient food for her growing numbers in the years ahead depends ultimately upon her farmers. Their performance is closely linked to what science develops and makes operative in rural life. Without continuing refinement of relevant knowledge and its efficient transference, especially to the poorer villages, the "Green Revolution" may foster more discontent than it satisfies.

In an age when radio-isotopes, a Gamma Garden and chemical mutagens are among the plant breeders' tools, Dr. Swaminathan is an originative follower of Gregor Johann Mendel, the Austrian monk and botanist who founded genetics over a century ago. A cytogeneticist, in the past sixteen years he has made major advances in breeding sturdier, more productive and better quality plant types at the Pusa Institute, as the Agricultural Research Institute outside of Delhi is popularly known. Included in the wide-ranging studies by him and his associates have been India's most essential food crops – wheat, rice, maize, sorghum, millet, pulses, potatoes and vegetable oils – plus cotton and jute. By purposeful manipulation of genes, he and his co-workers in 1967 developed a dwarf, non-lodging wheat variety – Sharbati Sonora, with amber grains – from Sonora-64 which has red grains and hence a low consumer preference in India.

An ability and enthusiasm for passing on his knowledge to others in the laboratory, classroom and field, and his prolific writing have earned him a reputation as a most lucid educator. In the past five years since he became Director of the Institute, Swaminathan has proven himself an equally gifted administrator.

Encouraged by him, scientists at Pusa extended their work to practical application in farmers' fields. University students were enlisted in this attack upon the limitations to a better life on the land. The primary demonstration arena for these efforts are villages around Delhi where tests of improved seeds – by farmers with whom the Pusa Institute cooperates – have won confidence in their productive potential. As part of a High-Yielding Varieties Program designed by Swaminathan one community was transformed into a "seed village" specializing in controlled multiplication of improved varieties to supply the needs of the entire state, and thousands of demonstrations were laid out by scientists in farmers' fields throughout India.

His particular combination of talents has made Swaminathan an acknowledged leader of India's community of agriculturalists. Now forty-six years of age, he is carrying forward his Madrasi family tradition of energetic personal emphasis upon professional excellence. That he is doing so with such broadly beneficial results for rural India is the mark of a first-rate scientist who is also a humanist.

In electing Monkombu Sambasivan Swaminathan to receive the 1971 Ramon Magsaysay Award for Community Leadership, the Board of Trustees recognized his contributions as scientist, educator of both students and farmers, and administrator toward generating a new confidence in India's agricultural capabilities.

Response by Dr. Monkombu Sambasivan Swaminathan

It is a very great honor and privilege to have been chosen for the Community Leadership Award instituted in the memory of one of the greatest world leaders and humanists of our time. The late President Ramon Magsaysay devoted his tremendous vision, wisdom and energy to the cause of helping his fellowmen build a better life for all. A satisfied stomach is a prerequisite not only for happiness but even to enable men to behave as human beings, a fact so well expressed by the Roman philosopher, Seneca, when he said, "A hungry person listens not to reason nor cares for justice, nor is bent by any prayers." Even as recently as the mid-sixties, the future of many of the developing nations appeared hopeless when assessed in terms of their ability to feed their fast growing populations. Drs. William and Paul Paddock, in their book *Famine Nineteen Seventy-Five!*, even went to the extent of comparing the fate of my countrymen after 1975 to that of sheep being led to the slaughterhouse. Then came the avalanche of scientific results from this great country, Mexico, Taiwan and India that opened up altogether new vistas in the yield potential of our major food crops. I consider this award as a tribute, which the Magsaysay Foundation would like to pay the entire community of agricultural scientists in my country who have helped me to provide a breathing spell during which efforts can be made to prove the prophets of doom false.

New concepts of crop planning and land use, designed to get the maximum benefit from the physical, biological and environmental endowments of tropical and sub-tropical agriculture, are being developed. Where there is water, four or even five crops can now be grown in a year in multiple, mixed and relay cropping systems, getting for the farmer the benefit of nearly 450 days of crop growth in a year of 365 days. New methods of water management and enrichment of soil fertility, coupled with crop varieties capable of avoiding the rigors of drought or floods through changes in their life length and coordinated pest control schedules, are opening up new hopes for peasants working in environmentally handicapped areas. The "Green Revolution" in cereals has paved the way for developing harmonious systems of agriculture and animal husbandry.

The scientific prospects for alleviating hunger, increasing the avenues for productive and remunerative employment, and banishing poverty through a radical transformation of cropping systems leading to the growth of agro-industries, are fascinating and immense. At the same time the magnitude of the problems of illiteracy, under- and malnutrition, under- and unemployment and population growth are truly staggering. In spite of all efforts, the number of illiterates in absolute terms is growing in many parts of Asia, including India. Leading scientific journals carry data indicating that protein calorie malnutrition in infants may affect the replication of the chemical substance of heredity, DNA, and lead to an irreversible underdevelopment of a child's intellectual potential, thus compounding the ill effects of the largely poor educational opportunities.

The Indian achievement in wheat production, leading to a near doubling of the total harvest from a little over twelve million tons to over twenty-three million tons in four crop seasons, has few parallels in recorded agricultural history and serves to illustrate what can be accomplished provided farmers, scientists, extension and communication experts and political and administrative leaders, all function like members of a symphony orchestra. Unless such an orchestration in effort is generated for all crops, a scientific breakthrough may not necessarily lead to a production breakthrough. This is illustrated by the yield stagnation in sorghum in my country where, although new hybrids and varieties capable of yielding two to three times more than the earlier ones have been available since 1964, the yield per hectare has hardly altered in the last decade. Wheat posed fewer pest management and marketing problems and the farmers responded with enthusiasm to produce as much or even more than what was harvested in the National Demonstration plots put up by scientists in the fields of poor peasants. In other crops like rice, sorghum, maize, millets and pulses, require sophisticated and coordinated efforts of a type which few developing nations have yet generated. Consequently, a genuine feeling that they have been bypassed by the "Green Revolution" is growing in the minds of many farmers.

We are thus faced on one side by great scientific possibilities and on the other by vast problems of organization, coordinated action, communication, and population growth absorbing the fruits of all advance. Those who have the power and capacity to serve their fellowmen – be they scientists, educationists, administrators or political leaders – have probably never had in human history so many challenging opportunities for service and for experiencing the thrill of fulfillment. What is needed is the will to act and the determination to learn and adopt the correct techniques of action, since Asian farmers have given ample evidence in recent years that they are ready for change if the change is for the better economically.

I would like to end on a personal note. When over ten years ago my colleagues and I at the Indian Agricultural Research Institute embarked on the relentless pursuit of high incomes from farming through high yields – without detriment to the long-term productivity of the soil – we hoped that the high yielding varieties would not only help raise production but would also act as catalysts in bringing about a total transformation in the outlook and agronomic methodology of our rural community. Where sights are limited, action is equally circumscribed and cynical comment is the only reception accorded to new ideas.

Mahatma Gandhi referred to this situation over forty years ago, when addressing those who wished to work in Indian villages. He said, "The fact is the villagers have lost all hope. They suspect that every stranger's hand is at their throats and that he goes to them only to exploit them. The divorce between intellect and labor has paralyzed our agriculture. The worker should enter villages full of love and hope, feeling sure that where men and women labor unintelligently and remain unemployed half the year round, he working all the year round and combining labor with intelligence cannot fail to win the confidence of the villagers." I have had the privilege of personally experiencing the wisdom of Gandhi's recipe. Hence, while accepting the Award for Community Leadership bearing the name of one, whose main characteristic was his passionate love of poor people, I plead in all humanity with the young men and women in the universities and scientific institutions of the developing nations to seize the opportunity and power, given them by science, to make real the possibilities of a truly human and meaningful life for millions of their fellow beings. It is to promote this cause that I propose to use the Award. ▯

51. MSS Launching the "Children for Happiness" Movement in India, 1999

7

Malthus and Mendel

The Malthusian picture of population growth outstripping the ability to meet the requirement of food is tending to become a grim reality. Significant advances took place during the nineteen sixties in improving the yield potential of major food crops like wheat, rice, pearl millet and sorghum. However, the scientific advances are yet to be converted into production advances through matching packages of services and of public policies. It is only in wheat that the genetic advance could to some extent get reflected in increased production. Genetic research, based on Mendelian concepts is showing the way for improving yield through better population performance and relentless research for raising the ceiling to yield is in progress. It is not only in plants but also in animals and fishes that considerable opportunities for improving production and productivity are now available. New coastal and inland aquaculture techniques can greatly increase fish production. Coincident attention to genetic upgrading, nutrition, health care and processing and marketing could help to accelerate the progress of improving animal productivity. The transmission of these research results to a large mass of poor and illiterate peasantry would call for an extension and educational effort of an unprecedented dimension. This is where every educated individual has an opportunity to assist.

Introduction

Famine leading to hunger, starvation, and so to death, was regarded by T.R. Malthus as one of the important natural checks on population growth. The Famine Commission of India in its report submitted in 1880 said, "famine which deprives an entire population of its customary food supply and arrests the ordinary employment of the wage-earning classes is one which, in a country such as India, wholly transcends individual effort and power of resistance." The Commission hence recommended that in time of famine "it becomes a paramount duty of the State to give all practicable assistance to the people." This recommendation has been implemented since then and widespread mortality due to famine has been, by and large, avoided, except during the Bengal Famine of World War II, caused by the wide-spread devastation of the rice crop by *Helminthosporium oryzae*. Over 728 crores of rupees have been spent on relief since the beginning of the First Five-Year Plan in 1951.

Presented as the Silver Jubilee Commemoration Lecture of the Indian Science Academy at Chandigarh, Punjab, January 2, 1973. First published in *Proceedings of the Indian National Science Academy* 39(6) Part B: 647-75 (1973).

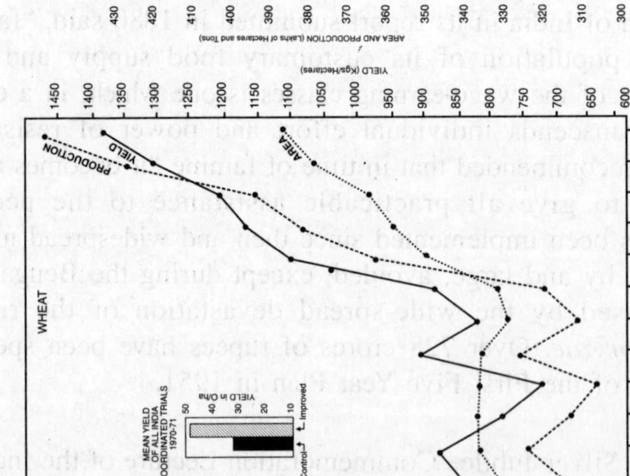

Figure 1

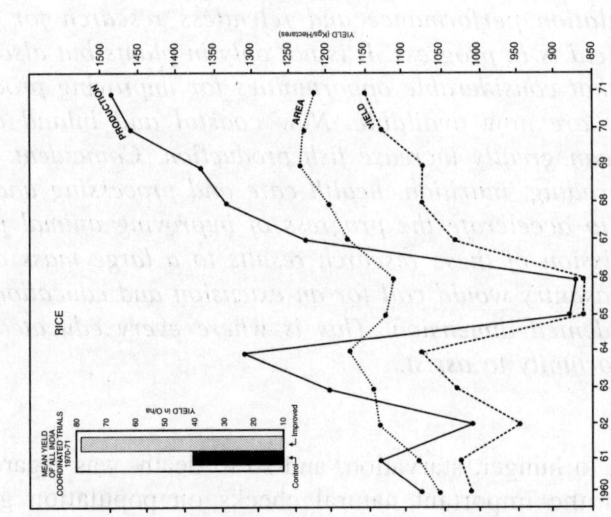

Figure 2

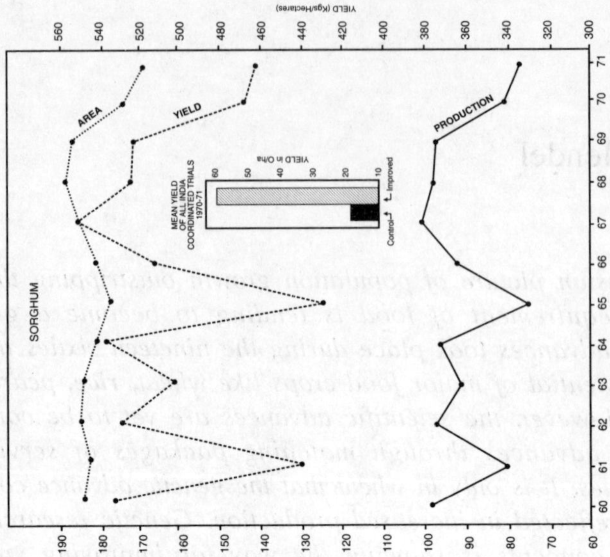

Figure 3

In the mid-sixties, India was considered by several experts to be a lost cause from the viewpoint of achieving food self-sufficiency. The prediction made by Paul and William Paddock in 1964 that our fate would be similar to that of "sheep going to the slaughterhouse", was followed by a doubling of wheat production during five crop seasons (1967-72), thereby dispelling the belief that the illiterate Indian peasant would not respond to new technology. If the optimism generated by this achievement has now given way to concern once again, it is partly due to our inability to capitalize on the scientific advances in rice, sorghum, maize and bajra (pearl millet) and partly due to the great impact which weather still has on the productivity of unirrigated, semi-arid areas which contribute over 40 percent of our total food supply. The year of doom predicted for us by the Paddock brothers – 1974 – is just one year away. Gunnar Myrdal, while reviewing the technological progress made since he wrote *The Asian Drama*, concluded that the unwillingness of "soft" States to take "hard" decisions would still lead to widespread misery in several of the developing countries of Asia, whatever the power and potential of scientific advances in improving biological productivity. I must express my agreement with Myrdal that while scientific work leading to the development of new technology to get more food from hungry and eroded lands is of crucial importance in facing the challenge of hunger, it would be mere self-deception to believe that a scientific breakthrough will automatically result in a production revolution and that a production revolution in agriculture will simultaneously result in a prosperity advance. This can only happen if there is a concerted effort to bring about the necessary changes simultaneously in socio-economic structures and policies and in education. These are basically dependent on "hard" political decisions. The experience of the 1960s has shattered two myths on which the Paddock brothers based their argument, viz., the fatalism and inertia of the Indian peasant, and technological stagnation. However, other factors still threaten danger.

While I am aware of the critical role which will be played by the political, educational and socio-economic factors, I propose as a scientist to confine myself here to examining the food-population equation from a scientific standpoint. To begin with, let me give as an illustration, the contrast between scientific advances in enhancing yield and the actual production figures in rice and sorghum. Due to a combination of favorable features, the wheat story has been different (Figures 1 to 3). The States in which wheat has been the dominant crop in the high-yielding varieties program have made the greatest progress in improving production (Table 1). The numerous intervening steps have been analyzed recently (Pradhan, 1972; Sivaraman, 1972; Swaminathan, 1972) and I shall only underline the urgent need to identify the causes for differential progress in different areas and in different crops, and to take appropriate remedial measures (Sivaraman, 1972).

I shall deal with the scientific possibilities largely from a Mendelian angle, although it is obvious that the genetic strain is only one component, though a pivotal one, in the farming system. Many new possibilities have been suggested for tomorrow's menu, such as algal soup, wood-based steaks, candied insects, grass pudding and plankton paste (Borgstrom, 1969), but I shall stick in my

Table 1. Increase in agricultural production in some states*

State	Percentage increase in foodgrain production during 1967-71	Area under high-yielding varieties as % of total area	Average yield of food-grains (kg/ha)	
	Over 1961-64	Under foodgrains	1961-62 to 1964-65	1967-68 to 1970-71
Punjab	87.2	36.1	942	1718
Haryana	64.9	13.3	770	1084
Uttar Pradesh	28.2	12.6	746	901
Gujarat	27.1	9.4	549	664
Tamil Nadu	7.5	21.6	1118	1202
Andhra Pradesh	4.7	5.3	773	770

* Data from Vyas, V.S. (1972). *New Agricultural Strategy and Regional Variations in Foodgrain Production.* New Delhi: Prices Commission.

Table 2. Projections of aggregate demand of foodgrains

Year	Cereals	Pulses	Total foodgrains (million tons)
Base period	84.76	11.39	96.15
1975-76	107.22	14.33	121.55
1980-81	120.96	16.70	137.66
1985-86	132.65	19.20	151.85
1990-91	146.65	22.20	168.69

Table 3. Requirements of foodgrains for human consumption

Year	Cereals	Pulses	Total foodgrains (million tons)
Base period	78.0	10.0	88.0
1975-76	93.6	12.4	106.0
1980-81	105.1	14.5	119.6
1985-86	116.0	16.5	132.5
1990-91	128.3	19.0	147.3

Table 4. Projections of milk demand

Year	Demand (million tons)
1975	32.3
1980	45.0
1985	55.0

analysis to the conventional food items. The problems are of immediate nature in our country and long-term solutions will offer no consolation to the hungry person, who, as Seneca pointed out long ago, "listens not to reason nor cares for justice nor is bent by any prayers".

Dimensions of the Problem

1) Population: The 1971 census revealed a total population figure of 547,949,809 out of whom 80.09 percent lived in rural areas. In contrast, the percentage of urban population is seventy-nine in the United Kingdom, seventy-four in Canada, seventy in the United States of America, sixty-eight in Japan and fifty-six in the USSR. According to the projections of the Registrar-General, our population is estimated to reach 657 millions by 1981 (Chandra Sekhar, 1972). While the Registrar-General has made projections of population only up to 1981, Dr. Thamarajakshi, Director, Agricultural Prices Commission, has estimated the figures at 703 millions on July 1, 1985 and 864 millions on July 1, 2000. These calculations are based on the assumption that the annual rates of growth of population would be 1.55 percent and 1.30 percent respectively, during 1981-1991 and 1991-2001. The Registrar-General, while emphasizing that we are still in the realm of conjecture owing to difficulties in assessing the likely impact of the massive family-planning program on the future population growth and size, has cautioned that we should be prepared for a population of one billion by 2000 AD.

2) Requirements of foodgrains: Estimation of requirements of foodgrains over the next few decades involves several assumptions regarding the growth of population, levels of income, income elasticity of demand, relative prices, changes in tastes, etc. Different estimates of these requirements have been worked out by different organizations. According to the Perspective Planning Division of the Planning Commission, India would need 1676 million tons of cereals and 18.6 million tons of pulses in 1981. [1] This requirement is expected to rise to 190 million tons in 1986, comprising 168 million tons of cereals and 22 million tons of pulses. [2]

The National Council of Applied Economic Research has projected aggregate demand of foodgrains at much lower levels, as will be seen in Table 2.

The figures for the base period are the average for the agricultural years 1967-68, 1968-69 and 1969-70, and are taken to correspond to the year 1968-69.

These estimates cover human consumption, as also seed, feed and other wastage. Requirements of foodgrains for human consumption as estimated by the NCAER are of the following order.

A Working Group on Demand and Supply Projections of the National Commission on Agriculture, chaired by Dr. A. M. Khusro has worked out some tentative projections of net and gross demand for foodgrains during 1975-2000. These projections (Table 5) indicate a gross demand of 228.50 million tons by 2000 AD on the basis of normative levels of per-caput requirements for urban areas and the 1985 projected levels of per-caput demand for rural areas.

---

[1] Quoted in the Fourth Plan Report

[2] Mimeographed (unpublished) publication of the Planning Commission

Table 5. Projection of net and gross demand for foodgrains, 1975-2000 (million tons)

Item	1975			1980			1985			2000*
	Rural	Urban	Total	Rural	Urban	Total	Rural	Urban	Total	Total
Rice	36.97	7.06	44.03	41.79	9.03	50.82	46.26	11.50	57.76	(Net demand) 172.80 (Assumption I)
Wheat	17.18	7.05	24.23	20.89	9.16	30.05	24.81	11.85	36.66	(Net demand) 182.80 (Assumption II)
Other Cereals	24.25	2.24	26.49	26.35	2.52	28.87	28.07	2.86	30.93	(Gross demand) 216.00 (Assumption I)
Pulses	9.30	2.58	11.88	11.18	3.47	14.65	13.21	4.68	17.89	(Gross demand) 228.50 (Assumption II)
Total net demand	87.70	18.93	106.63	100.21	24.18	124.39	112.35	30.89	143.24	
Total gross demand**			121.86			146.34			173.62	

* The estimates under Assumption I are based on normative levels of per-caput requirements for rural and urban areas, while those under Assumption II are based on normative levels of per-caput requirements for urban areas and the 1985 projected levels of per-caput demand for rural areas.

** Gross requirements are derived on the basis of the assumption that seed, feed and wastage ratio in 1975, 1980, 1985 and 2000 would be 12.5, 15.00, 17.5 and 20.0 percent respectively (of gross production).

Table 6. Indices of projection of per-caput demand for meat, egg, and fish*

Items	1971-00					
	Rural			Urban		
	1975	1980	1985	1975	1980	1985
Meat	113.1	141.0	182.1	108.2	122.8	141.6
Egg	115.42	149.28	201.16	126.48	183.34	278.44
Fish	107.60	123.00	143.5	110.6	129.6	155.0

* Khusro, A.M. and R. Thamarajakshi. "Demand projections for selected agricultural commodities: some preliminary results 1972". Working Paper for the Working Group on Supply-Demand Projections, National Commission on Agriculture.

With regard to milk, Dr. Thamarajakshi has worked out the following projections of demand. The indices of projection of per-caput demand for meat, egg and fish are given in Table 6.

3) Qualitative aspects of food consumption: On the basis of the overall average per-caput availability of different foods, it has been estimated that the per-caput availability of calories in our diets is about 2,000 calories and of protein a little over 50 g per day. These average figures tend to hide the existence of pockets of severe malnutrition, since a considerable proportion of our population live well below the poverty line and have a low purchasing power. The Task Force on Nutrition of the National Committee on Science and Technology has also drawn attention to the severe shortage of fats and oils, which are concentrated sources of calories, in the average Indian diet. In the developed nations, where excessive intake of fat (about 75 to 100 g daily) has aggravated the problem of certain kinds of heart disease, there is naturally an anti-fat movement. The current per-caput average consumption of fats is, however, only about 10 g per day in our country, and it has been suggested by the Task Force that the consumption rate should be increased to at least 30 to 40 g. This would call for a 3-to 4-fold increase in oilseed production.

According to Gopalan (1972), we will have to boost up our milk production by 205 percent and our meat, fish and poultry production by 388 percent to meet our demands by 2000 AD. There is thus the need for a simultaneous advance in the production of plant and animal products, including fisheries, if we are to assure a qualitative adequacy of food.

4) Nature and extent of agricultural endowments: The total cropped area was estimated at 164 million hectares during 1969-70. The net area sown was 139 million hectares, the intensity of cropping being 118. The net and gross irrigated areas were 30.34 and 37.22 million hectares respectively. Out of the net irrigated area of 30.34 million hectares, 12.25, 4.44, 11.14 and 2.49 million hectares were irrigated by canals, tanks, wells and tubewells and other methods respectively. The Irrigation Commission has estimated that it should be possible to develop a total irrigation potential of 81.7 million hectares by the end of this century by a planned expansion of major and medium projects and greater utilization of surface and ground water. It has been calculated that if all the projects now under various stages of execution are completed, we should have about 49.3 million

hectares of irrigated land (Ministry of Agriculture, 1972:208). While the progress in extending the area under irrigation has been impressive, intensity of cropping has increased from 114 in 1965-66 to only 118 in 1969-70. This suggests that there is great scope for improving the efficiency of water use.

5) Land requirements for non-agricultural uses: According to 1968-69 statistics, the net sown area was forty-five percent of the total area. Forests occupied 22.7 percent of the total land area. Although the forest area is about half of the cultivated area, the contribution of the forestry sector to the gross domestic product was only 1.6 percent in 1969-70. In the same year, agriculture contributed forty-six percent of the gross domestic product. The National Commission on Agriculture in a recent report on Production Forestry has pointed out the urgent need for paying scientific attention to forestry, since the estimated demands for wood and wood products will show a substantial rise in the coming years.

Land will also be increasingly needed for industry, housing, communications and non-food crops. For example, the demand projection for raw cotton to meet the cloth requirements in 1983-84 places the need at 90 lakh bales. Fortunately, there has been some significant recent advances in improving both yield and quality in cotton (Figures 4 and 5). We can be proud of being the first nation to exploit hybrid vigor commercially in cotton through Gujarat Hybrid 4 derived from the cross Gujarat 67 × American Nectariless and, more recently, Varalaxmi, developed at the University of Agricultural Sciences, Bangalore, by crossing Laxmi with SB 289 E. In view of the pressure on land for various uses, we will have to develop for each ecological area an integrated land-use plan, involving an appropriate admixture of agriculture, animal husbandry, fishery and forestry.

Recent Research on Improving Yield Potential

The credit for recent progress in improving the yield potential of the major food plants is due to the cooperative endeavor of scientists belonging to different disciplines and different organizations, all working towards the same end. The major scientific principles involved in this endeavor are: a) Development of varieties with a morphological architecture and growth rhythm that is conducive to making full use of good crop management, particularly the application of fertilizer and water; b) Capacity for good population performance; c) Wide adaptation arising from relative insensitivity to photoperiod; d) Higher productivity per day and suitability for multiple cropping; e) Quick-yielding habit, thereby helping in some cases to suffer less from the adverse effects of aberrant weather; f) Exploitation of hybrid vigor, additive genetic variance and other techniques by which the genetic potential for nutrient utilization and canalization of the total biological yield to the parts of economic value is enhanced; g) Development of broad spectrum resistance to pests and diseases and evolution of chemical and other methods of pest control; h) Standardization of agronomic practices which will bring out the yield potential of the variety.

The scientific principles underlying the improvement of the genetic potential for yield in wheat and rice are well known. Just to illustrate the principles employed and some of the current thinking on the methods of achieving further increases in yield, I give below a summary of the lines of research being undertaken by our scientists working in different agricultural universities and research institutes in coarse grains, sugarcane and tapioca.

Coarse grains

The research efforts during the past seven years have resulted in the release of five hybrids of bajra, four hybrids and a variety of jowar, and three varieties of ragi, and an integrated strategy of pest and disease control, specifically for rained areas.

1) New varieties: The new hybrids have a distinct advantage over locals during periods of drought, with an increase of 7 to 10 q/ha (Table 7). Two new male-steriles with acceptable grain quality in sorghum, and two male-steriles with tolerance to downy-mildew in bajra have also been evolved and are being tested. Two new male-steriles with acceptable grain quality in sorghum, and two male-steriles with tolerance to downy-mildew in bajra have also been evolved and are being tested.

2) Plant protection: Shoot-fly on sorghum and bajra and jassid on ragi can be controlled by seed treatment with carbofuron (5 g of active ingredient per hundred of seed) or ten percent phorate granules in seed furrows which improved the stands up to ninety percent (compared with less than ten percent in untreated plots as shown in Table 8).

Table 7. Relative performance of high-yielding varieties/hybrids of sorghum, bajra and ragi under rainfed conditions

Sorghum		Bajra		Ragi	
Hybrid and year of release	Average * yield q/ha	Hybrid and year of release	Yield q/ha (1962-72)	Variety and year of release	Yield q/ha (1969-72)
Local (H35-1)	0.2	Local (Maximum)	14.0	Local (Maximum)	20.6
CSH 3 (1972)	11.2	HB 5 (1972)	23.9	IE 28 (1972)	26.2
CSH 2 (1966)	11.0	HB 4 (1968)	22.3	EC 4840 (1970)	25.1
PSH 2 (1972)	6.9	HB 3 (1968)	23.8	Sarada (1970)	27.8
CSH 1 (1965)	8.9	HB 1 (1965)	20.7		

* Data at Bellary during 1971-72 rabi under extreme stress, when local failed.

Table 8. Effect of plant protection using carbofuron on grain yield (q/ha) in some coarse grains

Sorghum		Bajra		Ragi*	
Treated	Untreated	Treated	Untreated	Treated	Untreated
35.7	15.5	12.5	8.3	90-99%	32.8%

* Mortality of insect vector of ragi mosaic.

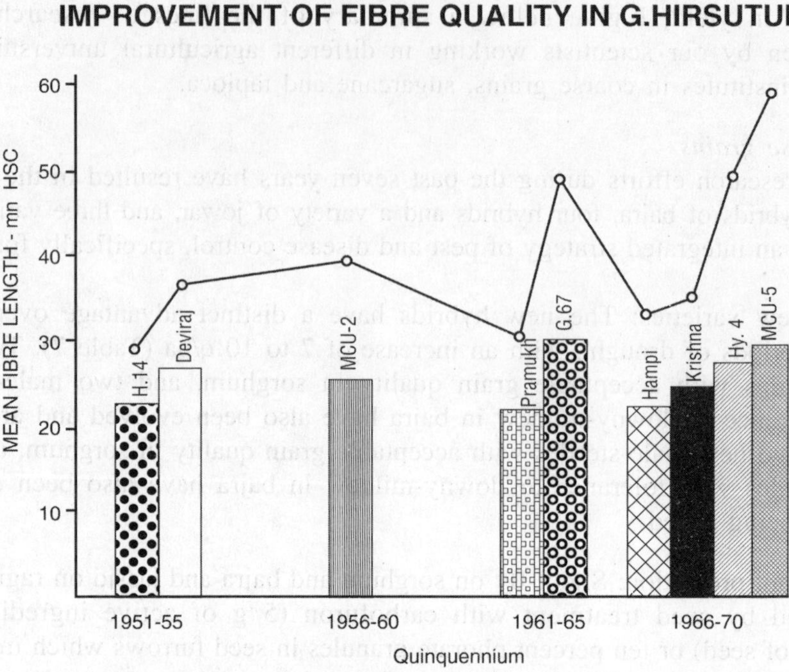

Figure 4

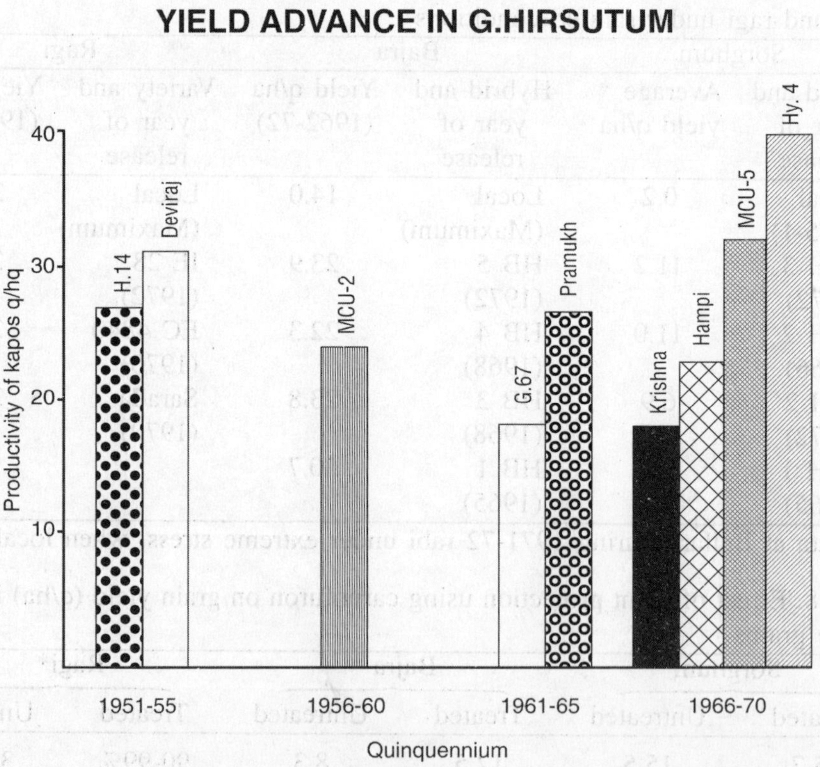

Figure 5

Sugary disease of sorghum and ergot of bajra can be reduced by spraying 0.2 percent Ziram or Thiram at flowering and can be combined with carboryl against application of lindane granules (2 kg/ha) controlled white grub on bajra by improving the plant stand from 1 percent in untreated plots to seventy-nine to ninety-seven percent in treated plots midge. Seed treatment with two percent salt solution and removal of ergot sclerotia are effective against seed infection in bajra.

3) Integrated strategy for disease control: Advancing date of planting in rabi sorghum to middle September together with an application of 60-80 kg N and 60 kg P_2O_5/ha, and carbofuron treatment of seeds is excellent in maintaining ninety-five percent stand. In bajra, sowing in middle of July at 2 cm depth in dry seedbed, with 60-80 kg N/ha and 175,000 plants/ha has been recommended.

4) Major difficulties encountered with the new strains: The grain quality in CSH 1 and CSH 2 is not comparable to locals and hence the hybrids suffered from this handicap in some areas. This drawback has been overcome by the new male-steriles with yellow endosperm and also by the new high-yielding varieties. Resistance to shoot-fly is yet to be achieved but a simple chemical control is available. Also, the problem can be minimized by changes in date of sowing where possible. A suitable hybrid for rabi-sorghum areas is yet to be released.

In bajra, the widespread incidence of ergot and downy-mildew is due to the high susceptibility of the female parent and a fall in quality of seed standards. Two male-steriles J 126 D_2A and MS 628 A with a fair degree of tolerance have been recommended for release. In ragi, mosaic and blast are serious in some seasons in Mysore and adjacent districts of Andhra Pradesh.

Hence, in all these crops greatly intensified research is necessary. I shall deal with in some detail the work in sorghum to illustrate what is being done.

Sorghum

1) The principles of genetics most effectively employed in improving crop productivity:

a) Exploitation of heterosis – Exploitation of heterosis employing cytoplasmic-genetic male-sterility has in the recent past resulted in the development of the commercial hybrids CSH 1, CSH 2, CSH 3 and CSH 4. The advent of the hybrids marked a genetic breakthrough in the otherwise stagnant yield levels. The hybrids yielded 60-100 percent more than the available improved varieties and several-fold when compared with national averages. The superior response of the hybrids to fertility and population levels has also been well demonstrated. While the first two hybrids are based on introduced male-steriles, the latter two are based on steriles developed in India.

b) Exploitation of additive genetic variance (varietal improvement) – The development and release of Swarna, the first high-yielding, variety, with yield levels equaling the commercial hybrid CSH 1, opened up further possibilities of line improvement. Swarna itself was utilized as a male parent for the hybrid CSH 4 (PSH 2).

Extensive studies involving selection in exotic × Indian crosses of sorghum resulted in the isolation of types like 604 (IS 3922 × Aispuri) and 302 (IS 3922 × Karad Local), which are on the verge of release for general cultivation. Such high-yielding derivatives, which combine the individual superiority of talls with the community performance of dwarfs, are representative of forms intermediate in height and maturity between the parental groups. Such intermediate forms satisfy the needs of food, fodder and grain quality.

c) Stability of performance – Besides high yields, stability of performance is of particular significance in breeding dry-land crops. CSH 1 has been observed to be a genotype that has performed well in several environments. It had an average stability index, i.e., b value close to 1.00 coupled with consistently higher mean yields. Even under extreme stress, CSH-F averaged 2500 kg/ha when the late locals failed to make any grain. In spite of the earliness and superior homeostatic properties of hybrids, barriers to their spread still exist, which appear to be more due to lack of appropriate extension efforts.

d) General and specific adaptation – The role of genotype × environment interactions has been critically analyzed and it has been well established that sorghums can be bred for a few large maturity zones rather than several small regions limited by State and geographic barriers.

2) Scope for further improvement in yield and quality:

a) New hybrids – Lack of variability among male-steriles, all of which are based on milo cytoplasm, is a major limiting factor in stepping-up the hybrid yield levels further. Sterile cytoplasmic sources different from milo have been discovered and attempts are in progress to convert lines known for their yield performance and combining ability into steriles using these new cytoplasmic sources.

b) Population improvement – There is increased interest in developing random mating populations of grain sorghum employing genetic as well as cytoplasmic sources of male-sterility. It is hoped that the use of such procedures involving wide germ-plasm resources may result in the breaking up of undesirable linkages and in generating enhanced variability on which selection could be practiced. Such population improvement programs could be useful in both varietal and hybrid development work.

It is now evident that gene-cytoplasm interactions could result in male-sterility or self-incompatibility, and each of these may in turn lead to apomictic seed development. Both self-incompatibility and male-sterility could be exploited to promote gene flow in populations and apomictic fixation could enable us to capitalize on the demands for diversity as well as uniformity.

c) Insect resistance – Shoot-fly has been the major pest for which absolute resistance is yet to be discovered. The tolerant sorghums which are predominantly Indian are tolerant because they exhibit non-preference for oviposition as against the preference of dwarf exotics. Studies are in progress to develop tolerant forms employing exotic × Indian cross-combinations..

d) Nutritional quality – With the exception of lysine, there is adequate ability of all other essential amino acids to match the FAO protein standards. The yellow-endosperm sorghums furnish sources for enriching the beta-carotene content. Two sorghum selections, IS 511 and IS 3797 have been established to be as well as stable for beta-carotene. Identification or induction of high-lysine character accompanied by a reduction in leucine content could provide the necessary opportunity for nutritional upgradation of sorghum to match high-lysine maize.

3) Theoretical limits to yield and production: While the higher-order yields of present-day Indian sorghums, have been of the magnitude of 8,000-9,000 kg of grain per hectare, they are yet not as high as those obtained in the United States (12,000 kg and above). One difference between tropical and temperate sorghums is that the former are single-stalked, while the latter tiller profusely. While tillering in sorghum is influenced by temperature at germination, we do not yet know whether this attribute would be desirable or not under Indian conditions.

The limit to progress from selection is a function of initial genetic variability, new genetic variabilities introduced and the selection scheme employed. However, at any given time the highest-yielding F_1 hybrid identified may be considered the visible limit to selection. Rao (1972) has pointed out that conversion of known parents as females and males to yield commercial hybrids, adoption of selection procedures to capitalize on additive and non-additive portions of the genetic variability and employment of tools like apomixis towards fixation of non-additive genetic variance could help in reaching higher yield levels.

Sugarcane

1) The principles of genetics most effectively employed in improving crop productivity: In the earlier years when systematic breeding work for sugarcane was started, interspecific hybridization was exploited to produce commercial varieties. Here, the early studies on phylogeny of the genus were of help in deciding the best source for incorporation of desirable characters. Co. 205, one of the important improved commercial varieties in the 1920s and '30s in this country, was obtained from a cross of *Saccharum officinarum* × *S. spontaneum.* Since then, interspecific hybridization has played an important role in the origin of several other commercial varieties.

Back-crossing also has been used, though in a somewhat modified form, to improve sugarcane varieties with success. Nobilization (i.e. a modified back-cross program, where a noble cane, *S. officinarum*, has been used as the recurrent parent) has given Co. 453, a variety which occupied for a long time considerable area. Here, different genotypes of *S. officinarum*, the recurrent parent, were used in contrast to the same recurrent parent being used in the conventional back-crossing program.

Studies on the genetics of quantitative characters have also been of use to the breeder. By use of parent-progeny regression, characters such as the percentage of germination, stalk height, brix, sucrose and purity have been studied at the Sugarcane Breeding Institute, Coimbatore. High heritability for quality characters such as brix and sucrose was found, showing that selection for these characters would give a greater genetic advance than other characters. In another experiment with two groups of varieties, the genotype × location variance component was found to have comparatively a low magnitude with respect to several economic characters. These data indicate that testing at a large number of locations may not be necessary for various genotypes. They also explain to some extent why selection at Coimbatore has been successful for eventual release of varieties in various parts of India.

By use of mainly bi-parental mating, a breeding technique not as powerful as some of the newly evolved breeding techniques, considerable genetic advance has been made in sugarcane improvement in India. In southern belt, from a base value of approximately fifty tons/ha and fifteen percent sucrose in juice, under average growing conditions, the yield could be improved to nearly one hundred tons per hectare and sucrose content to twenty-one percent. In northern India also from a considerably lower base value the yield potential can be improved to seventy-five tons per hectare and sucrose content in juice to nearly eighteen percent if research results obtained at Coimbatore and Lucknow can be transferred to the field.

2) Scope for further improvement in yield and quality: There is a considerable scope for improvement of yield and quality in sugarcane. For improvement of quality, the scope is high since payment by weight has led to the neglect of the factors governing sugar recovery. In view of this, only varieties which give high yield have been preferred by farmers. Varieties which could give high sucrose content, though they may be giving slightly less yield than the existing ones and are capable of giving more total sugar per hectare than the existing variety, have not become popular. Some of these varieties have been adopted in a few factory areas where an ad hoc premium for these varieties has been given. There is considerable scope for popularizing this practice among the various factories and starting a breeding program for producing varieties which could give more total sugar per hectare at different times of the crushing period of the factory. In the past, the breeding program has been mainly geared for producing varieties which give their peak recovery at the middle portion of the crushing period of the factory (Figure 6). There is a considerable scope for producing varieties which could give high recovery at the early and later parts of crushing

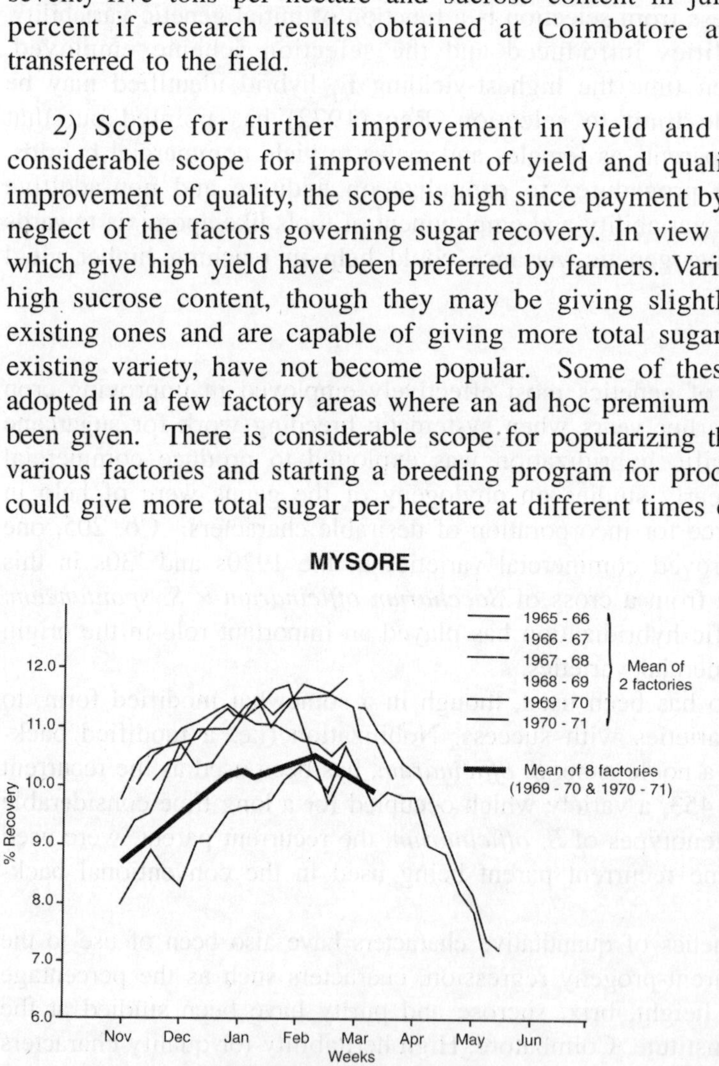

Figure 6

period of the factory. Compared with the all-India average of 9.5 percent to ten percent recovery, Australia has an average of fifteen percent recovery.

3) Theoretical limits to yield and production: As regards sucrose percent. Bull and Glasziou (1963) have worked out about twenty-seven percent of the fresh weight as an apparent physiological limit to sugar accumulation in the genus *Saccharum*. They derived this from data on fiber content, total sugar content, and moisture content for various species of genus *Saccharum*, allied genera, etc. As far as India is concerned, we are likely to get quick response to selection pressure in breeding programs where we have not very much exploited the genetic variability, as for example the breeding for early and late varieties and breeding for various physiological characters and disease resistance. So far as high yields are concerned, lodging will become a limiting factor and we will have to search for a non-lodging plant architecture.

Tapioca (Cassava)

1) Principles of genetics most effectively employed in improving crop production: Introduction – Exotic collections of tapioca have been obtained from countries like Brazil, Colombia, Madagascar, Fiji, Malaya, and Sri Lanka, and are being screened for their agronomic attributes for utilization in breeding programs. Susceptibility to mosaic disease and the presence of bitter principles in varying degrees in the tubers create difficulties in the isolation of promising types. There are strains capable of giving very high yields and high starch content. Though these could not be popularized for wide cultivation, they are being utilized in the breeding program for higher yield as well as for higher starch content in combination with disease resistance and better quality.

Hybridization – Intervarietal hybridization has proved to be useful. As a result of this method, H 97, H 165 and H 226 tapioca varieties have been evolved at the Central Tuber Crops Research Institute, Trivandrum. H 226 is suitable for culinary purposes. H 97, by virtue of having higher starch content (30.8 percent), is suitable for starch production, H 165 is found to be early maturing (seven to eight months) and may fit in a multiple-cropping pattern in different locations.

2) Scope for further improvement in yield and quality: Suitable breeding programs are being formulated to evolve varieties with high starch recovery, high protein content and low HCN. Some of the genotypes have shown desirable characters. But they could not be exploited for want of flowering in the material. Research is under way to induce flowering in these genotypes so that they can be utilized in the breeding program.

Tapioca mosaic is known to cause about thirty-five percent yield reduction. Inter-specific hybridization by making crosses and backcrosses with *M. glaziovii* has given types with mosaic resistance. Undesirable characters like lack of proper tuberization and high HCN have to be eliminated by further back crossing with tapioca. One of the important aims of current research programs is the development of early varieties (about six months) so that this crop can be grown more extensively and in new areas.

3) Scope for further genetic advances in yield and quality improvement: It is now becoming clear that through a greater understanding of production physiological processes and through an integrated application of inputs, particularly in the field of pest and disease control, yields higher than the plateau now reached in several crops can be attained. I give below a few approaches adopted in maize, a crop which has been studied intensively by both geneticists and physiologists.

a) Effective mobilization of new germplasm sources and creation of new variability through incorporation of more germplasm and induced genetic variation – This is now being done on a global scale.

b) Development of location specific strains – There is a need to identify varieties and cultural practices separately for intensive and rainfed farming. Under the former situation, yield levels can be further raised by selection of maize types that would be able to respond to a density of over 1 to 1.5 lakh plants per hectare. Varieties with smaller tassel, high nitrate-reductase activity and upper placement of the ear are responsive to high population density. However, for rainfed conditions selection of varieties with responsiveness to lower levels of fertilization and tolerance to moisture stress would help to stabilize yield.

c) Extending the filling period – It has been established that the photosynthate entering the ear is generated by the plant after pollination. Highly significant positive correlations between yield and the grain-filling period have been recorded. Obviously, the extension of grain-filling period will lead to higher yields. Extension of the grain-filling period can be realized independent of the other stages of growth and development.

d) Incorporation of resistance to serious disease and pest complex – This work has assumed particular significance after the devastation caused by *Heliminthosporium maydis* to maize crop in 1970 in the United States.

e) Improvement in quality – Further improvement in the level of lysine and tryptophan can be made by the development of double mutants involving Opaque-2 and other endosperm mutants fl_1, bt_1, bt_2, su_1, su_1. Increase in the level of lysine has also been shown by the multi-aleurone-layer stocks, particularly in the old Coroico maize from the eastern slopes of the Andes mountains.

Improvement in the level of total protein up to twenty-five percent has been demonstrated in Burrwhite. The problem of negative association observed between protein and lysine in normal maize is not serious when Opaque-2 is present. More work, however, is needed to ascertain the optimum levels of protein, lysine tryptophan that should be combined with high yields. Relatively little has been done with regard to selection for mineral content. Variability for Fe, Ca, K, Mg and others is available. Upgrading the levels of Fe and Ca, particularly in the background of modified protein stocks, would make maize a better food and feed.

f) Varieties for specific end-uses – Limited work has suggested that specialized types can be developed to meet industrial needs. Stocks like waxy, high amylose types, high oil varieties, and super sweets, on which limited work has been done, only suggest the possibilities for further development.

4) Theoretical limits to yield and production: Loomis and Williams (1963) on the basis of measurements of radiant energy and basic statistics on photosynthesis have calculated that 77 g/m^2, of dry matter can be produced (71 g of dry matter and 6 g of nutrient uptake) on a bright sunny day. Army and Greer (1967) calculated that 77 g/m^2 of dry matter would mean 690 lb/acre of dry matter and that if the entire converted energy went to grain, a yield of 12 bushel/acre (672 lb/acre/day) could be expected. In a 35-day active grain-filling period, a yield of 420 bushels/acre (26,362 kg/ha) could be reached. Goldsworthy and Palmer (1971) also observed that maximum yields of 27 tons/ha can be obtained.

In the calculations of maximum yield, the average measurements of radiant energy and relative efficiency of leaf canopies to convert energy have been used. It is observed that in certain environments higher radiant energy is received. Moreover, with a suitable modification of plant canopies higher estimates would result. Any increase in the grain-filling period will markedly influence the yield levels at the rate of three percent per day. Comparatively, maize is an efficient plant. Photosynthesis continues even beyond 5,000 ft. c. in contrast to several species where the saturation limit is reached around 2,000 ft. c. The CO_2 concentration compensation point for maize is very low. The crop can photosynthesize at as low as ten ppm of CO_2 concentration.

As against the maximum theoretical yield levels of 27 tons, yields of 19 tons have been recorded from Colorado (Cuanv et al, 1969). Yields of 14-16 tons have also been reported from the Arab Republic of Egypt, Mexico and a few other Latin American countries. Thus, there is much scope for further research in our country.

Need for Reappraisal of Breeding Procedures

The traditional methods of back-crossing and pedigree method and mass selection need to be supplemented using alternate systems of mating such as disruptive selection, biparental mating and inter-mating of intermediates with a basic yield and maturity level in the early segregating generations for diversifying the genetic base. The results so far obtained by scientists in Brassica, wheat and bajra have shown that by using such methods considerable variation is released, adverse association broken and genetic diversity enhanced with wider adaptability of the material generated by such selection. In *Brassica campestris* var. *sarson*, the rise in yield level over the best population by disruptive selection has been from about 700 kg/ha to 3,200 kg/ha with further possibility of improving the yield. In the case of bajra, biparental mating has increased the mean performance and variation for resistance to mildew and ergot, photosynthetic capacity as measured by effective leaf area, yield per unit area, grain size and grain quality. A similar system of mating using intermediates has improved the earliness, plant type and yield per plant in crosses between exotic and Indian sorghums with the new selections giving twenty-five percent more yield than the best Indian parent and 220% more yield than the best exotic parent.

The existing methods of breeding have to be modified for testing for wide adaptation in early segregating generations so that those genotypes with stability for maturity and better response to favorable environments can be identified and inter-crossed. Some of the hybrids developed by inbreds selected on this basis in bajra have a potential of more than fifty percent increase in yield over the existing hybrids.

Introduction of dwarfing genes should be accompanied by studies on crop geometry in pure stands and mixed crops, and estimates of competition effects should be related to leaf canopy, harvest index and rate of transfer of energy of leaf to gram under adverse environmental conditions like moisture stress. One such example in bajra is the new male-sterile J126 D_2A where a disease resistant but tall type has been dwarfened with simultaneous selection for synchrony of tillering, giving rise to an excellent combiner with wide-adaption and drought resistance. The transfer of energy of leaf to grain appears to vary with the genotype and also in different hybrids involving the same inbred. Therefore, selection for these physiological characteristics would be useful.

Seedling emergence is a major problem in saline areas. Recent studies at the Central Arid Zone Research Institute on different hybrids of bajra show that HB 3 is not only uniform with a high rate of germination under high concentrations of salt, but is capable of developing a good root system even under those conditions in the first four weeks. This capacity may be useful in its establishment much better than the local drought-resistant types of Rajasthan which are irregular in germination and slow and poor in root development.

1) Cooperation and competition in inter-cropping systems: The new technology for inter-cropping should be oriented to phase the maximum period of development of dwarf hybrids and varieties with the inter-crops. For example, hybrid bajra-soybean combination and hybrid bajra-groundnut are much better than other inter-crops involving the same crops with other cereals under rainfed conditions. An important need in such studies is the identification of inter-cropping practices where the companion crops "co-operate" and thereby increase the yield of each other. Sugarcane-moong and jute-moong inter-crops seem to have beneficial effects. Many other mixtures may show competitive interaction.

2) Population performance: Some evidence of the possible limits of productivity in the high-yielding varieties is available in hybrid bajra and brown-sarson. The new plant types in bajra can stand more than 250,000 plants/ha, compared with a maximum of 100,000 plants/ha with the traditional hybrids and much less for local varieties. The maximum yield levels in farmers' conditions achieved with hybrid bajra have been nearly 9,000 kg/ha with the lower limit of the above-mentioned population levels. Therefore, it can be assumed that the theoretical limit of about 150 to 180 q can be achieved with dwarf hybrids, provided the accompanying requirements for major and micronutrients are also met with and diseases are controlled. In the case of mustard, a plant population of 100 to 120 per square meter is likely to yield 50 to 60 q/ha compared with the traditional density of 15 to 30 per square meter with the local varieties. Selection for a longer period of maturity with early flowering can further enhance the yield level of oilseeds. It is expected that a theoretical limit of 60 q/ha of seed in oilseeds like mustard is attainable. In every crop, similar studies should be conducted to assess the maximum possible yield that can be obtained by promoting excellent population performance and good management.

Opportunities for developing new plant types are open in horticultural and other plants. Clear concepts of the numerous parameters controlling yield and of the most efficient plant architecture and growth pattern under a given agroecological milieu need to be developed, before much plant breeding advance can be expected.

Increasing Milk Production

Several programs such as the Intensive Cattle Development Project and Operation Flood have been introduced in recent years to increase milk production. The projects aim at paying coincident attention to genetic improvement, health care, nutrition and marketing. Progress in the genetic improvement of the milk-yield potential of pure-cattle breeds is necessarily slow. For example, the Sahiwal herd of the Indian Agricultural Research Institute, showed over a fifty-year period an improvement in milk yield from about 4.5 liters per cow per day to about 9.0 liters.

Cross-breeding Indian cattle with exotic dairy breeds for rapidly improving milk production became well recognized in the late fifties. In the early sixties, it was recommended that cross-breeding to exotic dairy breeds should be taken up in only such areas which have non-descript animals but which are suitable for maintenance of cross-breds. The results of cross-breeding in these attempts were so encouraging that, more recently, crossing of even high milk producing Indian breeds with exotic dairy breeds has been taken up. A cross-bred cow has the merit of higher milk yields without proportionately increasing the inputs in terms of cost of feeding her. From a recent study at the National Dairy Research Institute, Karnal, the average cost of maintenance per cow per day in a calving interval and the cost per liter of milk produced were reported from three Indian milch breeds in comparison with crossbreds between exotic and Sahiwal. The data are summarized in Table 9.

There is much scope for further improvements in breeding, health care and feeding procedures.

While the prospects for improving animal productivity are high, we should develop a long-term policy with regard to the relative proportion of plant and animal products in the food chain. Countries like the USSR which have increased their feed burden are forced to import large quantities of grains during bad seasons. China, on the other hand, has promoted the multiplication of only animals like hogs which can survive by scavenging wastes and excreta. Indiscriminate expansion in animal husbandry without long-term calculations on. the food and feed burden on both the nation as a whole and each ecological area will not be beneficial.

Increasing the Production of Fishes

1) Inland fisheries: Recent trends in aquaculture have indicated tremendous possibilities for increasing fish production in ponds, reservoirs and estuarine and brackish-water areas. In an All-India Coordinated Project on composite culture of Indian and exotic fishes, about 3,000 kg per hectare have been harvested in six months. This technique involves rearing in the same pond both Indian carps like catla (*Catla catla*), rohu (*Labeo rohia*) and mrigal (*Cirrhinus mrigala*) and exotic

carps like silver-carp (*Hypophthalmichthys molitrix*), grass-carp (*Ctenopharyngodon idella*) and common carp (*Cyprinus carpio*), in order to exploit all the water layers. Brackish-water fish culture has also made notable progress.

Table 9. Comparative economics of milk production by Indian pure-bred milch cows versus a cross-bred cow

	Tharparkar	Sahiwal	Red Sindhi	Cross-bred
Net cost of maintaining a cow per day (Rs.)	4.93	4.92	4.64	5.38
Average daily milk yield (kg)	3.96	4.45	3.86	7.48
Average cost of producing milk per liter (Rs.)	1.24	1.10	1.20	0.72

Source: Reddi, Y.V., B.K. Rastogi and B. Singh. 1972. *Indian Journal of Animal Science* 42(3): 157-163.

By suitable feeding techniques, an annual production of 2,400 kg/ha has been achieved in the culture of the mullet (*Mugil parsia*) and about 800 kg/ha in monoculture of the prawn, *Penaeus monodon*. Many problems such as the control of aquatic weeds, reclamation of swampy areas and the development of suitable culture techniques for air-breathing fishes are awaiting greater scientific attention.

2) Marine fisheries: During a recent symposium held in Cochin on the Pelagic fishery resources of the seas around India, the possibility of substantial increase of our fish production by increasing the harvest of the pelagic fishery resources was highlighted. It is well-known that in tropical oceans pelagic and bathypelagic fishes constitute the most important components of fisheries. In fact, some even believe that in the Indian ocean region, especially in the inshore waters of India, large-scale expansion of fisheries can only be based on the pelagic-bathypelagic complex, since the exploitation of demersal resources of the shelf areas has almost reached a saturation point at least in respect of some of the important groups. This, however, appears to be an extreme view because in recent years exploratory fishing operations have revealed the existence of several latent demersal resources and commercially exploitable fishing grounds off the east and west coasts of India. Therefore, there are possibilities for substantial increase in the yield both from the pelagic and demersal resources.

Various estimates based on exploratory surveys and primary organic production have shown that the Indian Ocean can easily provide an annual yield of nine to eleven million tons of fish and that, at this level of exploitation, the Indian Ocean would be on a par with the more intensely exploited Atlantic and Pacific Oceans. The total catch from the western Indian Ocean at present is about 1,090,000 tons and the potential yield from this area has been computed at 6,010,000 tons, consisting of about 4,150,000 tons of demersal fish, including crustaceans, and 1,860,000 tons of pelagic fish. The potential increase possible is thus about 4,920,000 tons per year, consisting of about 3,593,000 tons of demersal fish, including crustaceans and 1,327,000 tons of pelagic fish. In the eastern Indian Ocean, the total catch is about 895,000 tons and the corresponding potential yield has been estimated at 3,221,000 tons consisting of 1,400,000 tons of demersal fish and 1,821,000 tons of pelagic fish. Thus, from the Indian Ocean region the total estimated increase of yield would be of the order of 7,246,000 tons of fish from the present yield of 2,500,000 tons.

India alone accounts for 1,150,000 tons of fish out of the total catch of 2,500,000 tons from the Indian Ocean. Compared with the performance of other countries bordering the Indian Ocean, our achievements may be considered significant. But considering our country's locational advantage, the rich natural resources and availability of technical know-how for exploiting these resources, there is enormous scope for increasing the production. In this connection, it may be mentioned that out of 175,000 tons of tunas caught from the Indian Ocean, India's share is only 5,000 tons. Given proper facilities for harvesting these resources, it should be possible for our country to intensify the exploitation of tunas and market them at a rate cheaper than the marketing rates of Japan, Korea and Taiwan, which carry out fishing in the Indian Ocean far from their home bases.

Another aspect which requires research and development is that of coastal aquaculture. Although, the technique of culture of marine prawns is still in the laboratory stage, extensive culture of euryhaline fishes such as milkfish and mullets and molluscs would not be difficult. In countries like Taiwan, fish farmers are able to produce on an average 2,000 kg/ha/year of milkfish. Similarly, culture of molluscs, especially, oysters and mussels, has proved to be very profitable, and Japan produces more than 30,000 tons of oyster meat per year through aquaculture. The Central Marine Fisheries Research Institute, Cochin, has started experiments to culture mussels in the south-west coast of India by adopting the technique of rope culture. It is estimated from studies in the Galician Bays in Spain that the yield could be as high as 300 tons of mussels per hectare, as against the yield of about four tons per hectare from the natural beds. In our present state of knowledge, mollusc culture is known to yield the maximum quantity of animal protein per unit area. Therefore, large-scale culture of mussels and oysters would be highly rewarding. Nearly 2,000,000 hectares of open estuaries, brackishwater lakes and back-waters are available in this country. Out of these only about 10,000 hectares are now utilized for fishery purposes. Even assuming that we develop immediately only twenty-five percent of the available coastal areas for aquaculture (i.e., 500,000 hectares) and computing yield on the basis of a very conservative estimate of two tons per hectare involving culture of suitable varieties of fish, crustaceans and molluscs, an annual yield of 1,000,000 tons of edible animal products could be generated. Preliminary data on the economics of coastal aquaculture have shown that these practices could be economically feasible and viable.

One of our high-priority areas both in research and development should be the exploitation of the potential we have for increasing the yield of both inland and marine fisheries.

Some Challenging Scientific Problems

1) Fighting pests and diseases: The Committee on Genetic Vulnerability of Major Crops of the National Academy of Sciences of the United States has recently made a penetrating analysis of the risks to crop production arising from

the cultivation of the same genetic strain over large areas. In the past, breeding for disease resistance was largely based on, "vertical resistance" based on single genes (Plank, 1968). Can we produce strains with "horizontal resistance" which may be more stable? The natural accumulation of horizontal resistance in African maizes to *Puccinia polysora*, following the accidental introduction of this pathogen from Central America, suggests that it will be worthwhile to undertake intensive research in this field.

In fighting pests there is global awareness of the need to minimize the use of chemicals and to develop integrated control methods. Dr. S. Pradhan, a distinguished Fellow of the Academy, has been advocating the adoption of this approach in our country in crops like rice, cotton and sorghum, where the production advance is being hampered by pest damage. Integrated pest control would involve the use of two or more of the following approaches: a) selection of resistant varieties; b) use of parasites and predators, both native and introduced; c) development of microbial insecticides; d) ecological control involving modifications in cropping systems and agronomic practices; e) use of selective chemical insecticides; f) use of attractants and repellants, including sex hormones; g) development of autocidal methods including various genetic tools or induced sexual sterility.

The Indian Council of Agricultural Research is sponsoring integrated pest control projects in rice on a pilot scale in Andhra Pradesh, Orissa, West Bengal and Madhya Pradesh. We need urgently more research in this field.

2) Increasing the productivity of pulse crops: Pulse crops, particularly chick pea or Bengal gram (*Cicer arietinum*), are being pushed to even more marginal lands than before, since when irrigation facilities are developed wheat and other high-yielding crops are preferred. The short-term research strategy has hence been directed towards breeding strains suitable for multiple cropping and off-season cultivation, ruch as Bhaishakhi Moong (*Phaseolus aureus*) and Ageti and Prabhat Arhar (*Cajanus cajan*). The production trend is disturbing and the yield advance made in crops like pigeon-pea is not very striking (Figure 7). There is, however, vast scope for improving pulse production immediately by using the best available strains with fertilizer (particularly phosphorus), bacterial cultures, and pest control (IARI, 1971).

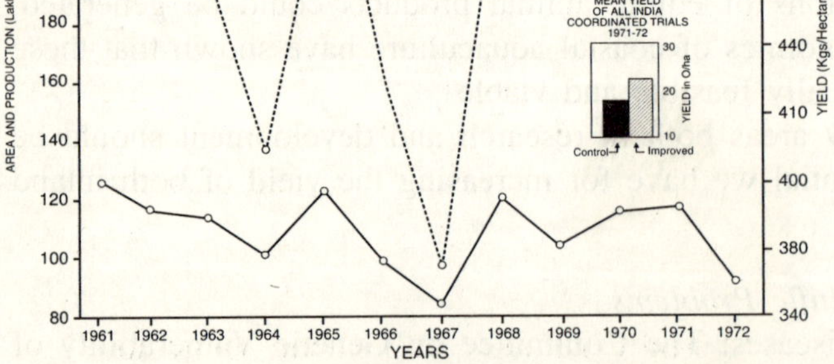

Figure 7

Physiological studies at the IARI by Dr. S. K. Sinha and his colleagues have brought out the following significant observations.

a) There is profuse flowering, thus a very high "sink" capacity, in most pulse crops. But as a result of heavy shedding of flowers, only a few fruits are formed.

b) In moong beans, cow-pea and pigeon-pea, the highest rates of photosynthesis and the activities of photosynthetic enzymes both C_3 and C_4 systems (i.e., RuDP carboxylase and PEP carboxylase) are observed in seedlings. At the time of flowering and fruit development, there is a sharp fall in all these characters. There is, however, a difference of magnitude in this respect in different varieties. This is in contrast to cereals and millets where the rate of photosynthesis is stimulated after anthesis.

c) The fall in photosynthetic systems coincides with the shedding of flowers and also a slight drop in the nitrogen level of leaves.

d) In cow-pea, the maximum number of nodules is observed at the time of flowering. There is disintegration of nodules during the fruit development period, indicating that the existing nodules are ineffective.

e) These results lead to the following theoretical considerations: i) for obtaining a yield of thirty q/ha of a pulse crop with thirty percent protein the amount of nitrogen required for seeds and fruit wall is 200 kg. N; ii) assuming a 0.2 harvest index and only two percent N in vegetative parts (dry weight) the amount of nitrogen required by the plant is 240 kg. N. Therefore, the total nitrogen requirement shall be approximately 440 kg. Since almost no nitrogen is supplied, the plant has to fix this amount of nitrogen through symbiosis.

These findings provide some clue as to the lines of work a breeder may undertake. We need more intensive studies for breaking the barrier to high yields in pulse crops.

3) New threats to plant products: Many plant products are increasingly facing competition from synthetics. For example, cotton is steadily losing ground. Cotton's share of the total textile market declined from about seventy-five percent in 1945 to about fifty percent in 1971. Every synthetic fiber plant which uses a few thousand employees displaces more than 32,000 agricultural workers, according to an estimate by the International Institute for Cotton. Cotton is an important commercial crop in the Semi-arid areas of our country and we therefore need tests and specifications which the cotton breeders will be able to use to assess their strains for easy-care properties. Also, new methods of opening, spinning, blending, weaving and knitting are being adopted by the industry, and we must make sure that cotton can be handled efficiently on these new systems. The situation is dynamic and the panorama of possibilities for creative research of immediate relevance to our people is vast. There is room for every one – whether working in a University or research institute or college – to contribute.

4) Crop life-saving research: A last urgent and challenging task is the development of alternative cropping strategies and crop life-saving techniques in order to ensure that crops are not totally lost during drought or floods. We must exploit the resilience which tropical and sub-tropical agriculture, unlike the farming systems of temperate countries, offers with regard to crop substitution.

During kharif 1972, sunflower gave an average yield of 8.2 q/ha, as compared to 4.7 q/ha by groundnut at six locations. Sunflower being relatively photo-insensitive is not very stringent with regard to sowing date requirements. As mentioned earlier, hybrids of jowar and bajra show greater stability of performance as compared to the earlier varieties. Aruna Castor developed by mutation breeding represents a good example of purposeful alteration of duration to overcome the adverse effects of poor rainfall. Hybrid cotton also shows promise under unirrigated conditions. Nutrient supply, pest control and water saving techniques are needed to fight total crop failure during aberrant weather. The preliminary studies carried out under the All-India Co-ordinated Dryland Farming Research Project as well as at the International Crops Research Institute for the Semi-arid Tropics located at Hyderabad, have given valuable data. Similarly, the results gathered at the Central Arid Zone Research Institute, Jodhpur, on the management of desert soils require widespread testing and adoption. More satisfactory weather models on the basis of rainfall reliability analysis would be of value in breeding work. If we do not step up our research on crop life-saving techniques and facilitate the adoption of the techniques already standardized, we will have to be prepared for considerable undulations in both food production, and the availability of oilseeds, pulses and cotton based on seasonal conditions.

Conclusions

I started this lecture by quoting the targets we have to achieve in the production of foodgrains, animal products and fish if we are to assure adequate nutrition to our population, both quantitatively and qualitatively. I then referred to the growing gap between what has been accomplished in the major crops in the experiment stations and in the field, except in the case of wheat, where research results and production records have found an honorable equation. The targets are not frightening, considering the irrigated area we have and the scope which exists for moisture conservation and integrated land use planning. Later, I entered the world of genes, their recombination, modification and selection, and the possibilities of tailoring the plant to our needs. Today's need is for population excellence in plants and farm animals as against individual and competitive virtue which was essential for survival both in nature and under stagnant agricultural conditions, where the unholy triple alliance among weeds, pests and pathogens made it necessary for the plant or animal to specialize more on survival than on production. This part of the scientific picture offers cheer and hope for striking a favorable balance between population growth and food supply at least till the end of this century, by which time we can expect a greater stability in population expansion.

When we leave the scientific world and enter the actual world of rural India and of urban slums, the scientific picture seems to be a mere mirage fit only to be admired with the eyes but not to be experienced in day-to-day living. The multiplication of population and the expansion of sub-human living conditions goes on relentlessly, and we tend to become numbed into inaction, overawed by the very dimensions of the problem. To explain our inaction and yet to satisfy our intellect, we become cynical, turn into destructive critics and try to paralyze even those few who are striving to end the divorce between intellect and labor in rural areas and thus to lay the foundation for the scientific transformation of our rural economy.

We are late starters in the economic rat-race. This has its advantages in the long-term, since we can try to create a new human and economic society based on the rational and symbiotic use of finite resources. On the other hand, we cannot be silent onlookers to the immediate human misery and suffering we see all around us. We therefore have to evolve a scientific strategy where the short- and long-terms goals are harmonized – the immediate aims being more and cheaper food, more purchasing power and more productive rural employment to arrest unplanned migration of landless labor to cities. The methods by which we aim to achieve these immediate goals should not be incompatible to long-term growth. You have to go and see the saline soils in some of the new irrigation project areas to realize what I mean.

I believe that every Fellow of our Academy, irrespective of the discipline of his specialization, can extend a helping hand in this task. Whether it be in integrated pest control or scientific water management or recycling of organic wastes including human excreta or post-harvest handling and storage, there are many small steps that can result in large beneficial effects, provided we can do to human society what plant breeders have done to wheat or rice, namely to generate collective and cooperative endeavor. This is more easily said than done; but there is no other way by which illiterate farmers with small holdings, who constitute the bulk of our agricultural community, can share fully the fruits of scientific work.

The immediate need is to convert the scientific work already done in cereals, millets, pulses, vegetables, animal husbandry and fisheries into wealth. Scientists in universities and research institutes can participate in this process in several ways. For example, in order to create a scientific outlook and temper among children and adults which would make the adoption of such innovations possible, there is need for scientists to become involved with school education and adult education in their vicinity. This need not take the conventional form of classes, but contribution can be in the form of development of science games, kits and toys for children, developing simple study projects on the immediate environment, demonstrations of scientific principles through hobby and study clubs, preparing of simple literature for neo-literates and school children, science writing at a simple level in Indian languages, and development of materials for communicating science through mass media. For those connected with agricultural science, steps that directly lead to new agricultural practices could be taken up, which I have spelt out elsewhere (Swaminathan, 1970).

We have in our country the strange phenomenon of having at the same time a small group of highly skilled and specialized scientists and a large population who may be described as scientific illiterates. In order to bridge this gap, evidently the initiative must come from the scientific community, which must make a special effort to interpret science and its applications in terms of the local environment and carry it to the users. The new educational technology, especially the mass media, can play a big part in this process; but the materials or the software that goes into it have to be developed by scientists. This can happen only if we begin to consider it as prestigious for a scientist to have developed a kit or demonstration technique or radio lesson for primary children or illiterate peasantry as to have published a paper in his own field of specialization. Similarly, the designing of simple short courses in technical skills for people with low educational qualifications will be necessary before we can launch a campaign of "technical literacy" for the illiterates. These courses will need to be closely geared to local needs and environment and can best be developed by scientists in close touch with the realities of the surrounding communities. The involvement of students in such work is also something which I have been pleading for some time (Swaminathan, 1970). In such small ways, the scientific community can contribute to the spread of knowledge of the new technology which will be one of the requirements for converting a scientific advance into a prosperity revolution. If the requirement of changes in socio-economic patterns and policies to bring about a better match between movements in science and in society are also brought about by the political leadership, then we can look forward with hope to a new future.

May I end by thanking the Academy for the Silver Jubilee Commemoration Award? While I appreciate the honor, I must emphasize that whatever we have been able to achieve in agricultural research is only because we have tried in a small way to emulate our biological creations – work cooperatively than competitively. I am only one member of a large army of research workers who toil in the sun, trying to understand how plants, animals and humans can live in a permanently productive and symbiotic biosphere. If you pick out an individual wheat plant from a good field of high-yielding wheat, it will look unimpressive. Its strength is in its partnership with its neighbors. I feel privileged to be in the same position and I thank you most sincerely.

References

Army, T.J., and F. A. Greer. 1967. Photosynthesis and crop production system. *Harvesting the Sun*, ed. by A. Santpiepro, F. A. Greer and P. J. Army, pp. 321-32. New York: Academic Press.

Borgstrom, G. 1969. *Too Many – An Ecological Overview of Earth's Limitations.* New York: Collier Books.

Bull, T. A. and K. T. Glasziou. 1963. The evolutionary significance of sugar accumulation in Saccharum. *Australian Journal of Biological Sciences* 16: 737-42.

Chandra Sekhar, A. 1972. *Population Growth and Agricultural Development in India.* India: Indian Society of Agricultural Statistics.

Cuanv, R. L., J. E. Swink and S. L. Shafer. 1970. Performance tests of corn hybrids grown in various regions of Colarodo in 1969. *Colorado State University Exp. Station General Series 905.* Colorado: Colorado State University.

Goldsworthy, P. R. and A. F. E. Palmer. 1971. *Proceedings of 1st Maize workshop*, pp. 93-94. Mexico: International Maize Wheat Improvement Centre (CIMMYT).

Gopalan. C. 1972. Meeting the increasing mankind needs. Mexico: IX International Congress of Nutrition.

Loomis, R. S. and W. A. Williams. 1963. Maximum crop productivity – an estimate. *Crop Science* 3: 66-72.

Plank, J. E. Van der. 1968. *Disease Resistance in Plants*. Delhi: Ministry of Agriculture.

Pradhan, S. 1972. In tropics protection research more needed than production research. *Indian Journal of Ent,* 33: 233-59.

Rao, N. G. P. 1972. Sorghum breeding in India. *Recent Development in Sorghum in Seventies.* Oxford and IBH.

Sivaraman, B. 1972. Scientific agriculture is neutral to scale – The fallacy and the remedy. *Indian Society of Agricultural Statistics.*

Swaminathan, M. S. 1970. Agricultural transformation and opportunities for a learning revolution. First Zakir Hussain Memorial Lecture, University Grants Commission, New Delhi.

-----. 1972. *Converting a scientific breakthrough in agriculture into a production advance*. Shri Ram Institute for Industrial Research.

8

Agriculture on Spaceship Earth

The expression "Spaceship Earth" was first used by Buckminster Fuller many years ago to dramatize the fact that the resources of Earth are finite. Today this viewpoint has become necessary for man. If one looks at it this way, one can see Earth as an extra large, very efficient, Nature-designed Spaceship, carrying its full load of three billion passengers through space at 18.5 miles per second, equipped with its own energy, refueling, recycling and life-support systems. A thin envelope of air, soil and water around the Earth together with energy from the sun provide the major ingredients for the evolution and survival of the large number of living things, including man on Earth, and all are closely interrelated. This complex and delicately balanced web of relationships together constitutes the biosphere within which man has his place. It is hence evident that man can ultimately make only a finite claim on the limited resources of Earth.

The utilization of Earth's resources as though they were infinite, particularly by the rich nations, has led to the present awareness of an impending environmental and energy crisis, if economic activity is continued on the same path and pace. Brown (1972), for example, has pointed out that in 1950 when the gross world product (GWP) was just over a trillion dollars, eco-catastrophes were few. In 1970, when the GWP became nearly three trillion dollars, rapid eutrophication of lakes, oil spills, fish kills, devastating floods arising from deforestation and similar disasters became more frequent.

Growth in the consumption of natural resources, both renewable and non-renewable, is proceeding at an alarming rate. Petroleum consumption is expanding at four percent yearly, fresh water at three percent, marine protein at four percent and the use of chemical fertilizers at seven percent. A three percent annual growth rate means a doubling in every twenty-four years, while a seven percent growth rate leads to doubling every ten years. In their study *Limits to Growth*, Meadows and co-workers (1972), therefore, stressed that the exponential growth in consumption of finite resources, however large the reserves may be, cannot continue indefinitely. Many of their assumptions may be open to question.

Taking agriculture, Meadows and co-workers have calculated that at the present level of productivity about 0.4 hectares of arable land will be needed to feed a person. The total world availability of arable land is about 3.2 billion hectares. As the population grows, more land is not only needed for producing

---

Presented as Coromandel Lecture, February 26, 1973. First published in *Coromandel Lectures: Platform for a Common Present and Future for Humankind*, pp. 37-66. Secunderabad, Andhra Pradesh, India: Coromandel Fertilisers Limited, 2001.

more food, but also for urban-industrial use. Calculations show that in 2000 AD, the land available for agriculture will fall short of the amount needed to produce the requisite quantity of food for the population at that time, at current productivity levels. This "doomsday" year will come twenty-five years later, if the productivity can be doubled and fifty years later, if the productivity can be quadrupled. There could be a difference of opinion about these dates and calculations but the basic thesis is clear – we just cannot continue to fondly hope that the Earth's resources can be indefinitely exploited on the assumption that some new discovery would always help to open up some hitherto unknown pathway of production.

The Ecologist (January 1972) in a much-discussed article entitled "A Blueprint for Survival" stated the choice available to man as follows: "The principal defect of the industrial way of life with its ethos of expansion is that it is not sustainable. Its termination within the lifetime of someone born today is inevitable – unless it continues to be sustained for a while longer by an entrenched minority at the cost of imposing great suffering on the rest of mankind."

In the consumption-oriented world of today, the rich nations consume most of the world's products, while within developing nations the richer sections of society, which may constitute only a fraction of the population, consume a large proportion of the goods produced. It is probable that the garbage cans of the richer nations may contain more protein than the total quantity of protein consumed by the poor of the developing world. One third of the world's people who live in developed lands consume nearly eight times more energy per capita than do citizens in poorer countries. The consumption revolution of the developed world obviously cannot go on at the current pace, particularly since the developing countries also wish to consume more energy now. This realization is leading more and more thinkers to begin to view the problems of the world as a whole, and not merely consider each part and aspect as an isolated problem. *Only One Earth* by Barbara Ward and Rene Dubos (1972) is an example of this new awareness.

We live in a world of sad scientific ironies and economic enigmas. The regions where man first settled down to cultivate plants and thereby initiated what we now call "agriculture" are also the regions which contain the greatest number of hungry people today. The regions which scientists designate as the center of origin of a crop plant are, strangely enough, the areas where the same plant is today giving one of the poorest average yields. The Jeypore tract of Orissa, which is believed to be an ancient center of origin of rice, is a good example. The regions where technical skills were of a high order in ancient days are today characterized by relatively poor quality of output and workmanship. One need only recall in this connection the past glory of India and China in the production of textiles, pottery, porcelain and metal goods.

This trend needs to be arrested, if the developing countries are to accomplish even partially their socio-economic goals. This is, however, more easily said than done. The compound interest law of growth rate operates without abatement. We are now in the second U.N. Development decade but by most economic yardsticks, the income and consumption gap between the rich and poor worlds is widening. Between 1967 and 1971, developed market economics increased their agricultural exports on an average by eleven percent per year, while the developing market economies lost ground by one percent (FAO statistics). Similarly in 1971, agricultural output went up by nine percent in North America and five percent in Western Europe, while the increase in the developing countries ranged from one to two percent. On the positive side is the fact that we have an opportunity to take to a pathway of development which does not lead to the degradation of both man and his environment. The "One Earth" concept is vital in any system of planning owing to the many tangible and intangible international linkages which influence national endeavor. However, we must accept the reality of the existence of sovereign nation states and the differences of interests and priorities between the two sharply divided groups in the world – the rich and the poor nations. Hence, within the overview provided by the concept of "One Earth" it will still be necessary for some time to come to think in terms of planning the rational and best use of resources for each country or region. I propose to confine myself, therefore, in this lecture to a discussion of the problems and possibilities for agricultural development in that sector of Spaceship Earth, which is India.

Table 1. Population Projections (Millions) – India

Year	NCAER*	Planning Commission	Registrar General	Dr. Thamarajakshi
1968-69	521	527		
1971			548	
1973-74		596		
1981	689	690	657	
1985	755	740		703
1991	827			
2000				864

*National Council of Applied Economic Research

Table 2. Food Demand and Supply Projections for India

	Gross Food-Grain Demand (Million tons)				Food-Grain Production (Million tons)		
	Demand				Supply		
Year	NCAER	Planning Commission	NCA	ESD	NCAER	Planning Commission	ESD
1968-69	99.8	94.4	-	-	96.2	98.0	-
1969-70	-	-	-	102.2	-	-	-
1973-74	-	126.9	-	116.1	-	-	115.6
1975-76	120.3	-	121.9	-	121.5	-	-
1978-79	-	-	-	145.4	-	-	134.8
1980-81	135.7	163.0	146.3	-	137.7	165.5	-
1985-86	150.3	190.0	173.6	-	151.8	192.3	-
1990-91	167.1	-	-	-	168.7	-	-
2000	-	-	228.5	-	-	-	-

Gross Demand: Net Demand (for human consumption) + Allowance for seed, feed and wastage

Population Growth and Food Supply

I have discussed elsewhere the various projections of population growth and food needs up to 2000 AD and also the possibilities for producing the needed quantity of food. The population and food requirement projections made in different studies are summarized in Tables 1 and 2. It should be possible for us to achieve the production targets, if we systematically adapt and extend the available scientific know-how to the different agro-ecological areas. But what we need from the agricultural sector is not only more food but also more income and more avenues for remunerative employment. Therefore, we have to develop for each ecological area a land and water use strategy which can, through a careful blend of agriculture, forestry, animal husbandry and fisheries, help to diversify the sources of employment and maximized income.

Potential for Further Improvement in Yield

I would like to take crop and fish production to illustrate current trends.

1) Crops: It has been estimated by crop physiologists that 12.2×10^5 cal/ha/day, which is equivalent to 124 tons/ha/year of carbohydrates, can be expected to be the maximum attainable biological yield in the tropics. Assuming a harvest index of fifty percent in crops like wheat and rice, we should expect a yield of 6.1×10^5 cal/ha/day or approximately 62 tons/ha/year of carbohydrates in the form of grains. There are reports that yields up to 80 q/ha and 100 q/ha have been obtained in wheat and rice respectively in 130 days under intensive agriculture. Due to cloudiness during the growing season, it may be difficult to expect yields exceeding 4.88×10^5 cal/ha/day in rice. In All India Coordinated trials, energy conservation to the extent of 3.05×10^5 cal/ha/day has been reported in rice which represents about sixty percent of the potential.

In wheat, the highest grain yield has reached nearly one third of the potential yield. In oilseeds, the single important factor responsible for poor yield is the low harvest index. Even with one irrigation and 50 kg. N/ha, it is possible to obtain eighty percent of the highest yield. Hence, striking improvement in the productivity of these crops is possible in areas with limited water availability.

Improvements in yield in most economic plants have taken place since the beginning of agriculture through small improvements and changes. Occasionally, jumps of larger quanta accelerate the pace of progress. Usually, such jumps are associated either with an important landmark in the genetic improvement of crop plants or in advances in water and pest management. Some significant landmarks in plant breeding during this century with reference to both yield and quality are indicated in Table 3. It should be emphasized that the role of the new seeds in improving yield has been primarily that of a catalyst and that the real yield advance has been brought about by associated changes in agronomy, plant protection and engineering.

Advances have also taken place in improving the yield of plants and animals in semi-arid and arid areas. The yields obtained in trials conducted under the All India Coordinated Research Project for Dryland Agriculture during Kharif 1972, when widespread drought occurred, indicate that there is much scope for improving production in these areas through an integrated plan for moisture

Table 3. Some Significant Landmarks in the 20th Century

Development	Plant Breeding: Years of widespread	Country where originally used	Genetic Principle(s) involved
1. Hybrid maize	1933	USA	Hybrid vigor
2. Hybrid sorghum	1957	USA	Hybrid vigor
3. Hybrid pearl millet	1959	USA	Hybrid vigor
4. Dwarf wheat	1961	USA	Dwarfing genes from Japanese "Norin" wheats
5. Dwarf wheat	1963	Mexico and later India, Pakistan & Middle East	"Norin" dwarfing genes & relative photo-insensitivity
6. Dwarf indica rice	1955	Taiwan	Genes for dwarfing and relative photo-insensitivity stiff and erect leaves
7. Dwarf indica rice	1955	Philippines, India & other countries in Asia	Genes for dwarfing and relative photo-insensitivity stiff and erect leaves
8. 'Opaque-2' maize	1965	USA	High lysine content
9. Hybrid barley	1969	USA	Hybrid vigor
10. "Hiproly" barley	1969	Sweeden	High protein and high lysine
11. Hybrid upland cotton	1970	India	Hybrid vigor

Table 4. Yields Obtained During the 1972 Kharif Season in Trials Under the All-India Coordinated Research Project for Dryland Agriculture

Rainfall Pattern and Center	Yield in Q/ha: Jowar	Bajra	Setaria	Arhar	Soybean
Seasonal rainfall deficient (-20 percent to -60 percent from normal)					
Hyderabad	17.8	11.8	16.3	14.1	
Indore	39.5				29.8
Dehra Dun					26.1
Rajkot	3.3	9.0		8.9	5.4
Varanasi			12.4	14.3	7.5
Seasonal rainfall scanty (-60 percent and below)					
Agra	6.0	19.9		11.9	

Table 5. Average Annual Yield of Different Animal Production Systems

Area	Product	Yield (kg/ha)
Pasture land	Cattle	6-308
Inland Ponds (India)	Fish	3000
Continental shelf	Ground fish	25-715
Humboldt current	Anchovies	375
Mariculture (Japan)	Oyster	57,500
Mariculture (Spain)	Mussels	300,000

conservation, fertilizer application, pest control and crop and varietal substitution (Table 4). Some imaginative and elegant experiments are now under way in achieving crosses between distantly related species through somatic cell hybridization, in using pollen cultures for getting haploids, in introducing the genetic capacity for symbiotic nitrogen fixation in non-legumes and in enhancing the "sink" capacity, or capacity for economic yield through the genetic manipulation of production physiologic processes. It is premature to guess what the ultimate impact of these researches would be on agricultural development. Meanwhile, we have enough work to do in converting the scientific breakthrough, which has already taken place in the major crops into production advance.

2) Fish: Among farm animals, fish is the most efficient converter of plant food. World fish production rose from 21.1 million tons in 1950 to over 69 million tons in 1970. Let me illustrate the position both in marine and inland fisheries in our country

.Marine Fish Production

The realistic estimate of the minimum potential catch per year (both demersal and pelagic) is of the order of 11-12 million tons as against the present annual production of about 2.8 million tons. From the western part of the Indian Ocean, the total estimate is approximately 6 million tons consisting of 3.9 million tons of demersal fishes, 1.9 million tons of pelagic fishes and 0.2 million tons of crustaceans. From the eastern Indian Ocean, 3.2 million tons is the estimated potential catch, which includes 1.4 million tons of demersal fishes and 1.8 million tons of other fishes. The Indian Ocean also has a potential of about one million tons of tuna and tuna like fishes of which less than 0.2 million tons are being caught, mainly using the long lines. There is, therefore, considerable scope for the development of fishery for tunas, particularly for the skipjack, smaller yellow fin tunas, frigate mackerel and the little tunny.

Continental shelf around India (The region of the sea up to 200m depth). The minimum estimated potential catch is about 2.4 million tons as against the present exploitation of about 1.2 million tons. Hence, the scope for increase is of the order of 1.2 million tons inclusive of both pelagic and demersal resources. The break-up for the west coast and the east coast is given below:

West coast of India: The estimated minimum potential catch is 1.6 million tons consisting of 1.02 million tons of pelagic catch and 0.58 million tons of demersal fishes. The present catch on the west coast is about 0.9 million tons and hence the new potential available is of the order of 0.7 million tons.

East coast of India: The east coast can produce at least 0.8 million tons (0.67 million tons of pelagic fishes and 0.14 million tons of demersal fishes). The present catch is of the order of 0.3 million tons.

Categories of New Resources Available in the Seas Around India. The new resources available have neither been categorized nor quantified in detail, although some general indications are available. It is now known that the catch

Table 6. Land Use Statistics (1968-69) ('000 Hectares)

Geographical Area	328,048
Reporting Area	305,814
	Percent Area
Forests	20.5
Not available for cultivation, Pastures, trees,	
Culturable waste and fallow	34.5
Net area sown	45.0

Source: Fertilizer Statistics, 1970-71

Table 7. Areas Under Different Soil Groups

Soil Group	Area (Million hectares)	Mean annual precipitation (mm)	Mean annual temp. (C°)
1. Red soils Red Sandy and Red Loamy	72.0	Extensive areas 600-1200 Small areas in West and North east 300 and 4000 respectively	25-28
2. Black soils (shallow, medium and deep)	64.4	500-1200	25-28
3. Alluvial soils (coastal, deltaic and plains)	63.3	Extensive area 500-1200 Eastern Plains 4000	20-28
4. Red and Yellow soils	33.5	Extensive area 600-1000 Eastern Part 3000-4000	20-28
5. Desert soils	21.9	100-400	25-28 &>28
6. Submontane soils	20.2	1200-3000	10-20
7. Mixed Red & Black soils	14.8	600-800	25-28
8. Laterite soils	11.7	Extensive areas 2000-3000 small areas-1200	20-25
9. Grey Brown soils	8.9	400-500	25-28
10. Saline and Alkali soils	7.0	200-1200	25-28
11. Mountain Meadow soils	6.6	200-1200	10

Table 8. Nitrogen Balance 1969 ('000 Metric Tons N)

	USA	Australia	India
I. N. in Yields Non-leguminous Crops	4,518	365	2,841
Leguminous Crops	1,996	6	460
Forage and Pasture	5,180	10,700	300
II. N. Input			
Legume Crops and forage	8,610	12,840	912
Free-living N-fixation	1,400	1,000	650
Fertilizer N	6,313	172	1,222
Difference II-I	+4,629	+2,941	-816

from traditional fisheries can be increased provided that the market conditions remain favorable for their utilization. During the time when the catches are high the prices go down to uneconomic levels, thus forcing the fishermen to exercise a voluntary restraint in landing more fish. The new resources of the Indian coast can be classified as follows:

1) Pelagic Fisheries: a) Mixed pelagic species: Recent studies have shown that about 200,000 tons of mixed pelagic species consisting mainly of anchovies, lesser sardines, ribbon fishes and silver bellies are available in the month of April along the south-west coast in the region extending up to forty meter depth. These are at present almost totally unexploited. The magnitude of total anchovies resource is probably much greater; b) Tunas: In Laccadives, there is traditional fishery for tunas. However, good concentrations of surface shoals have been noted beyond the traditional fishing range in these islands and also in the region between the islands and the south west coast. It is also known that the Central Bay of Bengal, especially the area near Andaman and Nicobar Islands, supports large population of tunas; c) Clupeoids off the east coast and off Goa and Maharastra: There are resources of sardines and other clupeoids available off the east coast and the coasts of Goa and Maharashtra. These have largely remained unexploited; d) Cephalopods (squids and cuttle-fishes): Recent work shows that the cephalopods are abundant in the Bay of Bengal and Gulf of Kutch. These are very poorly fished at present. However, the economics of their exploitation and utilization have not been worked out.

2) Demersal Fisheries: a) Prawns: Recent studies show that the Gulf of Kutch, south-east coast of India, Gulf of Mannar and Palk Bay and the coastal waters of Tamil Nadu and Andhra Pradesh have very rich grounds of prawns. The fishery in the grounds off the south-west coast has developed considerably, but the other areas are not fished intensively; b) Deep sea crustaceans and fishes: Off the south west coast in the depth range 75-450 meters between the latitudes 8° and 15° North (area 43,000 square km.), potential resources of the order of 100,000 tons are available. These resources have been discovered only during the last five years. Of these, at least 23,000 tons are crustaceans (deep-sea prawns, deep-sea lobsters and crabs) and the rest are fishes. The crustaceans are particularly abundant in the depth range 180-450 meters (the upper continental slope). Rich grounds of deep-sea lobsters have also been found along the upper continental slope of the Gulf of Mannar; c) Perch resources: Very good grounds of perches, fishable by hook and line have been found off the south-west coast between seventy-five and one hundred meter depths. The quantification of these resources has not yet been made, but the catch rates are expected to range between 200-300 kg per hundred hooks per hour, which is considered to be high; d) Other resources: Good grounds for sharks, catfishes, sciaenids, carangids, silver-bellies, pomfrets, and perches fishable by trawls have been found off the Indian coast, up to at least forty meters depth. These are only partly fished at present. Various forms of mariculture offer great opportunities for increasing food production (Table 5). In Japan, seaweed culture is also being exploited well.

Fresh Water Fish Production

In terms of quantity the world catch of fresh water fish is only $^1/_{10}$ of the total fish production, i.e., about 6.9 million tons. Nevertheless, freshwater fish include several varieties of high value which are in great preference in several countries. The inland fisheries have been grouped under capture and culture fisheries. Development of capture fisheries is somewhat expensive and has only a limited scope. As against this, culture fisheries can contribute significantly to food production. Freshwater fish culture has its longest history in Asia and Far East particularly in China. Stocking of streams, rivers, lakes and reservoirs, has been widely practiced for a multiplicity of reasons. Due to lack of adequate data, it is difficult to estimate the inland fish production, especially the quantum of fish produced through culture operations. According to a recent FAO estimate based on data from forty-two countries, the total world production by aquaculture is over five million tons. This estimate, however, does not include production of sport, bait and ornamental fishes. In the overall spectrum of inland fishery development, aquaculture plays an important role in the economy of many countries. In countries like China, India and Indonesia, nearly forty, thirty-eight and twenty-two percent respectively of the total production of fish is accounted for by aquaculture. In land-locked countries, fish culture is developing into one of the important sources of animal protein. Although reliable information on the area utilizable for aquaculture is not available, it is known that in addition to conventional pond culture, there are extensive swamps, estuaries and low-lying areas with assured source of water supply suitable for fish culture. Data so far collected show that only a small part of this is being utilized for aquaculture. With the development of a variety of techniques of fish culture such as closed

Table 9. Ultimate Irrigation Potential in Different States
(Data from the Report of the Second Irrigation Commission)

State	Area in million hectares	
	Ultimate potential	Area irrigated (1966-67)
Andhra Pradesh	10.3	3.9
Assam	2.2	0.6
Bihar	8.0	2.0
Gujarat	3.8	1.1
Haryana*	0.85	1.8
Jammu & Kashmir	0.6	0.3
Kerala	1.5	0.5
Madhya Pradesh	8.0	1.1
Maharashtra	4.5	1.4
Mysore	3.4	1.2
Orissa	4.0	1.1
Punjab	6.1	3.4
Rajasthan	5.0	2.1
Tamil Nadu	3.6	3.4
Uttar Pradesh	15.1	7.2
West Bengal	4.3	1.5
All India	81.7	32.8

*Area under Major & Medium Projects included under Punjab

water ponds, running water ponds, cage culture, paddy-cum-fish culture, hanging culture, culture in lakes and reservoirs etc., a manifold expansion of aquaculture in most parts of the world is possible. In many countries, the possibility of substantial increase in yield by the adoption of the technique of composite fish culture has been established. Similarly, it has been proved that integrated culture practices employing a variety of compatible organisms such as prawns, frogs, ducks and even plants like *Euryale ferox* and *Trapa bispinosa* along with fish are advantageous in making full use of the three dimensional aquatic environment. Further, rotation of crops involving fish, soybeans and rice are also being practiced with considerable success by some farmers. The labor-intensive nature of the present aquaculture operations also generates additional employment potential in the rural areas.

Water pollution and waterweeds pose some of the greatest threats to fish culture. A concerted attack on both these problems is vital to realize the potential fish offers to improve food supply and promote protein intake.

Our Agricultural Assets

Sunlight, land, water, plant, animal and human resources form the principal foundations of any agricultural system. The extent of advance made in enhancing both agricultural productivity and stability would depend upon the progress achieved in an integrated domestication of plants, soil and water based on the principles of synergy and symbiosis. In India, domestication of economic plants and animals evoked much interest since ancient days but equal interest in the domestication of soil and water was lacking. We see as a result large areas of eroded land and problems like salinity and alkalinity. Inadequate attention to the replenishment of soil fertility has also led to an impoverished nutrient status in most soils. One of our most urgent tasks therefore is to pay greater attention to the care and build-up of soil health and fertility and to the scientific utilization of water and sunlight. I would like to refer briefly to the opportunities we have for better land, water and sunlight utilization.

Land Use. The pattern of use of land for agriculture, forestry and other uses is indicated in Table 6. The soil types are of varied kinds, with red, black and alluvial soils occurring in almost equal proportions (Table 7). Forestry occupies a little less than half the area under agriculture but contributes only 1.4 percent of the gross domestic product. During 1969-70, the contributions of agriculture, animal husbandry and forestry to the gross domestic product were 42.4, 4.1 and 1.6 percent respectively. The poor contribution of forestry to GDP and the neglected state of this science have been highlighted in an interim report of the National Commission on Agriculture (NCA). NCA has pleaded that we must move away from the concept of mere conservation of natural forests to one of man-made forests. We need a dynamic program of research in forest tree improvement including imaginative attempts to shorten the duration by accelerating the rate of growth of trees which can provide fuel and raw material for the paper industry. Our ability to promote a better ecological balance and a desirable organic recycling process would depend to a considerable extent upon what happens in the field of forestry research and development in the coming years.

The authors of *Limits to Growth* have attempted to quantify the extent of fall in land availability for agricultural use that we will have to be prepared for with population expansion. They have assumed that each additional person would need 0.08 hectares of land for purposes like housing, roads, waste disposal, power lines and other uses and 0.4 hectares for producing the food he needs. If we use this formula in our country over five million hectares of land will be needed every year to meet the requirements of the newborn. Therefore, we have to aim at developing integrated land use systems in agriculture where the productivity of the soil can be continuously improved, without any impairment of its long-term production potential. To achieve this we have to learn both from the successes and failures of the developed countries in the matter of soil care and use. While there are several aspects to this problem, I would like to refer to two urgent ones relating to nutrient supply and soil structure.

Soil Care and Fertility. The recent advances in the genetic enhancement of the yield potential of several economic plants have stemmed from successes in improving their capacity to utilize more nutrients and convert them into the economic product. Thus, for getting a higher output, a greater input, particularly of nutrients, becomes essential. The first question that arises in this context is whether the continuous application of fertilizer is detrimental in the long-term. Excessive use of nitrogen and phosphorus has led to difficulties in some countries. Kanwar (1972) has discussed this question with reference to our country. He has shown that our problem is not akin to those of the developed countries for a variety of reasons, among which the following are some of the more important.

We are at present using about 6.15, 1.45 and 0.92 kg. N, P_2O_5 and K_2O respectively per hectare (according to 1969-70 statistics), which is about 1/47th of the fertilizer consumption per hectare in Japan. On a per capita basis, the amount is about 2.6 kg. Even if a fraction of this fertilizer is lost through leaching or other biological activities, the effect on the environment will be hardly measurable. However, in the monsoon season the loss of N can be high in rice. Drainage studies in heavily manured paddy fields at the Central Rice Research Institute, Cuttack, have revealed that during the Kharif season, 42 kg N, 29 kg phosphorus, 38 kg iron and 17 kg manganese were lost per hectare. No separate record is available on the nitrate content of the water. There are very few studies about the quantitative estimation of N losses in the gaseous form from submerged condition in paddy cultivation. Datta and associates at IARI using N^{15} technique concluded that gaseous N loss was 24.1, 30.2 and 53.7 percent in case of tagged ammonium sulphate, nitro phosphate (high water solubility) and nitro phosphate (low water solubility) respectively.

The loss of nutrients in gaseous forms as well as in drainage from heavily fertilized rice fields requires critical studies. Also, more studies are needed of the nitrate content of underground water in areas of intensive fertilizer application such as the grape gardens of Hyderabad.

From the available data, fertilizers do not seem to be significantly involved in raising the P levels in surface and ground waters. On the other hand, by promoting the growth of vegetation they help in reducing erosion and thus reduce the loss of P to waters. The contribution by human, animal and industrial waste and detergents to P content of streams, lakes and marine and also inland waters may be substantial as studies by the Central Inland Fisheries Research Institute, Calcutta, have revealed (Figure 1). Thus, organic sources of phosphorus may be a source of pollution in some areas.

Kanwar has stressed that the most important cause of pollution from the agricultural lands is soil erosion. This process not only leads to the loss of millions of tons of nutrients every year, but it is threatening our irrigation projects and cities. It is estimated that we are losing annually about 6,000 million tons of soil from 91 million ha, which are reported to be suffering from problems of erosion. Assuming that from a soil of 0.5 percent slope, the loss of nutrients is of the order of 5.8 kg. N., 10.7 kg. P_2O_5, and 42.8 kg. K_2O/ha, the annual total loss from Indian soils could be as high as 2.5, 3.8 and 2.6 million tons of N, P_2O_5 and K_2O respectively. This quantity is much greater than the present total consumption of NPK from fertilizer as well as organic manure sources. Moreover, there are millions of hectares which lose many times more nutrients per hectare because of greater slope and credibility. Obviously, such losses cannot be totally avoided, but we can certainly minimize them. Fortunately, a part of these leached nutrients must be going back to soil through water.

We have to assess the quantity of nutrients we need to produce about 228 million tons of food-grains by 2000 AD. We need nutrients not only for food crops but for all domesticated plants and for fish production. Research carried out under the All India Coordinated Project on the correlation between soil test values and response of crops to fertilizer has shown how we can maximize the return from fertilizer application. Based on this data, Dr. B. Ramamoorthy and his colleagues have calculated that to produce the food and other products we need by 2000 AD, we may need a minimum of 4.60, 3.94 and 1.92 million tons of the nutrients N, P and K respectively. These calculations are based on the assumption that the efficiency of fertilizer application will be very high. The present installed capacity for fertilizer production is of the order of 1.46 and 0.50 million tons for N and P respectively.

What pathways of nutrient supply should we take? The plant cannot give more yields from hungry soils. It is obvious that we should apply fertilizers judiciously and economically making full use of the research data relating to the potential for synergistic interactions. It is equally obvious that we should not use synthetic fertilizers as the sole means of nutrient supply. We have to evolve an integrated nutrient supply strategy for each cropping system, involving an appropriate combination of organic and inorganic manures and ensure that legumes find a place in the crop rotation.

1973 is a particularly suitable year to initiate the development and adoption of such an integrated procedure, since the availability of fertilizers is less than required.

Figure 1. Pollution load contributed by the various industries, estates and municipalities located on the banks of the Hooghly

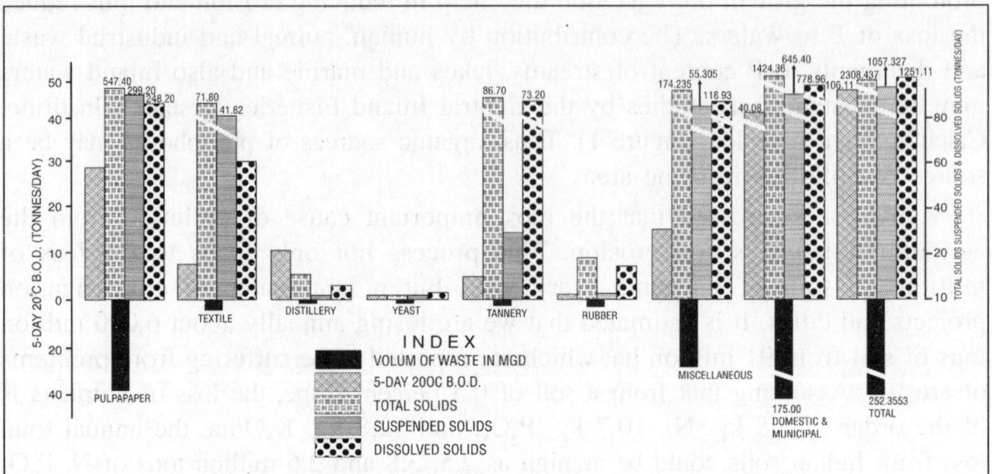

Pollution load contributed by the various industries, estates and municipalities located on the banks of the Hooghly.

Figure 2. Progressive demand, availability and utilization of India's water resources

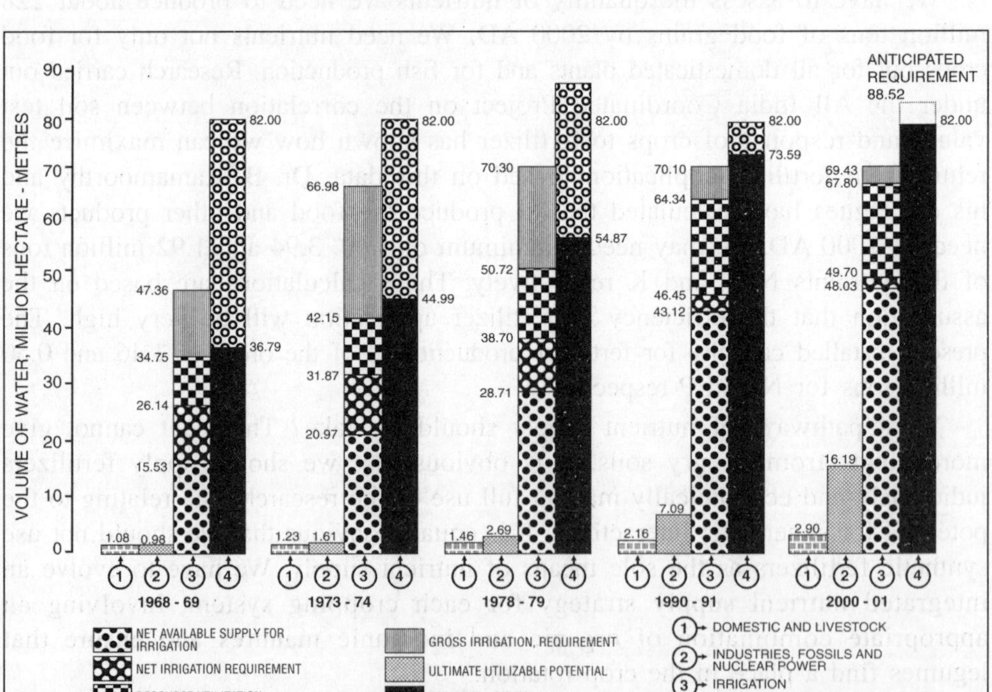

Progressive demand, availability and utilisation of India's water resources.

In its research bulletin on "Towards Self-reliance in Agriculture," IARI has advocated the tapping of the following sources of P to the maximum possible extent: a) Basic slag, a by-product of the steel industry wasted at present; b) Rock phosphate deposits in different parts of the country; c) Phosphorus-rich sands of the Arabian Sea Islands.

The following are some of the major sources of organic manure, which deserve widespread conservation and use: a) Urban solid wastes like garbage, night soil, sludge and slaughterhouse wastes; b) Urban liquid wastes like sewage, sullage and urine; c) Industrial wastes; d) All farm and crop wastes; e) Weeds, particularly water weeds like water hyacinth; f) Cattle-shed sweepings and all animal wastes; g) Inedible oil cakes.

If we are to conserve cowdung and bones for manurial purpose, we should popularize the widespread use of cowdung gas plants and bone digesters. If quick-growing fuel trees can be developed, the rural population need not have to rely so heavily on cowdung for fuel. In addition to the use of all such wastes, the cultivation of pulse and green manure crops both as a pure and as an inter-crop in rotations in irrigated areas will be very helpful both to get more pulses and to fix more nitrogen biologically.

There is a superabundance of nitrogen in nature, with about 87,000 tons in the atmosphere above each hectare. The contribution of legumes to nitrogen fixation in the soil is highest in Australia and least in India (Table 8). Of more than two thousand million hectares of meadows, pastures and rough grazing in Australia, much of it arid, about one hundred million hectares have been improved by introducing legumes like *Trifolium subteraneum*, inoculated with bacteria and usually sown with super phosphate, and sometimes with lime and minor elements. In the United States also, legumes provide the largest amount of nitrogen for primary production, soybeans alone contributing approximately 1.4 million tons of nitrogen. We will do well to emulate these examples.

In contrast to our country, where interest in organic recycling of all wastes has been generally poor in spite of Mahatma Gandhi's efforts to focus attention on its importance, much attention has been paid to the conservation and use of all wastes including human excreta in China. The following extract from a book by Han Suyin (1965) emphasizes the importance given to waste recycling in China. "Tobacco requires much manure, spraying with manure being done every five days in April and May when the plants grow. The good earth of Szechuan, moist and replete and rich, glistening as with oil is nourished with human and pig excrement, treated for some days in pits. It is only now after the Revolution, that chemical fertilizer plants are being built, but today man and pig are still the main providers of manure. And therefore in the city of Chengtu in the nineteenth century and up to 1949, some of the wealthiest families were those owning the public cloacae and selling excrement to the countryside. I remember Third Uncle on this subject, which rendered him prolix, for he was always aware of the earth, the sowing and planting and reaping of it; 'Three things are necessary, water, earth and manure. Water from the sky, soil of the earth, and manure of man and animal; the excremental waste confirms the harmonious cycle of life, since out of waste comes renewal.' In an agricultural society dung has its function and natural dignity; and even today, I can become engrossed in the relative merits of 'green lush' or 'brown gold.'"

In Europe and North America also human waste recycling systems are now under development and alternative methods which do not involve water to flush excreta, are rapidly being standardized. It is high time we also start seriously working on the development of efficient cycling and nutrient supply system which will lead to the proper maintenance of soil structure and fertility.

Pollution from Pesticides

As in the case of fertilizer, the average consumption of pesticides is so low that there does not appear to exist a situation akin to those of some of the developed countries. However, a good proportion of the pesticides used by us have the property of persistence in the plant material. Hence a more extensive surveillance and monitoring system is necessary. Also the right combination of pesticides is necessary for each region, as otherwise the pest which is not sensitive to the particular chemical used, would tend to survive preferentially and cause much damage. Entomologists now advocate the use of integrated pest control methods involving an appropriate admixture of genetic, chemical, biological, agronomic and other techniques. It is planned to get such an approach tested at the field level shortly in rice. So far we have not gathered operational experience in this field.

Water Requirements

On an average, we receive an annual rainfall of about 112 cm for 370 million-hectare meters. It has been estimated that about 120 million-hectare meters seep into the soil and about 170 million hectare meters remain in the top layers and contribute to soil moisture which support crop growth. The Irrigation Commission has made detailed projection of both water availability and scope for planned exploitation. It appears that the total ground water recharge available for utilization may be of the order of 26.75 million hectare-meters per annum, in contrast to the current utilization of about ten million hectare meters. Major, medium and minor irrigation works have gradually expanded the irrigated area.. The States, which still have less than fifteen percent of the net cultivated area with irrigation facilities are Madhya Pradesh, Maharashtra, Mysore, Gujarat and Rajasthan. The Irrigation Commission has made a broad assessment of the irrigation potential in the different States (Table 9).

The total water requirements of crops, animals and man and of industry up to 2000 AD have been calculated by Prof. A.M. Michael and his associates of the Water Technology Center, Indian Agricultural Research Institute, using the projections for domestic and industrial use made by Kathpalia and Varadan (1971) and by Planning Commission. These studies show that at present there is a big gap between the net available water supply and the amount required for intensive cropping. The demands on water resources, possible supply and progressive exploitation during the period 1968-69 to 2000-01 are shown in Figure 2. It appears that if the available resources can be tapped and husbanded properly, particularly through intensive efforts to increase the efficiency of water conveyance and use, we may be able to manage up to 2000 AD without having to go in for new sources of water such as desalination of seawater. The projections in Figure 2 have been worked by Prof. A.M. Michael of IARI. However, such an overall analysis of water budgeting hides the fact that even today there are considerable locational and regional gaps in the water demand-supply position. Rapid technological developments are taking place in areas like desalination of brackish water and weather modification and obviously in the choice of areas for conducting feasibility studies in the use of the new sources of water supply, the regions where the water budget shows greater deficit thereby causing much human hardship, should receive priority. There is, however, immediate scope in all parts of our country to capture more water through better water-shed management and by introducing water saving concepts such as design of houses which collect rain water, development and use of waste disposal system which do not involve the use of much water and adoption of improved methods of water conveyance and distribution (IARI, 1970).

Utilization of Sunlight

The need for using productively sunlight, which is one of our greatest endowments, has been stressed repeatedly by many research workers but little has so far been done in tapping for human use more of this valuable energy. Multiple cropping and inter-cropping provide an effective agricultural method of using solar energy (IARI, 1972). Green Plants can be made to harness sunlight throughout the year, wherever there is water to sustain crop growth. There is a vast potential for using solar energy in and semi-arid areas for household purposes, thereby saving the valuable organic matter for the soil. Garg and Krishnan (2001) have shown that a flat plate collector of two sq. meters at Jodhpur and three sq. meters at Delhi can help to heat 140 liters of water up to 55°C in the afternoon of January. This can provide sufficient hot water for a family of five persons. Such techniques should not remain only in the research laboratory.

Global Ecological Balance

I have deliberately refrained from discussing the problems of our spaceship in a global context, since there are many excellent books giving such an overview. It, however, hardly needs mention that technological changes particularly those which affect the planet balance of energy could trigger off catastrophic changes. For example, there is evidence to suggest that over the last decade, the release of CO_2 into the atmosphere, as a result of man's burning fossil fuels and of deforestation has been increasing by 0.2 percent a year. If this trend continues, the earth's temperature could rise by 0.5°C by 2000 AD. Industrial emission of dust, soot and gas as well as the development of supersonic transport may all influence earth's climate. Barbara Ward and Rene Dubos have hence pleaded for concerted world action in this field. According to them, "These concerns with global air pollution lie beyond the effective protection of individual governments. It is no use one nation checking its energy use to keep the ice caps in place if no other government joins in. It is no use the developed nations suggesting lower energy use just at the moment when the developing nations see increased use as their only exit from the trap of poverty. Man's global interdependence begins to require, in these fields, a new capacity for global decision-making and global care."

Dos and Don'ts

The environmental policy advocated in the richer nations is designed to protect the high standard of living resulting from the unprecedented growth in the exploitation of natural resources during the last century from serious damage by the very processes of such growth. It is of necessity a policy based on a series of Don'ts. This is inevitable since the aim is to undo some of the damage already done or to prevent further damage along the same lines. Thus, there is an outcry against some of the potent agents of improved agricultural productivity such as fertilizer and pesticide.

The poorer nations, however, are faced with the desire and need to produce more food from hungry soils, more clothing and more housing. They are aware that historically a rising standard of living has depended on the ability of agriculture to release manpower to other more industrial pursuits. They hence naturally wish to develop more industries and to find productive and remunerative employment for their growing population. For them, conditions of poverty and inadequate arrangements for human and other waste disposal may be greater cause of water and air pollution than the effluent from factories or fertilizer from the fields. Since the causes of pollution are by and large different, the solutions will have to be different too and it would be a grave mistake to attempt to copy the policies now being propagated in the developed world.

We are fortunately in a position to build a positive policy of economic-ecology based on a series of Dos rather than Don'ts.

The first Coromandel Lecturer, Sir Joseph Hutchinson spoke about the need for breeding soils for sustained high productivity and has in a recent book (1972) explained "the part that farming as a way of life has played in the preservation of good rural environments and the extent to which the future depends on this precious feature of agriculture." To the farmer, he says, the farm and family are one integrated unit. He remains poorer than his urban cousins who not only want him to produce cheap food but accuse him of destroying the amenity of the environment, if with purely economic motive he removes his hedges or uproots a tree. Too often his critics want it both ways. To Prof. Hutchinson the only hope for the future is that there are still many people who are prepared to live with less of the world's goods than they could acquire in an urban occupation.

The Second Coromandel Lecturer. Dr. Norman Borlaug, earned this theme further and reminded us that well-fed and over-fed human beings have no right to impose conditions which will preserve hunger and starvation for the majority. There is "no time to relax" in our efforts to eradicate malnutrition and poverty and as an illustration of choice of priorities and options in action programs, he said that if he were faced with the choice of having to watch the death of a few birds or the slow death of starving millions, he would "vote for the survival and feeding of man."

I have tried in this lecture to analyze how, within the concept of a careful husbandry of finite resources, we can move forward to produce more food and better rural living. 1 deliberately chose crops and fish for mention, since it is in these two areas that there is the greatest scope for rapid advance. In my view, it is only integrated land use involving a relevant combination of agriculture, forestry, fishery and animal husbandry, integrated strategies of nutrient supply and pest control, greater attention to organic recycling and water conservation and extensive tapping of sunlight through the agency of green plants and simple heat collectors that can help us to improve rural economy without ecological harm. We have a unique opportunity to evolve a policy for natural resource utilization based on the principles of economics and human welfare. To do this we have to learn how to deploy two of the most potent tools employed by Nature – synergy and symbiosis. Can we do this?

We can, but for this there are three pre-requisites. First, we have to shed single-discipline-oriented and one-sided thinking and develop a problem-solving and integrated approach. Integration is needed not merely in the planning process, but more urgently at the action level. Second, we must launch a techniracy (technical literacy) program aimed at importing technical skills to illiterate peasantry. Thirdly, we must shun the "band-wagon" and aping approach and generate a "do" rather than a "don't" atmosphere. We must help our farmers to apply fertilizer, pesticide, water and other inputs wisely and should not try to create confusion about the basic truth that inputs are needed for output. A "swadeshi" approach is even more relevant today than when Gandhiji first advocated it.

References

Brown, L.R. 1972. *World Without Borders*. New York: Random House.

Garg, H.P. and A. Krishnan. 2001. *Central Arid Zone Research Institute*.

Hutchinson, J. 1972. *Farming and Food Supply*. Cambridge, UK: Cambridge University Press.

IARI. 1970. Handbooks of Farm Irrigation Structures. *IARI Research* 3:82.

-----. 1972. Recent research on multiple cropping. *IARI Research Bulletin* 8:148.

Kanwar, J.S. 1972. Are fertilisers pollutants? *Fertiliser News* 17: 17-22.

Kathpalia, G.N. and N.S. Varadan. 1971. *Need for national water policy*. Symposium on Integrated Development of Surface and Sub-surface Water Resources, New Delhi.

Meadows D.H., D.L. Meadows, J. Randers and W.H. Behrens. 1972. *Limits to Growth*. New York: Universe Books.

Suyin, Han. 1965. *The Crippled Tree*. London: Jonathan Cape.

Swaminathan, M.S. 1973. Population and food supply. *Yojana* 17: 7-15.

Ward, Barbara and Rene Dubois. 1972. *Only One Earth*. Harmondsworth, UK: Penguin Books.

9 Our Agricultural Future

Lecture 1

Our Agricultural Balance Sheet: Assets and Liabilities

Food is a basic requirement in the hierarchical needs of man. The successful domestication of plants, animals, soil and water by settled human communities has hence been the starting point for the evolution of culture and orderly government. Sardar Patel often stressed this fact by emphasizing that the only culture he understood was agri-culture. Ironically, the very centers of civilization, where the domestication of plants and animals took place in the past, are the areas which are struggling today to find an honorable equation between population growth and food supply on the one hand, and population growth and opportunities for productive and remunerative employment, on the other. The secondary and tertiary centers of origin of agriculture have, in contrast, more than enough to eat and constitute the developed or rich nations of the world. It has been estimated that in the poor nations, which have contributed most of the domesticated plants and animals to the world, nearly 800 million individuals live under conditions of absolute poverty, resulting in protein-calorie malnutrition, squalor, illiteracy and disease. Again, ironically, the lands of the poor have often greater biological assets, the most important of which is abundance of sunlight throughout the year, rendering the historic concept of cereal inapplicable. This word refers to crops which grow during only part of the year, being derived from the myth of Ceres's happy seasonal reunion with her daughter. In our earth, appropriately compared to a spaceship by Buckminster Fuller, the first class compartment, which the rich nations as well as the rich of the poor nations occupy, is consuming a greater and greater proportion of the available resources, thereby making life for those living in absolute poverty more and more hard. Those living in the first class compartment of the spaceship have forgotten the meaning of the prayer "God, give us our daily bread", while for the vast majority of humanity squeezed into the small second class compartment, "God is bread", to use the words of Mahatma Gandhi. Why are we in this state of paradox? Why do plants seem to be unhappy and hence not so productive in their historic native homes?

Contrasting Trends

The state of food and agriculture in the world shows two contrasting trends. In one kind of agriculture, larger and larger farms are being farmed by fewer and

---

Presented as Sardar Patel Memorial Lectures, All India Radio, New Delhi, October 30, 31 and November 1, 1973. First published in *Sustainable Agriculture: Towards an Evergreen Revolution*, by M.S. Swaminathan, pp. 1-77. Delhi: Konark Publishers, 1996. Courtesy: All India Radio, New Delhi.

fewer cultivators. Such farms are highly automated and capital-intensive. A study on computerized farming in California revealed how a firm called Superior Farming Company, owned by a Texas Oil Company, raises twenty-six different crops on about 6,000 hectares, with land and equipment worth about Rs. 30 crores. In this agri-business, an irrigation system, called the drip method, brings individually piped water to every fruit tree and regulates the flow to a trickle, supplying exactly what each tree needs and no more. Under such systems of farming, not only is the efficiency of use of inputs like water and fertilizers high, but labor productivity also is very high. In one study conducted during the mid-fifties it was found that in many countries of Asia and Africa about 2.5 to 10 work days were needed to produce one quintal of grain and the extent of labor time needed to produce the same quantity of grain was about three hours in parts of France and only six to twelve minutes in parts of the United States. The gap in the relative productivity of 'farm labor' was, therefore, about one to eight hundred even over fifteen years ago and this gap, which also represents the relative earning and purchasing power of farm labor, has been growing ever since.

In many of the poor nations, including our own, smaller and smaller farms have to be cultivated by the same or even larger number of farmers. The percentage of work force employed in farming in India was 72.1 in 1971 and it is anticipated that this percentage will remain practically unaltered for a decade. The relative productivity of small farms, however, varies widely in the world. R.S. McNamara, past President of the World Bank pointed out: "If Japan could produce 6,720 kg of grain per hectare on very small farms in 1970, then Africa with its 1,270 kg, Asia with 1,750 kg and Latin America with 2,060 kg per hectare have an enormous potential for expanding productivity." (1973:16)

Agricultural Growth

Besides farm size and land and labor productivity, the other two major differences in the agriculture of the rich nations, as compared to the poor ones, are in the pattern of agricultural growth and the nature of the food chain. The mechanized and low-labor consuming agriculture has achieved increased productivity largely on the basis of a high consumption of energy derived from the non-renewable resources of the earth. Thus, while in 1964, about 286 kilo calories of energy were needed in countries like India and Indonesia to produce one kg of rice protein, 2,860 kilo calories of energy were needed to produce one kg of wheat protein and over 65,000 kilo calories to produce one kg of beef protein in the United States. Such a situation has now resulted in widespread awareness of the simple truth that any finite resource, if exploited in an exponential manner on the assumption that the resource is infinite, will some day or the other get exhausted, thereby bringing the pattern of growth based on its consumption, to ruin. It has also become clear that the tools of modernization of agriculture like pesticides, fertilizers, farm power and water, if indiscriminately used and excessively based on non-renewable resources, will end in crises, referred to as "ecological crises". For example, ninety-six percent of the energy input in the United States in 1970 came from oil, gas and coal, while in the same year non-commercial fuels like dung, firewood and wastes provided fifty-two percent of our energy needs. The reserves of the fossil fuels are expected to decline rapidly in the next thirty years and even now, we have started witnessing some shortages of the most desirable fuels. Our agricultural production process is still predominantly based on the use of renewable resources but our current productivity is very low. Hence, there is urgent need for the development of technologies where the productivity of land can be continuously increased with diminishing dependence on non-renewable components of energy, by deploying recycling processes more and more effectively. A consequence of the agriculture of the high-energy-consumption and low-labor-input pattern is the diversion of labor from agriculture to more industrial pursuits and a close linkage between farm and factory. Diversification of labor use, leading to a reduction in the number of people employed in the physical operations of farming, has historically been associated with a rise in the standard of living.

Contrasting Food Chains

A biological consequence of affluence has been a rise in the consumption of animal products. The poor and rich nations are hence characterized today by the former depending largely on the plant-to-man food chain and the latter, on the plant-animal-man food chain. Consequently, the average per capita consumption of grain is about one ton per year in the developed nations, out of which only about 70 kg are consumed directly in the form of bread, biscuits, cakes and other products made from flour. The remaining 930 kg are used to feed animals whose products like meat, milk and eggs are consumed by man. In contrast, the per capita consumption of grain per year in the developing countries is about 190 kg, most of which is directly consumed.

To summarize, the two major contrasting systems of agriculture differ in the size of farm, the proportion of work force employed in agriculture, the types of linkage developed between farm and factory, the extent of consumption of non-renewable resources of energy, management efficiency, per capita productivity and income and the extent of use of animal products in daily diet. Our need is an agricultural system where the benefits of a large human and animal population, robust soils, abundant sunlight, rich ecological diversity, availability of large quantities of organic wastes and a fairly extensive irrigation network are optimized in a manner that productivity is continuously increased without damage to the long-term production potential of the soil, stability is imparted to the production as well as prices of foodgrains and labor and land use are diversified, so as to increase real income and purchasing power.

During early sixties, some foreign experts believed that India could never become self-sufficient in its food requirements. They predicted the outbreak of widespread famine and hunger in India on the basis of theoretical calculations of the year when the food needs of the poor nations would exceed the capacity of the rich nations to meet them. Paul and William Paddock, for example, fixed this year as 1975. Some others felt that such a contingency would arise only in 1985. Our failure to achieve during the 1950s and early sixties the anticipated results with the Community Development and the Intensive Agricultural District Programs fanned the sentiments of the pessimists. It was widely believed that in our rural areas developmental action and achievement were being impeded by the limited sights of our farming community. Experts wondered why we should find it so difficult to become self-reliant in our food needs, when there appeared to exist several easy pathways of attaining this goal. An expert on rodent control would thus advocate the control of rats to wipe out the deficit in our food budget, while the agronomist would plead for using good seeds and some fertilizer for achieving the same end. The fact remained, however, that progress in improving production was slow and largely took place through an increase in the cropped and irrigated area, rather than through any appreciable increase in productivity. The economic plight of the small farmer remained unchanged since his real income would rise only if productivity was improved and marketing organized so that the small producer got a fair share of the price paid by the consumer for a commodity. While the Government was anxious to push up agricultural production and Pandit Nehru gave expression to this desire by his oft-quoted statement "Everything else can wait, but not agriculture", there was despair during the drought years of the mid-sixties concerning our agricultural destiny.

Famines and Scarcities

Historically, famines and scarcities have been known in our country from the earliest times. They are mentioned in *Mahabharata* and there is a Jain legend of a 12-year famine in the fourth century AD. It is recorded by the official historian in the *Badshah Namah* that in the famine of 1630-32, Emperor Shah Jahan opened soup kitchens, gave a lakh and a half to charity and remitted one-eleventh of the land revenue in the affected area. Subsequently, there have been numerous famines. But it was only during the famine of 1868-69, that it was clearly stated that "the object of Government was to save every life". The birth of agricultural departments in our States and of Famine Codes took place as a result of the recommendations of the Famine Commission of 1878. The Scarcity Manuals currently used by several State Governments are largely based on the Famine Codes of the last century. B.G. Verghese's penetrating articles entitled "Beyond the Famine" (1967) and "A Blessing Code Named Famine" (1973) on the droughts of Bihar and Maharashtra have highlighted the advantages conferred on the helpless rural population by famines, since it is only during such crises that at least some food and employment are assured to those living in absolute poverty.

Meanwhile a silent revolution has been taking place in the minds of our farming community as a result of the new production technologies introduced through the High-Yielding Varieties Programs, Intensive Cotton District Program including hybrid cotton which was introduced as commercial cultivation for the first time in the world and the Cattle Cross-breeding projects. Obsession with destructive criticism has blinded us to the meaning and implications of the increase in wheat production from a little over twelve million tons to over twenty-seven million tons in just six crop seasons beginning from 1967-68. There has not been much interest in studying how a small Government program in wheat was converted into a mass movement by our farmers. It is not only in Punjab or Haryana that wheat production went up dramatically but also in non-traditional wheat areas like West Bengal. While in the past it was difficult to induce farmers to take to a rat control operation, they took pains, in the wheat belt, to see that rats did not migrate from sugarcane fields to wheat. While in the past everyone knew that farmers were sitting over a large underground water resource in the Indo-Gangetic plains, it was only the introduction of high-yielding varieties of wheat and rice that provided the necessary motivation for them to take to the construction of tube wells – whether made of metal or of bamboo – on a large-scale. The truth of the saying "Necessity is the mother of invention" was proved to be true in the case of wheat cultivation, where farmers somehow managed to produce or get the seeds they needed, use diesel pumps, if electricity was not available, and get the requisite number of threshers and other implements as well as storage bins fabricated in record time. That our farmers may not wait for research workers to produce perfect tools but will themselves innovate and find a solution to a pressing need was shown by the Bihar farmer, Shri Ram Prasad Chaudhary Jaiswal of the village Lalpur who first put up a bamboo tube well in the Saharsa district. Farmers who learnt the economic value of good management in wheat also look to better practices in rice, potato and other crops, with the result that agriculture as a whole started moving forward in such areas. The sudden and steep spurt in fertilizer demand is an eloquent testimony to the credibility of the new varieties and techniques. The concept that agricultural advance in India would suffer due to the limited vision of the farming community was thus disproved. The social tensions between those who had access to the inputs needed for adopting the new technology and those who did not have similar access, only underlined the fact that those who have not derived economic benefit from such technology are equally anxious to take to the technology. The desire to change farming methods, thus, fanned both joy and sorrow in our countryside. This, in turn, generated considerable thinking and action on the part of the Government, resulting in programs for marginal and small farmers, expansion of credit facilities and more recently in integrated farmers' service societies, on the lines recommended by the National Commission on Agriculture.

Assets and Liabilities

Starting from the mid-sixties we have gone through waves of pessimism and optimism. We have been happy over good seasons and more than alarmed by bad ones. It is time that we look at our agricultural assets and liabilities critically and scientifically and evolve a pattern of growth which would help us to take full advantage of good seasons, be prepared for the worst and make agricultural growth yield more income and more jobs in addition to more food. Complacency at this stage and any relaxation of relentless efforts to achieve agricultural advance would be disastrous.

Assets. I would like to deal with our agricultural assets first. Sunlight, soil, water, plants, animals and human beings constitute the basic resources of our agriculture. Fortunately, sunlight is abundant except during limited periods in the monsoon season and generally does not constitute a limiting factor in production. In arid and semi-arid areas, the intensity of solar energy is fairly high and exceeds 650 cal/cm^2/day. Under such conditions evapo-transpiration rates become high. The average annual values of energies available in peninsular and north India are 473 and 460 cal/cm^2/day respectively. Apart from the use of sunlight by plants in the process of photosynthesis, solar energy is used in rural areas for purposes like drying of crops and grains, distillation, evaporation for salt production and heating. There is, however, considerable scope for improving the use of sunlight both by plants, through multiple cropping, and by converting solar energy into thermal energy.

Out of our total geographical area of 328.73 million hectares the net area sown during 1969-70 was about 140.15 million hectares. Forests occupied about 66.73 million hectares and uncultivated land was about 101 million hectares. Thus the net sown area under crop in 1969-70 was 45.9 percent of the reporting area and the area under forests was 21.9 percent. The net national product was Rs. 34,253 crores during 1970-71 at current prices and Rs. 18,876 crores at constant 1960-61 prices. At 1960-61 prices, the contribution of agriculture, forestry and fishing to net national product in 1970-71 was 44.4 percent. The contribution of agricultural products to the total export earnings in the same year was 37.1 percent.

Diversity in physiography: The physiography of our country shows great diversity. On the one hand, we have major mountain ranges with the Himalayas, at once one of the youngest as well as the mightiest of the world's mountain systems, in the north, and the Aravallis, the Vindhyas, the Satpuras, the Eastern and Western Ghats and the North-eastern ranges including the Garo, Khasi and Jaintia hills in the other parts of the country. Plateaus, ranging in elevation from 300 to 900 meters, constitute a prominent feature of our topography, the well-known among them being the Malwa, the Vindhya, the Chota-Nagpur, the Satpura, the Deccan, Ladakh and Meghalaya. A very large part of the country consists of extensive plains watered by great rivers where a considerable proportion of humanity lives. In the rivers originating from the Himalayas, the dry weather flow is generally good due to water coming from melting snows and glaciers. The lean period for these rivers is the winter months but at no time is the flow so reduced as in the peninsular rivers. The flow in the rivers of peninsular India undulates heavily, with big discharges during the monsoons followed by low discharges during the rainless months. Variations of the order of one to three hundred in the mean monthly flows of these rivers are common.

Our climate shows equally great diversity, ranging from continental to oceanic, from extremes of heat to extremes of cold, from extreme aridity and negligible rainfall to excessive humidity and torrential rainfall. The Himalayas present a barrier to the influence of cold winds from Central Asia and give the sub-continent the elements of a tropical type of climate. The variations in rainfall, temperature and humidity caused by the incursion of comparatively cool currents from the Indian Ocean across the Bay of Bengal and the Arabian Sea and by the movement of shallow depressions, which originate outside India, to the west, lead to extremely complex weather patterns. Thus, temperature, rainfall and the amount of vapor in the air, which influence greatly the growth of crops, show wide variations even within small areas.

An interesting feature of the wind system over the Indian Ocean and the adjoining sea and land areas is the seasonal reversal of the monsoon. During the late summer, the winds flow from the south-west over the sea towards India and

Burma, while during the winter, the flow of currents is from India and Burma over the Bay of Bengal and the Arabian Sea towards the Equator. The south-west monsoon season is responsible for over eighty percent of the total rainfall in most parts of the country.

On an average, we receive an annual rainfall of about 370 million hectare meters. It has been estimated that about eighty million hectare meters seep into the soil of which about forty-three million hectare meters remain in the top layers and contribute to soil moisture which supports crop growth. The ground water recharge available for utilization may be of the order of 26.75 million hectare meters, while the current utilization is about ten million hectare meters. As a result of the various major, medium and minor irrigation projects undertaken, the area under irrigation during 1969-70 was 30.3 million hectares. The Second Irrigation Commission has calculated that the ultimate potential for irrigation from conventional sources is 81.7 million hectares. It is premature to guess what the contribution of weather modification and desalination will be in the future to augment water availability in drought-prone areas. The States which have less than fifteen percent of the net cultivated area with irrigation facilities are Karnataka, Madhya Pradesh, Maharashtra, Gujarat and Rajasthan. Much of the knowledge we have on ground water resources is confined to the alluvial and semi-consolidated areas, whereas seventy percent of the total geographical area of the country is covered by hard rock. The geo-hydrological studies carried out by the Geological Survey of India as well as the studies carried out under the auspices of the Central Ground Water Board have revealed great opportunities for scientific ground water exploitation. Fresh ground water has been found in the heart of the great Indian desert near Jaisalmer, in the Narmada and Purna Valleys, the Rajamundry and Tirupati sandstone areas of Andhra Pradesh and the Cuddalore sand-stone and the Neyveli lignite areas of Tamil Nadu, in addition to the already well-known areas of Punjab, Haryana, Uttar Pradesh, Bihar, West Bengal and Assam. Opportunities for lift irrigation from perennial streams and rivulets are also yet to be fully utilized.

With the growth of vegetation and evolution of agriculture, what were once geological deposits, got gradually converted into soils. The topography, biology, physics and chemistry of soils provide an interesting account of past history of that soil. In particular, they indicate the extent of interest taken by man in the care and maintenance of soil fertility and productivity. Examined by such criteria, our record is dismal.

Major soil groups: We have four major groups of soils: alluvial, black, red and laterite; several other types like forest, desert, alkaline, saline and acidic soils occurring in smaller areas. The red soils occupy seventy-two million hectares and are found in both high and low rainfall areas. The stored moisture of a saturated red soil profile is generally sufficient to support a standing crop only for two to four weeks. Hence, mostly kharif crops are raised in such soils. Red soils are generally low in nitrogen, low to medium in phosphorus, and medium to high in potash. Drainage is not usually a problem in such soils and hence they can be converted into irrigated land without difficulty, if a water source is available.

Black soils which occur in about sixty-four million hectares are variable in depth and are generally found in regions with an annual rainfall ranging from 500 to 1,200 mm. While in the shallow black soils, only kharif crops are grown, rabi crops are raised in the deeper soils with residual moisture. Drainage and erosion are serious problems in such soils and hence there have to be appropriate arrangements for drainage and soil conservation, when such soils are brought under irrigation. Black soils contain forty to sixty percent clay and are plastic and sticky when wet and very hard when dry. They contain a high proportion of calcium and organic matter, magnesium carbonate, considerable iron and fairly large quantities of magnesia, alumina and potash. In black soils derived from ferruginous schists, there is generally a layer rich in nodules of kankar formed by segregation of calcium carbonate at lower depths. The most important difference between the red and the black soil lies in the amount and type of clay mineral present. The montmorillinitic type of clay predominates in the black soil and this imparts characteristic swelling and shrinking properties to these soils. On the other hand, the non-swelling kaolinitic clay predominates in the red soil. The black soils, also known as "grumosol", "regur" and "black cotton soils", have self-mulching properties, the repeated wetting and drying causing clods to crumble into small aggregates forming a surface mulch. These soils are also referred to as being "self-swallowing", since the surface mulch falls into the deep cracks and thus becomes incorporated into the sub-soil. Thus, on the one hand, the black soils have a good capacity for conserving soil moisture and, on the other, they need proper tillage and management to overcome the handicaps of their bad structure.

Alluvial soils occupy an area of about sixty-four million hectares and are generally deep and variable in structure, ranging from drift sand to loams and from fine silts to stiff clays. The main characteristics of these soils are derived from their having been deposited as silt by the rivers, particularly those of the Indo-Gangetic and the Brahmaputra systems. The physical characteristics and nutrients status of these soils vary considerably. In the south, for example, the Godavari alluvium is rich in lime, phosphates and potash, while the Cauvery alluvium is relatively poor in plant nutrients. In Assam, the old alluvial soils are acidic, while the new alluvium is usually neutral or alkaline. In West Bengal, the old alluvial areas are much less fertile and productive than the new alluvial soils.

Hard pans occur in the soil profile, particularly in the Indo-Gangetic alluvium of Uttar Pradesh, Punjab, and Delhi as a result of infiltration of silica or calcareous matter and the presence of kankar. Such hard layers impede root growth and percolation of water. In general, alluvial soils are ideal for irrigation and ground water exploitation.

In about thirteen million hectares, we have laterite soils, generally associated with undulating topography in regions with annual rainfall ranging from 1,200 to 3,000 mm. They are essentially composed of a mixture of the hydrated oxides of aluminum and iron with small quantities of manganese oxide and titania. These soils are poor in lime and magnesium and deficient in nitrogen. Generally, they are poor in available phosphorus and calcium and have a relatively low organic matter content. Such soils are well developed on the hills of the Deccan, Karnataka, Kerala, Tamil Nadu, Madhya Pradesh, Eastern Ghat region, Orissa,

Maharashtra and parts of Assam. The high and low level laterites of Tamil Nadu and Kerala are rich in plant nutrients, supporting paddy in the low levels and tea, coffee, rubber and cinchona at the higher levels. The higher the altitude, the more acidic is the soil. There is a good potential for exploiting ground water in many lateritic areas.

In addition to these major groups of soils, we have problem soils suffering from salinity and alkalinity in about seven million hectares. There are also peaty saline soils called kari soils which are heavy black soils, highly acidic and rich in organic matter. Marshy soils occur in the coastal tracts of Orissa, the Sunderbans and other areas of West Bengal, north Bihar, the Almora district of Uttar Pradesh and the south-east coast of Tamil Nadu. Forest soils containing a heavy deposition of organic matter derived from forest growth and desert soils occurring in parts of Rajasthan, Gujarat and Haryana, constitute the other important groups.

The main rivers of the Indo-Gangetic plains have extensive areas lying adjacent to them known as *khadir* in Western and Central Uttar Pradesh, *diara* in Bihar and eastern U.P. and *char* in West Bengal. The soils of these areas range from unconsolidated sand to silt and are rich in potash but low in phosphorus. Different management procedures are needed for the sandy, water-logged, saline and good areas of the riverine soils of Bihar, Uttar Pradesh and West Bengal. In U.P. and Bihar alone, such areas extend to about 2.4 million hectares and these are potentially high production soils.

We have also rich rock phosphate deposits both in north and south India and phosphorus rich sands on the beaches of Laccadives, Minicoy and related groups of islands in the Arabian Sea. Gypsum deposits are available in Rajasthan.

Floristic diversity: Consequent on our varied soil and climatic endowments, we have over 20,000 plant species occurring in our country, a number which is far higher than those found in countries with a larger land mass. Such a floristic diversity is obviously the consequence of the prevalence of tropical, sub-tropical, temperate and alpine areas, where the precise flora may vary depending on latitude, altitude, variation in mean annual day and night temperatures, and the severity of summer and winter. We have over 4,000 woody species of plants and among them about fifty are of major utility. The potential productivity of our forests has been estimated to be of the order of 490 million cubic meters of wood. The forest harvest works out to be very low in terms of gross revenue of only Rs. 21.50 per hectare.

Our wild plant wealth, used in various forms by people in tribal areas, includes about 500 species, mainly belonging to the families *Gramineae*, *Orchidaceae*, *Papilionaceae*, *Rubiaceae*, *Euphorbiaceae*, *Moraceae*, *Compositae* and *Labiatae*. If we total up plants under cultivation on some scale, their number comes to about 250, excluding the ornamental trees, shrubs and herbs. It is believed that about thirty-five of these species might have been first domesticated in India and our neighboring countries. There is a large genetic variability in our country in crops like rice, sugarcane, mustard, *arboreum* or Asiatic Cotton, *tossa* jute, various pulses, vegetables such as brinjal, cucumber, gourd, yam, black pepper, cinnamon, turmeric, mango, banana, citrus, jackfruit, ber, phalsa, amla, and jamun. Even in plants introduced from the New World, considerable variability appears to have been generated subsequent to their arrival in India. A few examples are maize, chilli, potato and cucurbits. We have areas in the country like the North Eastern Himalayas which have very rich variability in crops like rice, maize and citrus. A unique type of *arboreum* cotton with long bolls occurs in the Garo hills. There are trees of mandarin oranges in the Jowai area of Meghalaya which have been productive for over a century. The Jeypore tract of Orissa is also a center of considerable variability in rice, while in central India, there is vast variability in several pulse crops. It is only recently that serious attempts have been made to domesticate some of the medicinal plants which are otherwise directly collected from the wild state and used.

An example of such a recent domestication is the release for cultivation in the Bangalore area of strains of *Dioscorea floribunda*, a plant native to Central America and which contains Diosgenin, a steroid used in oral contraceptive pills.

Impressive animal wealth: Equally impressive is our animal wealth. According to the livestock census of 1966, our total livestock population was 344 million including 176.0 million cattle and fifty-three million buffaloes. Both in cows and buffaloes, we have an impressive array of hardy and productive breeds. Exotic breeds like Holstein, Brown Swiss and Jersey are being increasingly used in cattle breeding programs. More than sixty percent of the total milk production comes from about 23.4 million buffalo cows. The best dairy animals in our country have, however, an average production efficiency of only twenty-five percent of their counterparts in Europe and North America.

We have about 4.2 percent of the world sheep population, numbering, according to the 1966 census, about 42.0 million. In addition to wool, mutton and milk, sheep contribute at present 15.5 million pieces of skins and 2.1 million tons of manure. Sheep abound in semi-arid and arid areas and both the hot desert of Rajasthan and the cold desert of Ladakh offer great scope for sheep and goat improvement. Our poultry population is equally high, having been estimated at 115 million in 1966.

Animals also contribute over twenty-eight million H.P. of energy per day for agricultural operations. Good animals found in States like Punjab, Haryana, Andhra Pradesh and Tamil Nadu may give about 1 H.P. per pair per day of eight hours, while those in the eastern States may give only half as much. It has been calculated that to produce a yield of two tons per hectare, the average power requirement would be about 0.75 H.P. per day. The current power availability including those provided by man, animals, tractors and power tillers comes to only 0.30 H.P. per hectare. Thus, inadequacy of power is one of the basic causes for our inability to improve the efficiency of farming through timely agricultural operations.

Marine fisheries: Fisheries, both marine and fresh water, constitute one of our great assets. Over three million persons live on marine fisheries and the gross annual revenue to the country from fisheries is over 300 crores of rupees. Our marine fish catch has increased from about four lakh tons in 1947 to about 1.2 million tons in 1971. We are now the second biggest shrimp producing country in the world and the export of seafood during 1972 earned for the country over Rs. 60 crores. The estimated potential catch from the western Indian Ocean is about 8.9 million tons consisting of 4.4 million tons of pelagic fishes and 4.5 million tons of demersal fishes including prawns and other crustaceans. The potential catch from the eastern Indian Ocean is about 5.5 million tons which includes 2.4 million tons of pelagic fish and 3.1 million tons of demersal fish. Another index of the untapped marine fish resources we have is provided by comparative figures on catch in different oceans. The yield per sq km is about 233 kg in the Atlantic ocean and 196 kg in the Pacific Ocean, in contrast to 37 kg in the Indian Ocean. Even out of this low yield, we catch very little. For example, the total tuna catch from the Indian Ocean is about 175,000 tons, out of which our share is about 5,000 tons. The rest is caught mainly by Japanese, Korean and Taiwanese vessels.

Though the current contribution of inland fisheries, including both capture and culture fishes, is low and amounts only to about 690,000 tons per year, there is vast scope for improvement through modern aquaculture techniques. Water pollution and water weeds could become some of the greatest threats to fish culture and, hence, deserve serious attention. Domestic and municipal wastes contribute more in many areas to pollution than industrial effluents.

Any account of our agricultural assets will be incomplete without a reference to the excellent network of agricultural universities, research institutes and demonstration-cum-training centers which we developed. A recent analysis at the Yale University has shown that while investment in agricultural research has been very small in many parts of India in relation to other sectors of research, the payoff from investment in agricultural research in our country has been one of the greatest in the world. Among factors which have contributed to the effectiveness of our agricultural research system is the development by the Indian Council of Agricultural Research of a national grid of cooperative experiments conducted by scientists belonging to all the relevant disciplines and institutions. Such all-India coordinated research projects now number over seventy and cover all the major areas of crop, animal and fish improvement. The data collected in such projects are discussed at all-India workshops and decisions on recommendations to development agencies and farmers are made collectively by all the scientists concerned. Another important strength of our research system is the direct linkages which have been established between the research centers and the farmer through National and mini-kit demonstrations, *Krishi Melas* and traveling seminars. This feedback relationship assists in the fine-tuning of the research apparatus to the needs of its clients. Though impressive in terms of contributions, our research efforts are still very small in relation to the magnitude and diversity of problems facing us.

Let us now turn to some of our major liabilities.

Liabilities. The low productivity of our agricultural systems is well-known and comparative statistics place our country in the bottom group with reference to the yield per hectare of many economic plants like rice, wheat, jowar, maize, pulses, oilseeds and cotton. An important reason for our relatively poor average agricultural productivity is the vast area under the important crops, which includes considerable marginal lands. Historically, cropping systems have evolved partially due to ecological and pest and disease compulsions but more importantly, on the basis of the home needs of the farmer and his family. We now know that many low-yield environments for rice or wheat may constitute high-yield environments for some pulse or oilseed crops. An effective food distribution machinery can pave the way for re-adjustments in crop planning based on considerations of ecology and economics. If this is not done, our agriculture will become increasingly inefficient and expensive and our cost of production will increase, since wages tend to rise with or without coincident improvement in productivity. For example, our average rice yield by the end of the fifth plan was 1,170 kg per hectare. In contrast, the average yield of rice even in 1971 was over 5,000 kg per hectare in the Arab Republic of Egypt, Japan, Italy, the United States and several other countries. Thus, after the inadequacy of farm power, a major handicap is defective land and water use resulting in low yields.

Probability estimates: One of our great natural assets is variability in weather and soil conditions which fostered the domestication of a wide range of economic plants and animals. Aberrations in weather, however, also constitute a major handicap. The rainfall distribution is skewed, about eighty percent occurring during the south-west monsoon season. About thirty percent of our geographical area receives annually less than 75 cm rainfall and the occurrence of drought, floods, breaks in monsoon, cyclonic storms, thunderstorms and dust storms are common in one part of the country or the other. Although the periodicity of such weather aberrations cannot be determined with precision, it is possible to work out anticipatory measures and cropping systems on the basis of probability estimates. This is yet to be done systematically.

The next major handicap is the growing loss of soil and the damage that is being done to soil health and fertility. For reasons which are not clear, there has been a great neglect of the soil in our country in comparison with China or Japan. While it takes anywhere between one hundred to four hundred years for one centimeter of topsoil to be formed in nature, all this soil can be lost in just one year due to erosion. It has been estimated that nearly eighty million hectares out of the one hundred thirty-nine million hectares under cultivation require attention from the soil conservation point of view. A wide range of factors such as denudation of forests and vegetation cover, inappropriate tillage and cropping techniques and practices like shifting cultivation, are causing a considerable loss of valuable soil through water and wind erosion.

Shifting cultivation, known as *jhuming* in the north-eastern Himalayan region, involves cutting down all vegetation from hill slope, use of fire to clear the debris, growing a crop like hill paddy, millets, sweet potato or beans, abandoning the land after a few years and restarting the cycle at another place. According to an FAO estimate, this form of cultivation dates back to the Neolithic period. While this practice has gradually tended to disappear from States like Bihar, Orissa, and Madhya Pradesh, the area cleared annually for Jhuming may be about 100,000 hectares in Assam and Meghalaya. Observations in Assam hills indicate at least ten centimeters of soil may be washed away even for moderate slopes in each Jhuming cycle. One of the factors influencing such indifference to soil care is the fact that the land cleared for Jhuming is not owned by the cultivator, whose interest in the land is coterminous with the cropping cycle. Shifting cultivation as well as growing one crop once in two to three years were two of the ancient methods of overcoming the implications of the law of the diminishing returns of the soil in relation to crop yield. In areas with settled cultivation, application of organic wastes and the cultivation of pulse crops and other legumes which can fix atmospheric nitrogen in the soil were the common methods of restoring soil fertility adopted in the past.

Soil conservation: Soil erosion leads to an enormous loss of nutrients. Some calculations show that the annual loss of soil due to erosion is about 6,000 million tons and that of nutrients is 2.5 million tons of nitrogen, 3.3 million tons of phosphorus and 2.6 million tons of potash. A portion of these nutrients may get deposited elsewhere. In fact, such loss of nutrients may be the most important cause of water pollution in our country. A major approach to soil conservation has been the construction of contour bunds for the purpose of decreasing erosion, conserving water above the bund and increasing infiltration. During the first four Five Year Plan periods, we were able to treat about 13.3 million hectares of agricultural land and 1.2 million hectares of non-agricultural land with various soil conservation measures. While these programs have been very valuable both for diminishing erosion and providing employment, the agricultural benefits from such programs have not been commensurate either with the effort or expenditure involved. Often bunding has neither been followed up with other measures like providing a vegetation cover nor has it been carried out with an understanding of crop production technology. For example, a survey by the Madhya Pradesh Department of Agriculture on the effectiveness of contour bunds in increasing rabi wheat yields has shown that on the whole the effect of bunding was strongly negative. Similarly, a study on the effect of bunding on the yield of kharif crops in the Bellary area indicated that contour bunding decreased the yields of jowar, cotton and safflower both as a result of water stagnation and delay in cultural operations. It is obvious, therefore, that soil conservation measures need to be planned and implemented in a much more scientific manner taking into consideration not only the need to check erosion, but also the end purpose of promoting crop growth.

One has only to see how land is being used for brick making and road making in our country to understand the extent of indifference to the care of the soil. While planned use of land for brick making could help in building up permanent assets like tanks, unplanned use results both in the loss of good agricultural land and erosion. Similarly, there is always a danger that natural drains may be plugged if a total view on soil and water conservation is not taken while planning the construction of roads and railway lines. Closing natural drains leads to floods and thereby to considerable soil erosion. Thus, deforestation, shifting cultivation, overgrazing and improper cropping of undulating lands bunding without vegetation cover, plugging of natural drains and other kinds of poor land management are causing increased runoff, reduced ground water recharge and severe erosion resulting in the deterioration of soil, loss of valuable nutrients, lower yields, flooding of lowland areas, sedimentation of small tanks and large reservoirs and the runoff of precious water to the ocean. It is not surprising that in an article entitled "The Eleventh Commandment" (1966), Dr W.C. Lowdermilk stated a few years ago that "the use of land is a down-to-earth index of a civilization, for land has been the silent partner in the rise and fall of civilizations".

We can continue to neglect our soil only at the peril of our future. The authors of *Limits to Growth* (Meadows et al, 1972) have calculated that every child born today would need 0.08 hectare of land for purposes like housing, roads, waste disposal, power supply and other uses and 0.4 hectare of land for producing the food he or she needs. On this formula, we will need at least five million hectares of additional land every year to cater to the needs of those added to our population. In contrast even in 1969-70, the availability of agricultural land was only 0.34 hectare per person. Obviously, it is time we woke up and spread throughout the country consciousness of the value of soil and the importance of scientific land use.

Improper and inefficient water use, inadequate tapping of sunlight, poor utilization of biological nitrogen fixation, wasteful disposal of wastes, lack of understanding of recycling processes and poor integration of crop and animal husbandry on the one hand, and terrestrial and aquatic production systems, on the other, are some of our other major liabilities.

The slow pace of progress in getting the best out of our water resources is evident both from the relative stagnation in kharif crop production and the low intensity of farming. During the kharif season, when much of the rainfall is received, the production of foodgrains was 65.6 and 62.0 million tons in 1964-65 and 1971-72, respectively. In contrast, the rabi production during these two years was 23.7 and 42.7 million tons respectively. Thus, our major grains in production have come from the non-rainy season. With the development of irrigation facilities it should have been possible to bring more area under double and multiple cropping, thereby raising the intensity of cropping. The intensity of cropping is however still low, and has risen from 114 percent in 1965-66 to only 117 percent in 1969-70.

A low intensity of cropping even when water is available also implies a poor utilization of sunlight, since in the tropics and sub-tropics green plants can be made to produce food, feed and fodder continuously by photosynthesis from water and atmospheric carbon dioxide. The cellulose produced by photosynthesis on the earth is not only the chief basis of all fossil fuels, but is also the most abundant renewable raw material currently available.

Nutrient supply: In the area of nutrient supply to crops, organic manures like farmyard manure, compost, green manure, various oil-cakes and waste products of animal origin like dried blood, bones, fish manure and urine had been used in the past. With the growth in population, farmyard manure and other organic wastes have increasingly been diverted to fuel purposes. Though some work has been done on the generation of gas from such wastes in order to obtain both fuel and manure from the same material, such techniques have not come into use on any significant scale. Much of the urban wastes, sewage water, cattle and human urine and human excreta are not recycled in a manner that will promote productivity

A great marvel of nature is the way in which micro-organisms fix atmospheric nitrogen in the soil largely through leguminous plants. While synthetic nitrogen-fertilizer production requires very high temperature and pressure for combining nitrogen, hydrogen and oxygen, the nitrogen-fixing organisms like *Azotobactor*, *Rhizobium*, and blue-green algae are able to do this at ordinary soil temperature and pressure, with the help of the enzyme nitrogenase. Nitrogen is now being introduced into the earth in fixed form at the rate of about ninety-two million tons per year, whereas the total amount being denitrified and returned to the atmosphere is only about eighty-three million tons per year. The difference of nine million tons per year may represent the rate at which fixed nitrogen is building up in the soil, ground water, rivers, lakes and oceans. Unfortunately, studies show that while Australia, the United States and the Soviet Union are adding every year substantial nitrogen to their soils, we have a negative nitrogen balance to the extent of 8.2 lakh tons per year.

Before ending this discussion on agricultural assets and liabilities, we may refer to the most important factor which will determine our agricultural future, namely, our human population. The principal characteristics of our population are the predominance of youth, poverty, under-nutrition and illiteracy. We are mainly a land of youth and of rural people. Impressive statistics on poverty are available and there is also growing awareness of the implications of protein-calorie malnutrition on national development. Many authorities now compare nutrition education respectively to the hard and software components of technology. Although in terms of percentage, the number of literates rose from 24.03 in 1961 to 29.34 in 1971, the number of illiterates increased in absolute number during this period. Most of our agriculture is managed by illiterate peasantry and Mahatma Gandhi warned us forty years ago that unless there was a marriage between intellect and labor in rural India there would be neither agrarian advance nor rural prosperity. Our agricultural future will hence depend to a great extent on how successful we are in involving youth and the illiterate, both men and women, in rural transformation. The urgency of population stabilization hardly needs any emphasis, since all our efforts will have no effect in improving the quality of life, if population growth cannot be arrested and stabilized.

The second class compartment of Spaceship Earth is getting smaller and smaller, while the number of those who have to live there is getting larger and larger. Affluence is becoming a major claimant of world food and energy resources. Between 1967 and 1971, the developed market economies increased their agricultural exports on an average by eleven percent per year, while the developing countries lost ground by one percent. This is the net result of what is called the UN Development Decade. It is time the people living in the first class compartment realized that when the compartment of poor bursts due to excessive pressure, the whole spaceship will crash. But the poor have to begin helping themselves.

During the last few years, we have started making real progress in improving our agricultural capability. While food production dropped by seventeen million tons during the drought of 1965-66, the fall in production during the drought of 1972 was less than half of this amount. This is certainly an index of real progress and encouragement for greater effort.

To summarize the state of our agricultural balance sheet, we are endowed with considerable ecological diversity, a large volume of tapped and untapped sources of water, abundant sunlight, and large animal and human populations. Our liabilities, apart from those caused by population growth and aberrant weather, mostly arise from an improper use and management of these resources. Attempts to promote synergy, which is the only mechanism which can lead to rapid progress from small resources, have been few and halting. Our urgent needs hence are first, to develop and introduce in each ecological area an agricultural production technology which will lead to increased productivity based predominantly on the use of renewable resources and on the wise husbanding of non-renewable resources. We will have to learn to produce more and more food from less and less land. Secondly, we need to develop and introduce educational tools which will help to impart the latest technical skills to illiterate peasantry and which will enable educated youth to become catalysts of rural change. Finally, we need to develop and spread management and organizational techniques which will help those living in absolute poverty to overcome their handicaps and obtain their share of the fruits of agricultural advance.

Lecture 2
The Emerging Agricultural Technology

In the previous section an attempt has been made to give a synoptic view of our agricultural assets and liabilities and stress that in view of the inelastic nature of land availability and the competing demands for its use, we have to adopt an agricultural technology which will help us to achieve our immediate goals of more food, more jobs and more income for the present population, without at the same time endangering the long-term production potential of the soil. It is probable that extensive ocean farming, hydroponics or soil-less cultivation, test-tube cultivation, protein from petroleum and single cells and other similar techniques may contribute to our future food supply. In the immediate future, however, we will have to rely primarily on mother earth to feed us and our animals. As far as we can judge now, soil will continue to be the most important medium for crop growth, since several of the other techniques will have to rely heavily on the use of non-renewable resources of energy. We also do not know what the long-term consequences of certain trends in ocean farming, such as induced upwelling, will be. Upwelling is a physical, ocean-atmosphere process leading to the transfer of high concentrations of chemical nutrients from sea bottom to the zone where there is active carbon assimilation thereby producing food for fish. It has been estimated that about one half of the world's fish production occurs in the restricted coastal upwelling areas. If man can artificially create upwelling, it is possible that there may be richer harvests of marine life for some time. No one, however, can predict what will happen in the long-term. We already have the lesson of the Peruvian fisheries to remind us that if man in his greed tries to over-exploit a resource he will come to grief. In 1970, the catch of anchovies in Peru reached 12.3 million tons. It dropped dramatically to about four million tons in 1972 and is not yet showing any sign of recovery. Though the drop is partly due to a periodic ecological disturbance in the sea leading to the influx of warmer water, many authorities believe that over-fishing leading to the depletion of the basic renewal stock may also be a factor responsible for this big drop. A sad consequence of the drop in the availability of anchovies along the Peruvian coast is the dwindling in the population of the Guano birds, from about thirty million a few years ago to about a million now. If the Guano birds disappear, the finest natural fertilizer in the world formed by the deposits of these birds will also gradually get exhausted. Anchovies not only constitute the dominant food of the Guanos, but also of poultry and pigs in North America. The drop in the availability of anchovies hence escalated the demand for soybean as the main protein source in animal feed mixtures. The consequent sudden rise in soybean price led the Japanese to release rice for animal feed, rather than import soybean at very high prices. This illustrates the inter-dependence of biological and production factors and the consequences of one adverse event on other seemingly unrelated events.

The importance of the scientific management of the biosphere is thus obvious. This section would deal with some recent trends in the promotion and management of biological productivity, citing relevant situations in our country.

Attention to Economic Ecology

First among the emerging concepts of management of biological assets is attention to economic ecology. The prefix "economic" before "ecology" is used to underline the fact that what we need is as high an economic growth rate as possible through the use of the principles of ecology rather than the kind of ecology discussed frequently in the affluent nations which is of the conservationist or zero growth rate kind, intended to preserve the high standards of living already achieved. For the sake of convenience, our arable land is dealt in five groups: arid, semi-arid, humid, irrigated and hilly regions. It is obvious that there are numerous climatic variations in these groups. Kharif and rabi seasons have not the same significance in south India, as they have in the north. This is why agricultural technology becomes highly location and situation specific, necessitating a considerable amount of local research and testing work, before a new technology can be recommended to the farmers. While ideas and concepts can be transplanted from one region to another, the actual material and techniques will have to be tailored to suit the local agro-ecological and socio-economic milieu.

The arid zone occupies in our country an area of 3.2 lakh sq km of hot desert, mostly in Rajasthan, Gujarat, Haryana, and Karnataka and 0.7 lakh sq km of cold desert in Ladakh. In Ladakh, extreme aridity combined with low temperature limits the possibility of growing crops to about five months in a year. Hence, the strategy for agricultural growth in Ladakh has to depend largely on the cultivation of quick-growing cereals, oilseeds and fodder crops and the rearing of goats, giving Pashmina wool. The hot desert regions, in contrast, have an abundance of sunshine, land and soils capable of responding to management, well-adapted grasses and trees, excellent breeds of sheep, goat and cattle and considerable reserves of ground water. Water and not land, is the principal limiting factor and hence all attempts have to be focused on maximizing income per liter of water. This will be possible only if the ecological balance is not further disturbed and a proper land use pattern is adopted.

Three examples can be cited to illustrate the trends which are aggravating the unfavorable consequences of aridity. First, the area used exclusively for grazing in western Rajasthan has dropped from 13.09 million hectares to 11.04 million hectares during 1951-61, while the population of grazing animals increased during the same period from 9.4 million to 14.4 million. The same trend of diminishing grazing area and rapidly expanding grazing population is still persisting. Secondly, while most of the land in the arid zone is fit only for forestry or range management, land is increasingly being brought under cropping. The area cropped rose from twenty-six percent in 1960 to thirty-eight percent in 1970, thus extending cultivation even to sub-marginal areas. Thirdly, the area under forests is only two percent although the extent of land classified under barren uncultivable waste is twenty-eight percent and cultural waste is eighteen percent, all of which could be planted with tree species like *Acacia tortilis, Prosopis juliflora* and *Eucalyptus sp*. Fruit trees like ber and pomegranate can also be grown extensively.

Reclaiming the Desert

Afforestation has to be the focal point for reclaiming the desert. Techniques for soil and water conservation and for sand dune stabilization are fortunately available today. Large-scale planting of shelter-belts could help to minimize wind erosion and decrease the dust over the desert. Scope for the establishment of pastures and grazing land. is great and strip cropping involving the setting up of permanent grass strips to prevent wind damage will help to increase the yield of crops like bajra and moong substantially. If steps of this kind can be taken, it may be possible for the nomadic tribes like Banjaras and the Gadiya Lohars to start leading a more settled life.

Rajasthan has many fine breeds of sheep and research has shown that appropriate cross-breeding procedures can help to improve the yield and quality of wool as of mutton. Fortunately, techniques for artificial insemination have been developed and hence more widespread genetic improvement programs are possible.

An important need in the desert areas is the preservation of organic matter for the soil. Unfortunately, due to deficiency of fuel, the available organic wastes are generally used for burning and trees are cut in an unplanned manner. It is here that more extensive research and extension efforts on the utilization of solar energy for purposes like heating, cooking, lighting and distilling water will be of great value. Often in the desert areas, the water is saline and people have to walk several miles every day to fetch sweet water. Solar stills to produce water for drinking in such areas will be a great boon to the villagers.

Sophisticated techniques for raising crops in desert areas with very limited quantities of water are now becoming available These involve cultivation of vegetables and high value crops under air-inflated polyethylene houses using economic methods of water supply like drip or sprinkler irrigation. Since the Lathi aquifers of the Thar desert are rich, it may be possible to develop efficient water use systems at least for raising community nurseries of different crops so that transplanting can be done if there is a delayed onset of monsoon and for purposes like the cultivation of vegetables and seed production.

Unfortunately, we have no arrangements now either to mitigate the rigor of a bad season or to derive maximum advantage from a good season. For example, the year 1973 provided a unique opportunity for planting in the desert a large number of trees and for seeding large areas with grasses. While the average annual rainfall of Barmer and Jaisalmer districts is 277.5 mm and 164.0 mm respectively, 302 mm rain was received in just three days in Bormer and 377 mm in Jaisalmer in twenty days. In spite of the shallow nature of the soils and the calcium carbonate pans which occur in the subsoil, thereby increasing run-off and reducing infiltration, abundant moisture was available to facilitate the establishment of trees, shrubs and grasses. The district administration tried its best for getting the available seeds sown, but the magnitude of the program was small in relation to the opportunities that were available.

Good Weather Code. What we need in such areas is a Good Weather Code, which will spell out the types of activities that should be undertaken in the event of rainfall being good. While we have Scarcity Manuals which tell us what to do when rains fail, we do not yet have a corresponding action code to indicate what we should do to capitalize on good weather. It is obvious that a written code by itself will be of no value, if arrangements to implement its provisions, by way of extensive nurseries of trees, do not exist.

About seventy-five percent of our cultivated area is rainfed and since about forty-two percent of our food comes from such areas, there is variation in food production depending upon the amount and distribution of rainfall. Semi-arid areas where mean annual potential evapo-transpiration exceeds the mean annual rainfall occur in many of our States. Such areas have been plagued for centuries by periodic droughts, floods, soil erosion, instability in production, drinking water scarcity, unemployment, underemployment and other forms of human suffering. In about 101 million hectares, crop fortunes are closely linked to the behavior of the monsoon. These areas not only provide much of our millet and pulses, but are also the major suppliers of industrial raw materials like cotton and groundnut.

The earliest research effort on dryland agriculture was started by Tamhane in 1923 at Manjri. Subsequently, a rather comprehensive program was evolved by Kanitkar for improving the productivity of rainfed farming. This formed the basis for the Dry Farming Scheme started in 1933 by the Indian Council of Agricultural Research at Rohtak, Sholapur, Bijapur, Raichur and Hagari. The outcome of these projects was a series of dry farming practices commonly known as Bombay, Hyderabad and Madras Dry Farming Practices. The basic features of the dry farming practices developed in the past were bunding to conserve soil moisture, application of farmyard manure to supply plant nutrients, deep ploughing once in three years, shallow preparatory cultivation and interculture, low seed rate and wide spacing of crops. Official estimates placed the increase in crop yields due to the adoption of this package of practices at about fifteen to twenty percent over a base level of 200 to 400 kg per hectare. No wonder, the package failed to catch the imagination and interest of the farming community. Nevertheless, these programs did make a useful contribution in the areas of moisture conservation and erosion control.

Later in 1954, Soil Conservation Research and Demonstration Centers were established in different parts of the country and they gathered valuable data on the capability of land, rainfall intensity and distribution, run-off patterns and response to fertilizer application. In addition to developing conservation practices appropriate to a region, these centers also provided an insight into the factors limiting production in dry-farming areas. Lack of quick maturing varieties was a major handicap in the early programs. For instance, it was noted at the Bellary Research Center that in late December the soil, which has a high clay content, cracks heavily due to the loss of moisture. The flowering stage of the traditional varieties of cotton and jowar coincided with this period of rapid soil drying and hence their yields were poor. It was thus obvious that if cotton or jowar varieties maturing for harvest by early December could be introduced into cultivation, much higher yields could be obtained. We had to wait until 1965 when thanks to hybrid jowar CSH.l, H.B.I bajra and PRS cotton varieties, which could all mature before the soil lost its moisture, good yields could be achieved even in years of subnormal rainfall.

Intensification of Dryland Research

Experiences of this kind led to the intensification of dryland farming research in 1970 through the initiation of an All-India Coordinated Research Project on Dryland Agriculture with the collaboration of the Government of Canada. Meanwhile, international interest in upgrading the productivity as well as the stability of production of the dry farming regions of Asia, Africa and Latin America grew and resulted in the establishment of an International Crops Research Institute for the Semi-Arid Tropics in Hyderabad in 1972.

The new phase of dry farming research has already resulted in considerable data on better moisture conservation and use, new cropping patterns, crop life-saving techniques and mid-season corrections in crop planning in the drought-prone areas. The pilot project areas attached to the dry farming research centers are helping to identify the socio-economic and operational constraints in the transfer of technology from the research farm to the farmer's fields.

Since water is the major limiting factor, a priority area of research is the standardization of techniques by which as much of the precipitation as possible can be conserved for crop use, either directly in the soil profile through infiltration or through run-off collection and storage. Deep ploughing, for example, promotes a vertical rather than a tangential flow of water in red soils with a dense sub-soil. The cultivation of deep-rooted crops like castor, red gram or arhar and cotton further helps in improving the soil texture and in adding organic matter through root deposition. Run-off storage structures are being developed both for individual small farms and for larger watersheds. Obviously, the most effective method will be the cooperative management of an entire watershed. If sufficient water can be collected in community owned ponds, life-saving irrigation can be given at the time of grain formation, when the crop will benefit most from the supply of a little water.

In black soil areas with moderate rainfall, there is scope for double cropping or ratooning, provided suitable surface drainage can be introduced during the rainy season and appropriate tillage practices can be developed. Drainage and sound soil and water management will again need some degree of cooperative endeavor on the part of a watershed community. There is also great scope for introducing better inter-cropping practices in areas with annual rainfall ranging from 625 to 1,000 mm.

Drought-escaping varieties. Progress has been made in identifying crop varieties which are relatively photo-insensitive and which have a shorter duration and the resulting ability to escape droughts. Early seedling vigor and good population performance are the other attributes of such strains. Suitable varieties in rice, jowar, bajra, minor millets, sunflower, safflower, castor, mustard, groundnut, pulses like moong, urad, arhar and cowpea and cotton are becoming available. As a result, different cropping patterns can be developed to suit different weather models. For example, some of the common weather aberrations are early or delayed onset of monsoon, long breaks in the monsoon, and inadequate rainfall, and different crop schedules can be developed for each of these conditions. If the monsoon is very early, short duration legumes could be taken followed by regular season crops. For normal sowing, jowar, for late sowing, bajra, and for very late sowing, setaria, are some of the possibilities. When there are long breaks in the monsoon, the jowar or bajra crops affected by drought could be ratooned. Crops with indeterminate growth habits such as castor or arhar regenerate fast if given a urea spray after the receipt of rains during a prolonged drought period. The development of early maturing arhar varieties like Prabhat and Pusa Ageti has made it possible in north-western U.P to take a pulse crop before wheat. Since early varieties of arhar are now available, a crop of this pulse can be taken in north India between July and December. Then, with the help of dew or winter rains, a crop of sunflower can be raised in the same field from February to April. Much better use of dew can be made in the north India during rabi. Dr. R.D. Asana, one of our eminent plant physiologists, developed a model of a wheat plant which would be capable of retaining dew. We need research of this kind in all rainfed rabi crops. We should evolve for each soil and rainfall belt a series of alternative cropping patterns to match different weather probabilities. It would also be necessary to build the appropriate seed and fertilizer buffers and organize community nurseries in order to put into practice alternative crop schedules, if the weather proves truant. The National Commission on Agriculture envisaged a Drought Code which could help the developmental administration to be ready for any weather pattern and initiate anticipatory action to mitigate the effects of drought, rather than be content with palliative action after the misfortune has set in.

Multiple and relay cropping. The potential of irrigated areas to produce much higher quantities of food, feed, fiber and fodder plants through multiple and relay cropping is well-known. Due to various difficulties and shortcomings in the water conveyance and delivery system as well as due to shortage of power, fertilizer and good seed, the full potential of the irrigated areas is yet to be realized. If these constraints are overcome and a minimum of two good crops can be raised the production of about eight to ten tons of food and other grains per hectare per year, now being harvested in several of the experimental stations of agricultural universities, should not be beyond the reach of many small farmers. While taking up intensive farming practices, a continuous monitoring of the soil for major and micro-nutrients and for pathogens will be essential. Legumes should find a place in the rotation and as a rule, crops sharing common pests and diseases should not succeed each other. It would be better to alternate deep and shallow rooted crops so as to tap nutrients from different soil layers.

New systems of cropping are rapidly coming into existence in many of our irrigated areas. Thus, in the Punjab and Haryana, rice during *kharif* and wheat during *rabi* are gaining popularity. In West Bengal, wheat is becoming popular as a *rabi* crop in rice fields. Baisakhi moong and other strains developed in order to find a place for a pulse crop during a season when the land is normally fallow are becoming popular as summer crops. In the rice fallows of south India, jowar, bajra and cotton can be raised during the off-seasons. In the case of several of these new cropping patterns, we need much more data on the most appropriate cultural practices including fertilizer application and weed control.

Effective land use. Our knowledge of the most effective land use systems in the heavy rainfall areas is still fragmentary. Many of these areas, Assam, Meghalaya, Arunachal Pradesh, Nagaland, Mizoram, Tripura, Manipur and Kerala grow plantation crops and fruit trees in the medium and higher elevations. A viable alternative to shifting cultivation has to be found in the north-eastern Himalayan region. Similarly in Kerala, there is very little data on the optimum utilization of garden lands occupied by crops like coconut, arecanut, mango and jackfruit. Prof. K.N. Raj and his colleagues have pointed out that the current crop-mix in garden lands of Kerala and Mysore is the product of the technological possibilities as known until now to the farmers of these areas but that a fresh look is needed since the market conditions for some products, as for example arecanut, have considerably changed. By appropriate changes in spacing in coconut and arecanut, some other crops of high value like cocoa, cloves, and soybeans can be grown in between the palms. New varieties of grasses and legumes are available now which can also be grown as inter-crops in the arecanut or coconut plantations so that the farmer can maintain a few cross-bred cows for milk production. These and many other possibilities for capitalizing on the ecological possibilities of the humid tropic regions are yet to receive intensive attention.

High Altitude Research

High altitude areas have also received relatively little scientific attention. Ecological conditions vary greatly even within short distances in these areas depending upon aspect, latitude, altitude, slope, soil depth and distance from the plains. Large variations in temperature and rainfall occur and the length of the growing season is controlled by the time of occurrence of snow in winter and its disappearance in the spring. Erosion will be a serious problem and in several of these areas data are needed on minimum tillage, suitable implements to ensure timeliness of operations, fertilizer use and weed control. Both horticulture and nomadic sheep husbandry require considerable attention. For agriculture to start moving in these areas, marketing arrangements are pre-requisites, since in some of the Himalayan regions due to a variety of reasons the producer may get as little as ten percent of the price paid by the consumer.

Besides rainfall, altitude and latitude, the ecological conditions of soil and season need understanding for the improvement of biological productivity. Large areas are affected by salinity and alkalinity where crops either do not grow or give very poor yield. Soil salinity is also becoming a problem in black soils, where irrigation facilities have been developed recently. An understanding of the nature of the problems and of the physical and chemical characteristics of the soil profile would be necessary to introduce the most appropriate corrective measures like gypsum application, drainage, crop and varietal choice and fertilizer use. Fortunately, we now have techniques available for reclaiming such problem soils as a result of the work done at the Central Soil Salinity Research Institute, Karnal, and at other research centers and agricultural universities. Acid soils also require special care and, in particular, the most effective methods of phosphorus application will have to be worked out.

Strategies for ecological seasons. Detailed production strategies are yet to be worked out for each ecological season. In Assam, the period between September to May is usually free of floods, since the major floods have either taken place in May-June or more commonly in August. There is tremendous potential in this area for exploiting ground water and for arranging for lift irrigation. An assured rabi crop can be raised if arrangements can be made to supply water. There are areas in Assam with over 250 mm winter rains, where even pending the establishment of an irrigation system, good crops of sunflower, soybean, barley, moong and even a short duration rice variety like Bala or Pusa 2-21 can be raised. Similarly, coastal areas of West Bengal, Orissa and Andhra Pradesh are exposed to recurrent floods and cyclones during August-September. It is possible to tailor a cropping pattern which would escape or avoid the fury of floods. For such flood-prone areas a Flood Avoidance Code is needed which could provide guidelines for restructuring the agricultural system in accordance with the probability of occurrence of floods.

The entire riverine or Diara areas of Bihar and Uttar Pradesh will benefit greatly from arrangements for community tube wells. The Chota Nagpur and Santhal Parganas areas can grow fruits, vegetables and develop dairying so as to capture the emerging market in cities like Bokaro. Due to cloudy conditions and relatively greater incidence of pests and diseases during kharif, the yield of high-yielding varieties of rice is lower during this season than during rabi. Yet, while the total area under rice is over thirty-seven million hectares, the area under rabi rice is only about three million hectares. The major problems in extending rabi rice cultivation are the cold temperatures of December to February and scarcity of water. Cold-tolerant strains are currently being selected and with more economical water management for which techniques are available, it should be possible to take a good rabi rice crop in the delta region of the east coast, where ground water is available a few meters below the ground. Thus, the scope for restructuring cropping patterns on ecological and marketing considerations is vast. For this, we need to classify the existing cropping systems, which are largely the product of the home needs of the farmer and pest and disease syndromes into classes such as those with high yield potential and low risk, high yield potential and high risk, low yield potential and low risk, and low yield potential and high risk. Obviously, we should strive to eliminate the low yield potential crops from the cropping system of each area and reduce the risk elements in the high yield potential crops. Above all, effective purchase and marketing arrangements are vital, since the farmer is motivated not so much by yield per hectare but by net return per hectare.

Synergy in agriculture. Having discussed some of the broad features of crop planning on an ecological basis we may examine the application of three other important concepts in agriculture. The first of these is the concept of synergy in agriculture, one of the most fascinating scientific developments of recent years. Synergy is not a new concept, since this mechanism which results in the product being something much more than the sum of the parts, has been the most potent tool involved in natural evolution. Symbiosis, or mutually beneficial relationships and coordinated functioning are essential ingredients of synergy. For developing high synergy cropping systems, an understanding of population behavior and performance, rather than merely of the characteristics of individual plants or animals, is essential. Prudent observers of nature in the past had devised biological systems based on the release of synergistic interactions.

Chinese peasants had devised over a thousand years ago methods of producing high yields of fish with low inputs of money and technology. They did this on the basis of two principles. First, a body of water is a three-dimensional growing space; to treat it like a field by planting only one kind of crop is likely to result in wasting the majority of that space. Secondly, any fertile pond will produce a number of different fish food organisms. However, most fish are not omnivorous, but are rather selective in their diet. Thus, stocking single species in a pond wastes not only space but food. Chinese fish culturists took advantage of these two characteristics of the pond environment through polyculture or stocking several types of fish. Under an All-India Coordinated Project on composite fish culture, suitable species of fish have been identified which can give cooperatively about 3,000 kg of fish per hectare in about six months. The species involved are both Indian carps like catla, rohu and mrigal and exotic carps like grass carp, silver carp and common carp. Such high synergy aquaculture systems can revolutionize fresh water fish production in our country.

Terrestrial and aquaculture systems can also be devised, where each element would contribute significantly to the functioning of the whole, resulting in a profitable farm ecosystem. The Chinese have again taken advantage of the possibility for raising pigs and ducks in conjunction with the culturing of fishes. In a 4.4 hectare Chinese farm in Malaysia, pigs, fish and aquatic plants were raised together. About thirty tons of pig meat were produced per year, with the primary feed being the aquatic water spinach, *Ipomoea repens* which grew luxuriantly in the fish ponds fertilized by the pig manure. Besides the pigs, some 3,000 kg of fish were cultured in the small ponds. The main food for the fish was algae, which grew well in the enriched ponds. Ducks could complement a system like this, their wastes being added to the ponds.

Changing Plant Architecture

The search for synergy has led both to the re-patterning of plant architecture in many economic plants and to the development of cross-bred farm animals which are efficient in the conversion of feed into the product for which they are raised by man. What the breeders now look for are the plant types which can maximize production per unit of area, time and water. Thus, plant types which will not shade each other or fall over each other and which will promote better light interception and carbon dioxide fixation are now sought after. In the earlier strains, many of the characteristics had been selected for performance under adverse circumstances and not for high yield under good management. Because of their plant and leaf characters, a large number of productive tillers per square meter can be packed into the new varieties of wheat and rice, as compared to the old ones. Similarly, the distribution of the total dry matter produced by the plant between the part of commercial value and the remaining parts is much more favorable in the new strains. New plant types in bajra can respond to a population density of about 250,000 plants per hectare, in contrast to less than 100,000 plants in the case of the earlier strains. Similarly, in maize, dwarf varieties with smaller tassel, high nitrate-reductase activity and upper placement of the ear are under testing in several countries and these can respond to a population density of about 150,000 plants per hectare. The concept of population explosion in fields has also permeated horticulture. Growingly, emphasis is being placed in orchards on the selection of dwarfing root stocks. Some experts believe that the fruit orchards of the future may be "Lilliputian" in nature, facilitating a high management efficiency. Unfortunately, in our country root stock consciousness has not yet spread and citrus growers still raise orchards from seedlings although budded plants on appropriate root stocks like trifoliate root stock in drought prone areas can make a great difference in orchard productivity.

Apart from improving population performance and the yield potential of economic plants through a re-patterning of plant architecture, growth rhythm and allocation of dry matter between the commercial and non-commercial portions of the plant, the other areas of concern have been in introducing a broad spectrum of resistance to pests and pathogens and in improving nutritional and cooking or processing quality. Unfortunately, breeding for resistance to pests and diseases is often a never-ending task. The resistance tends to break down, following the build-up in nature of new races of the pathogen capable of attacking a variety earlier released for its resistance. This is being observed in the popular wheat varieties Kalyan Sona and Sonalika, which at the time of their release in 1967, had a high degree of resistance to brown and yellow rusts. Kalyan Sona has become susceptible particularly to brown rust and there is no assurance that the resistance of Sonalika may hold out for long. Seeds of new strains have hence to be multiplied and distributed speedily. Dynamic varietal diversification and seed multiplication programs are essential to sustain a good crop production program.

Outwitting the Pathogen

Will the struggle between the breeder and the pathogen be an endless one or will one outwit the other? This is an interesting question for which no clear-cut answer is yet in sight. The breeder is adopting a multipronged strategy. He is working on a type of resistance, technically referred to as horizontal or field resistance, which is likely to be of a more enduring type. Escape mechanisms are experimented to alter the growth phase of the plant in such a way that it does not synchronize with the peak multiplication and infectious phase of the pathogen. He is striving to develop strains called "composite varieties" which can help to reduce the pressure on the pathogen in building up new and virulent races. He is also trying to understand the biochemistry and physiology of resistance, hoping in this way to find effective and cheap chemical methods of control. While all these approaches need to be followed with vigor, our immediate hopes lie in the genes for resistance found in the primitive cultivars and other wild and cultivated genetic material occurring particularly in the centers of diversity of plants. Thus, the rich collections of rice made by our scientists from the north-eastern Himalayas contain genes for resistance to a wide range of pests and pathogens infecting rice. Areas where the pests are endemic, provide opportunities for screening for resistance. Under our All-India Co-ordinated Research Programs, such areas have been identified, and given an efficient seed multiplication and distribution machinery, we should be able to remain ahead of the pathogen in several crops.

Equally exciting are the opportunities now available for adding a nutritional dimension to crop improvement programs. According to several authorities, (including C. Gopalan and P.V. Sukhatme of our country) malnutrition in countries which have cereals as the staple food, is mostly the result of under-nutrition, arising in turn from inadequate purchasing power. Chronic illness and worm infection due to insanitary conditions, lack of community health care and ignorance about the feeding of infants resulting from poverty, all add to the problem. According to the report of the Task Force on Nutrition of the National Committee on Science and Technology (NCST), 90 percent of our pre-school children have heights and weights much lower than those of children of corresponding ages in the United States. Another important point made by the Task Force is that our nutritional problems are qualitatively very different from those of the richer nations and hence the remedies must be our own. For example, in richer nations, where excessive intake of fat of the order of 75 to 100 g daily has aggravated the problem of certain kinds of heart diseases, the emphasis has rightly been on reducing the fat content of the diet. The same situation holds good among the rich in our country. This section of the population, however, is very small and for a majority of our people, the inclusion of more fat in the diet is not only desirable but necessary. This is because fats and oils are not only concentrated sources of calories but act as important vehicles for some vitamins. The NCST Task Force has hence recommended that we should try to raise the per capita average consumption of fats from the current level of 10 g per day to at least 30 to 40 g per day. This would call for a three- to four-fold increase in oilseed production. Apart from opening up the possibilities for sunflower and soybean production, we have not had so far any striking advance in oilseed research. Scope however exists for improving the yield of groundnut, til, safflower and coconut, through better management and seeds. However, in the near future we may have to lean heavily on increasing the acreage under oilseed crops, if we are to substantially increase the production of these crops. More or less similar is the position with regard to pulses. Increasing the area under oilseeds and pulses would be possible only through multiple cropping and by increasing the productivity of crops like rice and jowar, so that some land can be released from them.

More Area for Pulses, Oilseeds

There is considerable scope for releasing areas from jowar and bajra for pulses or oilseeds in rainfed areas by improving the productivity of these crops, replacing the old varieties by high-yielding ones. For example, calculations show that in Gujarat, which produced during 1969-70 one million tons of bajra from 1.73 million hectares, it is possible to produce 1.12 million tons from 1.60 million hectares merely by increasing the proportion of the area under high-yielding hybrids from the original seventeen to thirty percent. Similar opportunities are available in jowar, as a result of release of varieties like 302, 604 and Swarna.

Genetic improvement of nutritive quality of basic staples is another venue of providing better nutrition at no extra cost. Chemical fortification with limiting amino acids, which used to be the major approach advocated during the sixties has helped to improve genetically the amino acid balance and nutritive quality of crops like maize, jowar, barley, potato, wheat, and rice. This field of research was stimulated by the discovery at the Purdue University in the United States of genes conferring a higher content of lysine, an essential amino acid, in the protein of certain strains of maize and jowar. The finding in jowar was made by an Indian scientist, Dr. Rameshwar Singh of Pant Nagar, while he was working at Purdue. This is of particular significance since the nutritional disorder Pellagra has been found by the National Institute of Nutrition, Hyderabad, in areas where jowar is the staple. Another important finding in this area is the identification in Sweden of a barley strain with high protein and lysine. It is interesting that both the jowar and barley lines originated in Ethiopia. This line of research may help to produce high yielding-cum-high nutritive quality strains. If it is coupled with the widespread preparation of home-made weaning foods such as the Hyderabad mix in villages on the basis of locally available cereals, pulses and jaggery, a dent on the malnutrition problem can be more easily made.

Harmony and Natural Balance

The next concept is harmony and natural balance. This involves the development of management concepts in areas like disease and pest control and the balanced growth of different components of a production system. "Prevention is better than cure" is the general motto in disease control and it is known that through appropriate immunization and sanitary and quarantine measures the health of farm animals can be maintained at a satisfactory level. Our scientists have been developing effective vaccines against the important diseases of cattle. Recently, a vaccine against lung worm of sheep has been developed using the principle of radiation attenuation. In general, however, the awareness of the need for quarantine and sanitation is not widespread.

Depending upon the cropping pattern, water availability and marketability of produce, profitable mixed farming systems can be developed. Where feed grain availability is low, ruminating animal could form a good mix with cereal and millet cultivation. Poultry rations usually contain as high as sixty to seventy percent cereal grains. However, several agricultural and industrial by-products like rice bran, wheat bran, slaughter-house waste and fishery wastes could be used as poultry feed. Similarly, the 1972 drought in Maharashtra provided an opportunity for introducing new rations like a mixture of bagasse, molasses and urea. Improved strains of fodder grasses and legumes are also becoming available and studies at the National Dairy Research Institute have shown that high-yielding cross-bred cows can be maintained largely on high quality fodder.

Growingly, concepts like pest control are giving way to procedures which help to raise good crops through better management with minimal assistance from chemical methods of control. Such integrated procedures of pest management involve an appropriate combination among methods like selection of resistant varieties, use of parasites and predators, both native and introduced, ecological control through modifications in cropping systems and agronomic practices, development of microbial pesticides, use of attractants and repellants including sex hormones, development of various methods of inducing sterility and use of selective chemical insecticides. Such an integrated approach has become necessary to avoid dangers arising from the destruction of beneficial insects and natural enemies of pests, the persistence of some pesticides in the food chain, the origin of pesticide resistant insects and environmental contamination. Safer chemical pesticides are being experimented and in crops like jowar, the control of shoot fly either through treating seeds with a suitable pesticide or an adjustment in the date of sowing could make a big difference in yield.

Biological Control of Pests

Biological control involving the control of pests through their natural enemies is receiving increasing attention. In the past, more emphasis had been placed on an inundation approach involving the mass breeding and release of indigenous natural enemies. Emphasis is now shifting to obtaining more effective parasites and predators from other areas where the same or related pests occur. In general, countries which have had the greatest successes in biological control are the ones which have imported the greatest number of natural enemies. A station of the Commonwealth Institute of Biological Control located at Bangalore has been helping in getting natural enemies of pests of apple, cabbage, castor, coconut, cotton, potato, sugarcane, rice and aquatic weeds from other parts of the world. Predatory snails and hermit crabs have been found to be useful in controlling the Giant African snail, which had become a menace in Andaman islands and in parts of Kerala and Assam. Attempts are in progress in some countries in Africa to control serious water weeds like water hyacinth and Salvinia through introduced parasites. If such parasites prove to be highly host-specific, we may also be able to initiate similar studies in our country, since these aquatic weeds are causing serious problems in tanks and waterways particularly in eastern India and Kerala.

Economy Through Recycling

The third concept is economy and maximization of farm income, which is being used with great profit in diverse areas like increasing the efficiency of nutrient supply, water use, energy release through recycling and effective use of all wastes and by-products, improved harvesting, storage, processing and utilization. Its major aim is to reduce waste in all possible ways and decrease to the extent possible the dependence on non-renewable resources for increasing production. An integrated nutrient supply system involves the development in each Block of an appropriate schedule of manuring with organic and inorganic manures. There is great scope for raising leguminous shrubs on bunds of irrigated fields or all along irrigation canals and rivers for providing green leaf manure. It would be wise to introduce pulses and fodder legumes in all rotations. Success in conserving cow dung and other wastes for use as manure will depend upon the availability of fuel.

Devices like gobar gas plants are yet to become very popular. Meanwhile, rapid technological progress is taking place in which all wastes which contain cellulose can be converted into pipeline gas or liquid fuels. The conversion of grains through fermentation into usable ethyl alcohol, the growing of trees specifically for generating thermal power, the development of new high-yielding crops for industrial energy and recycling animal waste by anaerobic fermentation are some of the possibilities under economic assessment in several countries. Unlike the situation in rich nations, non-commercial sources of energy comprising dung, firewood and crop wastes constitute the major sources of energy supply in rural India and we can profit greatly from recent developments on the efficient use of wastes to supply power.

Fertilizer use efficiency is very low at present and the common assumption of return from the applied nutrients is 10 kg of grain per kg of NPK nutrients. Scope for improving fertilizer use efficiency exists through procedures like adjusting fertilizer doses to soil test values, better water management, weed control, split application, placing the fertilizer a little below the soil surface, and use of nitrification inhibitors where leaching losses are likely to be high. Seed-cum-fertilizer drills help to achieve good seed germination and fertilizer distribution. When yields increase, deficiency of phosphorus and of micronutrients like zinc becomes important, as was seen in the Punjab after the introduction of high-yielding varieties of wheat. A continuous monitoring of the status of soil fertility is important.

There is apprehension that phosphorus reserves may get greatly depleted by the end of this century. Hence, every attempt should be made to use wastes like basic slag from steel mills, sources like rock phosphate and Laccadive sands and also promote the solubilization of insoluble phosphates through microbial secretion of organic acids which dissolve phosphates. Rock phosphates and basic slag have been found to be suitable for application in acid soils. Acid soils fortunately occur in the vicinity of steel factories and hence transportation costs can be minimized.

Reducing Post-Harvest Losses

Techniques are also becoming available for minimizing losses during harvest and post-harvest operations. A major problem in the safe storage of grains is the high moisture content often found, particularly in paddy. Studies carried out at the Paddy Processing Center, Tiruvarur, on the effects of salt sprays both before and after harvests have shown that this technique could help to bring down the moisture content to about twelve percent rapidly. We need much more work on the standardization of effective methods of preventing pre- and post-harvest losses in grains, fruits and vegetables under different ecological conditions. Bad storage not only results in a quantitative loss of all food material, but more importantly, there could be considerable deterioration in nutritive quality. The problem is acute in crops like groundnut which develop aflatoxins following fungal infection.

The situation in processing and utilization is dynamic. In some of the rich nations, where there is a move towards reducing the total per capita meat intake, so as to reduce the pressure on plant proteins which are needed to feed animals, food technologists are now developing textured vegetable proteins resembling meat products. Plant proteins which were not formerly of value in human nutrition are also now becoming usable due to the removal by processing of the toxic or anti-nutritional factor. For example, a milling technique called the Liquid Cyclone Process can help to remove gossypol from cotton seeds. Methods for removing the toxic glucosinolates present in rapeseeds and mustard seeds have been standardized in Sweden. Such detoxified rapeseed protein has still some anti-nutritional factors, which are currently being studied. A low cost sixty-five percent protein concentrate can now be prepared from fresh coconuts. The process gives at the same time a high grade oil. The Central Food Technological Research Laboratory at Mysore has standardized several useful milling and processing techniques and the improved methods of pulse milling particularly deserve popularization since they can save over ten percent of milling losses.

Changes in processing and utilization can completely change the market value of a commodity. For example, until some years ago research for the improvement of robusta coffee had been given less importance in our country on the assumption that the world demand would be only for arabica coffee, which has a superior flavor. The advent of instant or soluble coffee has, however, changed the position since robusta has good solubility. Thus, while the annual average prices secured in 1961 in export auctions per 50 kg of similar grades of arabica and robusta were respectively Rs. 144.50 and 70.50, these prices were Rs. 223 and 220.50 during 1970-71, thus, practically wiping out the difference. Similarly, lac was regarded until recently as a dying industry. However, the price of a ton of lac in the UK market in October 1973 was £1,450, while two years back the price was only £340. Obviously, some important new use for lac has been found. Demand for tapioca has also greatly grown and it is expected that within a decade countries of the European Economic Community would be importing at least 4.4 million tons of tapioca. These few examples would be sufficient to indicate how in order both to maintain and develop further markets for agricultural produce, we have to keep ourselves very well informed about trends in processing and utilization. Quality control and improved packaging and forwarding techniques are equally important. All this will call for an Agricultural Technology and Market Intelligence Unit which can keep our research workers constantly informed of the relative priorities they should assign to problems having a bearing on export trade.

Watching technological trends would also be important to stabilize the home market for labor-intensive cash crops in the dry farming regions, such as cotton. Globally, it is becoming clear that cotton may face serious threat from synthetics, unless cotton strains can be developed which can respond well to chemical finishing treatments capable of conferring on cotton "easy care" properties like drip-drying and crease retention. The preliminary work done so far indicates that there is scope for screening varieties for these attributes.

An exciting era. We are, thus, in an exciting era in agricultural and food science. Thanks to the growing involvement of more and more physicists, chemists, climatologists, mathematicians and engineers in studying and solving biological problems, we can hope for continued progress in unravelling new approaches to improving productivity. Remote sensing, satellite photography and satellite television have all great applications in soil and ground water survey, crop censussing, disease forecasting, prediction of floods and cyclones and mass communication. The National Commission on Agriculture in its report on Agricultural Research and Education has stressed that in our legitimate concern for results of immediate applied value, we should not dry up sources of basic research. Research in production physiology, molecular biology, cellular biochemistry, design engineering, construction of production models on the basis of a systems approach, all deserve to be supported and intensified. If some of the experiments designed to confer the ability for biological nitrogen fixation on cereal crops, thus making them behave like legumes, meet with success during the next decade or so, this will be one of the greatest boons that scientists can confer on the poor cultivators. There is evidence that the current highest yields obtained in wheat and rice represent less than half of what is theoretically possible. Scientists have hence much to do.

While the onward march of science relevant to human welfare must go on, it is our duty at every point of time to ensure that the people, for whose welfare the scientific work has been undertaken at public expense, benefit from the knowledge gathered and material developed.

Lecture 3
Social and Educational Requirements of New Technology

In the earlier sections, we reviewed several of the current trends in the improvement of the productivity of hungry soils in a way that the short- and long-term goals of development are harmonized. It was pointed out that the emerging technology is based upon an understanding of the principles of ecology.

A clear distinction needs to be drawn between the different meanings which this term can have for people in different situations. In the developed world, the growing importance of the science of ecology is largely the result of the efforts of the conservation movement. It is essentially a movement to use ecology to maintain, for the already rich people of those countries, the very high standards of living which they currently enjoy. It is a protective movement, aiming to prevent further depletion of resources already severely threatened or eroded and a further degradation of the environment.

However, for the developing world, ecology has to be understood in a very different sense. Most people do not have even the bare necessities of life and, therefore, hardly any standard of living to conserve or to protect. Rapid economic growth is a prime necessity. Human wastes and civic effluents form a far greater source of pollution in the poor nations than scientific and industrial pollutants. Mrs. Indira Gandhi made this point vivid at the United Nations Conference of Human Environment held in Stockholm in 1972 when she described poverty as the greatest pollutant. In our situation, knowledge of ecology should be regarded as an instrument of balanced and rapid economic growth.

The publication, *The Careless Technology* (Farvar and Milton, 1972), points out, with the help of fifty selected case studies, that a high proportion of the development efforts in the developing countries has been destructive, especially of precious natural resources and delicate but ill-understood natural balances. The book demonstrates that time and again not enough forethought was given to the ecological consequences of developmental projects. However, the authors of the book also admit that most of these projects were designed by experts from the developed countries, often with limited knowledge of the client countries, and based on the different and irrelevant experiences of their own countries. If such programs were accepted by the poor nations, it was for other reasons which need not be discussed here. What is needed, however, is a close and detailed understanding of local eco-systems, and its application to local situations by those who are fully informed of the parameters of the problem. If development is not to be destructive, then the developing countries must take responsibility for understanding their own ecological needs, resources and patterns and planning their development accordingly. Ecology has to be a positive force in the poor countries, one which supports the economic growth we so urgently need, and not just a conserving force. This can be achieved when the people of each area participate through Block and District level Land Use and Crop Planning Forums, in the formulation of detailed land and water use plans for their areas. Land use planning at the Block level, assisted by guidelines on market needs and opportunities by a suitable authority at the State level and on inter-State crop adjustments and export opportunities by a Central agency, would alone result in a dynamic agriculture, where the profits of progress accrue to all sections of the community.

Village Level Cooperation

For an efficient adoption of the technology at the field level it is necessary to develop appropriate drought, good weather and flood avoidance codes for each agro-ecological area. Such codes would be of immense benefit to the developmental administrators as well as to village panchayats, zilla parishads and other agencies which have to get projects translated into field accomplishments. It is obvious that tasks like recycling of all wastes, land use planning based on an appropriate admixture of agriculture, animal husbandry, fishery and forestry, require considerable understanding and cooperation at the village level. The only approach to rapid economic development which can succeed is to optimize the advantages of our large population, vast numbers of whom are now neither participants nor beneficiaries of development activity.

For example, let us take the case of the control of Pyrilla, a serious pest of sugarcane. Pyrilla is not only affecting sugarcane but is now becoming serious in other crops like wheat and jowar. The most economical and effective method of controlling this pest would be field sanitation involving the burning up of all trash after the harvest of sugarcane and the collection of eggs of the pest from the lower surface of the leaf during April-May. These tasks can be easily accomplished by even village schoolboys and girls. If such early steps are taken, the small population of Pyrilla that may still remain is likely to be attacked by natural enemies during the rainy season and the problem may come under control easily. If, on the other hand, simple steps like crop sanitation, removal of eggs from the under-surface of the leaves at the appropriate time and subsequent steps like attracting the nymphs and adults of this pest by light traps, are not followed, then the pest assumes a serious dimension towards September, as it did in some areas of north India during 1973. Since the sugarcane plantation would be full grown at that time it would be difficult to get into the field and spray with chemicals. At that stage, the method of aerial spraying or use of power sprayers becomes the only choice left. Though inevitable, when the pest has assumed menacing proportions, it is obvious that making a habit of such a method in our country would be tantamount to bypassing our major strength, viz., our large population, and to increasing the already high cost of cultivation.

The emerging agricultural technology is of two major kinds with regard to the ease of adoption. In one kind, as for example, the new technology of wheat cultivation, the economic benefits derived by a farmer by adopting the technology are not influenced by what his neighboring farmer does or does not do. In other words, the technology is capable of successful individual adoption, in economic terms. In the other kind, the economic benefits conferred by the technology on the farmer will be proportional to the extent of cooperative action generated on the part of an entire village or watershed community. Rice and cotton cultivation and prevention of epidemics in cattle are good examples. Even in Punjab, with one of the finest farming communities in the world, the average yield of cotton is low, being only 368 kg/ha. In contrast, in the Arab Republic of Egypt where cotton cultivation is managed cooperatively without any infringement of individual ownership, the average yield is 780 kg/ha. Pest control and water management in rice are best done cooperatively in a village. In fact, if this can be accomplished, even some of the fertilizer lost through leaching in drainage water can be recycled, by collecting such water in a pond at the lowest point and redistributing it. The need for achieving a doubling in fertilizer response from the present low level of about 10 kg of grain per kg of NPK nutrients was already stressed. It was also pointed out that the emerging concept of pest avoidance is pest management and not just chemical control. All these aspects of the new technology would need understanding and cooperation among neighboring small farmers. If this is not achieved, the risk element in crop production will be high. A low-cost and low-effort agriculture will be the poor cultivator's response to a high-risk farming situation, unless he has protection by means such as crop insurance. This is an important reason for the relative stagnation of kharif crop production in our country, where in spite of moisture availability, the farmer hesitates to invest on inputs in areas which are prone to pests and diseases, water-logging and operational problems in land preparation.

Cooperative effort is not only needed in the delta areas during the kharif season but even more importantly, in the dry farming regions. While an individual farmer can increase the storage of moisture in the soil profile of his own field through tillage, mulching and other measures, the possibilities for collecting all the run-off water and using it for a crop life-saving irrigation later, can be realized only if there is group action in accordance with the topo-sequence of the farms.

Changes in cultural practices. Several basic changes in cultural practices are also needed to get the best out of the new crop varieties. For example, in dwarf wheats, sowing has to be shallow, First irrigation has to be given at the time of initiation of crown roots and population density should be higher. In rice, traditional practices such as low seed rate under direct seeding, wider spacing, deep planting and bunch planting under transplanted conditions and drying the field to promote the establishment of seedlings are all intended to retard luxuriant vegetative growth in the earlier tall varieties, which have proneness for both excessive vegetative growth and lodging or falling down. In contrast, the new dwarf rice varieties need almost the opposite kind of treatment, if they are to reveal their full genetic potential for the yield.

It is obvious that to accomplish all these tasks successfully we need to develop educational and organizational procedure which would, on the one hand, make the literate better equipped to discharge the functions of gram sevak, gram sevika, credit agent, teacher and other agents of change, and on the other, make the illiterate highly skilled and efficient partners in adopting the latest scientific advances. It is not proposed to reiterate other obvious requirements like the timeliness of credit and input supply and appropriate pricing and marketing arrangements. I would, however, like to stress that in addition to the procedures currently in operation for product pricing, we should also pay serious attention to evolving long-term pricing for factors of production, such as fertilizer, pumps, pesticides and others. Product pricing alone without attention to the pricing of means of production will tend to result in controversies and problems.

It is not only in agriculture and dairying that new techniques need to be popularized but also in inland and marine fisheries. We have noted the possibilities of making full use of the three-dimensional space, and the availability of different kinds of fish food in a pond, through stocking different kinds of fish which can derive their food from the bottom, upper and middle levels. Similarly, there are opportunities for adopting new techniques in areas like the production of quality fish seed through induced breeding, the cultivation of air-breathing fish in the swampy areas and the culture of frogs for export. In marine fisheries also, culture techniques can be introduced to supplement the income from capture fisheries. Mussel culture, growing milk fish, mullets and certain species of prawns in coastal lagoons containing salt water and pearl culture can all improve the economy of the fishermen in our coastal areas.

To sustain high levels of biological productivity, we also need to generate a general awareness in the entire community of the crucial significance of conserving soil, water, flora and fauna. Like game and wild life sanctuaries, we need to develop "Crop Gene Sanctuaries" in those parts of our country where valuable genes are found in the primitive cultivars and native flora. Such a step can help to prevent the erosion of valuable genes. In contrast to soil erosion, gene erosion is not visible to the naked eye and hence does not attract attention.

Changes in educational outlook. Having referred to the extension needs of the new technology, let us examine the social and educational infrastructure which is a prerequisite for the successful adoption of the technology. I shall deal first with the educational aspect.

What kind of education do we need that would be relevant to our society dependent upon agriculture as its major activity? And what practical steps can we take to change our educational system? The changes required are of two kinds: at one end, we need to bring about a change in outlook, creating an awareness of biological surroundings, and a consciousness of the possibilities of synergy. This can be done only when there is an inundation, so to speak, at every level of education, of materials that will create such a consciousness. At the other end, we need to give technical skills to illiterate and semi-literate adults, which will enable them to understand and use efficiently the new technological package. Between these two, there is a wide range of activities which can be applied at every educational level.

During the early years of childhood, before the child begins to receive formal education, and in primary school, the major emphasis should be on living in nature and with nature, learning about nature through direct observation, and using the materials provided in nature to develop scientific skills, aptitudes and habits of thought. How often are children in balwadis and primary schools forced to remain within the limiting walls of a barren and bare classroom, because of the false impression that education is something that must take place within a building? How few children in balwadis are taken for walks and helped to learn about nature? On the other hand, there are commendable efforts in some places to use nature itself as a medium of education, as in the Meadow School of the late Tarabai Modak, which carried education to shepherd boys right in the meadows where they were grazing cattle. Rabindranath Tagore was an ardent advocate of this approach. There have been other similar ventures, but all have remained small, local and isolated. At a different level, the Avinashilingam College of Home Science in Coimbatore, for instance, has compiled a set of songs for preschool children incorporating knowledge about nutrition. The ICAR and the Extension Directorate of the Union Ministry of Agriculture have worked on a publication which will be a collection of games for use in classrooms and in recreation introducing information and concepts about agriculture. They can be played both by adults and by children. The details of the games will, however, have to be developed in each Block using data applicable to that region. We need many more such educational games and kits that will make use of the play way idea of teaching science in general, and agriculture in particular.

The traditional media of entertainment as well as the mass media of radio, cinema and television can also be harnessed to the cause. Even now many of the new crop varieties are known in villages at some places as "Radio varieties", since information on the new strains reached the illiterate cultivators through the radio. It is a happy sign that newspapers are increasingly assigning more space to items relating to agriculture and rural development. While the English newspapers play an important role in conveying messages to the policy maker, the language press, besides the radio and cinema, forms the most significant medium of conveying information to villages. Several of our leading newspapers have now whole-time agricultural correspondents and experts specializing in agricultural themes. Developmental journalism took to a new path when an integrated village transformation project was initiated in Chattera village of Haryana State by *The Hindustan Times*. Apart from its contribution to human happiness and welfare, this project is becoming one of the best documented accounts of rural change.

Role of Banks and Companies

Both banks and public and private sector companies can play a great role in combining their regular work with the art of communication. Banks can finance entire production programs, as for example, the production of a lakh of tons of sunflower oil or the production of a certain quantity of soybean in a suitable area, and thereby assure the credit and other infrastructure for the entire production chain and not merely for components of the chain. Similarly, pesticides firms can sponsor programs designed to ensure pest-proofing of important crops in the poorest villages in their area of operation. All such projects will provide excellent educational opportunities for the students of the area, besides serving as demonstrations of new techniques.

Work experience at the upper primary and secondary level too can be effective if carried out in a different manner. This idea of introducing work experience in schools seems rather artificial and unnecessary in our context. On an optimistic estimate, about seventy percent of the children aged six to eleven are enrolled in primary schools. If we consider children up to the age of fourteen, and take account of the wastage and stagnation in education, then the number of children between six to fourteen who are not in school may be as large or larger than the number of children in schools. Most of these children are already working, not indeed to gain work experience, though they are gaining it no doubt, but out of sheer economic necessity. In these circumstances, bringing work experience into schools seems redundant. Would it not be more meaningful to carry educative experience out to the work, where the children already are anyway? This means much more than open-air classroom or discussing arithmetic in terms of bags of manure or hectares sown or other agricultural operations. Of course, such activities are necessary and important. But, it would mean much more. It would imply a profound change in the school program, introducing such activities as allowing children to work in their own or others' farms for part or most of the day, planning for learning-oriented and not teaching-oriented education, and getting new kinds of teachers to deal with the new tasks.

Who is to teach work experience as conceived in this way? The traditional school teacher, who is mostly drawn from the white-collar class and has little practical experience of agriculture, may not always be the most suited to this task. In fact, it has already been observed in areas where the new technology has spread, that the technical knowledge of rural parents, who may be progressive farmers, is far ahead of the out-dated agriculture taught in the schools, even by qualified teachers. Why not allow such parents, or other local farmers, to teach part-time? Often even the children of farm families know much more about the latest developments in agriculture than the non-agriculturist school teacher. Even at the post-graduate level of education, the Indian Agricultural Research Institute, New Delhi, which started a few years ago the practice of inviting farmers with practical wisdom for brief periods as Visiting Professors, found the experience very rewarding to staff and students alike. Why not bring in more and more part-time teachers, people who have skills and knowledge in agriculture and related crafts or at least send students out to them? Some experiments along these lines are being carried out. For instance, in Bombay, at the city level, the Bombay Municipal Corporation has arranged for students from a number of schools to be placed for regular work in industrial establishments for two days a week. This gives the students real work experience, preliminary training for their future careers, and helps them to find placements after school. Here the same idea has been applied in a different sphere, the industrial one. In some districts of Maharashtra, secondary school students are being placed in villages in batches, residing with farm families, for several weeks at a time, under the supervision of their own teacher. The objective here is to enable non-farm students to learn more about agriculture and farm operations through direct participation and observation. The students live with the farm families and join in their daily work and routine. The difficulty has been that the teachers are often ill-equipped to guide the students with the information that they need. Such schemes are steps in the right direction. They need to be expanded and adapted to a wider variety of experiences and students.

Work Projects for Students

Another useful adaptation of the idea of work experience is to give academic recognition, through marks, to real economic work projects, self-chosen by students and carried out by them in their spare time, either alone, or with the help of their families. A boy or girl, who can run a simple project such as a poultry unit or a kitchen garden, would have learnt much by way of independence, responsibility and initiative, as well by way of arithmetic, science and economics. It is needless to add that the conventional literacy skills of reading, writing, and arithmetic will also be fully utilized in such projects. Of course, it means more time and effort on the part of teachers and administrators to identify and initiate such projects. Further, it requires the technical assistance of officials of agriculture departments and extension agencies. It is naturally much easier to provide artificial work in the classroom and call it work experience than to help children to engage in real work and to assess it as it deserves. Children become responsible when they are obliged to carry responsibility. The majority of children in our country are already working and carrying responsibility. It is up to us to devise an education that will encourage and support them in their efforts by providing a place for such tasks. Maslow, the humanistic psychologist, speaking on the development of responsibility stated (1972): "It looks as if one way to breed grown-up people is to give them responsibility, to assume that they can take it and to let them sweat and struggle with it. Let them work it out themselves, rather than over-protecting them, indulging them or doing things for them." It is time that we use this as a basic principle in education.

Yet another useful form of work experience would be the school project. Apart from the practical and economic value of the school farm, garden or dairy, it would teach another vitally necessary attitude which I have already spoken about. This is the necessity for cooperative endeavor. Even if each student had his own plot, or animal, or pond, he would, in the course of work, learn without being asked to combine with his fellow students for certain operations, the need for certain activities to be undertaken by all at the same time, the economics of sharing and so forth. In this way, cooperative endeavor would become a necessity for the individual, for the success of each individual's project would be inevitably linked to the performance of these joint operations at the right time and in the right way.

Involving university students. At the university stage, fewer people are involved, but they need to be involved differently. In 1969-70, about three million students were enrolled in higher education, about 2.5 million men and half a million women. Of this, only 1.6 percent of the men and 0.4 percent of the women were engaged in any kind of agricultural studies. Only 176 women were enrolled in agricultural universities. In a country with eighty percent of the population engaged in agricultural pursuits and living in the rural areas, it is sad that such a small fraction of men and women are engaged in studies which can contribute to rural development or agricultural productivity. It is specially sad that though women contribute equally to the labor force in agriculture, performing many of the key operations and most of the more tiresome ones, agriculture as a profession is not regarded as a suitable one for women at the higher stages of education. This reflects the gap between the educated classes, who are mostly drawn from the urban areas, and nowadays also from the rural elite, and the rest of the population. Apart from providing more opportunities for higher education and technical studies in agriculture, we have also to think of other practical ways in which the great majority of university students can be helped to become involved in the work of rural development and to acquire some understanding and sympathy for the problems of the greater part of this country's population. An example of how non-agricultural colleges can participate is the good farm run by the staff and students of the Madras Christian College, in what was until 1965 a scrub jungle.

Making work-experience an integral part of university education would help to generate a greater sense of self-confidence and self-reliance on the part of the student. One approach to achieve this aim may lie in making an in-built provision in each one of our developmental projects for students' participation.

Student work should not be regarded merely either as social service or training but must become a distinctive and advantageous part of the project. If this approach is accepted by the project authorities, students can become a source of great strength and dynamism to the project, if they are properly trained and deployed.

Every university student must be employed in an appropriate Plan Project for a period of two months every year, while he or she is in the university. A semester system of course curriculum organization would help in the implementation of this idea, although this is not absolutely essential. The summer vacation can be restricted to one month so that the formal teaching programs do not suffer. The assignment of students to various projects will need proper planning and adequate consultations between the universities and project authorities. Needless to say, the assignment of students will be based on the principle of learning through work and would hence involve a planned matching of the field of study with the field of work. Thus, students of history and archaeology may work for India Tourism Development Corporation, of zoology and medicine in the family planning and preventive medicine programs, of agriculture, botany, chemistry, physics, economics, engineering and home economics in agricultural development and child nutrition programs and of nearly all fields in pre-primary, primary and secondary education programs.

All scientific institutions in the country run by the Central and State Governments can provide opportunities for students to work in specific projects and can at the same time give a great fillip to the cottage industry movement by supplying detailed manufacturing drawings of new implements, machinery and processes. With these drawings, university students in engineering may be in a position to help village communities in starting small-scale industries, since rural credit is becoming more easily available now.

Technical Literacy for Farmers

The majority of our peasants are illiterate. Unfortunately, the proportion of illiteracy seems to be higher in areas such as the arid and semi-arid regions, where the scientific transformation of the rural economy requires a much greater understanding of the principles of economic ecology. Therefore, it is obvious that unless we seriously start programs to impart the latest technical skills to illiterate peasantry by the method of learning by doing, we will not be able to accomplish our goals. The Indian Council of Agricultural Research is supporting a number of Krishi Vigyan Kendras to impart technical literacy to practicing farmers, fishermen and others. These Kendras are designed to give major benefits to the poorest sections of the community. The sponsors of Krishi Vigyan Kendras will first have to make a survey of the agricultural potential of an area and then identify those aspects of growth which could help to improve the purchasing power of the poor. For example, in an exercise done in the Union Territory of Pondicherry, it was found that the following areas needed immediate attention, if the poor are to be benefited through Krishi Vigyan Kendras: a) Improving the income and reducing the under-employment of over 10,000 fishermen; b) Improving the average per hectare rice yield to about 2,000 kg as soon as possible from the existing 1,400 kg; c) Introducing a scientific animal husbandry program in order to produce more milk both for marketing in Madras city and for making milk powder.

A similar Kendra in Punjab should deal essentially with topics like tubewell technology, repair of implements and tractors and scientific multiple cropping.

The Krishi Vigyan Kendras are given infrastructure to provide training which will help the trainees to initiate change in their respective occupations. For example, in the case of the Pondicherry fishermen, training is envisaged in the area of coastal aquaculture, which will help them to undertake fish culture in coastal estuaries, in addition to improving the techniques of fish capture. Similarly, the animal husbandry programs impart skills in genetic upgrading, cultivation of nutritious grasses and legumes as well as the use of wastes enriched with urea as feed, health care through sanitation and immunization and better processing and marketing techniques.

Even if we are able to provide the educational infrastructure needed for agricultural reconstruction, we would still have to develop a social organization which can help small farmers to overcome the limitations arising from the small size of farm holdings, lack of capacity to purchase the requisite inputs and inability to get a fair price for the produce. To achieve this aim we will have to devise strategies to generate social synergy.

Social synergy. What is social synergy? Ruth Benedict, the anthropologist, who first applied the concept of synergy in social sciences, stated: "Societies where non-aggression is conspicuous have social orders in which the individual by the same act and at the same time serves his own advantage and that of the group. … Non-aggression occurs in these societies, not because people are unselfish and put social obligations above personal desires, but because social arrangements make these two identical. … Cultures with low social synergy are those in which the social structure provides for acts which are mutually opposed and counteractive, and cultures of high synergy where it provides for acts which are mutually reinforcing. In cultures with high social synergy, institutions ensure mutual advantage from their undertakings, while in societies with low social synergy the advantage of one individual becomes a victory over another, and the majority who are not victorious must shift as they can." (Benedict, 1970:72)

According to Abraham Maslow, "The high synergy society is the one in which virtue pays … High synergy societies all have techniques for working off humiliation, and the low synergy societies uniformly do not."

There is sufficient evidence in nature to prove that symbiosis, or the process of mutual assistance and support, is a necessary ingredient for synergy. However, it seems that the very concept of synergy has been very little used, or even understood, by social scientists in its application to man. Ruth Benedict was one of the few social scientists to apply this concept to societies, and it has been later developed by Abraham Maslow who is, significantly enough, not considered an orthodox psychologist. His definition of social synergy underlines a close correlation between synergy and non-violence, a fact which should be of particular interest to a peace-loving country like India. Further, the high synergy society seems to be another name for what Gandhiji described as the Sarvodaya society. It is also evident that our society as at present organized is a low synergy society. What steps can we take to ensure that we move from this to a better state of affairs?

Individual and social goals. Individual goals have to be made to coincide with social goals. This can be done in many ways. The communist countries have found their own approaches. Russia, China and Israel have all found their own ways of ensuring this merging of individual and social goals. Most of these methods are related to the practices of child-rearing and education. In the Soviet Union, for instance, the system of rewards and punishment in schools is such that the individual is rewarded only when the peer group succeeds. At every stage, the success and happiness of the individual is related to the success of the group in such a way that each individual strives to maximize the success of the group. Some of the approaches described earlier, such as encouraging children and students to take up economic projects in a responsible way, and linking them with academic success, would be good examples of social synergy at work.

This is not however the only means of achieving the goal. In a free economy, it is possible to conceive of a system whereby farmers could be motivated to produce what the nation considered necessary. This would involve a very complex set of arrangements including pricing, availability of inputs, technical knowledge, land ownership and a variety of other things. However, it is clear that slogans and appeals to the social conscience cannot work, except for short periods during national crises. "Grow More Food" or "Self-reliance" are not adequate as methods to persuade farmers even to grow anything, leave alone to do so in an ecologically balanced manner. It requires a far more widespread and dynamic movement of society to bring about such changes. In other words, it requires a release of synergy at the level of the farmer.

The farmer of India, who must be served by the new technology in such a way as to release social synergy, has three qualifications. He is essentially a small farmer, poor and illiterate.

A technology for the poor farmer. If slogans such as the transformation of rural life, or the achievement of growth with social justice are to be translated into reality, then the new technology must be such that it can be adopted by the small, poor and illiterate farmer. Do we have such a technology? Yes, but its adoption particularly during the south-west monsoon season requires a certain degree of cooperative effort and a changed outlook. These in turn imply far-reaching changes in education and social organization as pre-requisites.

We have only a few striking examples of the economic benefit that can accrue to small cultivators by cooperative action. The Bharatiya Agro Industries Foundation at Urli Kanchan near Poona has sponsored a whole set of cooperative activities like lift irrigation, dairying, poultry and joint farming based on the concept of uplifting the rural poor through the latest agricultural and animal husbandry techniques. The work of the Kaira District Cooperative Milk Producers' Union, Anand, is too well-known to need description. The Tudiyalur Cooperative Society has been providing total service to about 8,000 cultivators in Coimbatore district. The Syndicate Agriculture Foundation, Manipal; the Mahatma Gandhi Lift Irrigation Cooperative Society, Gaddipalli; Kerala Development Society, the Musahri Plan initiated at the suggestion of Shri Jayaprakash Narayan; Food and Marketing Center, Jamshedpur; the Agricultural Institute at Kosbad Hill; Shri Shivaji Council for Education and Extension, Godegaon; the Samanway Vidyapeeth, Bodh Gaya; Kishore Bharti, Hoshangabad; and several other organizations in different parts of our country are helping to uplift the economy of the poor. Various missionary organizations have been actively involved in educational and developmental work in the field of agriculture. The Aurobindo Ashram at Pondicherry runs a farm based on recycling principles.

The Archbishop of Trivandrum, for example, has personally tested and popularized new varieties of tapioca, hybrid napier grass and other strains on a large-scale. All those success stories need to be multiplied as fast as possible. This can be done, provided at the village level there is a combination between those who possess technical knowledge and those who are devoted to social service and together identify the most important need of the village in getting its economy moving. If the need is correctly identified, and then properly attended to, it will act as a catalyst in releasing synergy. An important task of social workers is hence the identification of catalysts which can bring home through tangible economic benefits the value of cooperation to the small and marginal farmers.

We have so far discussed about an agriculture in which the ecological advantages of the different parts of our country can be maximized and the hazards minimized through appropriate land and water use planning, in which high synergy crop and animal production systems are introduced, recycling procedures used to promote a high growth rate in productivity on the basis of renewable resources of energy, farm and factory are closely linked, maximum economy and efficiency effected in the use of all resources, export markets nurtured through continuous attention to the changing quality needs of the clients and a fair and remunerative return is assured to the primary producer. This is an agriculture based on ecological awareness, in which synergy, harmony and economy are the basic principles, and of which recycling is a tool. These are in essence the principles on which Mahatma Gandhi developed most of his concepts of rural reconstruction, though he used a different language and a different set of examples which were valid at that time. We should not make the mistake of rejecting the principles because some of the examples used by Gandhiji half a century ago may be out of date in the light of modem science. On the contrary, science and technology have in their search for a productive agriculture compatible with human growth and human welfare, uncovered the very same principles which Gandhiji arrived at by a different route. If we can take the steps necessary to evolve the kind of agricultural system based on the most advanced principles of biological science, we can probably claim to have developed a Gandhian Agriculture, because this would be an agriculture where Gandhian concepts become manifested in the form of an advanced rural economy, benefiting all sections of the community.

To achieve this, the tool has to be education, and the chosen method, cooperation. Social synergy has to be released by using appropriate educational and organizational methods as catalysts. In addition to official extension agencies, we have numerous workers belonging to the Gandhian and Sarvodaya movements, missionary groups, voluntary service societies and others, who are dedicated servants of the rural poor, struggling with zeal and devotion to transform the face of rural India. What they lack are the tools and the understanding of a modern agricultural technology which can help them to achieve their aims.

Transferring technology to the masses. The technology is now becoming available but the mechanism for transferring it to the illiterate and small users in an effective manner does not exist. Ironically, there is a global communication network which makes the latest findings of science available almost immediately to research workers in any corner of the world; but what is urgently needed is such a communication network at the service of the poor farmer in our country. It is not only knowledge that is needed, but an approach which will be able to supply the right knowledge and tools to the right people at the right time and place. The worker on the spot, whether he is an agricultural officer, a teacher, a gram sevak or a social worker, must be able to identify the local problem or need which will act as a catalyst for promoting cooperative endeavor in that area – it maybe lift irrigation in one place, dairying in another, pest control in a third, supply of farm machinery or fish fingerlings in a fourth and marketing in a fifth. In this network, the mass communication media, particularly the radio, television, cinema and the language press, have a very important role to play; so have the research and the extension worker, the education system and the drive for technical literacy.

In 1945, when the buffalo owners of Kaira District went to Sardar Patel to complain about their marketing difficulties due to exploitation by milk merchants, he replied that the obvious solution to this problem was to organize the dairy industry on a cooperative basis. He further advised the farmers to set up their own dairy processing unit. This was the seed for the growth of the Kaira District Cooperative Milk Producers' Union, which has brought prosperity to the formerly helpless owners of one or two buffaloes and has become a model for all parts of the country to emulate.

Our political freedom depends on rapid economic growth and this in turn depends on our performance in agriculture. The future of our agriculture in its turn depends on the success with which the educated can help the small and illiterate farmers to take the many small steps which alone can lead to a great agriculture. Science can only show the way; it is for us, the educated and the privileged class, to provide the will. ▯

References

Benedict, Ruth. 1970. Synergy: patterns of the good culture. *American Anthropologist*, p. 72.

Farvar, M. Taghi and J.P. Milton. 1972. *The Careless Technology*. New York: National History Press.

Lowdermilk, W.C. 1966. The eleventh commandment. *Selected Readings to Accompany Getting Agriculture Moving,* Chapter 5. ed. by Raymond E. Borton, New York: Agricultural Development Council.

Maslow, A.H. 1972. *The Farther Reaches of Human Personality.* London: Pelican.

McNamara, R.S. 1973. Address to the Board of Governors, World Bank Group.

Meadows, D.H., D.L. Meadows, J.Randers and W.B. Behrens. 1972. *Limits to Growth.* New York: Universe Books.

10

Science and Integrated Rural Development

This is the first session of the Indian Science Congress when, in addition to discipline-centered papers and symposia, there will be a cross-disciplinary examination of a topic of national relevance. The aim is to utilize the vast inter-disciplinary expertise present at the Congress for a deeper understanding of a specific national problem. "Science and Integrated Rural Development" was chosen at our last session held at Delhi, as the focal theme for this session. Scientists, both from our country and abroad, and our Sectional Presidents have taken much trouble to articulate their thoughts on what scientists can and should do to promote rural development and agrarian prosperity. For the first time, we will have a Forum on Home Science and Nutrition and will give special attention to Fisheries. The major suggestions arising from our discussions will be summarized on the last day when the Deputy Chairman of the Planning Commission, who is also the Chairman of the National Committee on Science and Technology, has kindly agreed to receive them personally.

The various initiatives taken by our Central and State Governments since 1947 have now placed our country in the third position in the world in scientific and technical manpower. Yet we were included at the World Food Congress held in Rome in 1974 among the "most seriously affected" (MSA) countries with regard to food, which is the first requisite in the hierarchical needs of man. Why do we find this mismatch between our position in the world of science and the quality of the lives of a majority of our people?

A Lesson from the Gene

It seems to me that a basic deficiency of our developmental system is our inability to emulate our genes, in each one doing his or her specific task properly and well and everyone working together in a coordinated manner. Let me elaborate this further. I wonder how many of us realize that the total weight of the chemical substance of heredity, de-oxyribose nucleic acid (DNA), for our entire population is only about 4.2 mgs, if we add up the weights of DNA present in the single cell embryo of each individual from which we grow. By the same method of calculation the present population of Homo sapiens would have a DNA content of about 28 mgs. In these 28 mgs of DNA, the specifications for the heritable characteristic of our entire species are inscribed. How has the genetic system combined economy with such a remarkable efficiency and precision?

Genes have three essential properties. Each gene has a specific function: each is capable of replication or making exact copies of itself; and each is capable of mutation of heritable change. The function of a gene is regulated both in terms of

---

Presidential Address at the Sixty-third Session of the Indian Science Congress, Waltair, India, January 3-7, 1976.

its particular location on a chromosome and also in terms of a particular time sequence in development. During the reproductive phase, genes are able to combine in different ways and this process of recombination followed by segregation generates wide variability. Thus, in practice, except for identical twins, no two individuals are exactly similar. Mutation, or heritable change, adds to the capacity to respond to changing selection pressures.

Another remarkable property of genes is the system of vertical and horizontal coordination to which they subject themselves. Thus, each gene performs its specific function and is able to replicate itself, to change and to work cooperatively and sometimes competitively with other genes. Integration of vertical and horizontal coordination has invested the heredity system with precision and power. If only everyone of us was aware of the way in which our own bodies function and applied the lessons learnt from it in our daily life, we might be more successful in achieving what individually and collectively we wish to accomplish.

Human Resource Development

What is it that we want to accomplish? In my view, the most important task is to draw the greatest benefit from our existing human resource and to limit its further unplanned growth. Past experience suggests that human resource cannot be effectively used by traditional approaches to "creating jobs", which often tend to degenerate into doles. What is needed is the growth of employment policies from an overall strategy of resource utilization designed to convert sunlight, soil, water, mineral, plant, animal and other resources into wealth meaningful to people. Instead of devoting undue attention to brains drained from their countries, we should pay serious attention to the utilization of brains within the country. The starting point is obviously the village, which is where both untapped assets and native brains exist. This at why the Science Congress has chosen "Science and Integrated Rural Development" as its first focal theme.

The Rural Scene

Eighty percent of our population now live in rural areas. Some experts have calculated that this percentage will be seventy-one in 2000 AD. Even if the percentage goes down, the rural population is expected to grow in number from 441 million in 1971 to 662 million by 2000 AD. Of the existing rural population, nearly fifty percent is believed to suffer from poverty. Rapid rural development based on the scientific utilization of all our resources, both natural, and human, is therefore a must.

It is worth noting that countries which have gone too far on the road of urbanization are now repenting this choice. Rapid urbanization has generated a steep rise in the consumption of non-renewable forms of energy. For example, the transport needs of large cities have grown so vast that the number of motor vehicles in the world increased by 120 million during the 1960s, generating problems of atmospheric pollution and human temper. There is today a marked movement back to the countryside, but with the integration of some of the basic benefits of urban life with rural living. A planned reclustering of jobs, services and amenities more widely throughout the country in accordance with the agro-ecological potential and socio-economic needs of each area will facilitate this movement.

Rural-Urban Relationship

Gandhiji stressed that only a marriage between intellect and labor could lead to rural regeneration. Education, in the past, particularly at the university level, unfortunately tended to promote the concept that rural jobs, which were mostly related to agriculture, required only brawn and not brain. The exodus of the educated caused neglect of the hinterland. Yet we now know that unless life in a rural community is made tolerable for all, the problem of poverty cannot only not be solved but will get worse. Therefore, a national policy of scientific rural-urban development as an integrated package is essential.

So far we have generally tended to pay only lip service to the cause of rural development. People living in cities talk about the freshness and beauty of nature but would not like to go and enjoy them even after retirement. Similarly, public policies have tended to promote an interest in urban living. For example, today if a scientist or a public servant moves from a Class "A" city to a smaller place, he loses in his total emoluments. It is only in the case of the Defense Services that the principle of serving all parts of the country has been developed with appropriate provisions for non-family postings. Recently, the Indian Council of Agricultural Research has introduced the concept of compulsory service for a specific period in a tribal or neglected area by all members of the Agricultural Research Service. But compulsion alone cannot yield the desired results. What is important is to incorporate, in personnel policies, appropriate provisions for the education of children and for medical facilities to enable qualified scientists and technologists to work in areas where these facilities do not yet exist.

Rural Assets and Liabilities

An important task for this year's Science Congress is to review our rural assets and liabilities and the present state of the art of harnessing science to improve rural economy and living. On the basis of such a review, certain broad guidelines can be developed for using the tools of science to enhance rural living.

Though our assets are well known, it is worth repeating them because they are so impressive. First, we have the second largest human population in the world, most of whom are young. Although a considerable proportion of the adult population is classified as illiterate in the formal sense, they have shown a great capacity to absorb, adapt and benefit from modern technology. There are abundant recent examples to justify this statement. The doubling of wheat production within a span of five years, the progress made in improving rice and wheat production in areas where they were unimportant before, the spread of maize, formerly regarded as a kharif crop, in the rabi season along the Indo-Gangetic Plains, the spurt in long staple cotton production, the availability of apples everywhere in the country, the progress in the production of potato, tapioca and other tubers, the development of low-cost ground water exploitation technology like bamboo tubewells, the spread of gobar gas plants and the growing diversification of export products from the rural sector are all indices of rural capability. Striking changes are visible in several parts of our country in farming systems, and animal husbandry is beginning to get more efficient. Our farmers have thus shown their readiness to adopt new technology, provided it is economically viable and low-risk in character and if appropriate packages of services and public policies help to ensure a reasonable return for their labor and investment.

Our animal wealth is also vast. We have over sixteen percent of the world's cattle population, forty-five percent of the buffalo population, over sixty-nine million goats representing the largest population in the world, over forty-three million sheep occupying the sixth position in the world, and rich populations of poultry and inland and coastal fisheries. We are also endowed with excellent wildlife resources. Our soils are by and large robust and productive, although in some areas lack of care has led to considerable erosion and the development of salinity, alkalinity and other problems. Shifting cultivation is still a major source of living for nearly two million people in the north-eastern Himalayan region. This method of handling the soil was developed hundreds of years ago as the only then available answer to the law of diminishing returns from the soil. There is, however, no place for it now.

Our water resources are vast and varied, and the National Commission on Agriculture has calculated that the area under irrigation can almost be doubled during the next twenty-five years from the present forty-two million hectares. In the gross sown area of about 164 million hectares, regions with high (1150 mm and above), medium (750 to 1150 mm) and low (less than 750 mm) rainfall occurs in almost equal proportion. Thus, over forty-five percent of the net sown area in the country has a reasonably assured water supply. Even in the arid zone of Rajasthan there is fortunately a good underground water reserve. The water quality is by and large good, although there are areas where the groundwater is saline. An integrated strategy for the utilization of ground and surface water and for harvesting all rainwater in each ecological and topographic area will help to transform our agriculture and rural economy.

The high priority accorded to irrigation, command area development and groundwater exploitation in the Fifth Plan should help to make our agriculture self-reliant and provide considerable resilience in crop planning. If we consider only light duration (energy), water availability and temperature, which are three of the major factors regulating crop production potential, the period of maximum insolation in the tropics and sub-tropics unfortunately coincides with low availability of water and high temperature. The latter would lead to high evapor-transpiration and consequently greater demand for water. In the absence of irrigation, water becomes the chief limiting factor in crop productivity. Thus, the period of potential maximum yield unfortunately becomes in reality a period of minimum productivity. In contrast, in the temperate zone the period of maximum day length fortunately coincides with periods of precipitation and temperatures conducive to growth (see Table 1). The high and stable yields obtained with irrigation during *boro* and summer seasons in our country with several crops have fortunately focused attention on this problem.

Table 1. Estimated yield potential and percent achieved in four rice-growing countries*

Country	Approximate duration of effective grain-filling period (days)	Average sunlight (cal. cm^{-2} day^{-1})	Estimated yield potential (t ha^{-1})	Average yield** (t ha^{-1})
Spain	35	500	17.9	6.27
Australia	35	600	21.1	6.25
Japan	35	350	12.1	6.02
India (Kharif)	25	300	7.3	1.85
(A.P.) (Rabi)	25	500	13.4	2.20

* Efficiency for solar energy utilization is assumed to be 2.5%.
** FAO Production Year Book 1973.

Our plant resources are vast and we have nearly 20,000 plant species, a greater number than in countries with a much larger geographical area. This is yet another index of the varied agro-climatic conditions prevalent in our country which offer scope for a wide range of plant species to thrive.

Thus, if one draws an agricultural balance sheet purely in quantitative terms, our assets are great. It is an irony, therefore, that we should still find it difficult to provide for the basic minimal needs of our population and that we should face problems of unemployment and under-employment both in rural and in urban areas. Since I consider the population-food supply equation and the population-employment equation equally important and interdependent, I would like to dwell briefly on certain scientific aspects of these two equations.

Food Requirements

According to the estimates of the National Commission on Agriculture, our population by 1981 may be about 668 million. The total food needs for the present population and estimated population of 945 million by 2000 AD are about 122 and 220 million tons respectively. The major nutrition problem of our country is inadequacy of calories in the diet of the economically handicapped. Under-nutrition, in turn, has been attributed in many instances not so much to lack of food in the market as to lack of purchasing power in the hands of the urban and rural poor. Therefore, the food problem in many areas needs to be stated not just in terms of a certain quantity of foodgrains alone but also in terms of certain person-years of jobs which would provide the wherewithal to buy food. If we separate the problem of increasing food production from the ability of the market to absorb it at remunerative prices, we will find that there should be no difficulty in producing the food we need. In several countries of Asia, an increase or decrease in food production by a margin of about five percent make all the difference between an uncomfortable glut and acute scarcity. Consequently, price fluctuations tend to be high, thus making it difficult for poor farmers to decide how much to invest in inputs. Some form of crop insurance will help, but this presents many operational difficulties. Ideally, there should be a global solution to this problem. The World Food Congress had recommended the establishment of an International Food Security system through a nationally or regionally held but internationally financed grain reserve of about eighty million tons. If implemented, this will provide a mechanism for channeling adequate external resources for the purchase of home-grown food. Both an uneconomic depression in prices and considerable loss of surplus produce in poor home storage structures can be prevented in this way.

Improvement in Productivity

Ultimately, the only real mechanism for achieving improved standards of living is increased productivity both in farms and factories. In the area of crop productivity, we occupy an unenviable position in the world.

The National Commission on Agriculture has calculated that even by 1985 the average yield of rice and wheat in our country would be only of the order of 1.6 and 2.1 tons per hectare respectively. The position with regard to jowar, pulses and oilseeds is even worse. For these crops the National Commission feels that the average yields in 1985 will range only between eight and nine quintals per hectare. We have only to compare this with the average yields of rice of five to six tons per hectare prevalent even now in countries with small holdings like Japan and Taiwan to see the large gap between what seems to be possible and what we are able to accomplish. Unless we take speedy action to identify the major constraints to productivity in each cropping pattern and remove them, ours will be a very inefficient farming system. Further, there is a positive correlation between productivity and stability of yield – the higher the average yield the greater is the stability. In my view, the keys to achieving a comfortable position on the food front in our country in the near future are rice and jowar (sorghum), which occupy over fifty million hectares. I would hence like to refer briefly to some institutional arrangements which will have to be made to achieve a higher growth rate in productivity in these crops, and to some problems affecting rice production during the south-west monsoon period: a) Inability of farmers to raise nurseries and transplant at the optimum time; b) Lack of availability or application of improved technology for direct-seeded and upland rice; c) Difficulties in efficient water management, resulting either in too much or too little water; d) Inability to control pests effectively and in time; e) Poor fertilizer use efficiency; f) Poor post-harvest technology.

Collective Solutions Needed

In several of our rice-growing areas and more particularly in the tribal areas, a farmer spends over two months in preparing rice seedlings for transplanting. He collects cowdung and other organic refuse and often burns this material in the place intended for raising seedlings. He uses a thick seed rate and thereby gets thin seedlings. As a result, he plants a bunch of weak seedlings together and places them rather deep in the puddle. Such practices need to be almost reversed if the yield potential of the new strains is to be realized. Community nurseries provide an institutional solution to this problem. The timely supply of healthy seedlings to farmers enables not only transplanting at the optimum time but also correct varietal choice, according to the particular situation of the farm in the village and the supply of nutrients like phosphorus and zinc at the seedling stage.

Scientific water and pest management would also need community endeavor. The new plant varieties form a dense crop canopy. This kind of crop canopy also promotes greater incidence of some pests which were not important before, such as the brown plant hopper, which has played havoc with rice production in Indonesia and also in some parts of our country like Kerala. It would be difficult for each smallholder to undertake the necessary tasks himself, even if he has the requisite will. According to the National Sample Survey, farm holdings below one hectare increased from 19.9 million in 1961 to 26.1 million in 1971. In fact, the Agricultural Census of 1971 estimates that holdings below one hectare are 35.7 million. Whatever be the correct figure, the trend is towards an increase of smallholdings, which is in accord with the national policy. It would be necessary to match the national policy on size of holding with appropriate institutional arrangements for helping small farmers not only to overcome their economic handicaps but also the biological handicaps beyond the control of an individual farmer. Devices like farmers' service societies, small and marginal farmers' agencies and, more recently, rural banks have been set up to tackle the economic and input supply aspects of this problem. However, in my view, unless we approach the problem in its totality, it will be difficult to achieve substantial jumps in productivity which are otherwise well within our reach purely on scientific and ecological considerations. This is particularly urgent since the technology of the future will be increasingly based on recycling principles and integrated approaches which will demand collective action by farmers in a village or watershed for efficient adoption.

As in rice, there is immense scope in jowar, a major food crop of unirrigated areas, for improving yield and production. Our average yield today is only about 500 kg per hectare, while in many countries where jowar is grown for feeding cattle and pigs, the average yield is about ten times higher. A breakthrough in rainfed agriculture can be expected only by planning for large quantum jumps rather than for slow and graded annual targets which are within striking distance of environmental fluctuations. The predominantly black soil belt of jowar, as in Madhya Pradesh, Maharashtra and parts of Karnataka and Andhra Pradesh, usually grows tall varieties of five to six months duration. By substituting such varieties with a three-month hybrid or variety, assured yields can be expected even during years with sub-normal rains. However, if the rains are plentiful as happened during 1975, a ratoon crop of jowar can be taken up or alternatively a second crop like safflower, sunflower or chickpea can be grown. It has been estimated that nearly four million hectares out of about eighteen million hectares under jowar would provide opportunities for taking a double crop in years of normal rainfall, provided short-duration varieties and hybrids are grown. The establishment of single maturity jowar zones with regard to varietal distribution would help to minimize pests like midge.

Scientists should make, block by block, a detailed analysis of the factors impeding biological and industrial productivity. Even in a crop like sugarcane, which is one of our great botanical assets, we find that our yields are low for a variety of reasons. The National Commission on Agriculture has calculated that even by 1985 the gur yield will only be six tons per hectare as against about five tons per hectare now. Sugarcane is the most efficient natural quantum converter that we know today as a collector and storer of solar energy in a useful form. The efficiency of quantum conversion to sugar is about 0.25 percent, which is very good for a field crop. Countries like Brazil where land is not a limiting factor are trying to exploit the botanical efficiency of sugarcane, even to produce ethanol and thereby cut down the import of petroleum products. For example, it has been calculated that in the San Francisco river region of Brazil, where yields are high, only 150,000 hectares of sugarcane will be needed to meet ten percent of the annual gasoline requirements of Brazil. Why are we lagging behind in exploiting this wonderful crop?

Some answers to this question have been provided by the district of Visakhapatnam, where our Congress is being held. The yields in this district were formerly poor. But by a correct varietal choice and by extending the needed help to growers through cooperative societies which run all the sugarcane factories in the district, there has been a steady increase in sugarcane production. Similar cooperative endeavor in the supply of disease-free planting material, scientific nursery and planting practices and pest management will have an immediate beneficial effect on sugarcane yield in States like Uttar Pradesh and Bihar.

A Policy for Prosperity

In my view, some basic steps are needed if we want to convert the scientific breakthrough now taking place in most crops, farm animals, and in inland and coastal aquaculture into a production advance, and a production advance into improved prosperity for all sections of the rural community. First, all groups must accept the concept of productivity. For this purpose, minimum productivity targets will have to be fixed and a farmer who does not achieve the target continuously for a few years without valid reasons should stand every chance of losing his land. Minimum targets such as three tons per hectare per year in areas with over 1,000 mm rainfall are neither ambitious nor unrealistic if proper institutional arrangements are made for supporting farmers. Appropriate Agricultural Productivity Acts by our State Governments could provide the social compulsion for cooperative efforts in areas like soil and moisture conservation and pest control. As an ingredient of such legislation, consideration deserves to be given to minimum limits to operational land holdings, below which the holding should not be fragmented as a unit of management, whatever be the ownership pattern.

The second need is to match the obligations of farmers under Agricultural Productivity Regulations with corresponding obligations on the part of State Governments. For this purpose, the Gram Sabhas or Panchayats may have to be restructured in such a manner that every member of the Panchayat assumes specific responsibility to organize the social support necessary for the effective adoption of technological innovation. Unless there is a link between those who move society and those who move science and technology, it will be difficult to achieve the ends we desire. Every prospering experiment in our country in rural transformation owes its impact to a combination of these two factors.

Unlike in industry, no depreciation can be allowed on the basic assets of agriculture. The land-man ratio has already reached critical limits in several parts of our country. The history of agricultural advance of the past century has shown that in many countries productivity has continuously risen through scientific farming without any harm to the long-term production potential of the soil. As productivity increases, the area under a specific crop can go down and more land could become available for energy plantations and silvi-pastoral systems, which are essential for meeting rural fuel and feed needs. In a few States like Karnataka, there is evidence that total production is going up in a few major crops side by side with a reduction in area. This trend should become national.

Need for Accelerated Research

I have talked so far of the institutional and policy packages needed for increasing the feasibility and efficiency of adoption of new technology in our villages. Emphasis on this does not mean that all that needs to be done on the scientific front has been accomplished. On the contrary, every change produces several reactions, some favorable and some adverse. Scientific vigilance and vision will be needed to maximize the beneficial effects and minimize the negative consequences of new technology. I have already referred to the undesirable association between a "high-yield environment" for plants and a corresponding favorable atmosphere for pests. Unscientific multiple cropping and monoculture of the same genetic strain of a crop over large and contiguous areas compound the problems of pests, which are even otherwise serious in the tropics and sub-tropics. Not only annual crops but perennial crops like citrus, mango, coconut and sandal, are affected by serious disease problems which are still evading solution. Besides pest management, I shall refer to a few more obvious areas where research work needs to be stepped up speedily.

a) Of the highest priority is more intensive work on the preparation of an integrated inventory of land, water, mineral and other natural resources, area by area, and the development of scientific land and water use plans. Such an inventory should indicate the steps needed to reclaim large areas now stricken by salinity, alkalinity and acidity, as well as to conserve water, soil fertility and germplasm of plants and animals.

b) A more careful examination of weather in relation to crop and animal productivity is needed. Fortunately, a close coordination has of late been developed between meteorologists and agricultural research and extension workers. Day-to-day decisions by farmers require the correct interpretation of short-range weather prospects. We have a good mass media network and we can easily develop the capability to convey accurate and timely advice to farmers. Meanwhile, research on soil-crop-weather correlations, weather modification, hail dispersal and efficient forms of utilizing dew should go on.

c) The efficiency of conversion of cultural energy (i.e., all forms of energy introduced by man after he domesticated plants and animals) into digestible energy has to be improved. The most important components of cultural energy are implements and farm power, water and nutrients, particularly fertilizer. In the technology developed so far in the affluent countries, the ratio of conversion of cultural energy into digestible energy became less coincident with an increase in productivity. For example, David Pimental and co-workers in a recent article[1] point out that the use of US agricultural technology to feed a world population of four billion a high-protein calorie diet for one year would require the equivalent

[1]*Science,* Vol. 190, 754-61.

of 5,000 billion liters of fuel. If petroleum were the only source of energy for food production and if we used all petroleum resources solely to produce food, the estimated reserve of 66,053 billion liters would last a mere thirteen years. The challenge therefore lies in developing techniques which will help to improve productivity continuously without a concurrent loss in the efficiency of conversion of cultural energy. This will need a stepping up of research on tillage, water use, integrated nutrient supply and recycling of recoverable sources of energy. Also this area needs support from well-defined public policies. For example, the conservation and use of all organic wastes would need the introduction of a rural fuel supply policy based on a combination of steps such as raising energy plantations, organization of community bio-gas plants, supply of coal or lignite at subsidized prices and supply of electricity and other forms of energy at reasonable rates.

d) Research relating to the control of the menace posed to crops through the triple alliance of pests, diseases and weeds must be stepped up. It is obvious that for the foreseeable future chemical pesticides will continue to be used. For applying them with due care and caution, surveillance and early warning systems need to be developed more extensively which will facilitate the timely application of the minimum quantity of pesticide needed. Simultaneously, work on genetic resistance through screening at "hot-spot" locations and through collaboration with International Agricultural Research Centers needs to be stepped up. Diverse genes for resistance will have to be identified since biotypes occur in many serious pathogens and insect pests, leading to the breakdown of host resistance.

e) All aspects of production physiology aiming at the improved utilization of solar energy and reduction of photo-respiratory losses by genetic or chemical means will have to receive greater attention.

f) The whole area of biological nitrogen fixation both by symbiotic and non-symbiotic mechanisms requires more attention. Nitrogen Fixation Research and Development Centers might well be established at State level jointly by Agricultural Universities and State Departments of Agriculture. Here, collection of nitrogen-fixing micro-organisms, both symbiotic and free-living, could be maintained. Such Centers could distribute pure cultures of recommended strains for further multiplication and also administer quality control programs. Since considerable basic work involving sophisticated instruments is also needed, the establishment of a National Nitrogen Fixation Research Center supported by all the major scientific agencies might be desirable.

g) We need more work on fertilizer technology in order to prevent the loss of fertilizer taking place during the kharif season through leaching. Some work has been done on reducing losses through blending nitrogenous fertilizers like urea with neem and other cakes and by using shellac-coated urea and sulphur-coated urea. What we need is a low cost method invoving locally available material which could simultaneously help us to reduce leaching losses and provide the needed micronutrients. Fertilizer technologists should concentrate on local problems and local solutions. Fertilizer is the most effective as well as most expensive input, and our requirement will grow not only for crop production, but also for animal and fish production. A national grid of small, medium and large fertilizer plants, compost units and an expanded soil testing service will hence be needed for sustaining agricultural advance. Since data on the fertilizer needed for specific yield targets are becoming available for major crops, fertilizer could be used as an effective trigger in any national policy of stabilizing food production.

h) Forestry research, including the development of quick-yielding fuel and pulp trees such as annual strains of bamboo, needs more intensive attention and support.

i) Increasing the production of domestic animals, poultry and fish by breeding, reduction of infertility, better nutrition and health care requires much closer study. The exploitation of hybrid vigor in cattle and fish deserves more attention. Disease problems of new genetic strains of cattle and sheep, and the harnessing of non-conventional sources of feed, particularly call for more scientific work. Recent research on fish culture in both inland and coastal waters has opened up new vistas of products. This will have to be supported by vastly expanded research on fish diseases and nutrition.

j) Genetic engineering involving the transfer of genes from one species to another is being increasingly regarded as a potential tool for achieving recombination of favorable genes. Basic research in this area requires expensive facilities. A recent symposium on basic research, as relevant to agriculture, organized by the Indian National Science Academy, has stressed the advisability of creating such facilities.

k) A serious lacuna in our present research and development efforts is the inadequate attention to problems of post-harvest technology. An uneven match between production and post-harvest technology has resulted from our relative neglect of the latter. Even in areas where rice is cultivated with improved varieties and fertilizer, the harvested crop has to be dried most times by small farmers on paved roads. The high moisture content of grains at the time of harvest and inadequate drying prior to storage, particularly during the kharif season, are resulting in increasing problems of food toxins. Studies by scientists of the Central Food Technological Research Institute in Gurupur village in coastal Karnataka revealed a heavy fungal growth and high levels of aflatoxin in rice consumed by the population. Hepatitis outbreaks due to toxins in food have been reported at the National Institute of Nutrition.[2] Thus, appropriate post-harvest technology is essential both for the farmer to get the maximum return for his labor and investment and the consumer to get food of good quality.

Population Growth and Opportunities for Gainful Employment

The next area to which I wish to refer is the population-employment equation, which, in my view, will assume even greater importance in the coming years than the population food supply equation. The obvious first step is a vast intensification of our research and developmental efforts in the area of population stabilization. Recently opened areas of research, such as vaccination against fertility, male contraceptives and immunization with specific placental proteins, need to be extended full support. The cultivation of improved strains of plants

[2]*Proceedings of the Nutrition Society of India,* No. 12, 1975.

containing certain steroids is now part of our strategy for reducing the cost of oral contraceptives.

The Bhagwati Committee had estimated that total employment opportunities of the order of 22.52 million man-years would have to be created in rural areas in 1969, if full-time employment was to be provided to all the available labor force. Other calculations have shown that at least ten million man-years of employment need to be provided immediately in the rural areas. If we measure unemployment by the criterion of productivity per person-day, the figure becomes very high, as several economists have pointed out. Ultimately, it is only productivity per person-day that can provide the basis for true prosperity. Thus, rural employment, productivity and prosperity are all interdependent.

The National Commission on Agriculture has calculated that to generate full employment, at least thirty percent of the rural labor force may have to be employed in the non-agricultural rural sector, including processing, textiles and other village industries. The Commission has hence suggested both agricultural and non-agricultural programs for generating the needed number of jobs. Rapid generation of non-agricultural, though agro-based, sources of employment is also essential if a minimum limit to operational holdings is to be introduced.

Among non-agricultural occupations in the rural sector, the development of infrastructure such as roads, housing, irrigation works, electrification and agro-service centers, has been important in the past. As agriculture advances, marketing, trade, processing, storage, distribution and transportation will demand more labor as has already been observed in north-west India, where the rice-wheat rotation and mixed farming practices are making substantial progress. There has been much talk of the growth of village and small-scale industries and the planned allocation of labor-intensive industries to rural areas, but in practice it has been difficult to make such projects viable. The non-availability of improved management and marketing techniques has rendered sustained advance difficult. Thus, wherever sericulture or honey production or any other such occupation makes progress in a village, marketing problems soon become overwhelming and farmers are unable to sustain their interest in the new vocation.

There are, of course, outstanding exceptions to this trend and these are usually associated with either well-organized cooperatives or exceptional selfless individuals who command the respect of rural society and are able to organize the entire production system efficiently. Considering the fact that about seventy-five million people may have to be employed in the non-agricultural sector during the next two decades, we need to pay urgent attention to the development of self-replicating models of rural growth. All the calculations mentioned by me take it for granted that children need not look for jobs and that they can be spared for schooling. Today, child and woman labor, much of it unpaid, constitutes a dominant component of the rural work force. While steps have been taken to ensure that women and men get equal wages, only economic progress can result in the child being freed for school.

Steps have been taken recently through the establishment of rural banks to channel more credit to rural areas. Similarly, rural agro-industrial complexes are being developed. In addition to the various pilot employment schemes of the

Central Government, different State Governments have initiated programs to meet the objective of full employment. Thus, Maharashtra is operating an Employment Guarantee Scheme and Gujarat a "Right to Work" Project. Other projects include the Labor-cum-Development Banks of Kerala, organization of a Land Army in Karnataka and mobilization of laborers in the Rajasthan Canal Project area. The Small Farmers, Marginal Farmers and Agricultural Laborers Development Agencies aim to generate more opportunities for self-employment through a diversified farming structure. A weakness of some of these programs is the lack of detailed scientific attention to resource use. Little benefit has generally been derived from the data available from different surveys including the recent rural engineering surveys. Scientists, technologists, educationists and rural communities have not been involved in an organized manner in such exercises. In the Maharashtra Employment Guarantee Scheme, a nutritional dimension has been incorporated and this process of integrated scientific approaches to human resource use needs to be fostered.

Education for Rural Development

The ultimate success in achieving balanced rural-urban growth will depend much upon the nature of our educational system and its integration with other developmental inputs. Different countries have approached this problem in different ways and case studies from seventeen countries[3] indicate that the greatest success by any criteria has been achieved where there is a clear and overpowering national ideology and strong leadership committed to this ideology. Also, success has been assured where education, formal or non-formal, has been developed as an ingredient in a package where economic initiatives are central. Two other points emerge: a) Women as a group have been almost universally neglected, and so too the landless, since most of the agriculture-based education programs benefit only those with ability to farm land; b) The danger of internal stratification developing between formal and non-formal education is very great. This could further strengthen undesirable urban-rural, elite-masses and educated-illiterate distinctions.

Our country has been a leader in many areas of educational thought and endeavor. What are our assets? We have: a) a huge and by and large well-organized infrastructure of formal education which can be reoriented to new needs and objectives; b) a large reserve of educated manpower, especially at the high levels, though insufficient at the middle levels; c) well-developed mass media which can be geared to educational purposes. Our major liability is that the imperial legacy has tended to withstand too well the encroachments of innovation.

When we became independent in 1947, it was generally assumed that quantitative expansion of the formal system of education at all levels, with suitable diversification, and qualitative improvement, would be sufficient to take care of the needs of both economic development and social transformation. A radical change of consciousness is today evident for it is now recognized that the vast mass of human beings outside the formal system of education constitutes

[3]Manzoor Ahmad and P.H. Coombs, International Bank for Reconstruction and Development, 1975.

and will continue to constitute the majority of our citizens for a considerable time to come. As a result, non-formal education is now seen as the major task for the future. The new role for formal education in this larger context will be to underpin, support and contribute to the development of non-formal education in a variety of ways. The blend of both approaches will depend upon local needs.

Our attempts to equalize educational opportunities for children of all strata of society have also not been fully successful. While incentives like mid-day meals have been of some help, the main tool for equalization has been the scholarship schemes. A drawback of these schemes is that the child who has dropped out is not considered, while selection by tests tends to leave out the very persons whom it should help. Unfortunately, culture-free tests to evaluate natural ability by unconventional approaches in villages and urban slums are yet to be devised.

Vocationalization as a Response to Unemployment

The assumption behind vocational education has been that it will equip a person with certain technical knowledge and skills to enable him to make a livelihood through a vocation. The absence of a close linkage between the available avenues of employment on the one hand and the training on the other has invalidated this approach. For example, though agriculture, forestry and fisheries account for the employment of over eighty percent of the labor force, the fewest number of vocational courses are available in these areas. There is a shortage of teachers who can effectively teach these subjects even if courses were offered, and there has been no serious effort to train teachers on a large-scale in these skills. Agro-industries, too, have been neglected.

Another lacuna is training for self-employment. Very few courses are offered in the skills needed for self-employment on a small or medium scale at the middle levels of education. Only recently, the Indian Council of Agricultural Research has started establishing Krishi Vigyan Kendras which will cater to the needs of practicing farmers and fishermen. Nearly all non-formal educational programs for women undertaken by welfare agencies in the last twenty-five years have concentrated on teaching skills like sewing, embroidery and tailoring. Women tailors, however, are notoriously few. More recently, non-formal education programs for women have emphasized home-based skills related to health, nutrition, childcare and kitchen gardening. Schemes for training women in skills related to agriculture, forestry and fisheries are still few, despite the fact that fifty percent of the agricultural labor force is women, and eighty percent of all women in employment are engaged in the production and post-harvest aspects of agriculture.

Expansion of Higher and Professional Education

The pattern of higher education and professional training in the past has evolved from colonial models on the assumption that quantitative expansion and high "quality" (quality being measured according to hypothetical "world" standards rather than in relation to national needs or objectives) would automatically take care of our scientific manpower needs. It is now recognized that the "brain-drain" is in part an outcome of a mismatch between the type of specialized education offered and the types of employment available. In retrospect, it appears as though unconsciously we have trained our scientific manpower to meet the requirements of Western industry, sciences, technology and society. That "content" is an important ingredient in "quality" is now being realized. The recent committee on the restructuring of medical education has made a breakthrough in the subject by beginning from the needs and goals of the country for medical care and analyzing the ingredients of a medical education suited to meet them. Similar exercises are needed in other areas. Agricultural education has generally tended to remain close to the field. Such nearness to field problems has a pay-off in research, as is evident from the fact that we are the first nation to cultivate hybrid cotton on a commercial scale, using labor-intensive technologies. Agricultural Universities which have not yet done so should develop experimental and demonstration farms which will inspire farmers and serve as windows into our agricultural future.

Social Responsibility of Science

Though lip service is often paid to the ideals of social commitment, our system of selection, training and evaluation of scientific personnel at all levels, or of the teaching profession does not provide for it. For instance, selection and later promotion are mainly based on academic records and attainments and publication of papers. Job requirements rarely include such elements as participation in developmental activities. These could take a variety of forms, ranging from the writing of textbooks and preparation of kits, toys, games and educational materials to participation in mass media programs, demonstration and extension campaigns for the popularization of science, and work in teacher training and lower-level education and training programs. Such elements could be built into the system if the relevant activities with "social" value that a scientist could legitimately engage in were spelt out.

Some Significant Innovations in Education

Lest the foregoing should give the impression that we are in a hopeless situation, I would like to cite a few significant recent innovative approaches: a) The recent introduction of compulsory work experience in schools, which has its roots in earlier aspirations expressed in Basic Education, has a great potential for making education more meaningful and relevant to life. For its proper handling, however, the active assistance of all scientific agencies in the country will be essential; b) The introduction of 10 + 2 + 3 pattern has as its major objective the task of making secondary education terminal and reducing the rate of expansion of higher education. Earlier attempts in this direction have not met with success chiefly because of the lack of a positive correlation between educational content and employment possibilities. Detailed planning will have to be done on this if this scheme is also not to fail for the same reason. Until this gap is bridged, there may be some merit in linking education and employment in the same Ministry of Government; c) The Farmers' Functional Literacy Project and similar schemes in adult education and their success in the last decade have indicated the tremendous potential for need-based and functional adult education; d) The National Social Service for youth and the experimental mass programs for youth, such as Youth Against Famine and Youth Against Dirt and Disease, are others with potential for increasing the contact of students with the realities of our rural life.

Some smaller programs which are still in an experimental or pilot stage but which need to be watched carefully for their possible use as prototypes for change are: a) The establishment of Nehru Yuvak Kendras in one hundred districts to channelize the energies of rural youth; b) The Bhumiadhar Experiment in non-formal education carried out by the National Council of Educational Research and Training; c) The Science Education Center of the Bombay Municipal Corporation launched and supported by the Tata Institute of Fundamental Research and the Community Science Center, Ahmedabad; d) The program for improvement of primary science education, primary curriculum development and community participation in education being developed in certain areas with UNICEF/UNESCO support; e) The establishment of Krishi Vigyan Kendras by ICAR for imparting the latest technical skills to practicing farmers, fishermen, rural youth and women.

In the non-official or voluntary sector, some significant experiments have been made. They have many lessons to teach in regard to planning, supervision, methods and objectives, and a careful evaluation of such programs would be rewarding. Often, such programs tend to be linked inevitably with certain dedicated individuals and their fate also becomes linked with the fate of those individuals. We should hence evolve an organized way of institutionalizing such innovative procedures.

Criteria for a Suitable Infrastructure

Taking the existing institutional and educational patterns as given, I shall try to indicate the directions in which they need to move to become a force for the regeneration of our rural economy and life. A satisfactory system should: a) Enable the formal education system and all scientific laboratories to become instruments which can further the non-formal education of the masses. Ways can be found, but the first essential is that institutions accept the servicing of the non-formal sectors as one of their basic objectives; b) Encourage group effort and group development in contrast to the attempt to lift individuals from one level to another; c) Relate vocational education and training to the job requirements of the present and the future through intensive local studies; d) Spell out the goals of professional education at all levels – from paraprofessional through middle level to high-level specialists – in terms of the objectives and needs of our country; e) Spell out the social responsibilities of scientists (in both research and education) and of academics in all disciplines in concrete terms of job requirements. People should be enabled to understand and fulfill their responsibilities to development through professional contributions and not merely by being associated in their leisure time with political, social welfare or educational organizations.

Steps Needed to Build the Infrastructure

With these as guiding principles, let me now indicate briefly the necessary mechanism to get the system moving: a) Every institution of higher education (including university departments, colleges, research and training institutions, junior colleges and schools) should incorporate some form of development activity into their regular teaching-learning program. This could take the form of placements of students in projects as part of their field practical work, or as apprentices; involvement of the staff or the institution itself in "action or operational research" and other possible studies. Such participation should be a condition for the receipt of financial assistance from Government. I would like to stress that this involvement be not thought of merely in terms of "adopting a village", holiday work camps and other sporadic activities, or different forms of "Sunday social work" but be a part and parcel of the regular program of work and be regularly budgeted for. As an example of what can be done, ICAR had submitted to Vice-Chancellors of our Universities a program for student and faculty involvement in rodent and pest management, control of noxious weeds like "carrot-grass", and tree planting. The response has so far not been encouraging, except from a few Agricultural; b) Projects taken up will have to be extremely localized in character so that in each case it would be possible to have some kind of coordination between the various agencies concerned; c) Grant-giving agencies and Government departments funding scholarship schemes should introduce more group awards as recognition of contributions of interdisciplinary groups to society, and group scholarships to enable institutions to strengthen themselves and provide better education and services to the whole community; d) Academies and research organizations should start honoring those who have done good work under difficult circumstances, for example, in tribal and neglected areas. e) Curriculum planning at +2 level should incorporate a period of fieldwork placement as an integral part of the course for both academic and non-academic streams. Several approaches are possible, including: i) Employment for specific period in an appropriate development project in the area; ii) Survey or field study; iii) Apprenticeship in a working agency.

These may sound fairly small steps to be taken now. However, the task is enormous and requires intensive work in localized studies, planning, working out of details, field surveys, coordination of a number of agencies, and learning new ways of working together. The involvement of students in the planning exercise is vital, as unless they are convinced of the utility of such an approach for their future, the attempts are unlikely to prove successful. Few of the steps that I am advocating, for instance, are new. They have been advocated many times before, separately and by many persons, including myself. Yet, movements in this direction seem pitifully small in comparison to the magnitude of the effort needed.

Spreading the Innovative Procedures

All over our country, there are hundreds of little programs pursuing innovative approaches. We are not short of ideas or the men to carry them out. But to transform these myriad efforts into part of an organized plan and to involve the entire academic community in this effort is the task I see before us now. It is a task the order of magnitude of which is probably unprecedented in the history of our education or scientific endeavor. Time is short. We may have forestalled the prophets of doom who predicted a future of starvation for our country, but shall we be content to drag the spectacle of our misery into the next generation? How much longer shall we drift, making valiant uncoordinated individual efforts here and there to hold back disaster but seemingly unable to gear up the total organism to the challenge? This is the question to which the scientific community must address itself.

We need to fully exploit our human ability to create rational patterns of collective activity. Such patterns will continue to be colored by personalities, but the basic structure should lend itself to replication, like the DNA molecule. It is easy to state this principle but the paths to achievement are by no means within easy reach, since they encompass our entire working culture. Nevertheless, we should begin by applying ourselves to the question of the basic common requirements that would enable us to move in the right direction, and to evolving the appropriate legislation that would serve as a framework for action.

Legislation Can Make a Start

Legislative measures for integrated rural and urban development designed to promote the symbiotic growth of the village and the city can take into consideration the following needs: a) The setting up of scientific and administrative consortia for each block and town which can help to develop and implement ecologically sound rural works and urban growth programs (in the scientific consortium, all scientific and technical institutions in the area as well as colleges and schools should be involved, and in the administrative consortium appropriate representatives of all rural community, mass media, industry and input supply agencies will have to be members); b) Minimum limits for land productivity and operational holdings for irrigated and unirrigated land in each area to provide the social stimulus for cooperative endeavor and proper use of land; c) Reservation of specific industries, credit and energy for the rural sector; d) An employment guarantee scheme as an integral part of an overall resource utilization strategy; e) Reservation of unproductive land for non-agricultural uses like brick-making, construction of building, etc. and avoiding the use of good soil for such purposes; f) A rural drinking water and fuel supply policy; g) A scientific plant-animal-man food-chain policy for each agro-ecological area based on long-term considerations of the fertilizer, feed, water and land requirements of each production system; h) Soil and water conservation and tree plantation; i) An integrated formal and non-formal educational system involving participation by students and teachers in appropriate rural/urban development programs.

A beginning towards providing ecological guidelines for development has been made through the Water (Prevention and Control of Pollution) Act of 1974. Comprehensive legislation for integrated rural-urban development will be a continuation of this process. However, we should also guard against the tendency to feel that once a law has been enacted all that is necessary has been done. Good administration and implementation of policies can make legislation worthwhile. ■

11

Agricultural Research in an Intellectual Property Rights Environment

1995 marks the fulfillment of the mission launched by Mahatma Gandhi in the beginning of this century for ending skin color based apartheid. It is appropriate that President Nelson Mandela is the Chief Guest of India's Republic Day this year, since it is a striking symbol of one of the greatest transitions in human history initiated by Gandhiji. This is therefore an appropriate year to intensify our efforts to achieve the other great transitions for which Gandhiji worked and died, namely promotion of tolerance and love of diversity and pluralism in human societies, eradication of hunger, poverty, gender discrimination and achieving harmony between humankind and nature.

The United Nations has declared 1995 as the International Year of Tolerance. This year is also the first year of the International Decade for the World's Indigenous People. The UN Conferences held at Rio de Janeiro, Vienna and Cairo in 1992, 1993 and 1994 respectively on the themes of environment, human rights and population have all recognized the need for ending unsustainable life styles and unacceptable poverty. Eradication of poverty and unemployment will be the major themes for the UN Social Summit scheduled to be held at Copenhagen in March 1995. Issues relating to the empowerment of women will be discussed in September this year at Beijing.

The recent international conventions on climate, biodiversity, desertification and the law of the sea are all indicative of the emerging trend towards cooperative and concerted global action in arresting further damage to our basic life support systems of land, water, flora, fauna and the atmosphere. To this list of multilateral mechanisms for the management of the global commons and economy, the most recent addition is the World Trade Organization, which was established on January 1 this year with its headquarters in Geneva.

While the above represent positive achievements in the area of global partnership, the most serious negative aspects of contemporary developmental pathways are: the growing rich-poor divide both among and within nations, jobless economic growth, environmental degradation and numerous mini-conflicts arising from a sense of injustice, either perceived or real, on the part of distinct ethnic or minority groups. Skin color based apartheid is being replaced by technological and economic apartheids. The built-in seeds of discrimination against the poor in the present developmental paradigm are evident from the following data on the distribution of annual global income.

Presidential Address delivered at the Second National Agricultural Science Congress, Hyderabad, India, January 19, 1995.

Nearly thirty percent of the world's poor are in India. Despite a gradual decline in fertility our country annually adds more people (seventeen million in 1991 and 1992) to the global population than any other country. Our total population of 846 million, according to the 1991 census, grew to over 900 million at the end of 1994. The population is likely to cross the billion mark by 2001. During 1991, over thirty-six percent of our population were aged below fifteen. Even if all young Indians decide to have no more than two children each our population would continue to grow for the next sixty to seventy years. Population growth coupled with enhanced purchasing power would lead to a considerable escalation in the demand for food and other agricultural commodities in the coming years.

The demographic profile of our country dictates that agriculture has to be intellectually satisfying in addition to being economically rewarding, if we are to attract and retain youth in farming. The share of land in the total assets of rural households was 60.3% according to the agricultural census of 1981-82. The same census indicates the following distributions of ownership of land holding (figures in percentage).

Since most of the jobs in rural areas are in the primary sector, the problem of rural poverty can be solved only through the intensification and diversification of farming. The smaller the farm, the greater is the need for increasing the marketable surplus. Thus productivity improvement and producer-oriented trade are essential for improving the income of households with small farm holdings. It is clear that the flagship of a job-led economic growth strategy rooted in the principles of ecology, equity and renewable energy utilization, has to be agriculture. It is in this context that the prediction of experts like Lester Brown and Hal Kane that India may have to import forty-five million tons of foodgrains annually by the year 2030 has to be viewed seriously. Such predictions are based both on diminishing land and water resources per capita and the near stagnation of scientific efforts in improving aquatic and terrestrial productivity. Brown and Kane (1994) in their analysis of the agricultural situation in India state as follows: "Unfortunately, it is becoming much more difficult to expand food production in India. The grainland area that peaked in 1983 at 107 million hectares has since fallen to 102 million, a decline of five percent in ten years. With the projected addition of 590 million people, the non-farm claims on both cropland and water will be heavy. In a country where overpumping is already widespread, satisfying the residential needs for water of the additional population will come in part at the expense of irrigation. Against this backdrop, it comes as no surprise that India's grainland area is shrinking, and that it is likely to continue to do so as its population climbs toward 1.44 billion in 2030. It may be optimistic to project a growth in grain output in India at close to one percent per year, matching the global rate of growth from 1984 to 1993, given the cropland losses in prospect. Nonetheless, if it succeeds, India would boost its harvest from 158 million tons to 222 million tons. Achieving this will not be easy because some of the existing production is based on the unsustainable use of land and

Table 1. Income distribution

Years	Income going to the richest 20% of population (%)	Incoming going to the poorest 20% of the population (%)
1960	70.2	2.3
1970	73.9	2.3
1980	76.3	1.7
1989	82.7	1.4
1993	4.7	1.4

Source: UNDP Human Development Reports.

Table 2. Land holding

Land holding	Households	Area
Owning no land	11.33	—
Sub-marginal holdings (below 0.5 ha)	36.88	2.75
Marginal holdings (1 ha & below)	18.43	9.47
Small holdings (1 to 2 ha)	14.70	16.49
Medium holdings (2 to 5 ha)	4.68	38.03
Large holdings (over 5 ha)	3.98	33.26

water. If it happens, India would still need to import forty-five million tons of grain in 2030."

In a predominantly rural and agricultural country like ours, importing food is equivalent to importing unemployment. According to the Union Planning Commission, the number of new jobs created during the first three years of the Eighth Five Year Plan has been of the order of 6.5 million per year, as against the target of ten million per year. There is general agreement that most of the new jobs will have to come from land- and water-based occupations, i.e., crop and animal husbandry, forestry, fisheries and agro industries. In spite of this realization, there is no political will to enforce some discipline in land and water use. Prime farmland continues to be diverted for non-farm uses and those who have money have no restriction on the unsustainable exploitation of ground water. In contrast, in another great democracy, the United States of America, the US Food Security Act amended in 1990, makes it mandatory that farmers must use erosion predictions to develop and implement soil conservation measures. Farmers who do not control erosion will lose all subsidies, which are substantial in that country. The step was taken to stem the current erosion rate of roughly one billion tons of topsoil that ends up in US waterways each year. Even if we cannot develop and implement a National Food Security Act as in USA, we should at least arm the State Land Use Boards with powers to prevent the abuse of prime farmland. Unsustainable pumping of the static component of ground water should also be stopped. It is time we wake up and take steps to conserve and protect the ecological foundations essential for sustainable advances in agricultural productivity.

Compounding the problems arising from shrinking per capita land, water and biodiversity resources and expanding biotic and abiotic stresses, is the new situation arising from the spread of an intellectual property rights (IPR) regime in agricultural research. For over ten thousand years, agricultural innovations and development have progressed under an environment free of considerations of ownership of knowledge and skills. With the coming into existence of the World Trade Organization and the World Trade Agreement on January 1, 1995 we are entering a new era where competition will replace cooperation, possessiveness will substitute sharing, and private profit rather than public good will provide the prime motivation for innovative research. The World Trade Agreement in the field of agriculture has four main components, which need our attention. These are: i) market access, ii) domestic support and export subsidies, iii) sanitary and phytosanitary measures, iv) and trade related intellectual property rights (TRIPS). In the area of market access, non-tariff border measures will be replaced by tariffs that prove substantially the same level of protection. Tariffs on agricultural products are to be reduced by an average of thirty-six percent in the case of developed countries and twenty-four percent in the case of developing countries. Reductions are to be undertaken over six years in the case of developed countries and over ten years in the case of developing countries. Least-developed countries are not required to reduce the Total Aggregate Measure of Support (AMS) including export subsidies, there will be a real opportunity for India to export more farm goods. This will however call for a massive investment in post-harvest technology, including much greater attention to sanitary and phytosanitary measures. The new "sun-rise" industries like horticulture, aquaculture, medicinal plant products, feed grains, hybrid seeds, biological software products like biofertilizers and biopesticides and different kinds of organically grown and tissue culture raised plant products, need integrated attention at the levels of technology, training, techno-infrastructure and trade if initial successes are to be sustained over time.

An area of great concern in our country has been the potential impact of TRIPS on the easy availability of the fruits of modern plant breeding and biotechnology to resource poor farming families. The position in this matter is as follows.

There is a general obligation to comply with the substantive provisions of the Paris Convention (1967) with reference to patents. In addition, the agreement requires that twenty-year patent protection be available for all inventions, whether of products or processes, in almost all fields of technology. Inventions may be excluded from patent ability if their commercial exploitation is prohibited for reasons of public order or morality: otherwise, the permitted exclusions are for diagnostic, therapeutic and surgical methods, and for plants and (other than micro-organisms) animals and essentially biological processes for the production of plants or animals (other than microbiological processes). Plant varieties, however, must be protectable either by patents or by sui generis system (such as the breeder's right provided under the UPOV Convention). Detailed conditions are laid down for compulsory licensing or governmental use of patents without authorization of the patent owner. Rights conferred in respect of patents for processes must extend to the products directly obtained by the process.

Issues related to the sui generis system of plant variety protection have been discussed at many forums in our country during the last year. At the Center for Research on Sustainable Agriculture and Rural Development in Madras, an

International Dialogue was held on this topic last January. The result of this dialogue was a carefully prepared draft legislation, which provides for: a) The recognition of the intellectual contributions of both breeders and farmer-conservers; b) The protection of the rights of farmer-cultivators to retain seeds for their own use and for limited exchange; c) Conversion of the license of rights from exclusivity to non-exclusivity where public interest warrants such a change.

The Madras Draft Act provides a transparent and administratively implementable mechanism for recognizing and rewarding the contributions of tribal and rural women and men in the field of plant genetic resources conservation and enhancement. Since the draft Plant Variety Protection Act prepared by the Union Ministry of Agriculture has not been made public, we can only hope that it has been drafted in a manner that the cause of plant breeding will be advanced and not retarded. It should not be forgotten that we have one of the world's largest and dynamic plant breeding establishments in the public sector, solely devoted to public good. Similarly, there is a large network of research institutions under the Consultative Group on International Agricultural Research (CGIAR), all devoted to the cause of international public good and the food security of developing countries.

The World Trade Agreement is a reality; so also is TRIPS and all other agreements. They provide opportunities for enlarged trade. We should act now to derive maximum benefits from new trade opportunities. At the same time, we should take steps to avoid damage to: a) The ecological security of the nation, arising from the over-exploitation of natural resources; b) The food and nutrition security of the country through disproportionate emphasis on crops for export; c) The livelihood security of the poor due to their being bypassed by technological innovations and losing their traditional sources of income.

These should be the basic considerations guiding our post-World Trade Agreement, agricultural research and development policy.

Fortunately, the GATT Agreement provides several provisions, which can be used to insulate the poor from further marginalization and to expand investment in scientific research, intended to promote public good. Such provisions should be clearly understood and woven into the policy fabric. For example, investments in scientific research and extension are excluded from the calculation of the subsidies and AMS. Therefore, it will be prudent to create a National Trust Fund for Science and Sustainable Food Security with an initial allocation of at least Rupees 1,000 crores for the purpose of expanding strategic, anticipatory and participatory (with farming families) research designed to ensure that not only predictions such as those of Lester Brown and Hal Kane do not come true, but also that India becomes a major agricultural power in the world.

I am not pleading for greater investment in national and international agricultural research devoted to public good just because I have worked in a tradition where the sole purpose of agricultural science was the eradication of hunger and rural poverty. I plead today for greater investment in public sector agricultural research mainly because location specific varieties and resource management technologies are essential for converting the concept of sustainable agriculture from rhetoric to reality.

The commodity-centered research characteristic of private companies often tends to bypass the poorly endowed and fragile resource systems such as mountain and coastal region, tribal belts and arid and semi-arid zones. In the absence of the needed biophysical resource base, these areas are unable to take advantage of the fund of knowledge and material created through commodity driven research. These are the very areas, which account for much of the poverty and household food insecurity prevalent in our country. Therefore, research programs for such resource-scarce areas have to be reformulated on the basis of optimum land and water use. Resource scarce areas need high value agriculture, so that the people of the area can derive greater income. Both high yield-low risk areas and low yield-high risk areas need specific strategies based on concurrent attention to resource management, productivity improvement and poverty alleviation. If the research agenda is entirely market-driven, orphans will remain orphans and the urgently needed resource management-cum-commodity improvement orientation to agricultural research may never come about. This will be disastrous from the point of view of integrating the dimension of ecological sustainability with those of economic viability, social equity, employment intensity and energy efficiency in technology development and dissemination.

The plea for adequate investment in public sector research through ICAR institutions and Agricultural Universities does not imply that private sector research is either not necessary or should not be encouraged. In fact, the dynamic growth of private sector research will stimulate the growth of relevance and excellence in research supported from public funds. The proposed Plant Variety Protection Act will help to attract greater private investment in plant breeding. However, the ability of resource poor farmers as well as of farmers living in less favorable environments, such as arid and semi-arid areas, mountainous terrain and flood-prone regions, to derive benefit from private sector plant breeding will be proportional to the help they receive from ICAR institutions and Agricultural Universities in the improvement and sustainable management of their natural resource endowments. Thus, public and private sector research can become mutually reinforcing.

Private sector companies should also enter into a new social contract with resource poor rural families, particularly with women and men belonging to landless labor families. This will help to convert unskilled workers into skilled ones, thereby adding economic value to the time of the poor. There are attractive opportunities for symbiotic partnerships between private seed companies and the families of agricultural labor and marginal and small farm holdings. Such partnerships will help to fight the spread of technological and economic apartheid. An excellent example of the vast opportunities existing for the technological and economic enlistment of the poor through cooperative empowerment is provided by the work of the National Dairy Development Board in the dairy sector.

In conclusion, I would like to share my personal conviction with reference to three major questions facing us today in the farm sector.

First, *can we produce enough food to feed the growing population*?

My answer is yes, provided we take steps both to enable resource poor farming families to derive advantage from the untapped yield reservoir existing even at currently available levels of technology and to defend and enlarge the gains in the most favorable areas (MFA) and in high yield-low risk environments and crop seasons. Particular attention needs to be paid to assist farmers in rainfed and semi-arid areas as well as in the entire eastern region.

Second, *can we produce food in a sustainable manner, without damage to the basic environmental capital stocks of land, water, forests, biodiversity and atmosphere*?

My answer is again yes, provided political leaders are willing to enforce discipline in the use of natural resources, while at the same time generating the necessary public opinion and awareness with the assistance of the mass media in favor of nature conservation. The numerous central and state Land Use Commissions and Boards, which today exist on paper, need to be empowered and placed under professional leadership.

Third, *can we ensure that food is accessible to all*?

Yes, provided we promote higher consumption through both protective social security measures and greater opportunities for skilled employment in the farm and off-farm sectors. An aggressive program of market-driven knowledge and skill dissemination among the poor is urgently needed.

Thus, I am cautiously optimistic about our agricultural future. My optimism stems from the receptivity and innovativeness of our farm families and creativity of our young agricultural scientists. Science is an exciting adventure. The greater the challenge, the greater is the impetus for creativity. Let me cite an example.

Dr. Gurdev Khush and his colleagues at the International Rice Research Institute in the Philippines have recently developed a prototype of a new rice variety capable of yielding twenty-five percent more than the existing high yielding varieties. The new plant type is based on the bulu rice of Indonesia. Bulu rices have very large panicles, which is one of the important characteristics of the new plant type. The new material is susceptible to several diseases and pests and hence may be available in a useful form only at the end of this century. Progress can be faster, if the prototype rice strains are used in location specific varietal breeding and also in achieving a further rise in the yield potential of hybrid rice.

Such scientific developments have three major lessons for us. First, strong south-south collaboration in germplasm exchange should continue. Second, a dynamic national research capability, particularly of a participatory research mode with farmers, is a must to take advantage of new opportunities opened up by science. Finally, only through ecotechnologies based on blending traditional wisdom and technologies with frontier technologies such as bio-information, space and renewable energy technologies, can sustainable advances be achieved in improving the productivity, profitability, stability and sustainability of our major farming systems.

Jawaharlal Nehru's statement, "everything else can wait, but not agriculture", is even more relevant today than when he made it forty-seven years ago, since only the rapid spread of ecological agriculture, can help to achieve the national goals of food and jobs for all. Deceleration of damage to our environmental capital stocks and acceleration of progress in the eradication of poverty, illiteracy and gender inequity should be the most important transition we should strive to accomplish during the remaining years of this century. ❑

Reference

Brown, L.R. and Hal Kane. 1994. *Full House: Reassessing the Earth's Population Carrying Capacity*. New York: W.W. Norton and Company.

12

Benjamin Peary Pal

52. Dr. B.P. Pal

The gate of life swings to and fro
and soon, too soon, it closes
and that is why, beside my door,
I grow red roses.

This quotation from Benjamin Peary Pal's article "The rose - its beauty and its science" (1987) is a vivid expression of his personal philosophy. His life was a combination of dignity, courtesy, compassion and love of beauty. He was a scientists' scientist and through his dedication and scientific innovations earned the respect and admiration of not only fellow professionals but also political leaders and the general public, particularly farm men and women.

In science, he was one of the earliest to demonstrate the value of heterosis breeding in a self-pollinated cereal like wheat. This was the forerunner of similar work in rice, which ultimately led to the commercialization of hybrid rice in China. In the 1930s he launched a search for new genes in wheat and potato in order to achieve pyramiding of genes for resistance/tolerance to a broad spectrum of biotic and abiotic stresses in these crops. He succeeded in breeding the bread wheat variety NP 809 possessing genes for resistance to the then important biotypes of stem, stripe and leaf rusts. In life, he remained a bachelor wedded to the breeding of new varieties of wheat, rose and bougainvillea and to painting. In addition, he looked after with great affection and care his two sisters, who stayed with him until their death. His wit and wisdom gave cheer and hope to all who came into contact with him. He shaped the course of Indian agricultural research for over three decades and designed instruments for achieving interdisciplinary and inter-institutional cooperation in basic and applied research. He strengthened educational infrastructure and programs at the postgraduate level. His goal was excellence and relevance in all fields of agricultural research, education and extension of knowledge from the laboratory to the field. He ranks among the most eminent biological scientists India has produced.

Early Years

B.P. Pal was the youngest of nine children (three of whom died in infancy) of Dr. Rala Ram Pal and Mrs. Inder Devi. Although originally belonging to the village Sarhal Qazian in Jullundur District of Punjab State, Dr. Rala Ram Pal settled in Burma as a Medical Officer. Pal's name at birth was Brahma Das Pal but this was changed to 'Benjamin Peary Pal' when he joined St Michael's School at Maymyo in Burma in 1914. This school had a lovely rose garden and several of the teachers in the school were fond of both gardening and painting, hobbies which Pal also acquired at a young age.

First published in *Biographical Memoirs of Fellows of the Royal Society*, London, Vol. 42, pp. 267-74.

Pal was always first in his class at the school. Once he was presented with a paintbox as a gift and this led to a life-long passion for painting. As I write this memoir, I have in front of me a lovely painting of the River Cam in Cambridge which Pal painted when he was working for his Ph.D. at the University of Cambridge with Sir Rowland Biffen, F.R.S., and Professor Sir Frank Engledow, F.R.S., and which he later presented to me.

Pal did his B.Sc. and M.Sc.degrees at the University of Rangoon between 1924 and 1929. For his M.Sc. degree in Botany, he worked on Burmese Charophyta. In 1929, he was awarded, by the Government of Burma, a scholarship to pursue postgraduate studies at Cambridge. It was in Cambridge that Pal's interest in agriculture and plant breeding was born. Following the retirement of Biffen in 1930, Pal started working for his thesis with Engledow on hybrid vigor in wheat. At Cambridge, Pal attended courses given by eminent scientists like Professor F.T. Brooks, Dr. R.N. Salaman and Dr. H.G. Sanders. He returned to Burma in December 1932 after taking his Ph.D. degree. His Ph.D. work established that there is scope for the commercial exploitation of hybrid vigor in wheat provided suitable methods of large-scale production of F_1 seeds can be developed.

In March 1933, Pal was appointed as Assistant Rice Research Officer in the Burmese Department of Agriculture. However, in October 1933 he moved to Pusa in Bihar in India to become the Second Economic Botanist at what was then known as the Imperial Agricultural Research Institute (renamed the Indian Agricultural Research Institute in 1947). In April 1937, Pal was appointed as Imperial Economic Botanist at IARI.

Professional Career and Honors

Following a severe earthquake which damaged the main buildings of IARI at Pusa, the Institute was shifted to New Delhi in 1936. Pal became the Director of IARI in September 1950 and served in that capacity until May 1965, when he became the first Director General of the reorganized Indian Council of Agricultural Research (ICAR.). He served as Director General of ICAR from May 1965 to January 1972. Pal considered the period 1933 to 1965 when he worked at IARI as the best years of his personal and professional life. He loved the Institute so much that he left all his property, including two lovely houses in New Delhi and Simla to IARI in his will.

Pal generously gave his time to a wide variety of public and professional causes. He was Chairman of the All India Fine Arts and Crafts Society for several years. He was the founder President of the Rose and Bougainvillea Societies of India. He founded the Indian Society of Genetics and Plant Breeding and edited the *Indian Journal of Genetics and Plant Breeding* for twenty-five years. Pal was elected President of the Indian National Science Academy in 1975. He served as the President of the Indian Science Congress in 1974. He was elected a Fellow of the Royal Society in 1972. Pal was the recipient of numerous awards, including Padma Vibhushan, which is the second highest civilian award conferred by the President of India. He was elected a Fellow of the Science Academies of France, Japan and the former USSR. A highlight of his career at IARI was the establishment, in 1958, of a postgraduate school offering educational programs leading to the degrees of M.Sc. and Ph.D. in a wide range of subjects in agriculture. IARI in its turn showed its gratitude to Pal by conferring its first honorary D.Sc. degree on him in February 1986.

Scientific Contributions

Pal's major research contributions are in the area of wheat genetics and breeding. At the time Pal joined IARI at Pusa, the Institute had already released a few good wheat varieties possessing excellent grain quality, as a result of the selection work done by Sir Albert and Dr. Gabrielle Howard. Pal observed that rust diseases were mainly responsible for low and uncertain wheat yields and hence initiated systematic breeding efforts to develop varieties possessing resistance to stem, stripe and leaf rusts. In this work he collaborated with Dr. K.C. Mehta, an eminent mycologist.

Special facilities for screening for rust resistance were created at Simla. An off-season nursery was established in Wellington in southern India. This research led to the release of a series of varieties under the New Pusa (NP) 700 and 800 series. Varieties were bred for resistance to the then prevalent biotypes of all the three rusts, both separately and in combination. The variety NP 809 released in the early 1950s combined resistance to most of the common biotypes of stem, stripe and leaf rusts as well as to loose smut. This achievement received much national and international acclaim.

Realizing that the improvement of wheat yields in India would depend largely on our ability to control the severe losses caused to the crop by rust infection, in 1934 Pal initiated a breeding program for evolving rust-resistant wheat varieties. In collaboration with Mehta and others he bred several varieties individually resistant to different rusts. He combined the genes for rust resistance with those controlling good grain quality and high yield and the varieties NP 710, NP 718, NP 761 and NP 770 were thus evolved.

These varieties were widely grown in many states in India. Later, he developed the highly rust-resistant varieties NP 797, NP 798 and NP 799 which are in addition high-yielding, early in maturity and resistant to loose smut. During the severe rust epidemics in Bihar and West Bengal in 1956-57, these varieties were healthy and gave good yields while the local varieties were destroyed.

But for the rust-resistant wheats of the NP 700 and the early NP 800 series developed under Pal's guidance, there would have been no wheat crop worthy of harvest in years characterized by severe rust epidemics. K.M. Munshi, formerly Union Minister for Food and Agriculture, in a talk delivered on the occasion of the Golden Jubilee of the Indian Agricultural Research Institute in April 1955, estimated that the improved rust-resistant wheat varieties bred by Pal were responsible for an increased annual income to the farmers of about Rupees 300 million.

Pal was chief architect of the Coordinated Wheat Rust Control Scheme, a project under which varieties resistant to the different rusts and suitable for cultivation in all the wheat growing states were bred. He initiated work on several aspects of the project, such as the preparation of a National Register of Genetic Stocks of Wheat, the use of exotic varieties either as breeding material or for direct cultivation, and the study of the milling, baking and chapati-making qualities of Indian wheat varieties. He also established a germplasm bank of wheat and initiated work to take advantage of the genome approach to wheat breeding.

As mentioned earlier, the variety NP 809, which was the first wheat variety to possess resistance to all three rusts, was released in 1954. This variety was the result of many years of breeding work by Pal with the collaboration of plant pathologists. With the setting-up of fertilizer factories within the country and the consequent possibilities for undertaking intensive cultivation, Pal anticipated the need for varieties that would respond well to heavy fertilization. He consequently reoriented the wheat breeding program so as to include good response to manuring as one of the criteria for selection. As a result, several new varieties, such as NP 823, NP 824 and NP 830, which responded well to the application of about 60 kg of nitrogen per hectare and at the same time were resistant to rust and loose smut diseases, were evolved by wheat breeders at IARI. Another variety, NP 825, gave high yields under drought and became popular with farmers cultivating under rain-starved conditions.

Lack of precise knowledge concerning the inheritance of characteristics of agronomic interest is a handicap in breeding work. To remove it, Pal intensified studies on the genetics of wheat. Several lines of approach were followed. First, by using classical methods, the genetics of rust resistance and of several morphological characters were studied in many wheat varieties. This knowledge helped in understanding the nature of genetic control of rust reactions more fully. In a joint paper with his colleagues presented at the International Genetics Symposium held in Japan in 1956 (Pal and Sikka, 1957), Pal clarified the genetic situation concerning rust resistance in wheats. He showed that two sets of genes – one conferring rust resistance and another inhibiting the resistance reaction are present in wheat varieties and that the dominant or recessive nature of rust resistance may vary in different crosses depending upon which set of genes is involved. He thus clarified a complex genetic situation arising from the variable dominance relationships found with reference to the same characteristic in different crosses.

A second line of approach encouraged by Pal for the study of the genetics of Indian wheat varieties was monosomic analysis, i.e., use of plants deficient in individual chromosomes to locate the genes situated on the missing chromosomes. The chromosomes on which genes conferring resistance to several races of stem, stripe and leaf rusts were present were identified in this way in different varieties and experiments were started to transfer selectively those chromosomes to susceptible wheat varieties.

Soon after starling his work on the breeding of improved varieties of Indian crop plants, Pal realized the need for a thorough reclassification of the varieties of different crop plants and for the formulation of suitable keys for identifying the different varieties. He himself undertook this work in wheat and potato and published comprehensive papers on the classification and identification of the varieties of these crops. Thus he brought to an end the chaotic situation that existed in the field of varietal nomenclature in India. He also took an active interest in the revision of the International Code of Nomenclature of Agricultural and Horticultural Plants. In pursuance of his interest in evolving suitable keys for the identification of wheat varieties, he explored the use of anatomical characteristics in classification and found that the structure of leaf hairs was useful for this purpose.

In 1957, Pal made an important contribution to the available knowledge concerning the taxonomic relationships among hexaploid *Triticum* species when, in collaboration with one of his colleagues, H.B. Singh, and the eminent geneticist, Dr. Edgar Anderson, he published a paper on the genetics of *Triticum vavilovi*, which is one of the lesser known hexaploid wheats (1957). By studying the progeny of crosses between *T. vavilovi* and the cultivated species *T. aestivum* and *T. aestivum* subsp. *sphaerococcum* he showed that *T. vavilovi* is more closely related to *T. aestivum* than might be inferred from a superficial study. He suggested that *T. vavilovi* and the other cultivated hexaploid *Triticum* species should all be considered as subspecies of *T. aestivum.*

In addition to breeding for disease resistance, Pal initiated work in breeding wheat varieties capable of responding to irrigation and good soil fertility. NP 824 was one such variety capable of yielding over three tons of wheat per hectare. A further breakthrough in improving the yield potential of bread wheat had to await the introduction in 1963 of semi-dwarf varieties possessing the Norin genes for dwarfing from the wheat breeding program of Norman E. Borlaug in Mexico. Pal gave whole-hearted support to the semi-dwarf wheat breeding program, which resulted in wheat production going up from twelve million tons in 1964-65 to seventeen million tons in 1967-68. Such a quantum jump in wheat production led to the coining of the term "Green Revolution" in 1968 by William Gaud of the US Department of Agriculture.

As Imperial Economic Botanist, Pal initiated breeding work, using the tools of modern genetics, in a wide range of economic plants. The crops on which Pal himself worked included potato, tobacco, maize, tomato and okra.

Pal emphasized that the extent of availability of genetic variability in a crop determines the scope for recombination breeding and for pyramiding of genes for desirable traits like resistance to diseases. In his paper "The search for new genes" published in *Agriculture and Livestock, India* (1937) he advocated a systematic search for genetic variability coupled with genetic conservation, evaluation and use. This led to the establishment of a Division of Plant Introduction at IARI for the purpose of genetic exploration, collection and introduction. This division later resulted in the establishment of a National Bureau of Plant Genetic Resources in 1974.

Leadership of The Indian Council of Agricultural Research (ICAR)

A very important contribution of Pal during his tenure as both the Director of IARI and the Director General of ICAR was the organization of all-India coordinated projects first in wheat and later in most of the important crop plants and farm animals. The coordinated project methodology became a powerful instrument for fostering interdisciplinary and inter-institutional collaboration in testing new varieties and in monitoring the dispersal of biotypes of fungal diseases and insect pests like the brown plant hopper in rice. The multi-location testing procedures developed in the coordinated projects helped to accelerate the pace of progress in varietal identification and release. Results from many locations gave scientists confidence to recommend to the Government of India the import of 250 tons of seeds of the Mexican semi-dwarf varieties Lerma Roja 64 A and Sonora 64 in 1965 and of 18,000 tons of seeds of these two varieties in 1966. Such large-scale import of seeds of high-yielding wheat varieties helped to improve wheat production very rapidly in India. Although the Mexican wheat varieties were deficient in their chapati (unleavened bread) making properties, they found acceptance among farmers because of their high yield. Using the Mexican wheats as parents, Indian wheat breeders soon evolved a wide range of high-yielding varieties possessing the requisite degree of disease resistance as well as good grain quality. Wheat production consequently has gone up from twelve million tons in 1964-65 to about sixty-five million tons in 1995-96.

Strategic Research

Pal encouraged basic and strategic research in major crop plants. For this purpose, he established a School of Basic Genetics at the Division of Botany at IARI. Induced polyploidy and mutagenesis, aneuploid analysis in wheat and rice, and radio-tracer experiments to improve the efficiency of fertilizer use were all encouraged by Pal. As mentioned earlier, he himself was interested in identifying the location of genes for resistance to rusts and loose smut in wheat, using the monosomic series developed by Dr. E.R. Sears in the wheat variety Chinese Spring. He compiled the results of his extensive research in the form of a comprehensive monograph on wheat published by the Indian Council of Agricultural Research (Pal, 1966).

Pal was deeply interested in novel genetic combinations. He therefore promoted alien gene transfer involving embryo rescue techniques. This approach resulted in commercial varieties of tomato and okra resulting, respectively, from crosses between *Lycopersicum esculentum* and *L. pimpenellifolium* and between *Abelmoschus esculentus* and *A. tuberculatus*. Similarly in potato, he and his associate Pushkarnath initiated an extensive program of hybridization between *Solanum tuberosum and S. demissum* possessing genes for resistance to the late blight fungus, *Phytophthora infestans*.

Pal played a major role in getting Dr. R.N. Salaman interested in the organization of an Empire expedition to South America during 1938-39 for collecting wild relatives of the potato. This collection, which was later known as the Commonwealth Potato Collection, proved to be a source of many valuable genes for resistance/tolerance to a broad spectrum of biotic and abiotic stresses in potato, and laid the foundation for rapid progress in potato improvement and production in India.

Pal's interest in germplasm collections extended to roses and bougainvillea. His comprehensive monograph on roses remains a classic. He also wrote a book on *Beautiful Climbers of India* (1960). Both these books were published by the Indian Council of Agricultural Research.

Education

Pal was deeply interested in human resource development for agriculture. He used to emphasize that what India needs is not just agricultural education but education for agriculture. This approach led him to establish a postgraduate school at IARI in 1958, for imparting education leading to the degrees of M.Sc. and Ph.D. in a wide range of subjects basic to progress in agriculture. The IARI Post-Graduate School soon rendered it unnecessary for Indian scholars to go abroad for their Ph.D. degree. In fact, most teachers in Indian agricultural universities are IARI alumni. The establishment of a postgraduate school at IARI granted by the University Grants Commission the status of a university, ranks among the most significant of the many contributions of Pal to the modernization of Indian agriculture.

During his tenure as the first Director General of the reorganized Indian Council of Agricultural Research (ICAR), Pal took special interest in imparting college curricula with greater relevance to the needs of farm families. ICAR is a unique organization, since it has concurrent responsibility for research, education and extension education in crop and animal husbandry, fisheries and agroforestry. Pal fostered the growth of agricultural universities in India and helped them to integrate in a symbiotic manner their triple functions of education, research and farm-level demonstrations and training. Agricultural universities like those of Punjab, Haryana and Pant Nagar in Uttar Pradesh played a dominant role in spreading the message and methods of "Green Revolution" technologies in wheat, rice, sorghum, maize and pearl millet (*Pennisetum typhoides*). Pal exhorted the agricultural universities to function in such a manner that farm families consider them as their own. When he was Director of IARI, he brought about a meaningful partnership between the farm families of the Khanjawala Block of Delhi State and the scientists of IARI. Pal's tenure as Director General of ICAR between 1965 and 1972 was marked by the spread of the coordinated project approach to animal sciences and fisheries. He promoted a farming systems approach to agricultural research, since in India mixed farming involving crop-livestock integration is both a way of life and a means to secure livelihoods. Above all, he raised the status of agricultural scientists in the country and helped to create the scientific backstopping essential for a sustainable Green Revolution. Many of the professional societies in agriculture in India owe their origin to his encouragement and support. He firmly believed that the health of a professional society is an index of the health of that particular profession. He gave his time generously to attending meetings of students in agricultural colleges and universities, since he felt that the future of India's agriculture depends upon the country's ability to attract and retain youth in farming. His humor and humility endeared him to the younger generation. His self-effacing nature won for him universal admiration and respect. He became a role model for agricultural scholars and scientists, both young and old.

After retirement from ICAR in 1972, Pal concentrated on issues relating to environment protection in addition to his "hobbies" of rose and bougainvillea breeding and painting. He was the first chairperson of the National Committee on Environmental Planning and Coordination of the Government of India. Following his death, the Ministry of Environment and Forests of the Government of India established a prestigious award in the name of Pal. Similarly, ICAR established a B.P. Pal National Professorship in Genetics and Plant Breeding. Pal loved his family, country and IARI. When I succeeded him as Director of IARI, I felt that the best tribute we could pay him was to establish, in front of the IARI Library, a B.P. Pal Rose Garden. In a letter, in his lovely handwriting written some months before his death, Pal wrote: "As I write this letter, my terrace garden is full of beautiful roses, bathed in the cool winter sunshine. I recall you named the rose garden in front of the IARI library after me. No one else could have thought of such a warm and affectionate gesture."

It is appropriate to end this memoir with a quote from one of Pal's last papers (1987): "In a world which is becoming rapidly over populated, where the environment is continuously degraded, where forests are being destroyed, where beautiful areas are being threatened by 'development', when vulgarization is proceeding apace, and advertisements of various consumer items are being blatantly displayed even in the most remote areas of the globe, the message of hope and beauty symbolized by the rose is all the more needed."

It can be truly said of Pal that his life was his message. He remained a bachelor to help his unmarried sisters to have a meaningful life. He took pleasure and pride in providing unfettered opportunities for creative young minds to blossom. He brought cheer and hope to fellow agricultural scientists as well as farmers. By creating the research and educational backstopping essential for the origin and spread of the "Green Revolution". Pal served as a bright affirming flame in the midst of the sea of despair then prevalent in India on the food front.

Keats wrote: "You may break, you may shatter the vase if you will,
But the scent of the roses will hang round it still."

Pal's life, like the scent of roses, will continue to inspire generations of Indian agricultural research scholars and scientists. ◨

References

Pal, B.P. 1937. The search of new genes. *Agri. Livestock, India* 7:372-78.

----- and S.M. Sikka. 1957. Breeding crop varieties resistant to diseases and pests in India. *Cytologia Supplement* (Proceedings of the International Genetics Symposium held in Japan in 1956), 435-41.

-----, H.B. Singh and E. Anderson. Studies in the genetics of Triticum Vavilovi Jakub. *Agronomy Journal* 49:4-10.

-----. 1960. *Beautiful Climbers of India*. New Delhi: Indian Council of Agricultural Research.

-----. 1966. *Wheat - A Monograph*. New Delhi: Indian Council of Agricultural Research.

-----. 1987. The rose - its beauty and its science. *Interdisciplinary Science Review* 12(i): 77-88.

13

Science and Sustainable Development

Until three decades ago, the concept of sustainability was considered only in economic terms. Rachel Carson's book, *Silent Spring* (1962), generated widespread awareness of the need to assess technological and economic progress from the ecological angle. The facts contained in her book shocked the world, since it was only in 1948 that Paul Hermann Muller of Switzerland had been awarded the Nobel Prize in Physiology and Medicine for his discovery of the insecticidal properties of DDT (dichlorodiphenyltrichloroethane). The very properties of DDT, such as broad-spectrum action and long residual toxicity, for which the Nobel Prize was awarded, became its principal weaknesses. Concern over the ecological consequences of post-World War II economic and technological progress led the United Nations to convene the first UN Conference on the Human Environment in 1972 at Stockholm. At this Conference, Indira Gandhi, then Prime Minister of India, stressed that, unless considerations of economic and gender equity were integrated with those of ecology and economics, development will not be socially sustainable.

Thus, today, the concept of sustainability has enlarged to include the dimensions of economic viability, environmental soundness, and social and gender equity. In every sphere of human activity, new definitions and monitoring tools are being developed to reflect this paradigm shift in our approach to sustainable development.

The UN Conference on Environment and Development held at Rio de Janeiro in June 1992 adopted the following conventions and action plans for ensuring that our earth remains green forever: Convention on Climate; Convention on Biological Diversity; Agenda 21; Forestry Principles.

Subsequently, a UN Convention on Desertification and the UN Law of the Sea Convention have come into force. Thus, we have a framework for ensuring that economic development is both environmentally sustainable and socially equitable in the future.

In spite of the growing public and political awareness of the dangers of unsustainable development, we are continuing to clamor for unsupportable lifestyles and excessive consumption of natural resources. A recent book (Calborn et al, 1997) indicates how many new synthetic chemicals harm our hormone and immune systems. The tobacco industry has at long last accepted that serious health risks are involved in cigarette smoking and have offered the

---

Presented as Dr. A.L. Mudaliar Endowment Lecture at Anna University, Chennai, September 7, 1997. First published in *A Century of Hope, Harmony with Nature and Freedom from Hunger*, by M.S. Swaminathan, pp. 57-69. Chennai: EastWest Books (Madras), 1999.

State and Federal Governments in the United States huge sums of money to improve the health of tobacco victims. Unfortunately, the same tobacco companies, who have offered nearly three hundred billion US dollars to help undo the damage they are doing to human health, are trying to shift their marketing tactics to developing countries, particularly to countries in South Asia as well as in Eastern Europe.

The special session of the UN General Assembly, held at New York in June 1997, to review the global environmental balance sheet five years after Rio, provided proof of the growing gap between environmental rhetoric and reality. Similarly, the rich-poor divide is widening on the economic front. Social disintegration now being witnessed in many societies, including our own, is the result of the current pattern of economic growth, which is not rooted in the principles of equity and ethics. What then is the remedy? I would like to begin with rural India, the backbone of India's culture and agriculture.

Over sixty years ago, Mahatma Gandhi made the following remarks while addressing young scholars who wished to work in the area of rural development: "The fact is the villagers have lost all hope. They suspect that every stranger's hand is at their throats and that he goes to them only to exploit them. The divorce between intellect and labor has paralysed our agriculture. The worker should enter the village full of love and hope, feeling sure that where men and women labor unintelligently and remain unemployed half the year around, he working all the year round and combining labor with intelligence, cannot fail to win the confidence of the villagers."

Gandhiji's words are by and large still true. However, during the last fifty years, villagers have lost their original ethics of self-help because of the patronizing approach of government agencies, who refer to the women and men toiling in sun and rain to feed the country as "beneficiaries" and "target groups". In recent years, voluntary organizations have been struggling to change this attitude and develop methods of including the excluded in development programs. There are nearly one hundred forty to one hundred fifty schemes in many States for helping the poor and there is, therefore, no political or economic excuse for the persistence of poverty and deprivation on such a large-scale.

Since agriculture is the basic industry of rural India, I would like to discuss rural transformation in the context of population and natural resources.

Ashish Bose (1996) has recently summarized the present demographic scenario in India. The desired transition to low birth and death rates is yet to take place in most parts of the country, excepting in the states of Kerala, Tamil Nadu, Goa, and Mizoram. Based on considerations of population, illiteracy, percentage of malnourished children, and per capita income, Mahbub ul Haq has developed a measure of human deprivation (1997). He concludes: "The overall extent of human deprivation is simply colossal in the SAARC region. The deprivation of human capabilities far exceeds the deprivation of income alone and affects over

Table 1. Poverty and Literacy Profile of Major Indian States (rural areas)

State	Per capita Income (Rs. per annum)	Persons below absolute poverty line(%)	Literacy rate, age 7+ (%)
Punjab	6,380	32	60
Haryana	6,368	27	55
Kerala	5,778	30	90
Maharashtra	5,525	34	58
Gujarat	5,288	39	59
Andhra Pradesh	5,046	21	50
Rajasthan	4,229	40	41
Uttar Pradesh	4,185	40	47
Himachal Pradesh	4,168	45	68
Bihar	3,691	42	44
West Bengal	3,157	51	59
Orissa	3,028	55	55
Average for rural India	4,485	39	54

Source: NCAER 1996 (personal communication).

Table 2. Disaggregated HDI for India

Indian States	HDI Value
Madhya Pradesh	0.341
Uttar Pradesh	0.343
Tamil Nadu	0.432
Kerala	0.597
West Bengal	0.452
Haryana	0.476
Maharashtra	0.513
Punjab	0.516
India (Average)	0.436

Source : Mahbub ul Haq 1997.

Table 3. Paradigm Shift Recommended by the Swaminathan Committee (1995)

Existing	Swaminathan Committee
Strategy	
Target- and technology-driven	Human- and social development-centered
Approach	
Think and plan centrally, and act locally	Think, plan, and act locally, and support nationally
Awareness generation	
National slogans, symbols, and educational strategies	Sensitisation and self-awareness of rural and urban communities concerning the population supporting capacity of their ecosystem
Planning tool	
Five year Plan of the Department of Family Welfare, Government of India	Socio-demographic charter for the village/town prepared by the people
Delivery services	
Contraceptive Services	Integrated health security including reproductive health and user-preferred family planning devices

five hundred million people in South Asia." Of this five hundred million children, women, and men, nearly fifty percent are in India.

The human deprivation characteristics of the SAARC region (India, Pakistan, Bangladesh, Nepal, Sri Lanka, Bhutan, and Maldives) enumerated by Haq are:

Most illiterate region:	Forty-eight percent adult illiteracy
Most malnourished region:	Fifty percent children are underweight
Least gender-sensitive region:	Gender Equality Measure of 0.23
Highest human deprivation:	Human deprivation affects over five hundred million people

Inter-state variation in India in human and social development indicators sheds light on the pathways we should adopt for achieving a balance between human population and the supporting capacity of the ecosystem.

A Government of India committee for drafting a national population policy statement for the consideration of Parliament recommended a paradigm shift in our population stabilization strategy, with a view to achieving a Total Fertility Rate (TFR) of 2.1 by the year 2010.

Unfortunately, a well-defined population policy is yet to be adopted by Parliament on the basis of an all-party consensus. I can only repeat what we said in our report: "If population policies go wrong. nothing else will have a chance to succeed."

Energy, in many ways, is a critical component in meeting the challenges of food security. The major input of energy into food production is in the form of mineral nitrogenous fertilizers, besides energy that goes into operating farm machinery and for pumping water. Such input of energy in agriculture has been continually on the increase since the 1950s. An estimate by Postgate suggests that the energy required to transport nitrogenous fertilizers in 2000 would be the same as the energy needed to produce the entire quantum of fertilizers used in 1975. The utilization of electrical energy in pumping of water has been on the increase, particularly since power is available free to farmers in many States of India. Nonetheless, it is essential that measures to improve the efficiency of input energy are implemented. Such measures in the past required considerable amount of capital, which acted as a disincentive. However, it is possible to improve efficiencies with critical input of knowledge.

Wood fuel is an important source of rural energy, although the exact proportion of cooking and other requirements met by this fuel has not been estimated with certainty. With land under forest cover diminishing, this important extraction has been on the decline. Therefore, there is need to create a strategy for the promotion of forestry practices that help us meet the increasing demand for wood fuel in rural areas, and such an approach must be an integral aspect of an agricultural policy designed to meet the challenge of food security.

Promoting improved efficiency is an essential step, but only a first step in sustainable development. A critical way that current inefficient patterns of energy utilization can affect food security and sustainable development is through forcing climate change. The work carried out by the Intergovernmental Panel on

Climate Change (IPCC) has established that the occurrence or onset of climate change due to emission of greenhouse gases is a near certainty. The implications of climate change for Indian agriculture have been documented in detail in a workshop (December 1995) organised by the MSSRF. Some of the probable adaptations predicted are: a) Warming by 1° C would require major rearrangement of crop calendar in low latitude regions; b) In the tropical regions, the frequency and extent of area over which losses of agricultural output could occur from heat stress are likely to increase. This is particularly liable for the wheat-producing regions in North India; c) The studies carried out by the International Rice Research Institute (in twenty-two countries over eight years) indicate that rice yields are significantly affected by increase in temperature and solar radiation.

While carbon dioxide enrichment will have positive impact on yield, changes in temperature and precipitation will negate this advantage. On the whole, it is estimated that under the changing climate scenario, it is considerably more expensive to meet food security requirements. Thus, attention will have to be paid to increase efficiencies and effect newer fuel-mixes in the context of global food production as well.

In building a paradigm of sustainable development, it is necessary to link concepts and procedures at the macro-level (such as the concepts and measurements relating to climate change or the international conventions /frameworks) with field-level practices to arrive at a way of translating a conceptual response into a reality. MSSRF has been engaged in the development of integrated, intensive farming systems (IIFS) as an example of sustainable agricultural development that integrates notions of equity, efficiency, and conservation. Whole villages are being enabled to adopt such farming systems. This approach is essential for meeting the triple goals of more food, income, and jobs from available land and water resources.

We have excellent opportunities today to achieve the goals of food, health, literacy, and work for all in rural India, provided we take advantage of the power of both grassroots-level democratic institutions and the tools of the information age. To take advantage of such opportunities, we need a change in our approach to poverty alleviation and the conservation of natural resources. The poor are poor only because they have no assets, such as land or livestock or special skills. Hence, they tend to work as unskilled labor. Asset building and technological information, and organizational empowerment are, therefore, essential to add value to the time and work of the poor.

Past experience shows that the trickle-down approach in relation to helping the poor does not work in economics or in extension or in environmental protection. If the poorest farmer is helped with new technologies, all farmers tend to benefit.

Similarly, if women are helped, all members of a household benefit. In both these cases, the reverse does not happen. The same is true in relation to ecological security. Unless the entire population participates, protection of the environment is not easy.

If the past fifty years led to the Green Revolution in agriculture and impressive progress in industry and commerce, the next fifty years should witness the emergence of a nation known for tolerance and love of diversity and pluralism in terms of religion, caste, gender, language, ethnicity, and political belief and for achieving an ever-green revolution in agriculture and industry. That these two goals are inter-related was pointed out by Mahatma Gandhi when he said over sixty years ago: "How can we be non-violent to nature, if we are going to be violent to each other?"

How do we achieve the twin goals of "unity in diversity" in day-to-day life and "green productivity" in every area of economic activity? In my view, the following seven steps are necessary to help us to arrest the current trend towards social, political, and economic disintegration and help the nation to become one.

a) A demographic transition to low birth and death rates is vital for achieving the goals of food, drinking water, health, literacy, and jobs for all. Population stabilization should be a shared concern and responsibility of both men and women. With over a million women in elected panchayats, it should now be possible for the elected members of panchayats and municipal councils to draw up socio-demographic charters for their respective village/town, which can lead to achieving a balance between human numbers and the population supporting capacity of the ecosystem.

b) Globally and nationally, the rich-poor divide is growing. Today, we have as many people below the poverty line as the entire population of India in 1947. Globally, 358 billionaires have more income than 2.5 billion of the world's poor. Unsustainable lifestyles and unacceptable poverty co-exist in every city, town, and village. This situation must change if a *sarvodaya* society is to emerge. The transition from destitution to secure livelihoods can happen fast, if the rich consider themselves as trustees of their surplus wealth and adopt a caring and sharing ethic. It will be in their self-interest to do so. Gandhiji's concept of *antyodaya* as the pathway to *sarvodaya* must become the guiding spirit in developmental planning.

c) India is one of the least gender sensitive nations in the world. The sex ratio is very adverse to women. A beginning has been made to reverse this trend through the sociopolitical empowerment of women at the level of grassroots democratic institutions. This process should extend to Parliament and must be coupled with the technological and skill empowerment of women. A gender code must guide all developmental and socio-economic and sociopolitical activities.

d) We should eschew the path of jobless growth of today's industrialized and rich nations. We should instead promote a job-led economic growth strategy based on a pro-nature, pro-poor, and pro-women orientation to technology development and dissemination. This will call for a movement to spread market-driven micro-enterprises supported by microcredit.

e) Our population is predominantly young. We should actively promote teaching happiness in schools. Teaching happiness involves inculcating a love of nature – flora, fauna, rivers, oceans, forests, mountains. Young people should know that it takes over a hundred years in nature to make one cm of top soil and that we destroy several meters of top soil within a few days in brick kilns. Again,

Gandhiji's dictum that "nature provides for everyone's needs but not for everybody's greed" should be internalized in the habitual behavior of every citizen. At the same time, it must be realized that it is not adequate to just oppose unsustainable development, it is equally important to just oppose unsustainable development – it is equally important to propose sustainable options. We need an ecology of hope and action.

f) Industry should move towards adopting ISO 14000 standards in relation to environmental management. In agriculture, more food and other agricultural commodities will have to be produced without ecological harm from less per capita arable land, irrigation water, and fossil fuel based energy sources. This will call for a massive effort in developing and disseminating ecotechnologies based on a blend of frontier technologies such as biotechnology and space, information, and renewable energy technologies, and the best in traditional wisdom and ecological prudence.

g) The growing marginalization of rural economies is leading to the unplanned migration of the rural poor to urban areas, thus inducing the proliferation of urban slums. Strengthening rural livelihoods through the effective use of panchayats in the technological empowerment of socially and economically underprivileged families should receive attention. Innovative approaches are needed to create more wealth and jobs in rural areas. For example, all along our vast coastline "green health" tourism can be promoted, based on our traditions of Ayurvedic and herbal systems of health management.

On the basis of a recent study, the International Food Policy Research Institute (IFPRI), located in Washington DC, believes that by the year 2020 significant reductions can be made in poverty, food insecurity, malnutrition, and micronutrient deficiencies and significant advances made towards efficient, effective, and low-cost food and agricultural systems that are in harmony with the natural resource base. The fact remains that, although today the world produces enough food to feed each and every person, nearly 800 million children, women, and men remain food insecure. The reasons for this tragic irony are to be found in the non-food sector, such as lack of purchasing power and entitlements, poor sanitation and health care, and non-availability of clean drinking water. It is therefore time that we redefine our concept of food security as "livelihood security for the households and all members within, which ensures physical and economic access to balanced diets including the needed micronutrients, safe drinking water, environmental sanitation, basic health care and primary education." Such a denotation will help to articulate the public policy package needed to ensure food security at the level of each individual in a household. The emphasis on individuals rather than on families is important, since, in many societies, women and girl children tend to suffer more from under- and malnutrition and there is a trend towards an increasing feminization of poverty.

In view of the increasing globalization of economies, we also need action at the international level. I would like to refer to a few of these: a) Implement both in letter and spirit the plans of action agreed to at recent UN Conferences, particularly at Rio, Cairo, Copenhagen, and Beijing, and take speedy action to implement the provisions of the Desertification Convention and the UN Law of the Sea Convention; b) Ensure adequate support for research and development efforts designed to promote national and international public good, since the fast spreading Intellectual Property Rights (IPR) environment is likely to promote greater investment only in research that is of a short gestation period and profit-driven. Integrating the dimension of sustainability in research objectives, however, involves location specific and participatory research with rural families. This kind of research is unlikely to attract adequate investment from the corporate sector; c) Bring about the rapid dissemination of environmentally sound agricultural technologies, particularly among resource-poor farm families who may have difficulties in gaining access to green technologies as a result of the provisions of the Trade-related Intellectual Property Rights (TRIPS) of the World Trade Agreement; d) Promote increased trade in value-added agricultural commodities produced in developing countries by assisting them in improving post-harvest technology and sanitary and phytosanitary measures; e) Contribute generously to global emergency food security reserves and to food for livelihood security programs through the World Food Programme and other appropriate global, regional, and national social security programs and organize a global coalition for the sustainable end of hunger so that every child, woman, and man has the requisite entitlements for a healthy and productive life; f) Mobilize the power of the information age to spread timely details among farming communities on meteorological, management (land, water, pests, and post-harvest technology), and marketing factors.

To achieve these goals, however, we have to go back to Gandhiji who wrote in his autobiography: "The most important lesson I learnt in my life was from my illiterate mother who taught me that all rights accrue from a duty well done." ▯

References

Carson, Rachel. 1962. *Silent Spring*. Boston: Houghton Mifflin Co.

Calborn, T., D. Dumanoski and J.P. Myers. 1997. *Our Stolen Future*. New York: Penguin Books.

Bose, Ashish. 1996. *India's Population Policy: Changing Paradigm*. Bangalore, India: Bangalore Urban Art Commission.

Haq, Mahbub ul. 1997. *Human Development in South Asia 1997*. Oxford: Oxford University Press.

14

Indira Gandhi Prize for Peace, Disarmament and Development

Citation of Award by Shri K.R. Narayanan, President of India, New Delhi, November 19, 2000

The International Jury of the
Indira Gandhi Memorial Trust
herewith awards
The 1999 Indira Gandhi Prize
for Peace, Disarmament and Development
to
Prof. M.S. Swaminathan,
eminent agriculture scientist
Chairman, M.S.S. Research Foundation,
for his outstanding contribution
in the domain of plant genetics
and
ensuring food security
to hundreds of millions of citizens
in the developing world.

Acceptance speech by Prof. M.S. Swaminathan, November 19, 2000

On behalf of my wife, and family and co-workers in the scientific and farming world, I express my sincere gratitude for the privilege of receiving this prestigious Prize bearing the name of one of the unique leaders and visionaries of the 20th century. I particularly value this honor, since the Prize is linked to peace, disarmament and development. The futility of war as a means of solving problems either between nations or among individuals has been emphasized in our history by both kings and saints. The life and message of Emperor Ashoka serve as a classic example. Even successful war leaders in industrialized countries have expressed similar convictions at the end of wars. For example, General Dwight D Eisenhower stated in 1953: "Every gun that is made, every warship launched, every rocket signifies in the final sense a theft from those who are hungry and are not fed, from those who are cold and are not clothed. The world in arms is not spending money alone. It is spending the sweat of its laborers, the genius of its scientists, the hopes of its children."

The International Peace Research Institute in Oslo, after a careful analysis of the causes of armed conflicts within nations during the decade of the nineties, concluded that the reasons for conflicts were invariably associated with struggles for entitlements, particularly to food, water, land and shelter. It may be worth mentioning that when I joined the Union Planning Commission in March 1980, as the Acting Deputy Chairman, Indiraji mentioned that her priorities for the Sixth Five Year Plan were "work, water, food and power". The Oslo Peace Institute has suggested that investment in agriculture, which is the backbone of the livelihood security system in a majority of developing countries, is the most effective strategy for preventing future wars. The International Commission on Peace and Food, which I had the privilege of chairing, also stressed in its Report submitted in 1994 that numerous mini conflicts arising from hunger and poverty are now taking the place of mega wars. Mahatma Gandhi often stressed that *sarvodaya* or a society characterized by high social synergy, is the means to achieve lasting peace and non-violence, and that *antyodaya* is the path to *sarvodaya*.

Respected Rashtrapathiji, it is a signal honor to receive this Prize from your hands. In your introspective analysis of the state of our nation on the occasion of the Golden Jubilee of our Republic, you rightly emphasized, "Fifty years into our life in the Republic, we find that justice – social, economic and political – remains an unrealized dream for millions of our fellow citizens. We have one of the world's largest reservoirs of technical personnel, but also the world's largest number of illiterates; the world's largest middle class, but also the largest number of people below the poverty line, and the largest number of children suffering from malnutrition. Many a social upheaval can be traced to the neglect of the lowest tier of society, whose discontent moves towards the path of violence."

The Roman philosopher Seneca expressed over 2000 years ago a similar view when he said, "a hungry person listens not to reason nor cares for justice, nor is bent by any prayers". Hunger in our country today is not related to the availability of food in the market but is caused by inadequate opportunities for sustainable livelihoods resulting in insufficient purchasing power. Since a famine of purchasing power is responsible for the famine of food at the household level, food security is best expressed in terms of million person years of jobs, rather than in million tons of foodgrains.

From the beginning of the Industrial Revolution in Europe, technology has been a major source of economic inequity among nations and among communities within nations. Fortunately, thanks to the support and encouragement extended by successive Prime Ministers, from Pandit Jawaharlal Nehru to Shri Atal Bihari Vajpayee, our national capability in frontier areas of science and technology, such as biotechnology, space, nuclear, information and communication and renewable energy technologies as well is impressive. Our national strength in biotechnology opens the door to the conversion of our rich bio-resources into economic wealth, in a manner which is meaningful to the poor. Therefore, if technology has been a cause of economic and social inequity in the past, we now have an opportunity for enlisting technology as an ally in the movement for social, gender, and economic equity. Modern information technology provides this opportunity. Knowledge and skill empowerment can now be achieved at a fast pace, if we shift our approach from patronage to one of genuine partnership with the poor in the design and operation of rural knowledge centers.

While we should admire the prospects of progress and prosperity promised by the virtual world, it would be foolish to overlook the state of póverty, hunger, malnutrition and environmental degradation prevailing in the real world. In her historic address to the UN Conference on the Human Environment held at Stockholm in 1972, Indiraji explained with clarity and conviction the linkages between poverty and degradation of the environment.

My work carried out in collaboration with numerous colleagues, students and farm women and men during the last fifty years relating to a hunger-free India and world, has been greatly influenced by the concept, "the test of our progress is not whether we add more to the abundance of those who have much; it is whether we provide enough for those who have little." Translated into action, this has meant concentration on improving the productivity of small farm holdings and the well-being of the economically and socially underprivileged women and men who toil day and night and in sun and rain to produce food for their fellow human beings. I am firmly convinced that hunger and deprivation can be substantially reduced by August 15, 2007, which marks the 60th anniversary of our Independence, provided we launch a community led and controlled food and water security system with the help of Gram Sabhas and elected local bodies. I am glad such a system will be an integral part of the Gram Raj movement being launched in Madhya Pradesh on January 26, 2001.

My experience in India and in many countries of Asia and Africa and the Caribbean has taught me that the poor take to new technologies like fish to water, provided they are enabled to do so through opportunities for "learning by doing" and for producer-oriented marketing. Imparting greater economic value to the time and labor of the poor through technological and knowledge empowerment should, therefore, receive overriding priority. The United Nations has designated the first decade of this century as the International Decade for a Culture of Peace and Non-violence for the children of the world. Every third child born in our country is characterized by low birth weight, which in turn affects adversely brain development. Low birth weight can be due to several causes, but the most common cause in our country is maternal and fetal malnutrition and under-nutrition. Denying a child the opportunity for the full expression of its innate potential for mental and physical development is the cruelest form of inequity and violence. The French mathematician Marquis de Condorcet, a contemporary of Thomas Malthus, said over two centuries ago that population will stabilize itself if children are born for happiness and not for mere existence. Can we therefore undertake a well-planned program for the eradication of maternal and fetal under-nutrition by August 2007? With over forty million tons of grains in government stocks, this is not an impossible task. Also, this will be an investment of profound significance for the future intellectual capability of our nation.

I mentioned earlier that agriculture, comprising crop and animal husbandry, fisheries, forestry and agro-forestry and agro processing, is the backbone of our livelihood security system. Our agriculture is now at the crossroads. Prime farmland is going out of agriculture all the time and the average size of holdings is becoming smaller each year.

Ground water table is receding at a fast pace in many areas. The smaller the farm, the greater is the need for marketable surplus, so that small farmers can have a reasonable income. Unfortunately, markets are also becoming volatile, causing acute distress to small and marginal farmer families. Let me cite one example. Eastern India is regarded as the sleeping giant of Indian agriculture. However, thanks to the steps taken by the Central and State Governments to promote irrigation through shallow tube wells during the non-flood season, the small farmers of Assam have brought about a silent revolution in the productivity and production of rice. The price of rice has however crashed and farmers will have hard time in repaying the loans they have taken. Hence, a systems approach giving concurrent attention to all the links in the production-processing-consumption chain, or what the late Prime Minister Rajiv Gandhi termed "Technology Mission Approach" is essential, if farmers are to benefit from their success in improving production.

At the global level, we face a confrontation between products resulting from mass production technologies, and those produced by the Gandhian concept of production by masses. For example, the over eighty million tons of milk produced now in our country are the result of the hard work of nearly eighty million women. In contrast, the over seventy million tons of milk produced in the United States of America involve probably less than 200,000 persons. The total number of farm families in USA is about 900,000 in contrast to over one hundred and five million in India. Since the onus of livelihood generation falls on the farm sector, it is important that policies relating to the import of agricultural commodities are based on a careful analysis of their impact on rural and urban livelihoods, particularly of those living below the poverty line.

The recipe for poverty alleviation proposed by international organizations and bilateral agencies is micro-enterprises supported by micro-credit. National macro-economic policies and global trade policies, however, often threaten the survival of micro-enterprises. There is no level playing field between the technology, capital and subsidy driven macro-enterprises of the industrialized countries, and micro-enterprises functioning under conditions of poor infrastructure, low investment and high risk. The fast expanding transnational supermarkets are threatening the livelihoods of small-scale traders and vendors. The absence of venture capital for micro-enterprises further compounds the problems of the poor. The culture of eternally "talking poor and living rich" characteristic of many international and national organizations will only lead to the perpetuation of poverty and not to its eradication.

I applaud the Government of Bhutan for its choice of Gross National Happiness as the index of progress. The major components of this index are: environment protection, economic growth, cultural promotion and good governance. The component of cultural promotion includes spiritual and ethical values. Without ethics, technological progress may become a curse rather than a blessing. Without the same love for diversity and pluralism in human societies, as for biodiversity in plants and animals, lasting human peace will not be possible.

Experience indicates that human population tends to stabilize itself if children are born for happiness. If we wish to improve Gross National Happiness, children constitute the natural starting point for action. I would like to end by quoting a poem on children by Gurdev Rabindranath Tagore, which brings out the beauty of a child's value system.

Child, how happy you are sitting in the dust
Playing with a broken twig all the morning;
I smile at your play with that little bit of broken twig
I am busy with my accounts adding up figures by the hour
Perhaps you glance at me and think what a stupid game
  to spoil your morning with,
Child, I have forgotten the art of being absorbed in sticks and mudpiles
I seek out costly playthings and gather lumps of gold and silver
With whatever you find, you create glad games;
I spend both my time and my strength over things I can never obtain;
In my frail canoe I struggle to cross the sea of desire;
And forget that I too am playing a game ...

Children living in poverty are like blossoms in the dust. The fund associated with this Prize will be dedicated to enable at least a few such blossoms to flower and bear fruit. ▯

53. With Prime Minister Indira Gandhi and farmers, 1968

15

Food Security and Sustainable Development

In 2001, the M.S. Swaminathan Research Foundation and the United Nations World Food Programme brought out a *Food Insecurity Atlas of Rural India*. This important publication reveals that every State in the country has its strengths and weaknesses in relation to the five major dimensions considered in the analyses. These are: availability of food, which is a function of production; access to food, which is related to purchasing power; absorption of food in the body, which is determined by the availability of safe drinking water, environmental hygiene, primary health care, and primary education; vulnerability to transient hunger, which is related to natural and man-made calamities and disasters; sustainability of production, which is influenced by the extent of attention given to the ecological foundations essential for sustained advances in production.

The Atlas makes clear that non-food factors, like livelihood and income-earning opportunities, health care facilities, education, sanitation and environmental hygiene are as important for food security at the level of every individual. It also provides an opportunity for State governments to draw up food security balance sheets based on strengths and weaknesses, and to identify "hotspots" with reference to endemic and transient hunger as well as to open (i.e., protein-calorie under-nutrition) and hidden (i.e., micronutrient deficiencies) hunger.

While we should give the highest priority to improving food consumption and equitable distribution, we should not decelerate our efforts in improving agricultural production through yield improvement, higher factor productivity, and better post-harvest management. Agricultural production, factor productivity, and investment in irrigation and post-harvest and rural infrastructure are all declining in India. Prices of many agricultural commodities have collapsed and Indian farm families are in deep economic and psychological distress. This trend, if not arrested immediately, will lead to social chaos, since agriculture (crop and animal husbandry, fisheries, forestry, agro-processing, and small-scale agri-business) is not just a food-producing machine, but the backbone of the livelihood security system for nearly seven hundred million people in the country. There is no time to relax on the food production front, just because the major problem today is in the area of marketing and distribution. At the same time, we should not continue to remain silent spectators to the co-existence of mountains of grains and millions of hungry.

Keynote address presented at the "Third Asian Conference of Agricultural Economists on Sustainable Agriculture, Poverty and Food Security in Asia: The Perspectives for the 21st century," Jaipur, India, October 18-20, 2000. First published in *From Rio de Janeiro to Johannesburg*, by M.S. Swaminathan, pp. 109-28. Chennai: EastWest Books (Madras), 2002.

From Green to an Evergreen Revolution

The first sixty years of the twentieth century were marked by a sense of despair and frustration regarding India's capability of achievin a balance between human numbers and the production of foodgrains and other agricultural commodities. In 1968, this mood of despondency and diffidence gave way to one of optimism and self-confidence in relation to our agricultural potential and our farmers' ability to adapt and adopt new technologies, a phenomenon which was christened in that year as the Green Revolution. This agricultural transformation helped to strengthen national sovereignty in many areas, including the capacity to remain non-aligned in foreign policy.

Our agriculture is now at the crossroads. On the one hand, our national capability in frontier areas of science and technology – as, for example, in biotechnology, information, communication and space technologies, nuclear and renewable energy technologies, and in management science – has opened up uncommon opportunities for achieving an evergreen revolution. I use this term to denote sustainable advances in crop productivity per units of land, water, and time, without associated ecological harm.

Thus, sustainable food security will have to be defined as "physical, economic, social, and ecological access to balanced diets and safe drinking water, so as to enable every individual to lead a productive and healthy life in perpetuity". A life-cycle approach will have to be followed in the case of nutrition, ranging from in utero to old age. Achieving such a form of food security will require synergy between technology and public policy (Figure 1). There are, on the other hand, both internal and external threats to our agricultural progress. The most important among the internal threats is the damage to the ecological foundations essential for sustained agricultural advance – land, water, forests and biodiversity. The other major internal weakness is the mismatch between production and post-harvest technologies, and the consequent need for the Government of India to undertake "trade relief" operations on lines similar to those of cyclone, flood, and drought relief.

Figure 1. **Whole Lifecycle Nutrition**

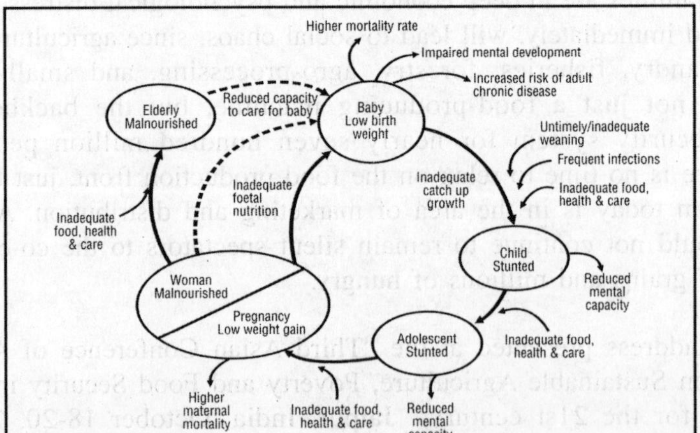

Nutrition throughout the life cycle

The external threats include the unequal trade bargain inherent in the WTO Agreement of 1994, the rapid expansion of proprietary science, and potential adverse changes in temperature, precipitation, sea level, and ultra violet ß radiation.

The global threats to the agricultural destiny of developing countries can be overcome only if industrialized countries, particularly the United States of America, are willing to take the following steps: a) Ensure that the Kyoto protocol relating to the Climate Convention is implemented; b) Extend adequate support to public good research at the national and international levels, thereby fostering a new social contract between science and society (Swaminathan 1999); c) Revise the Agreement on Agriculture of the World Trade Organization in a manner that trade becomes a powerful tool for poverty eradication. At present, there is no level playing field between the produce emanating from factory farming and farmer's farming (i.e., small-scale producers).

Sustainable Development in Agriculture

The smaller the farm, the greater is the need for marketable surplus to ensure cash income. The gap between potential and actual yields is high in most farming systems. Even in the case of rice and wheat, the present average yield is just 40 percent of what can be achieved even with technologies currently on the shelf. Therefore, a massive effort should be made to launch a productivity revolution in farming. An integrated approach is necessary to remove the technological, infrastructure, social, and policy constraints responsible for the productivity gap and, in some cases, productivity decline. Reducing the cost of production through eco-technologies and improving income through efficient production and post-harvest technologies will help to enhance opportunities for both skilled employment and farm income. Precision farming methods which can help to enhance income and yield per drop of water and per units of land and time need to be standardized, demonstrated, and popularized speedily, if a reduction in the cost of production is to be achieved without reduction in yield.

As an immediate measure for strengthening food security at the level of individuals and households, there is no better option than initiating a systematic effort in each agro-climatic zone to identify and remove the constraints responsible for the prevailing yield gaps. This is true not only of crop plants but also of livestock and fisheries. The local panchayati raj institutions or other forms of local bodies should be fully involved both in identifying constraints that limit production and in removing them.

Some of the measures needed to conserve land for agricultural purposes are: a) Arresting land degradation and the loss of the biological potential of the soil (desertification); b) Promoting land and water use on the basis of agro-ecological, meteorological, and marketing factors; c) Restoring degraded and wasted land through agro-forestry and other appropriate methods of restoration ecology; d) Launching community-centered water harvesting, conservation, and use programs to ensure the efficient harvest of rainwater and the sustainable use of groundwater. The adoption of conjunctive water use practices ensures integrated use of river, rain, ground, treated sewage and seawater, in such a manner that every drop yields more crop and income; e) Introducing public policies to prevent the diversion of prime farm land for non-farm uses and the unsustainable exploitation of groundwater.

There are currently several Central and State government programs dealing with wasteland development, rainwater harvesting, watershed development, command area management, shallow tubewell construction, social and agro-forestry, and prevention of damage to hydrologic cycles in hill areas. There is an urgent need for convergence and synergy among these programs so that land and water conservation and use can be dealt with in a scientific and holistic manner. Land and water management problems (the term "management" is used to denote concurrent attention to conservation, sustainable use, and equitable sharing of benefits) are multi-dimensional. So, uni-dimensional approaches through numerous independent schemes implemented by separate departments of Central and State governments will only result in inefficient and ineffective use of financial and technical resources.

Schedule 11 of the Constitution Amendment 73, relating to panchayati raj institutions, entrusts to panchayats/local bodies, responsibilities for the management of land, water, and common property resources. If these bodies, in which a third of the members are women, are legally, technically, and financially enabled to discharge the functions listed in Schedule 11, a beginning can be made in fostering land and water management in an environmentally and socially sustainable manner. There are some legislative, administrative, and legal hurdles in the way of decentralized planning. Hence, it is important to address these issues immediately.

The existing State Land Use boards should be revitalized and reorganized in such a manner that they can give pro-active advice to farm families on land use during the south-west and north-east monsoon periods based on the following factors: a) Farming systems (crop, livestock, fish, and agro-forestry) which will be most efficient under the given soil, water, and climatic conditions; b) Short- and medium-range weather forecasts (the country has developed considerable capability in this area); c) Projected market demand (both home and external markets); d) Cost of production, risks involved, and expected return; e) Potential for on-farm and non-farm livelihood generation, so as to maximize income and employment per units of land and water

If such advice is given at least several weeks before the sowing season, a proper match can be achieved between production and potential market demand. Uneconomic market interventions can then be avoided. The agro-ecological potential of every village can be utilized in an ecologically and economically optimum manner. The "Blue Box" of the Agreement on Agriculture of the World Trade Organization provides for expenditure on achieving a balance between demand and supply in farm products. Seed banks of alternative crops will have to be established at the local level.

Figure 2. Proactive Advice on Land Use (Land use decisions are also Water use decisions)

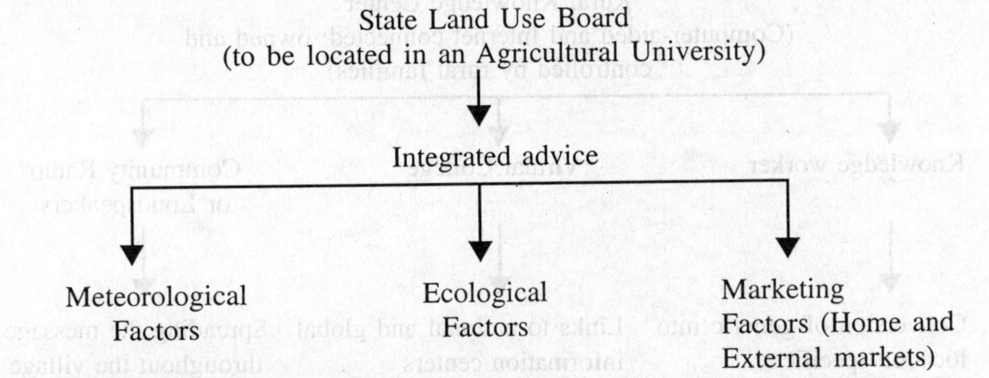

The Land Use Board through a virtual college should give proactive advice on the choice of crops and farming systems, so as to achieve a match between demand and supply in farm commodities and to ensure that the most efficient crops are grown in different agro-climatic and agro-ecological regions.

The reorganized Land Use Boards should also be able to develop contingency cropping patterns to suit different rainfall and water availability patterns. Thanks to the long-range weather data available with the Meteorological Department, it is now possible to develop computer simulation models of likely deviations in monsoon behavior. These can be used for formulating land use advice based on GIS maps, which also take into consideration the moisture holding capacity of soils, physiological efficiency of crops, home needs, and market demand (Figure 2).

If these steps are taken, we can promote land use based on considerations of both ecological sustainability and economic efficiency. Since land use decisions are also water use decisions, land and water care and use are best dealt with in a simultaneous and interactive manner. For example, if the ongoing Technology Missions in crops like oilseeds, pulses, maize, and cotton are linked to the watershed development and dry farming programs, these missions will become more effective.

Reorganization of Extension Services

Advances in information technology also provide opportunities for farm graduates to establish computer-aided and Internet-connected rural knowledge centers. These centers should help to convert generic into location-specific information. The present extension service has outlived its utility. It can be replaced over time by farmer-owned-and-operated knowledge centers. A virtual college linking such village knowledge centers to agricultural universities and research institutions can be established, so that farm women and men are able to get up-to-date and authentic technical advice. Nearly a million farm graduates (both men and women) can be involved in establishing and operating such rural knowledge centers based on modern information and communication technology. Such centers can also operate local community radio stations. Such

Figure 3. Restructuring and Retooling Extension Services for an Era of Precision Farming

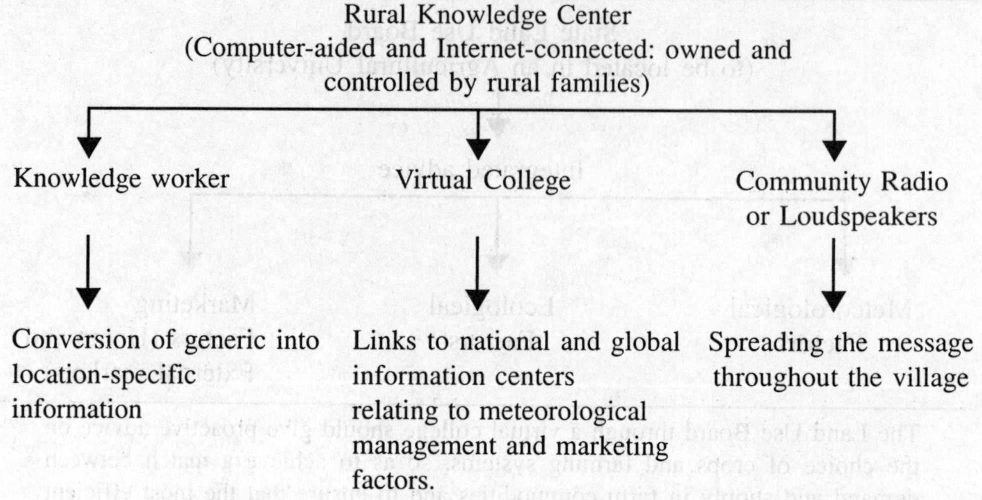

All existing extension workers can be retrained to serve as Rural Knowledge Workers under the control of Local Bodies

a restructuring and retooling of extension services will help to provide farming systems of specific value to farmers. They will trigger a knowledge revolution in agriculture and will lead to an efficient and eco-sensitive precision farming movement. This great opportunity for achieving a transition from unskilled to skilled work and for designing a new extension service for the new economy should not be missed (Figure 3).

Linking Food and Ecological Security

The provisional figures for Census 2001 reveal that our population is higher by nearly twenty million than expected and that sex ratio continues to be adverse to women (933 women per 1,000 males). The sex ratio is even more adverse in the 0-6 age group, suggesting the possibility of increasing female feticide. Fortunately, female literacy is improving and Madhya Pradesh has shown the way for achieving a quantum jump both in male and female literacy through its education guarantee program. While we must relentlessly pursue the goals of literacy and health for all as well as gender justice and equity, we should take advantage of a rare and unique opportunity in the history of independent India provided by the growing grain stocks in government godowns to strive in our effort to realize Gandhiji's vision of a hunger-free India.

Current government stocks of wheat, rice, and other grains exceed sixty-five million tons. A considerable proportion of these stocks remains in gunny bags and temporary storage structures. The Government of India has announced a scheme for the construction of large numbers of rural godowns. Severe drought in several parts of Rajasthan, Gujarat, Madhya Pradesh, and other States is compounding the problems of poverty-induced endemic hunger and drinking water scarcity.

The time is, therefore, opportune to launch an imaginative movement. On an average, one ton of wheat or rice supports the food needs of five individuals in our country over a year. Community Grain Banks (CGB), each with two hundred tons of wheat or rice or other locally acceptable staples like ragi, jowar, bajra, and maize could be established, to begin communication, as the desert areas of Rajasthan, and hill, tribal, and drought affected areas, can be given priority in starting the CGB movement. Twenty-five thousand grain banks can be established during the next few months if the Government of India will approve immediately the release of five million tons of grains for this purpose. Because large quantities of government stocks are in gunny bags, it is easy to move them to CGBs where they can be stored using the low cost technology standardized by the Food Corporation of India. It will be sad if government sits over sixty million tons of foodgrains, allowing some of it to rot, rather than take them to places where, in Gandhiji's words, "God is bread".

Based on the experience of the initial 25,000 village level grain banks, another 25,000 can be established later this year, thus using ten million tons of the surplus stock in a socially meaningful manner. Let us go into the new millennium with decisive action and resolution to provide every individual in the country an opportunity for a productive and healthy life.

CGBs can be sustained with locally procured grains, wherever feasible. They should be linked to the rural godowns scheme. The banks could function under the overall umbrella of the Gram Sabha, and can be operated by local self-help groups of women and men. This will ensure their relevance to local conditions in addition to involving low transaction costs. They could be used for initiating at the local level food-for-work, food-for-nutrition (i.e., distribution of food among pregnant and nursing mothers, infants and old and infirm persons), wasteland and watershed development, ecological restoration of common property resources, and for establishing community water banks. They can also be the vehicles for operating the targeted public distribution, Antyodaya Anna Yojana, and other Central and State Government schemes. Thus, CGBs can become instruments of eco-restoration, water, harvesting, and hunger elimination.

We should link conversation, cultivation, and consumption in a mutually reinforcing manner. For this it will be useful to foster the establishment of community gene, seed, water and grain banks in every village (Figure 4).

Increasing Production and Productivity

Future agricultural production programs will have to be based on a strategy designed to foster an evergreen revolution, which leads to increased production without associated ecological and social harm. There are four major elements in this strategy for producing more in an environment-friendly manner.

Defending the gains already made. This will call for conservation and enhancement of soil and water resources as well as forests and biodiversity through an integrated package of government regulation, education, and social mobilization (through panchayats and local bodies). The traditional "Green Revolution" areas are in urgent need of such an integrated natural resources

Figure 4. Community Food and Water Security System

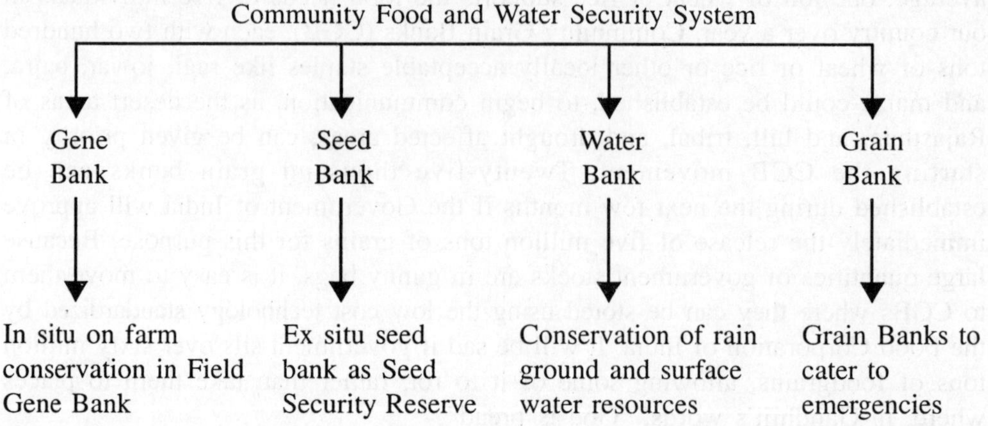

management strategy so that the pattern of present production does not erode future prospects. The Punjab, which is India's granary today, will become food insecure in fifteen to twenty years from now, if the current unsustainable land water use practices continue. Defending the gains already achieved will also need stepping up maintenance research for ensuring that new strains of pests and pathogens do not cause crop losses. Special steps are needed to prevent the introduction of invasive alien species that are coming into the country along with imported food and agricultural commodities. These invasive alien species, like new and aggressive weeds and nematodes, can cause incalculable harm to the future of Indian agriculture.

Conservation and enhancement of land and water resources is important. Water harvesting, watershed development, and economic and efficient water use can help to enhance productivity and income considerably. Conjunctive use of different water sources should become the rule rather than the exception. As I have repeatedly pointed out, unless there is equity in water sharing, there will be no cooperation in water saving. Therefore, equitable methods of water sharing should be promoted. Where water is scarce, high-value but low water-requiring crops should be grown. In this context, the organization of pulses and oilseed villages should become a national movement. This should be a major aim of the Pulses and Oilseeds Technology Missions. Solving internal shortages of pulses and oilseeds through imports will only add to the economic woes of dryland farming communities. Pulses and oilseeds are important income-earning and soil-enriching crops in dryland areas.

There are various estimates of land degradation. The Ministry of Rural Development has also published a *Wasteland Atlas of India*. The following kinds of soil degradation have been quantified:

Wind erosions	19.7 million ha
Salinization	4.1 million ha
Water logging	3.1 million ha
Water erosion	69.6 million ha
Soil fertility decline	13.7 million ha

Thus, there are vast opportunities for launching wasteland development enterprises by local self-help groups through strategies like identifying the precise nature of soil degradation and developing scientific restoration measures. Based on agro-ecological conditions, tree species that can help to initiate suitable enterprises can be chosen. For example, a plant pesticide model of wasteland development could involve the planting of neem and melia. Appropriate species can be chosen and planted, depending on soil and water conditions, for undertaking the preparation of furniture, doors, windows, etc., or raw material for paper, fiber, or fruit packaging industries.

The aim is to add value to wasteland development through an integrated strategy of restoration and commercialization. Such a twin approach will impart greater momentum to wasteland reclamation particularly in peri-urban areas.

Extending the gains. Regional imbalances in agricultural development are growing, based largely on the availability of assured irrigation on the one hand and assured and remunerative marketing opportunities on the other. North Bihar is an exception, where water is plentiful but agricultural growth is slow. Eastern India has a large untapped yield reservoir and, by and large, falls under the "green but no Green Revolution" category. West Bengal has made impressive headway during the 1990s, while more recently Assam has started making progress, thanks to a large shallow tubewell program designed to tap groundwater during rabi and summer (boro) seasons. The introduction of eco-regional technology missions to provide appropriate packages of technology, techno-infrastructure, services, and input and output pricing and marketing policies will help to include the excluded in agricultural progress.

Technologies for elevating and stabilizing yields are available for semi-arid and dry farming areas, as a result of the work done by the Indian Council of Agricultural Research institutions, State Agricultural Universities and the International Crop Research Institute for Semi-Arid Tropics. Agro-forestry and animal husbandry are extremely important in arid and semi-arid regions. Livestock and livelihoods are closely linked in such areas. A major effort in water conservation and management and land use planning is needed in all areas that have been by-passed by scientific agriculture. Attention to horticulture, with particular emphasis on post-harvest technology, will help to optimize income and employment from every drop of water. Both livestock and tree farming will provide opportunities for downstream employment. Therefore, the emphasis should be on farming systems that can optimize the benefits of natural resources in a substantial manner and not merely on cropping systems. Also, in coastal areas, there is need for a massive program of coastal systems research and development, involving capture and culture fisheries, coastal forestry and agro-forestry, and integrated crop and animal husbandry. Coastal agro-forestry and forestry can provide much of the fuel wood needs of inland areas. As in the crop sector, there is a vast untapped production reservoir in the fisheries sector. With the coming into force of the UN Convention on the Law of the Sea, India's exclusive economic zone in the oceans around the country extends to over two million sq. km.

Dry farming areas are also ideal for the cultivation of low water-requiring but high-value pulses and oilseeds. The Pulses and Oilseeds Technology Missions should be revitalized and linked to the water harvesting and watershed development projects during the Tenth Plan (2002-2007). As emphasized earlier, taking the easy option of importing large quantities of pulses and oilseeds forecloses the great opportunity for improving the economic well-being of farm families in dry farming areas through improving the production and productivity of pulses and oilseeds.

Making new gains. During the past decade, discoveries in information and biological technologies have contributed in unanticipated ways to fundamental changes in the global economy and to unprecedented economic growth, particularly in industrialized countries. There are also growing bonds of partnerships between universities and industries. In the US, for example, industrial research parks surround leading research universities. We have more than forty agricultural, animal sciences, fisheries, and rural universities, in addition to numerous agricultural and forestry research institutions in our country. Universities and research institutions should serve as the engines of growth in a knowledge-based economy. They should also address, through their research and training agenda, the great challenges that confront our country in terms of poverty and the lack of basic human needs.

Detailed agro-climatic and soil maps are available for the country. Watershed and Wasteland Atlases are also available. We have considerable capacity in remote sensing and GIS mapping. These should be used for developing improved farming systems, which can provide more incomes and jobs. Value addition to primary products should be done at the village itself. Integrated crop-livestock-fish production systems should be fostered. Opportunities for non-farm employment will then improve.

Institutional support. Higher production can be sustained only if there are opportunities for assured and remunerative marketing. A major challenge relates to reducing the cost of production by improving productivity. This will call for appropriate institutional structures which can help to provide key centralized services to small and marginal farm families and to equip them with the power of scale in eco-farming (i.e., integrated pest management, scientific water management, integrated nutrient supply, precision farming, etc.) as well as in marketing. The role of the Small Farmers Agri-business Consortium (SFAC) that was established for this purpose should be reviewed and appropriate institutional structures, owned and controlled by farm families, should be promoted. Federations of self-help groups, farmer controlled cooperatives, corporate business entities, and other socially relevant institutional structures should be promoted. Without enhanced efficiency in the production and post-harvest phases of agriculture, Indian farmers will not be able to face the challenge of globalization in terms of cost competitiveness, quality of produce, and consistency of supply.

It should not be forgotten that but for the existence of a very capable and professionally run National Dairy Development Board as well as a dairy farmers' cooperative movement, we would not have been able to achieve the first position in milk production in the world. User controlled and driven institutional structures characterized by low transaction costs are essential to provide the needed assistance in post-harvest technology, like drying, storage, processing, and marketing.

Adequate food availability is necessary both for stabilizing prices and ensuring the operation of an effective public distribution system. There is therefore no time to relax on the food production front. There is particularly an urgent need for greater investment in irrigation, power supply, rural roads, cold storages, godowns, and food processing units. By extending the benefits of technological transformation and institutional reform to more areas and farming systems, India can become a leader in world agriculture.

Policies for Improving Economic Access to Food

As early as 1856, Colonel Baird Smith, who investigated the causes of a serious famine in north-west India, wrote: "Indian famines are famines of work, and not of food. Where there is work, there is money. Where there is money, there is food." This situation is as relevant today as it was one hundred and fifty years ago. Food security in India is best described in million person-years of jobs and livelihoods rather than in million tons of foodgrains.

Agriculture, comprising crop and animal husbandry, inland and marine fisheries, forestry and agro-forestry, agro-processing and agribusiness, constitutes the backbone of the livelihood security system of India, particularly in rural areas. Our agriculture is still "farmers' farming", and not "factory farming" as in industrialized countries. This is our great strength, since the health of plants and animals and other hazards associated with factory farming are now becoming evident. Therefore, jobs/livelihoods for Indians must be the bottom line of all our economic and development policies. Unfortunately, modern industry is not labor absorbing and usually enhances its efficiency by downsizing of staff to improve output per person. The "new economy" based on information technology and knowledge industries is also by itself not employment intensive, but could lead to "new employment", if intelligently used.

Farming is the largest private sector enterprise in India. Nearly 58.9 percent of the workers depend upon agriculture for their income and livelihoods. Now that the share of agriculture in total workers is declining, we must think of increasing their productivity and alleviating rural poverty. Rural poverty is greater than urban poverty. Nearly fifty percent of the rural population, belonging to labor families, is engaged in unskilled low-wage work in several parts of the country.

Opportunities for new employment include the production of eco-foods, "biological-software" for sustainable agriculture like bio-fertilizers, bio-pesticides, and vermiculture, bio-processing, health foods, herbal medicines, recycling of solid and liquid wastes, and agriculture and agro-processing machinery. In the new knowledge-based economy, good ecology will be fundamental to good business.

Thanks to both the ongoing technological revolution, particularly in molecular genetics, information and space applications, and the spread of democratic systems of governance at the grassroot level, it is now possible to work towards achieving a substantial reduction in chronic and hidden hunger by the year 2007, which marks the 60th anniversary of India's independence. What is important to ensure is that the means adopted to achieve this end are not at the expense of the prospects for sustainable food security for the generations yet to be born.

References

Department of Statistics, India. 2000. *55th Round of the National Sample Survey (NSS)*. Delhi: Government of India.

-----. 2001. *Provisional Population Totals.* Paper I of 2001. Delhi: Controller of Publications, India.

Swaminathan, M.S. 1999. Science in response to basic human needs. *Current Science* 77(3): 341-53.

-----. 2002. An evergreen Revolution. *Biologist* 47(2): 85-89.

Vepa, S.S., R.V. Bhavani, et al. 2001. *Food Insecurity Atlas of Rural India.* Chennai: M.S. Swaminathan Research Foundation and the World Food Programme.

16

Towards a Sustainable Human Future

Introduction

I am grateful to the Trustees of the Lovraj Kumar Memorial Trust for giving me the privilege of delivering this lecture. I have known Lovraj since 1952 when he married Dharma and we were close family friends. I have throughout admired Lovraj's commitment to national development, environment protection and social equity. He and Dharma constituted intellectual powerhouses and were full of dreams for a happy and prosperous India. Lovraj was an achiever and he was never content with just floating good ideas, but was equally active in getting them implemented. He was a multi-splendored personality and his life and message will always continue to inspire us.

The UN Summit on Sustainable Development now scheduled to be held at Johannesburg, South Africa, from August 26 to September 4, 2002, was originally to be held from September 2 to 11, but the dates were brought forward. The reason was that the dreadful date, September 11, marked the entry of humankind into an era where the fear of one another has reached unprecedented heights. The Johannesburg Summit is being convened ten years after the Earth Summit held at Rio de Janeiro in 1992 and thirty years after the UN Conference on the Human Environment held at Stockholm in 1972, to take stock of the current status of our common ecological future and to examine how far the concept of sustainable development has been converted from a desirable goal into practical accomplishments.

The Stockholm Conference resulted in the establishment of the UN Environment Program (UNEP) at Nairobi, as an instrument for shaping global governance of environmental isues. Unfortunately, UNEP could not live up to its early expectations due to constraints of financial and political support. It is high time UNEP is developed into a World Environmental Organization, on the lines of the World Trade Organization. Even if the name is not changed, UNEP's mandate and political backing need to be enlarged.

At Rio, several International Conventions like those relating to Climate and Biodiversity as well as an Agenda 21 containing guidelines for sustainable development were adopted. Later, a Convention on Desertification was taken up. The UN Convention on the Law of the Sea also came into force. More recently, an International Treaty on Plant Genetic Resources for Food and Agriculture was ratified in the forum of FAO. Fortunately, the Global Environment Facility (GEF) set up just prior to Rio has been receiving good financial support and is rendering extremely valuable service. GEF has saved many biodiversity and

---

Presented as the Ninth Lovraj Kumar Memorial Lecture, New Delhi, August 23, 2002.

hydrologic hotspots from total destruction and has been effectively serving as the financial mechanism for the Climate and Biodiversity Conventions. It is a matter for satisfaction that thirty-two Governments from developed and developing countries have pledged US $2.92 billion for GEF's operations during 2002-2006. This is the highest replenishment so far for GEF.

In spite of all these positive signals, sustainable development is still a far cry, particularly if the social dimension is added to those of ecology and economics. There is a growing violence in the human heart, to some extent due to a feeling of social exclusion and injustice on the part of those who feel that they are losers in the present pattern of development. Mahatma Gandhi asked many decades ago: "How can we be non-violent to nature, if we are going to be violent to each other?" It is obvious that the principles of ethics and equity enshrined in the Convention on Biological Diversity have not been extended to human diversity. The increasing intolerance of variance and pluralism in human societies underlines the need for a Convention on Human Diversity that will help to foster understanding and appreciation of differences in terms of gender, color, religion, language, ethnicity, or political belief.

We see today increasing conflicts in regions characterized by gross social and economic inequity. Thousands of young men and women in Johannesburg, as in many other cities of the world, have no jobs and no hope for a healthy and productive life. However, they often have guns, leading to a threatening climate of violence. It is likely that at the Johannesburg Summit more money will be spent on protecting the participants than what the wealthy nations are likely to pledge for protecting the planet. Like other UN meetings, Johannesburg has poverty eradication as one of its goals. If speeches and statements would have led to poverty eradication, we will not be witnessing the increasing rich-poor divide today. According to UNDP, two hundred ultra-rich persons in the world currently earn more than the two billion living in poverty.

Damage to biological and human heritage is also continuing unabated. Since Rio, some highly disturbing and anti-conservation and anti-sustainable development terms like "ethnic cleansing", "bio-piracy", and "bio-terrorism" are frequently appearing in the mass media. The threat of nuclear warfare is again looming large on the horizon, although this had receded to the background after the end of the Cold War and the breaking of the Berlin Wall. While until recently the eradication of smallpox was considered a triumph of modern immunization technology, steps have recently been taken in the United States to inoculate several groups of professionals against smallpox as a precaution against a possible terrorist attack. Similarly the "dirty bomb" scare has led to severe restrictions on the handling of radioactive materials in research laboratories.

What then is the role of the Johannesburg Summit? Can it shed some light where only darkness prevails now? What should be the outcome of the Johannesburg Summit?

"Do" Ecology: A Message of Hope for the Johannesburg Summit

If I were to describe my hope in two words, it is "do ecology". In 1973, soon after the Stockholm Conference on the Human Environment, I elaborated this concept in my Coromandel Lecture, "Agriculture on the Spaceship Earth":

"The environmental policies advocated in the richer nations are designed to protect the high standard of living resulting from the unprecedented growth in the exploitation of natural resources during the last century. It is of necessity a policy based on a series of 'don'ts'. The poor nations, in contrast, are faced with the desire and need to produce more food from hungry and thirsty soils, more clothing, and more housing. They hence need a 'do ecology' and not just a 'don't' philosophy."

In operational terms "do ecology" involves the creation of new eco-jobs. Since eco-jobs are knowledge intensive, the knowledge era, initiated by the digital, space and biotechnological revolutions, provides uncommon opportunities for realizing the goal of jobs for all.

The M.S. Swaminathan Research Foundation (MSSRF) organized an interdisciplinary dialogue on the theme "Environment, the New Economy, and New Employment" at Chennai from January 28 to 31, 2002, with support from the Global Environment Facility. Its main purpose was to identify opportunities for fostering environmentally and socially sustainable job-led economic growth. Today we know that the famine of jobs or livelihood opportunities resulting in inadequate family income is the basic cause of the famine of food at the level of individuals. Where hunger rules, peace cannot prevail. Sustainable employment or livelihood opportunities hold the key to peace and progress.

Though our planet is politically divided by frontiers and concepts of sovereignty, our fates are ecologically intertwined. While ecology and communications unite us, economics is unfortunately serving as a divisive force. Trade is becoming free but not fair. Globalization should be designed in a "win-win" mode for all and should not result in creating a large number of losers and some winners. This can be achieved if we make opportunities for sustainable livelihoods/employment opportunities for youth as the yardstick for measuring the beneficial impact of globalization. Jobs for all should be the bottom line of trade arrangements, including export and import policies. WTO should subject its agreements to all livelihood impact analysis.

Thanks to the onset of the age of innovation and invention, we have exceptional opportunities for fostering job-led economic growth. For this promise to be realized, political will and action as reflected in priorities in investment in human resource and infrastructure development are a must. There is no single or simple solution to the problem of overcoming the growth of unemployment in the world. The problem is so vast and yet the opportunities are so great that we should foster the fusion of political will, professional skill, and national and global partnerships, to achieve the goal of sustainable livelihood opportunities for all. We can learn from successes, as for example from the integrated approach to on-farm and non-farm employment adopted in China through the Rural Township Enterprises program: The cooperative dairy movement in India, based on small-scale production and on animal nutrition based largely on agricultural residues, is another good example of production by masses, since this provides over fifty million women with secure livelihoods. Global, regional, and national partnerships will be necessary to achieve a paradigm shift from jobless to job-led growth.

Youth constitute the majority of population in most developing countries. If governments acknowledge the opportunity to work and thereby earn one's livelihood as a basic right as well as a basic prerequisite for internal and international peace and human security, they should design strategic interventions which will help to build the self-esteem and competence needed to undertake self-employment. Unity of goals, but multiplicity of approaches, will be needed to achieve success. The complex problem of youth employment and its multiple dimensions will have to be dealt with in a desegregated manner, as for example, first generation learners (i.e., the children of illiterate parents) as well as "school push-outs" (often girls) and handicapped youth will have to receive special attention. Without peace and security, there can be no progress in any field of human endeavor. Therefore, if youth employment efforts go wrong, nothing else will have the chance to go right.

Framework for Action for Fostering Sustainable Livelihoods

I would like to suggest a framework for generating sustainable employment and livelihood opportunities for all. The basic strategy would include enhanced opportunities for skilled employment in the primary, secondary, and tertiary sectors of the economy; an integrated approach to technology, training, techno-infrastructure, and domestic and export trade; public policies for enlarging the space for remunerative and market-driven self-employment; and, consortiums of public and private sector companies for fostering assured marketing opportunities through buy-back arrangements.

To operationalize this strategy we would need to: a) promote location-specific jobs, based on the natural and human resource endowments of the area; b) carry out livelihood/job impact analysis of all technology-driven development projects; c) form in each compact area a "Jobs for All Consortium" consisting of appropriate representatives of government and non-government organizations, business and industry, service oriented civil society associations, women's groups, financial institutions, bilateral and multilateral donors, and the mass media; d) organize a network of training and capacity-building institutions capable of imparting market-driven skills; e) develop marketing infrastructure needed to link primary producers with domestic and international consumers, and promote an Employment Infra-structure Fund as well as Venture Capital Fund for small-scale decentralized production enterprises; f) encourage, whenever appropriate, a "production by masses" approach, rather than "mass production" and labor-displacing technologies; support such decentralized small-scale production with key centralized services in order to reduce transaction costs; g) develop institutional structures which can confer the power of scale to small producers both in the production and post-production phases of the enterprise; h) facilitate the organization by young entrepreneurs of effective structures like Eco-enterprises Parks, Biotechnology Parks, Food Parks, Horticulture, Poultry, and Dairy Estates, Renewable Energy Parks, etc.; i) assist in creating more non-farm employment opportunities in rural areas and provide opportunities to young farm, commerce, home science, veterinary, fishing, and engineering graduates to organize socially relevant and economically viable services; j) help unemployed youth to acquire the necessary self-confidence and skills to embark upon initiatives which can help to bridge the digital, genetic, gender, and other divides; k) devise short-term non-degree training programs at appropriate universities and technical institutions to help build the capacity of youth to harness both traditional and frontier technologies. The aim should be to assist the country to progress rapidly in the technological transformation of crop and animal husbandry, fisheries, agro-forestry and forestry, agro-processing, and small rural and urban enterprises.

The report of the Brundtland Commission was titled *Our Common Future*. The period from Rio to Johannesburg has amply demonstrated that there can be no happy common future for humankind without a better common present. However, extending the extrapolation domain of successful experiences and examples will be possible only if we understand the basic principles underlying such efforts. We should emulate the example of the DNA molecule (the basic unit of heredity and life) and design programs capable of replication, recombination (i.e., marriage of successful experiences), and mutation (mid-course correction when needed, or even total recasting of strategies where appropriate). Thus, unique successes can be converted into universal progress.

Action Today - Not Just Promises for Tomorrow

For every human being, today is a reality and tomorrow is a possibility. The hungry need food today and not just promises for tomorrow. India is a case in point where food for all can be ensured today and hence I have chosen this uncommon opportunity in this lecture dedicated to the memory of Lovraj Kumar, a man of action.

Both in terms of incidence and absolute numbers, India has the largest number of children, women and men going to bed undernourished in the world. Although the Food Corporation of India has currently over sixty-five million tons of foodgrain stocks, and there are numerous safety net measurers of the Government of India for the economically and socially underprivileged sections of the society, the FAO has listed India as one of the countries which has not been able to achieve the target of the World Food Summit of 1996 in terms of reducing the number of undernourished. This calls for serious introspection of both our strategies for fighting hunger as well as of our ethical commitment to the cause of hunger-free India.

The prevalence of endemic hunger should no longer be accepted as unavoidable. The social security systems for the able-bodied poor should include a food component in the form of an open-ended National Food Guarantee Scheme on the model of Maharashtra's Employment Guarantee Scheme. The public distribution system should be strengthened and should be designed to reach the unreached. On the basis of ICMR norms, about one hundred and fifty-eight million persons, belonging to about thirty-two million households, fall under the category of ultra-poor and need immediate assistance to help them to lead a healthy and productive life. They are best identified by Gram Sabhas and local bodies. They can be issued with Food Entitlement Coupons, which should entitle them to be provided with work under the Food Guarantee Scheme. The total requirement for such an open-ended employment-cum-food security project may need at the maximum about ten million tons of foodgrains per year. Meeting this need is well within our national capacity.

Consumption inequality decreases as income increases. Diversification of diets also happens with increased purchasing power. Unfortunately, inequality of income distribution is growing resulting in increasing marginalization of the poor, with children and women suffering more. A job-led economic growth strategy will help to reduce protein-energy malnutrition, and at the same time stimulate farming systems' diversification. Public investment in rural infrastructure should be greatly increased.

A whole lifecycle approach should be introduced by providing a horizontal dimension to the numerous on-going vertically structured programs supported by the Central and State governments as well as by bilateral and multilateral agencies. At the local level, the management of a lifecycle-based nutrition security system, beginning with pregnant women and extending up to old and infirm persons, is best left to the one million elected women members of the Panchayats.

Decentralized procurement and decentralized storage will help to minimize transport and transaction costs. Uttar Pradesh, Madhya Pradesh, West Bengal and Tamil Nadu have accepted the principle of a decentralized procurement system. The storage can be done through a national grid of Community Food Banks (CFBs) managed by self-help groups. The construction of CFBs can be done at the local level under the food for work program. CFBs can help to implement this social safety net program in an effective economical and transparent manner. We must end the prevailing mismatch between production and consumption as soon as possible.

Short-term problems should not blind us to long-term realities. Population is growing and per capita land and water availability is decreasing. We have to produce more but produce it differently – in a manner that high yields can be obtained in perpetuity without associated ecological or social harm. Farming systems' intensification, diversification and value addition are extremely important to generate the needed on-farm and non-farm employment. There is also an urgent need for greater attention to the nearly one hundred million hectares of rainfed land. Better dry farm techniques are available. They offer great opportunities for increasing the production of pulses and oilseeds. Taking the easy option of importing large quantities of pulses and oilseeds will only increase poverty and unemployment in dry farming areas. Productivity improvement in both irrigated and rainfed areas will help to increase marketable surplus at the household level and thereby the cash income. The gap between potential and actual yields even with the technologies currently available is high in most farming systems. Livestock husbandry and livelihood security tend to be closely correlated in the case of poorer households. There is greater equity In livestock ownership as compared to land. Hence, support services should be organized for livestock farming families in the form of Fodder and Feed Banks and health care and insurance facilities. With increasing urbanization the demand for animal products will grow. Accelerated agricultural progress is the best safety net against hunger and unemployment.

There is an urgent need for spreading quality literacy throughout the country, which will help to sensitize both producers and consumers on issues relating to food safety, *Codex Alimentarius* standards and sanitary and phytosanitary measures. In addition, there is need for greater research in the field of breeding crop varieties having the quality characteristics needed for food processing and exports. If productivity and quality are improved, farm income will go up substantially.

Sustainability of food security measures should be ensured. PDS at the local level should provide reasonable income to those operating the ration shops. Community Food Banks may receive their initial food supply from government but subsequent replenishment should come from local communities. A wide range of grains and wild foods can be stored in CFBs. This will help to widen the food security basket. Also, the silos can be locally manufactured, thus providing another source of rural employment in the non-farm sector. Further, the silos will help to reduce post-harvest losses considerably. Both private and public sector institutions and companies have the necessary technical capability to launch the CFB movement.

The organization of a National Consortium for Sustainable Food Security will be useful to mobilize the necessary technical, financial and managerial resources to eliminate endemic and transient hunger. Indian farmers have shown that they will produce more if they are given opportunities for assured and remunerative marketing. Agriculture is the backbone of India's food, livelihood and ecological security systems. Therefore, India's agricultural image within and outside the country needs to be refurbished. At the same time, chronic, hidden and transient hunger should be eliminated as soon as possible.

There is no time to relax, both on the production and consumption fronts. The stock of sixty-five million tons should not lull us into complacency. Concurrent attention to production and consumption is important. Enhancing production through an evergreen revolution is a priority task. However, eliminating widespread endemic hunger is both a national responsibility and an ethical obligation. A universal and user-sensitive public distribution system, food guarantee scheme community food banks and various other food entitlement projects like Annapoorna, Antyodaya Anna Yojana, Integrated Child Development Service, Gramin Rozgar Yojana, etc., need to be implemented in an integrated manner, so that the goal of hunger-free India can be achieved. Nutrition status will, however, continue to fall unless the purchasing power of the poor is increased. Hence, livelihoods for all should be the bottom line of all national development and import and export policies.

Emerging Scientific Revolutions and an Ecology of Hope

Fortunately, we are now experiencing three major revolutions in science and technology, which will influence agriculture and industry in a fundamental manner. It will therefore be appropriate to make a brief reference to them: a) The gene revolution which provides a molecular understanding of the genetic basis of living organisms, as well as the ability to use this understanding to develop new processes and products for agriculture, industry, the environment, and for human and animal health; b) The ecotechnology revolution which promotes the blending of the best in traditional knowledge and technology with frontier technologies such as biotechnology, space and information technologies, renewable energy and new materials; and c) The information and communication revolution which allows a very, rapid growth in the systematic assimilation and dissemination of relevant and timely formation, as well as a dramatically improved ability to access the universe of knowledge and communicate through low cost electronic networks.

In principle these three types of advances – when coupled with improvements in the management and governance – greatly increase the power of a scientific approach to genetic improvement, the integrated management of natural resources and ecosystems, and the management of local and regional development strategies. Eco-technologies enable the adoption of ISO 9000 and ISO 14000 standards of environmental management. These scientific revolutions seem to be proceeding at an ever-increasing pace, with most of the action occurring in industrialized nations. Also, new discoveries of great relevance to sustainable food and health security are coming under the purview of proprietary science, since they are covered by Intellectual Property Rights. How then can we mobilize recent advances in science and technology for meeting the basic needs of the economically and socially underprivileged sections of the human family?

a) The gene revolution: The past twenty years have seen dramatic advances in our understanding of how biological organisms function at the molecular level, as well as in our ability to analyze, understand, and manipulate DNA molecules, the biological material from which the genes in all organisms are made. The entire process has been accelerated by the Human Genome Project, which has poured substantial resources into the development of new technologies for working with human genes. The same technologies have been applied to the mapping of Arabidopsis, rice and other plant genomes. Thus, new scientific disciplines of genomics and proteomics have emerged and are expanding at a rapid pace.

Several large corporations in Europe and the United States have made major investments in adapting these technologies to produce new plant varieties of agricultural importance for large-scale commercial agriculture. The same technologies have equally important potential applications for addressing food security in the developing world, provided objective methods of risk-benefit analysis are put in place.

Let me cite one example from the work of MSSRF scientists to illustrate the value of the new tools. As a part of the anticipatory research program to meet the consequences of sea level rise arising from global climate change, genes responsible for conferring the ability to withstand seawater intrusion were identified in a few mangrove species through molecular mapping. They have been transferred to annual economic plants like rice through recombinant DNA technology.

b) The ecotechnology revolution: Knowledge is a continuum. There is much to learn from the past in terms of the ecological and social sustainability of technologies. At the same time, new developments have opened up new opportunities for developing technologies which can lead to higher productivity without adverse impact on the natural resources base. Blending traditional and frontier technologies leads to the birth of ecotechnologies with combined strengths in the following areas: Ecology, Equity, Employment, and Energy.

c) The information technology revolution: New communication and computing technologies are already influencing life on our planet in a profound manner: i) Access to the Internet will soon be universal, and it can provide unrestricted low-cost access to information, as well as highly interactive distance learning. The Internet will not only facilitate interactions among researchers, but also greatly improve their ability to communicate effectively with the potential users of their research knowledge; ii) Computing makes it possible to process large-capacity databases (libraries, remote sensing and GIS data, gene banks) and to construct simulation models with possible applications in ecosystem modeling, preparation of contingency plans to suit different weather probabilities and market variables; iii) The software industry is continuously providing new tools that increase research productivity and create new opportunities for understanding complex agroeco systems; iv) Remote sensing and other space satellite outputs are providing detailed geographic information useful for land and natural resources management.

The promotion of ecotechnology development and dissemination, the effective adoption of integrated systems of gene and natural resources management, and the effective harnessing of information technologies should become essential elements of the "science and technology for basic human needs" movement.

Converting Calamity into Opportunities

I would like to cite one example of a potential application of 1CT to help rural families to manage aberrations in precipitation. This year, there is a severe drought in several parts of India. It will be possible to mitigate the adverse impact of irregular rainfall, if farm families can have access to dynamic information on crop life saving techniques and alternative cropping strategies. Hence, we are planning to establish a Virtual University for Climate Management of the type described below.

Establishment of a Virtual University for Climate Management

Recent advances in information and communication technologies, when combined with application of contemporary instructional methods of open and distance learning offer unprecedented opportunities for a large number of people to gain access to education over an extended period of time. The need for continuous learning while pursuing an occupation is being effectively addressed in the OECD countries, especially in North America, through such a combination. The recent enhancement in technology availability for connectivity in India and other developing countries can now be combined with distance learning in higher education to establish a new paradigm of virtual education for large masses of rural women and men. To make it useful in the immediate context, we need to identify topics that are in demand by the rural communities who stand to gain in a considerable measure by such a combination of approaches. The Virtual University for Climate Management will exemplify such an approach and will enable rural families and local administrators and policy makers to prepare effective responses to the emerging problem of climate-induced disasters.

The Methodology. India has been viewed as a hotbed of experiments to overcome the digital divide. Recent analysis of over thirty such field projects in India indicates that sustainability of digital access in rural and semi-rural areas is to a considerable extent dependent upon popular appreciation of the manner in which the felt needs of the local population are fulfilled. Connectivity alone, by itself is not sustainable unless content creation and its management to meet changing and enlarging requirements of rural families are met in an accurate and in a timely manner. Thus, the model we are adopting is a hub-and-spokes one, where a local hub performs key content management functions, while maintaining two way links with its heartland, namely, villages (the spokes), and with the larger external world. This function of credible and intelligent intermediation is a crucial leadership function that is the core principle of the entire operation. This model, conceived ten years ago by Dr. V. Balaji and other staff members of MSSRF, has been put into practice successfully during the last five years in the Pondicherry region. With refinements that come with such an experience, this model is potentially the most suitable for the establishment of a Virtual University.

Technology. The primary technology will be the direct satellite access technology called DTH-RCS which is licensed for use in India. The State Level hub will have a high bandwidth (T1 level) to the Internet using a fiber optic cable (for assured high quality service to users). The local hubs will connect to the State Level Hub using DTH-RCS technology whose costs are declining rapidly. The estimated capital cost for connectivity at the local hub is US $4,000 (dish antenna plus three computers and a UPS) and the recurring cost for bandwidth for local hubs will vary from US $1,000 to 2,000 depending upon the bandwidth required. The new business models in this area permit purchase of additional bandwidth in a dynamic way against specific requirements. The principal advantage with DTH-RCS is the speed with which it can be deployed, and it does not require specially trained server side administrators. The cost of setting up the State Level Hub may be estimated at about US dollars 500,000.

While Internet will be the primary mode of communication, the Virtual University will also use media such as radio (terrestrial or satellite) and television, besides print media based on specific requirements. The possibility of using satellite radio is high in the last mile area that is between the local hub and a village.

The establishment of a Virtual University for Climate Management by local communities is one example of the uncommon opportunities now available for taking the benefits of technology to the rural poor.

The people of Afghanistan have undergone extreme suffering during the past ten years. There have been recent reports, particularly by the Pulitzer Prize awardee, Barry Bereak, that hunger is so acute that parents have no option except to sell children for bags of wheat. It is under such circumstances a grid of Community Food Banks located in all remote areas where transient and endemic hunger prevails will be of great value. About one thousand CFBs each containing five hundred metric tons of foodgrains will insulate the population from the pangs of hunger. Unfortunately such simple and easily implementable solutions do not get support, while proposals like setting up a Food Security Authority, etc., get implemented. Again, this is reflective of a mindset which accords priority to promises for tomorrow, than action today.

My work during the last fifty years relating to a hunger-free India and world, has been greatly influenced by the concept, "the test of our progress is not whether we add more to the abundance of those who have much; it is whether we provide enough for those who have little." Translated into action, this has meant concentration on improving the productivity of small farm holdings and the well-being of the economically and socially under-privileged women and men who toll day and night and in sun and rain to produce food for their fellow human beings. I am firmly convinced that hunger and deprivation can be eliminated sooner than most people consider feasible, provided there is synergy among technology, public policy and social action. This is the major lesson we can learn from the farm revolution that transformed India's agricultural destiny during the last part of the last century.

The World Summit on Sustainable Development scheduled to be held at Johannesburg in August-September this year provides a unique opportunity for strengthening existing partnerships and promoting new ones. It is in the long-term enlightened self-interest of the rich that they should match the words they will speak and the sentiments they will express at Johannesburg with action as expressed by a firm commitment of financial and technical resources for fostering sustainable human security of which sustainable food security is the foundation.

17

It's Time to Have a National Biotech Policy

The country has well-defined policies in the fields of atomic energy, space applications and information technology. No further time should be lost in developing a national food and agricultural biotechnology policy through political consensus.

Otherwise India will experience a serious genetic divide. A similar policy is also needed in the area of medical biotechnology, which involves ethical issues with reference to both human and animal experiments. The urgency stems from the fact that the twenty-first century will belong to those who help to advance the frontiers of science and technology in the areas of functional genomics, proteomics, bioinformatics and molecular breeding (i.e., genetic modification).

Scientific leap-frogging in both the theoretical and applied aspects of the new genetics will take place mostly in industrialized countries. China is fast becoming a developed country in respect of achieving mastery of techniques relating to genomics and recombinant DNA technology for improving human nutrition and health.

Equally important is to formulate a well-defined implementation plan for the national food and agriculture biotechnology policy. Because, any policy without an appropriate and effective implementation framework will have no value. Therefore, the policy should provide the terms of reference to an autonomous and professional Biotechnology Regulatory and Advancement Commission.

The aim of regulation should be to help in harnessing this powerful technology in a risk free and responsible manner. The Commission should not only develop and enforce a code on "don'ts", but should also propose "dos", which will help to gain benefits without risks. It should build on the Cartagena International Protocol on biosafety and introduce a system of regulation and monitoring which inspires public, political and media confidence.

The term, "biotechnology" encompasses a wide range of technologies – both traditional and frontier. For example, the production and use of biofertilizers, biopesticides, vermiculture, and bioremediation agents are essential for fostering ecologically sustainable farming methods. Bioprocessing and bioprospecting offer new opportunities for skilled jobs and livelihoods. These areas of biotechnology also offer scope for decentralized village level enterprises operated by self-help groups.

Regulation for the responsible advancement of biotechnology for public good should be the motto. The National Biotechnology Regulatory and Advancement

---

First published in *BioSpectrum Today* (Mumbai, India), pp. 28-29. (March 2003)

Commission, which could be attached to the Ministry of Agriculture for administrative purposes, should be headed by an eminent professional known for objectivity and credibility.

There should be a multi-stakeholder representation on the Commission and its Standing Committees, including scientists, concerned government officials, representatives of public and private sector industry, consumer and women's associations, farmers' associations and the mass media. Such a Commission will be effective only if it is created on the basis of consensus among political parties.

The following could be some of the important responsibilities of the Commission: a) Create the expertise and infrastructure needed to undertake a critical and transparent scientific assessment of the food and environmental safety of Genetically Modified (GM) crops; b) Coordinate the work of the Ministries of Agriculture, Health, Environment and Science and Technology, Department of Biotechnology (DBT) in the area of biosafety assessment; c) Provide guidelines for research collaboration between public and private sectors in areas such as functional genomics, proteomics and bioinformatics as well for priorities in public investment, as, for example, research on drought tolerance, water use efficiency and salinity assistance; d) Assist in fostering linkages among biodiversity, biotechnology, biosafety management; conservation of habitats rich in agro-biodiversity should receive special attention; e) Monitor the role of the biotechnology industry in assisting the effective implementation of the Protection of Plant Varieties and Farmers' Rights Act and the Biodiversity Act; f) Help to upgrade patent offices and develop expertise in dealing with issues connected with the Trade Related Intellectual Property Rights (TRIPS) regime of World Trade Organization (WTO); g) Promote regional and international biosafety collaboration, particularly among South Asian Association for Regional Cooperation (SAARC) and Association of South-east Asian Nations (ASEAN) countries; h) Serve as a single window regulatory and monitoring agency and develop and introduce a time bound regulatory process; i) Help to create public awareness and understanding on issues relating to biosafety and biotechnology in local languages and vernacular media; j) The Commission should submit an annual report to Parliament on the state of Food and Agricultural Biotechnology in India.

Building National Capacity

To build national capacity in all areas of risk assessment and biosafety valuation and monitoring, it will be useful to set up a National Research Center for the safe and responsible use of GM crops.

Such a National Research Center could provide the scientific and technical support needed by the proposed National Biotechnology Regulatory and Advancement Commission.

The Center should maintain a global database on biosafety assessment procedures and legislation. It should undertake training, capacity building and networking in the field of biosafety evaluation. Ultimately, considerations of human health and environmental safety should be the bottom line in risk assessment.

The policy has to deal with the apprehensions relating to molecular genetics and genetic engineering in many broad categories: a) Issues relating to science itself, such as its ethical implications and the problems associated with the antibiotic markers used, etc.; b) Issues relating to the control of science such as the probability of the control of global food security falling into the hands of a few transnational corporations; c) Issues relating to access, such as the implications of IPR for the poor, technologies becoming more exclusive than inclusive leading to a further expansion of the rich-poor divide in terms of technological empowerment; d) Issues relating to the environment, such as impact on biodiversity, possibility of "genetic pollution" in the centers of origin and diversity of crop plants emergence of super weeds.

Finally, issues relating to human and animal health and food safety and allergenicity, which are extremely important in the case of food, feed and fodder plants.

Of the above, the ethical issues assume greater importance in medical biotechnology in areas such as human cloning. A disaggregated approach to the study of the above issues will be important for a rigorous analysis of risks and benefits.

If a disaggregated approach is not used to analyze the issues involved, the conclusions arrived at in international meetings will tend to deal with them in a composite manner as will be clear from the following statement made by NGOs and civil society organizations at the World Food Summit meeting held in Rome in 2002. "GM organisms represent a threat to family farmers, other food producers, the integrity of genetic resources and human and environmental health. They will affect particularly the rural poor, who cannot afford this costly alternative."

The benefits of molecular breeding techniques like the use of molecular markers and undertaking precision breeding for specific characters through recombinant DNA technology are immense. The work already done in India has revealed the immense potential for breeding new GM varieties possessing tolerance to salinity, drought, some major pests and diseases and improved nutritive quality. A new era of Integrated Mendelian and Molecular Breeding has begun.

This is the only way we can face the challenges of the future, particularly in the context of the growing water scarcity as well as the urgent need to step up productivity in semi-arid and dry farming areas. Denying ourselves the power of the new genetics will be doing great disservice to both resource poor farming families and to the building of a sustainable national food and nutrition security system. Food self-sufficiency is essential for preserving our national sovereignty in foreign policy. ▯

18

Towards Hunger-Free India: Countdown from 2007

During the last three years, scientists of the M. S. Swaminathan Research Foundation at Chennai and of the UN World Food Programme, New Delhi, have been working together on the causes of food insecurity in rural and urban India using multiple indicators relating to food availability, access and absorption in the body. The Food Insecurity Atlases of Rural and Urban India provide valuable guidelines for public action to end chronic and under- and mal-nutrition in the different States of the country. Using these guidelines, an action plan was developed at a multi-stakeholder consultation held at New Delhi on April 4-5, 2003, under the joint auspices of MSSRF, WFP and FAO.

Releasing the *Food Insecurity Atlas of Rural India* on 24 April, 2001, the Prime Minister of India, Shri Atal Bihari Vajpayee, called for concerted action to achieve substantial freedom from hunger by August 15, 2007, which marks the 60th anniversary of India's Independence, thereby helping to fulfill the hope of Mahatma Gandhi that no child, woman or man should go to bed hungry in Independent India. To quote Shri Vajpayee: "The sacred mission of a 'Hunger-Free India' needs the cooperative efforts of the Central and State Governments, local self-government bodies, non-governmental organizations, international agencies and above all, our citizens. We can indeed banish hunger from our country in a short time. Let us resolve today to make this mission substantially successful by 2007, which will mark the sixtieth anniversary of our Independence."

The Challenge

The quantitative and qualitative dimensions of the challenge of achieving a hunger-free India are daunting. The incidence of poverty, endemic hunger, communicable diseases, infant and maternal mortality rates, low birth weight children and stunting and illiteracy, is high. There are however many examples where progress in the elimination of poverty-induced hunger has been rapid because of a symphony approach in dealing with the multi-dimensional problem of hunger and malnutrition. Successful experiences in the elimination of hunger and poverty have shown that synergy between political will and action and strategic partnerships can help local communities to achieve seemingly impossible tasks. Such "messages and methods of hope" should therefore be documented and spread widely, since they not only inspire confidence that the goal of a hunger-free India can be achieved, but will also help to build the self-confidence of all engaged in the mission of overcoming under- and mal-nutrition.

First published in *The Hindu*, April 27, 2003.

Basic Approach: Food with Human Dignity

Food with Human Dignity should be the basic approach. The poor should not be subjected to a patronage approach and referred to as "beneficiaries" but should be treated as partners in achieving the aim of ensuring that every child, woman and man in the country has an opportunity for a productive and healthy life. The right to adequate food and clean drinking water should be regarded as a basic human right.

Thrust of the Tenth Five Year Plan (2002-2007)

The Tenth Five Year Plan has shifted the emphasis from food security at the household level to nutrition security at the level of each individual. Emphasis has been placed on employment, education, health and nutrition, which are all important for poverty eradication and hunger elimination. The inter-sectoral nature of chronic hunger has been recognized. By shifting the attention to individuals, the strategies adopted will be based on the principle of social inclusion and will help to foster a lifecycle approach in nutrition interventions. For example, pregnant women will need special attention, since maternal and fetal under-nutrition leads to the birth of babies characterized by a weight of less than 2.5 kgs at the time of delivery. Such low birth weight (LBW) children suffer several handicaps in later life and may not be able to express their innate genetic potential for mental and physical development. Such inequity at birth is inexcusable since we are now entering a knowledge-based economy. Similarly, old and infirm persons need special attention. Thanks to advances in preventive and curative medicine, we are now adding years to life. However, we should pay equal attention to adding life to years through nutrition and health care. We should recognize that we are now entering a new chapter in human longevity. There is therefore need for a proper match between nutrition requirements and nutrition support at different stages in the life of an individual.

Hunger is the extreme manifestation of poverty, since the poor spend a high proportion of their earning on food. The elimination of hunger is thus the first requisite for eradicating poverty. Without adequate nutrition, the energy needed for higher work output will not exist in malnourished individuals.

Guiding Principles for Converting Goals into Accomplishments

Decentralization: The desired goal can be achieved speedily and surely only if a decentralized approach to implementation is adopted. "Think, plan and act locally and support at the state and national levels" should be the motto. Elected local bodies, together with the concerned Departments of Government (Health, education, women and child welfare, rural and tribal development, etc.) should prepare Micro level Action Plans. They should form a local level "Alliance for a healthy and productive life for all". Elected members of local bodies, particularly the one million elected women members can be empowered to spearhead the freedom from hunger movement, since they are more aware of the problems of nutrition and drinking water. Decentralization will enhance accountability, reduce transaction costs and remove corruption in delivery systems.

Life-cycle Approach: For ensuring nutrition security at the level of each individual, a life-cycle approach is necessary so that the nutrition needs of an individual can be met from birth to death. Special programs for adolescent girls, pregnant women, nursing mothers, infants (0-2 years) and old and infirm persons should continue. What is needed is the horizontal integration of numerous vertically structured programs. Such a functional integration will help to create a symphony at the level of each village/town/city to ensure that all links in the food availability – access – absorption chain function at a high level of efficiency and effectiveness. Management tools and not additional monetary support will be needed to bring about at the field level such convergence and synergy among ongoing programs.

Information, Education and Communication: There is need for launching a Nutritional Literacy movement to spread awareness of the adverse consequences of malnutrition-induced intellectual and physical dwarfism among children. The Nutritional Literacy movement should include issues relating to food safety, *Codex Alimentarius* standards, sanitary and phytosanitary measures, etc. Mass media, particularly those in the public sector like Doordarshan and All India Radio can play a very important role in making the Hunger-free India movement a success. Community radio stations, giving location-specific information, should be encouraged to assist other mass media in spreading messages of hope. It will be useful to set up Media Resource Centers for a Hunger-free India. Such centers can provide credible and timely information to the print, audio, video and new (i.e., Internet) media.

Household Entitlement Card: It will be useful to provide every family with an Entitlement Card, giving information on the various government projects which they can access. The information may be disaggregated by gender, age, religion, caste and class, and precise addresses of contact persons and offices may be given. Such information will enable everyone to make the best use of their entitlements. A single step of this kind will help enormously to ensure the effective utilization of all the schemes of central and state governments and bilateral and multilateral donors.

Asset Building and Community Development: The poor are poor because they have no assets like land, livestock or fish pond. They often are illiterate and lack proper dwelling. They survive on wage employment, which particularly in the case of women, does not reach the level of even the prescribed minimum wage. A massive effort is needed to help them to shift from unskilled to skilled work through training in market-driven skills. The on-going micro-finance led self-help revolution will be the speediest way to help them to rise above the poverty level. This will call for establishing effective forward and backward linkages, particularly with technology sources and markets. Insurance and Venture Capital support should also be available to micro-enterprises. We have now an opportunity to leapfrog in achieving our goal of enabling every one to earn his / her daily bread.

Capacity Building: Since a decentralized approach involving the empowerment of over three million women and men members of local bodies holds the key to the success of this national movement for food and clean drinking water for all, it is essential that a national consortium of agricultural, rural and women's universities as well as government and non-governmental training and research institutions is formed for undertaking capacity building in areas such as management, communication and organizational skills with reference to the implementation of the hunger-free area program. The capacity building programs can be organized on a Trainers' Training model, in order to achieve a multiplier effect.

Initiation of a National Food for Social Capital Programme: The social capital of a country is the product of interaction between the human capital and the cultural, political, economic, nutritional and natural environments. Human and social capitals constitute the most precious wealth of a nation. Mahatma Gandhi and Vinobha Bhave advocated the principles of *antyodaya* and *sarvodaya* for achieving high social synergy and capital. A society committed to building its social capital will try to promote programs which represent a "win-win" situation for all, thereby avoiding winners and losers and the consequent social conflict and disruption. During the last few years, the Government of India as well as some State Governments have initiated many programs like Sampoorn Gramin Rozgar Yojana, Annapoorna, Antyodaya Anna Yojana, Universal Noon-meal Programme for School Children, etc. It is now clear that our farmers will produce more if we can enhance consumption and thereby opportunities for assured and remunerative marketing. Therefore, the initiation of a National Food Guarantee Scheme will help to ensure that all who are hungry today due to lack of livelihood opportunities or other constraints are able to have food for a productive life. Such a National Food Guarantee Scheme can serve as an umbrella for all ongoing projects like those mentioned earlier. In addition, it can provide foodgrains for initiating a Nagarpalika Rozgar Yojana as well as for a wide variety of social support initiatives like Food for Health (TB, HIV / AIDS, Malaria etc.), Food for those employed in ICDS, Nutritious Noon Meal and other similar projects. In other words, food can become a powerful currency for achieving the goal of a hunger-free India. Using food as a currency has twin advantages, namely, there could be greater off-take of food-grains from farmers, thereby providing them with an incentive to produce more, and secondly for meeting the immediate needs of the poor, destitutes, migrant labor and all who are undernourished today.

The Urban Food Insecurity Atlas released by The President in October 2002 clearly brings out the urgent need for attending to the hunger problems of the bottom 10% of the poor (ultra-poor) in towns and cities. Therefore, the setting up of an umbrella program combining the principles of the Employment Guarantee Scheme of Maharashtra and of various Food for Work Programs under a National Food for Social Capital Programme will be timely. This could serve as the hub of a series of activities. It can start with a total allocation of 15-20 million metric tons of foodgrains during 2003-04. Such a block grant of foodgrains can be managed by a Malnutrition-free India Trust, headed by the Deputy Chairman of the Union Planning Commission. The Trust can sanction small projects to meet specific local requirements as well as to fill gaps in the ongoing programs. At least five million metric tons of grain should be available to support local level Community Food Banks in "hunger and hydrologic hot spot" areas and specific programs designed to improve maternal and fetal nutrition as well as to provide nutrition support to those affected by TB, leprosy and other diseases. Such a Food Guarantee Initiative will be psychologically an important index of the capability of Indian farmers to produce more on the one hand, and the political commitment to achieve Prime Minister's goal of a hunger-free India by August 15, 2007, on the other.

Monitoring and Evaluation: This could be done at various levels starting with Gram Sabha and Citizens Groups (like the Right to Food Group) and Government Agencies and Research Institutions. An effective and transparent monitoring system will also help to ensure the implementation of the directives of the Supreme Court of India.

Consultative Group for Freedom from Hunger: Both at the national and state levels multi-stakeholder Consultative Groups for "Agenda 2007: Hunger-free India" could be organized comprising representatives of the concerned Government of India Ministries and Departments, Professional Experts, National Commission for Women, Civil Society Organizations, Business and Industry, Mass Media and bilateral and multilateral donors, with the Union Planning Commission serving as the nodal agency for such a Consultative Group. Such a CGFFH would help to foster strategic partnerships as well as synergy among political leaders, professionals and peoples' organizations.

Standing Committee of the National Development Council: A Standing Committee of NDC could be set up for monitoring progress and ensuring the success of Agenda 2007. The NDC Committee chaired by the Prime Minister could include Chief Ministers of States where there is wide spread under- and mal-nutrition as well as of food secure states, so that there could be lateral sharing of experiences among states. Such a Standing Committee could provide the political guidance and support needed for implementing this important program.

Immediate Action during 2003-04: Besides the announcement of the twenty million metric tons Food for Social Capital program and the setting up of a Malnutrition-free India Trust immediate action needs to be taken to end poverty-induced chronic hunger and the transient hunger caused by drought and natural calamities through a series of Community Food, Fodder and Feed Banks. Such Banks may be established in all the "hunger hot spots" of the country. CFBs managed by local Self-help Groups preferably of women, would save considerable transaction and transport costs. They will also help to widen the food security basket through inclusion of local grains like millets, pulses, oilseeds and tubers. It will also be advisable to reclassify coarse cereals as "nutritious grains" in order to underline their desirable nutritive properties.

No Time to Relax on the Food Production Front: While the alleviation of hunger by improving access to income and balanced diets and safe drinking water should receive high priority, there is no time to relax on the food production front. We need to bring about productivity, quality (including food safety), profitability and sustainability revolutions in farming based on a Farming Systems Approach. There is an urgent need for enhancement of investment in agriculture and rural infrastructure development. There is also need for conferring on small producers the power of scale through cooperatives, self-help groups and other socially viable methods of group endeavor both at the production and post-harvest phases of farming. Our spectacular progress in the Dairy sector is largely through such management innovations. There is need for a movement for trade and quality literacy including an understanding of sanitary and phytosanitary measures and *Codex Alimentarius* standards. There is also need for launching a Jal Swaraj and Water Literacy Movement. Above all, there is need for a paradigm shift from jobless to job-led growth in order to ensure that every poor person is enabled to earn his or her daily bread.

Unless our agricultural competitiveness improves, our earnings from the farm sector in international markets will not grow significantly. In a predominantly rural and agricultural country like India, agricultural progress (i.e., crop and animal husbandry, fisheries, forestry and agro-forestry and agro-processing) is the most effective social safety net against hunger and poverty. Hence, the ongoing fatigue of the Green Revolution in wheat, rice and other major crops should be converted into an ever-green revolution designed to promote productivity improvement in perpetuity without associated ecological harm. A beginning has been made in the Union Budget for 2003-04 to focus attention on precision farming, economic use of water and nutrients through fertigation techniques and green house horticulture. Such a focus on improving factor productivity in agriculture should be further enlarged to cover all the major agro-ecosystems. Agricultural and rural development, if given adequate and appropriate attention, will help the country to take to the path of job-led economic growth.

Our substantial grain and foreign exchange reserves and the three million elected women and men members of local bodies have provided us with an uncommon opportunity for launching a frontal attack on hunger and poverty. It will be a tragedy if we don't act, when we are in a position to act.

To conclude, the "Agenda 2007: Hunger Free Area Programme" should keep in mind the following advice of Gandhiji given before his death: "Forget the past. Remember every day dawns for us from the moment we wake up. Let us all, every one, wake up now." ◼

I.B
Toward a Hunger-Free World

54. MSS Portrait, 1999

55. With Syed Ahmad Marei, Norman Borlaug, A.H. Boerma at Rome World Food Forum, 1974

56. With Pope John Paul II for discussion on the Sahelian drought, 1985

57. With President Nelson Mandela for discussion on eradicating hunger in Africa, 1988

58. With World Food Laureate He Kang, Former Minister of Agriculture, for discussion on improving agricultural development in China, 1992

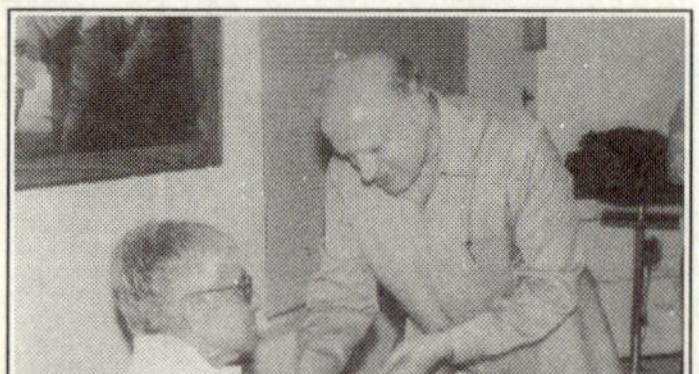

59. With C. Subramaniam, pioneer of Green Revolution in India, 2001

19

The Green Revolution Can Reach the Small Farmer

The term "Green Revolution" has been used in different senses by different people throughout this book. Some have used it to imply that the technologies associated with the "Green Revolution" have built-in seeds of discrimination against small farmers, because they are capital-intensive, labor-displacing, and ecologically damaging. These statements have been made, I believe, in ignorance of what has really happened. For instance, in the Punjab or the western part of India, more jobs have been generated. The capital required to grow the new varieties of wheat or rice is there; but the extent of investment needed certainly varies. In the Indo-Gangetic basin, where some of the first new varieties of wheat were planted, the capital required may still be beyond the means of a small farmer, but it is not really so high as to make it difficult to organize a reasonable credit-supply system. Some farmers have standardized the construction of a tube-well made of bamboo and, including a one or two horse-power pump, it costs about forty to fifty dollars.

With a self-pollinating plant like wheat or rice, the farmer can keep his own seeds and with a little input of fertilizer he can reap a better harvest. These are the kinds of changes which enabled the northern part of India to double its wheat production within a few seasons. Had the technology been confined to a few farmers, it would have been impossible for production to go up from about twelve million tons to twenty-six million tons within four crop seasons. If a technology is capital intensive or labor displacing or ecologically damaging, certainly it must be got rid of. But again taking the Punjab as an example, as the small farmer gained a little more money he was able to invest more in purchasing gypsum to reclaim his alkaline soil. The proportion of salinity and alkalinity in the soil has been a big problem in that state, but much soil is now improved.

Clearly what is needed is a technology that is relevant to the socioeconomic and agro-ecological conditions of a place. In our discussions of the Green Revolution, we should either clarify the issues or avoid referring to it; disparagement has already done a lot of damage in terms of development planning. And we must stress the fact that output cannot be increased without some inputs. But how to get them? We should not confuse the small farm with the small farmer. A small farm is capable of high productivity, but a small farmer (being a euphemism for a poor farmer with no capital and low risk-taking capacity) is not able to benefit. There is a need for arrangements to help him, whether it is in terms of crop insurance or credit policies.

Presentation at the Rome Forum (preceding the World Food Conference) organized by Barbara Ward at the FAO, November 1-2, 1974. First published in *Hunger, Politics and Markets: The Real Issues in the Food Crisis,* ed. by Sartaj Aziz, pp. 83-86. New York: New York University Press, 1975.

As for the social tensions arising from the Green Revolution, in the case of India, one finds that farmers who have been unable to enjoy the economic benefits of technological innovation are eager to do so today. In other words, there is a desire for change, there is a desire to get higher yields and income. In my view, this represents the most important positive change that has taken place in the Indian rural scene.

To capitalize on this desire, we have to promote the technology which is relevant to each country. In India, Mahatma Gandhi pointed out that the technology which is relevant to our circumstances is the technology of production by masses in contrast to the mass production technology adopted in the affluent nations. For example, if someone can develop an inexpensive solar-operated water pump, this will be the greatest gift to the Indian farmer of the Indo-Gangetic basin. The water is there, just thirty feet below. Diesel fuel has become expensive and electricity is scarce, with the result that the farmer is not able to operate the pump. The old Persian wheels operated by camels or bullocks are reappearing.

My second point is that, while naturally we are concerned with the population/food equation, the population/food/employment equation is really the more serious one in most parts of South Asia. The malnutrition problem could be more appropriately stated in terms of a certain number of man-years of needed jobs and not just in terms of a shortage of so many tons of grain. Employment is most important. Nobody wants a large segment of humanity to live on a permanent dole. The question is now: Do you develop agriculture in a way that will create more jobs? I believe that to develop an employment-oriented technology for rural areas we have to go much more deeply into the scientific planning of land and water use. The international effort in this field is lagging. We need better methods of using the data from Earth satellites. Today, many governments believe resource satellites are also gathering information for military or para-military purposes. But if some of these information-gathering and resource-planning systems can be depoliticized, they could be of great value internationally for land use surveys.

Another area where research and assistance is urgently needed is plant nutrition. In the main monsoon season in Asia, as much as thirty to forty percent of the fertilizer used in the rice paddies is lost due to leaching. This is one reason why farmers are reluctant to buy fertilizers. But there is now an emerging technology by which these leaks and losses could be minimized. Similarly, work is needed on both biological and non-symbiotic nitrogen fixation. ▯

20

Science and Agricultural Progress in the Third World

Agriculture began in what is now called the Third World. The ancestral homes of the world's most important crops are in the Third World countries. This is not surprising because agriculture revolves around the harvest of solar energy, and abundant, year-round sunshine is the Third World's greatest asset.

Culture was cradled by the prehistoric shift from food-gathering to food-growing in the Third World. Domestication of crop plants and farm animals and the development of water conservation and irrigation systems led to the civilizations of the Nile Valley, Mesopotamia, the Indus Valley, the Andean highlands and Central America, and, of course, China. Thus, the world's most experienced farmers are found in the Third World.

Asia is today the home of more than half of the world's people. But some five hundred million Asians live in absolute poverty and go to bed hungry each night. About ten million or more children die every year from causes associated with hunger and malnutrition.

We thus live in an unequal and paradoxical world. It is an irony that "God is bread" (to quote Mahatma Gandhi) to children, women, and men in many parts of the "South" while the prayer "God give us today our daily bread..." has lost historical relevance in many countries of the "North".

Why should an abundance of natural resources coexist with low farm productivity and consequent hunger? Can't we take steps to convert the natural blessings of the Third World - its basic life support systems of land, water, sunshine, flora and fauna - into wealth that is vital to its people?

The answer to this question was also provided by Mahatma Gandhi when he said that the divorce between intellect and labor is the bane of rural India. So long agriculture was regarded as a profession requiring only brawn, production and productivity remained stagnant. However, in the last twenty years, the situation started changing. Thanks to the introduction of science-based technologies. The result was a rapid increase in the production of wheat, rice, potato and several food crops.

How did the farmers of Bangladesh, Burma, China, India, Indonesia, the Philippines, and other Asian countries raise rice production more in the past twenty-five years than in the preceding 7,000 years? Five major scientific advances helped Governments to bring about a doubling and tripling of rice production through appropriate extension and input supply measures and public policies.

1). The modern semidwarf plant type that increases the efficiency of nutrient, water, and sunshine use. The traditional varieties "lodged" or fell over when soil

---

Based on the "Third World Lecture" delivered at Beijing on April 4, 1983.

fertility exceeded the capacity to produce more than two or three tons of rice per hectare.

Modern semidwarf rice varieties yield five tons or more because their strong stems - inherited from Chinese parents - enable them to stand erect and convert nutrients, water, and sunshine into grain more efficiently. Many of the newer varieties carry built-in genetic resistance to a host of insects and diseases.

2). Short growth duration and insensitivity to daylength. Teams of researchers selectively bred these traits into the modern rice varieties to make it possible for farmers to grow two or three crops where they previously grew only one crop, and to grow superior varieties at different latitudes.

3). Cultivation of F_1 hybrid rice. China is the world's pioneer in hybrid rice production. China now grows more than six million hectares of hybrid rice; yields are about 30% higher than those of ordinary semidwarfs.

4). Improved irrigation and water management. Better availability of water gives farmers yet another opportunity to grow two or more crops per year in the same field.

An old Chinese saying vividly portrays the green power of tropical rice land: "a field can look gold, black, and green on the same day." The gold represents the mature rice crop, which the farmer and his family harvest early in the morning. The same rice field looks black around noon because the family has already plowed the field in preparation for the next crop. Late that afternoon the field looks green because the farm family has transplanted a new rice crop!

In the Rice Garden of the International Rice Research Institute, a crop is harvested every Monday and. another planted every Friday. Good water management and early maturing varieties allow the planting and harvesting to be staggered throughout the year. The system spreads labor, cash inputs, and risks, and enhances both income and food production for the family. And the intensive production more fully exploits human inputs, water, sunshine, and soil.

5). Introduction of new farm management techniques. Yields are higher and farm incomes more secure because farmers now have improved implements, mineral fertilizers and better methods to apply them, integrated procedures for pest control, and improved postharvest technology.

Last year, I asked Mr. Lu Shen Rong, an old Chinese farmer who has cultivated rice for over fifty years, "What are the most important farm problems today, and in what areas do you feel we should focus research in the future? "The farmer's immediate reply was, "The care of the soil and the health of the plant."

I have worked in agricultural research for almost thirty-five years, but I could not have better summarized the most urgent challenges for the future.

Asia has little unused land, and agricultural land is a shrinking resource as population multiplies and urbanization forces its way into the countryside. Asia's ratio of land to people - 0.21 hectares per person - is the lowest in the world. Future increases in agricultural production must come mainly from higher productivity and cropping intensity per unit of land, time, water, energy, and labor. The Food and Agriculture Organization estimates that rice production must increase by more.than 3% per year during the remainder of the 20th century. This will require a continuous improvement in productivity, and it must be done without harming the long-term production potential of the soil. This challenge requires the intensification of research on all aspects of plant and soil care.

Proper attention to plant health not only prevents crop damage from the unholy triple alliance of insects, diseases, and weeds but also reduces risk and uncertainty in agriculture. Institutes such as IRRI offer unusual opportunities to develop varieties with multiple pest resistance.

In the International Rice Germplasm Center of IRRI, more than 65,000 varieties of rice are preserved for the use of today's rice scientists and farmers, and, for generations yet unborn.

But the IRRI collection probably includes little more than half of the world's genetic variability in rice. With the cooperation of national research systems, we propose to develop a global strategy to collect the remaining genetic strains as rapidly as possible. Many irreplaceable rice strains - the fruits of thousands of years of natural and human selection - grow in endangered habitats and face the threat of extinction. Agricultural organizations everywhere must join forces to collect and preserve this invaluable germplasm for posterity.

But preserving the world's rice germplasm is not enough. More important, we must know the diversity of genetic traits these strains possess so that we can put those traits to work for the benefit of rice farmers everywhere.

Through the multidisciplinary Genetic Evaluation and Utilization (GEU) program, IRRI and worldwide cooperators carefully test this vast seed collection for the expression of more than fifty inherited characteristics, such as resistance to the diseases and insects that ravage farmers' crops and tie them to costly chemical protection, and tolerance of adverse soils.

The best rices from all nations are selected under diverse environments through the International Rice Testing Programme - a worldwide cooperative network. The power that such global cooperation gives all scientists is illustrated by the success of IR36, a rice variety now grown on almost eleven million hectares in Asia. To develop IR36, IRRI scientists crossed thirteen parents from six countries including a wild species *Oryza nivara* from India. Scientists in the Philippines, Indonesia, and India cooperated in the selection of IR36.

We have no time to relax. Eternal vigilance is the price of a stable, prosperous, and productive agriculture. A new biotype of brown planthopper recently attacked IR36 in North Sumatra. Fortunately, Indonesian scientists were prepared and immediately arranged a new "genetic barrier". In February 1983 IRRI airlifted more than twenty tons of seeds of a newer variety, IR56, which is resistant to the new biotype, to Indonesia for multiplication and distribution to farmers.

Most experts agree that fighting the famine of jobs is as important as fighting the food famine during the remainder of the 20th Century. The nutrition problem in many parts of the Third World can be improved even more by an increase in person-years of employment than by an increase in grain production. Rapid agricultural development is important to generate the employment necessary to purchase food and increase its availability. In fact, rural development in Asia implies largely the development of land- and water-based occupations such as crop and animal husbandry, horticulture, fisheries, and forestry.

Asia is a continent of small-scale farmers; more than 75% of its farmers cultivate less than two hectares of land. Earlier I mentioned that recent scientific advances have enhanced the production potential of small farms in the Third World. But we should recognize that although new technologies have in many cases been used economically by all farmers, regardless of their farm size, the technologies themselves are not "resource neutral". Farmers need cash inputs to enhance agricultural output. Modernization of agriculture involves the increased use of purchased inputs and a greater dependence on marketing opportunities. Public policies that simultaneously help small farmers raise production and help poor consumers increase consumption are vital to advancing agricultural well-being in the Third World.

Because of risks, farmers largely base their decisions to adopt new technology on the likely net returns per hectare as well as on security of income. In many Southeast Asian countries, an increase of one million tons in rice production requires the active participation of about two million farming families. Hence, attention to their problems is vital for success in increasing production.

Inadequate attention to small farm management has increased production costs and risk. Agricultural scientists must find more ways, using the techniques of biotechnology, to enhance the resource neutrality of technology by substituting non-monetary or inexpensive inputs for high-cost inputs and minimizing yield fluctuations caused by weather aberrations and pest epidemics. New technologies must ensure that labor demand increases faster than its supply and that food supply increases faster than its demand. These objectives require that social, biological, and physical scientists work as teams.

New systems of small farm management must marry individual initiative and group endeavor. Japan has pioneered yield increases in agriculture through integrated systems of research, extension, input supply (including credit) and remunerative pricing. In China, a household responsibility system of crop production has been superimposed over social land ownership. This provides a mechanism for blending collective management of farm operations such as irrigation, plant protection, and postharvest operations with individual enterprise and incentive.

In India where land is individually owned, can social or group management of key farm operations be superimposed on individual initiative and ownership rights? Obviously, farmers have often done this themselves. Agricultural research organizations can help by demonstrating how enlightened self-interest demands that farmers living in a watershed or a village participate in the management of farm operations that elevate and stabilize yield per unit of cash input. In particular, there is immediate need for increasing returns from irrigated agriculture through equitable systems of water distribution and better on-farm management of water. The major irrigation systems of the world are in Asia, but there is considerable scope for improving productivity per unit of water.

Farmers live in a world of action, not of words, so personal experience is the only meaningful source of conviction. Therefore, they must be shown how blending of Individual ownership and group action can better serve the interests of the individual and the community in the same manner that China is demonstrating how social ownership and individual initiative can be combined in a meaningful way.

The most precious resources of the Third World are its intelligent and hardworking rural populations and its abundant sunlight. These assets can be purposefully married through a photosynthetic pathway of development that gives greater attention to green plants. Poverty persists when human resources are undervalued and land and other physical and material resources are overvalued. Human resource development holds the key to national prosperity as well as to peace and progress.

Priority to the farm sector does not imply neglect of industry. In fact, there are several major linkages between agriculture and industry. Agriculture supplies the raw materials for employment-intensive industries. Agriculture stimulates and sustains industrial output through rural household demands for consumer goods and services. Agriculture influences industry through government savings and public investment. Agricultural growth requires fertilizer, machinery, tools and energy.

Science and technology are important components of the wall dividing poverty and prosperity. Today there are unusual opportunities for developing countries to improve the quality of life of their people and to reduce drudgery in rural areas, particularly in women-specific occupations, through an integration of traditional and emerging technologies. Recent advances in microelectronics and microprocessors, biotechnology, satellite communication and imagery, and solar and renewable energy technology can be harnessed to upgrade traditional skills and occupations. In the past, industrialization was synonymous with urbanization, centralization, automation, and pollution. We can now promote in rural areas sophisticated agro-industrial complexes based on decentralized production techniques supported by appropriate centralized services particularly in marketing.

History shows us that man causes civilizations to blossom - and to decay. Some historic centers of agriculture are useless deserts today. In contrast, some areas that were considered hopeless wastelands in the early 20th Century are now fertile farming areas. A proper combination of political will, scientific skill, and people's action is essential to harness the power that science-based technology gives us for increasing human happiness and welfare. ▯

21

Eulogy for Shouichi Yoshida of IRRI

As I rise to join others in paying tribute to an esteemed colleague and a great scientist, I am reminded of a beautiful proverb which likens the soul to a migratory bird that flies through a lighted room. It comes from darkness, we know not where, and flies once more into the darkness, we know not where.

The same thought is expressed in other words in the Bhagavad Gita, and I quote:

> All beings have their beginnings unseen and unknown,
> their middle alone is seen,
> their end is again unseen and unknown,
> why then this lamentation?
> The self that dwells within each body is indestructible;
> therefore do not grieve.

So we should not grieve for the departed, but rather reflect on the passage of the soul through the world, and strive to learn what we can from the brief sojourn of the soul in the light. Each one of us has had the opportunity, for shorter or longer periods, to come in contact with the self, embodied in the person of Shouichi Yoshida, during his stay at IRRI. I should like to dwell for a few moments on the many things that we can learn by reflecting on that contact.

Dr. Shouichi Yoshida's life and work as a scientist serves as a shining example for all of us to emulate. In a broad sense, science is search for the truth. This implies that qualities of openness and receptivity of the mind and the ability not to be carried away by dogma have to be among the major endowments of a scientist. That Dr. Yoshida possessed these attributes in rich measure will be evident from the depth and width of his scientific contributions, all made during a brief span of twenty years.

Much has already been said about his seminal role in advancing the frontiers both of knowledge and of production in rice. I would however like to pick up ten of his scientific contributions to illustrate their enviable blend of practical relevance and scientific excellence.

First is his work on zinc deficiency leading to the understanding of its importance to rice production.

Second is his work on sulfur deficiency and the demonstration that irrigation water plays an important part in the supply of sulfur to rice.

Third is his classic work on nutritional disorders, which led to a greater understanding of the importance of phosphorus.

Read at the Memorial Service held at the International Rice Research Institute, Feburary 1, 1984.

Fourth is his demonstration that there is no optimum leaf area index in dry matter production, a finding which helped to impart greater clarity to the design of ideotypes for higher productivity.

Fifth is his finding that about 20% of grain starch comes from accumulated carbohydrates before flowering, which led to dispelling many mistaken notions in this field of knowledge.

Sixth is his study on root systems in relation to drought resistance which helped to develop indices for the choice of low or high tillering varieties depending on when the plant experiences water stress.

Seventh is the demonstration of genotypic differences in the extent of sterility caused by high temperature in experiments carried out in the Phytotron. The importance of *Oryza glaberrima* in work relating to heat tolerance won for him an award from the Japanese Society of Tropical Agronomy in 1982.

Eighth is his observation that oxygen released from rice roots is a physical process and does not involve the enzyme system, a finding which helps us to understand better the performance of rice plants under anaerobic environments. This study also helped to explain the tolerance of some rice varieties to iron toxicity.

Ninth is his adaptation of the calcium peroxide coating technique for direct sowing into puddled soils, a technique which proved useful at several locations in the Philippines.

Finally, as a result of continuous field experiments, he showed that annual rice production in the tropics could be 23-24 t/ha if water and the necessary inputs are not limiting. He frequently stressed the fact that the tropics have a greater potential for food production because of the possibility of harnessing solar energy through photosynthesis throughout the year, although on a per crop basis the temperate regions have greater potential for yield than the tropics. Based on an understanding of the physiological potential for yield, it was his conviction that countries in the tropics can become the future bread-baskets of the world.

In recent years, Dr. Yoshida was deeply interested in exploiting the potential offered by tissue-culture-induced somaclonal variation for increasing tolerance to adverse soil factors like salinity. As recently as January 16, 1984 he dictated a letter expressing his hope to participate in the Inter-Center Seminar on IARCs and Biotechnology.

He took great interest in training scholars and students. A great teacher is always identified by the performance of his students. Several of Dr. Yoshida's colleagues and scholars have achieved distinction in their subsequent careers. To cite one example, Mr. J. S. Ahn was given a Special Presidential Award in Korea for his work on the role of zinc deficiency in reducing rice yield in Korea.

Dr. Yoshida combined great scientific capability with both humility and humanism. He added wisdom to wit and even a most devastating comment would be accompanied with a smile and laughter. True to the innate Buddhist culture he was reared in, he would never hurt but would try to heal. He believed in shedding light rather than generating heat during scientific discussions. He was one whose judgement could always be depended upon. I have had the good fortune during the last two years to have received his advice and assistance on many occasions.

The esteem, respect, and affection with which Dr. Yoshida was held by the Board of Trustees of IRRI is evident from the recent decision of the Board to appoint him as a Principal Scientist, the highest honor that the Institute can confer upon its scientists. It is proposed to develop a Shouichi Yoshida Memorial Fund for initiating appropriate activities in his memory.

Dr. Yoshida took immense pride in pointing out that the dominant force behind his accomplishments was the love, consideration, and help of his wife, Yoshiko, who gave up her personal scientific career to support him. In fact, he used to derive pleasure in pointing out that the eminent scientist and a Founder-Trustee of IRRI, Dr. Hitoshi Kihara used to ignore him whenever he came to their house, but used to ask only Yoshiko the botanical names and history of the different plants growing in the Los Banos-IRRI campus. Our sympathies and prayers go to Mrs. Yoshida, her two gifted daughters, Ayako and Tomoko, and the other members of the family.

Though we will miss Shouichi in flesh and blood, his example and life will continue to guide us in our work and in our day-to-day life. May I conclude with a quotation:

Though a good life have but a few days
Yet a good name endures for ever
The righteous shall be in everlasting remembrance
And the memory of the just shall be blessed
The people will tell of their wisdom
And the congregation will show forth their praise. ▮

22

Future Horizons and Advances in Biotechnology

Recent research in cell biology, molecular genetics, recombinant DNA and tissue culture have opened up vast new vistas in agriculture, medicine and industry. New developments in the management and manipulation of the biological system are broadly grouped under the term biotechnology.

There has been considerable public and political interest in this field of technology because of reported breakthroughs possibly of great significance to life on earth. Some examples of these are: the development of new nitrogen-fixing plants, single-cell edible protein, new crops resistant to pests, bacteria for use in waste recycling and pollution control, and petrochemical substitutes; the use of gene therapy to correct monogenic diseases such as sickle-cell anemia; the use of genetic screening for the isolation of genes responsible for birth defects; the microbial production of human insulin; the production of interferon and growth hormones; and a general increase in the knowledge base of imunological processes including cancer.

It is widely believed that biotechnology will probably have its most important impact on agriculture and health. Bollinger has called tissue culture "the botanical equivalent of the laser, in that there are more potential applications than originally conceived." The vast panorama of possibilities for the use of tissue culture techniques in agriculture, animal husbandry, aquaculture, and forestry is just unfolding.

FAO has estimated that world cereal production may amount during 1983 to 1,605 million metric tons or six percent less than in 1982. In more than thirty low-income countries, food production rose at a slower rate than population during 1982. Developing countries already import more than one hundred million metric tons of food a year. The situation is especially serious in Africa where average per capita food production has fallen over the last fifteen years. Therefore, developing countries should strengthen their research and extension efforts in plant breeding and agriculture.

Most farmers in developing countries cultivate small holdings. They have few resources to invest and little capacity to take risks. Although new technologies are scale-neutral with regard to their relevance to farms of varying sizes, they are not resource-neutral. Hence, adding the dimension of resource-neutrality to scale-neutrality should be a priority task of genetic engineers. This implies: a) developing varieties with multiple resistance to pests and diseases, b) improving biological nitrogen fixation, c) developing varieties which can use soil nutrients more efficiently, d) breeding varieties which can tolerate adverse soil stresses, and e) tailoring varieties for multiple cropping and higher yield per day.

Presented at Pakistan Agricultural Research Council Seminar on Science and Technology, Islamabad, March 28, 1984.

In short term, the following areas of biotechnology offer promise in agriculture: a) Tissue culture for clonal propagation and germplasm conservation; b) Haploid culture techniques; c) Somaclonal variation and cell culture screening; d) Embryo culture and rescue; e) Protoplast fusion; f) Monoclonal antibodies and hybridoma techniques in disease diagnosis and control; g) Gene transfer; h) Regulation and control of gene expression. Depending on the nature of the problem, the appropriate technique will have to be chosen.

The world population may reach about ten billion within the next fifty years. If world population is stabilized by then, the new tools of genetics should help us to see a world without hunger. The famine of jobs will still be a difficult one to vanquish particularly if some of the developments in genetic engineering lead to substitution of microbes for human input in agriculture. An employment impact analysis is hence important while developing national policies in the choice of technologies.

I would like to briefly enumerate some of the opportunities open now for basic and applied research using the new techniques. I shall use rice as an example to illustrate the potential for the application of biotechnology in crop improvement.

Tissue Culture

Tissue culture techniques include the in vitro culture of meristems, haploid and somatic cells, embryos, protoplasts and fusion of protoplasts, which together with other technologies such as monoclonal antibodies and genetic engineering techniques constitute what is now being called biotechnology.

To meet the additional demand for rice by the year 2000, the developing countries should have a harvest area of 125 million hectares, an increase of 25 million hectares over the area harvested in 1974-1976. Similarly, yield should increase from 2.0 to 3.2 tons per hectare.

To increase production, either yield per unit area or the area planted to crop should be increased, or both. Rice tissue culture could help in the breeding or improvement of varieties with increased yield and pest and disease resistance, and adapted to problem areas such as saline and aluminum toxic soils or cold regions. Furthermore, tissue culture technique could be applied in the production of mutants with better grain quality such as brown rice with high-lysine, high-protein content.

Recognizing the potential application of biotechnologies such as tissue culture in the production of new rice varieties, the International Rice Research Institute in 1980 initiated a tissue culture program. IRRI's rice tissue culture program includes somatic cell culture, anther culture and embryo rescue.

Meristem Culture

Meristem culture is the most advanced and widely used technique in plant propagation since it allows the rapid production of genetically uniform plants and virus-free planting materials. Disease-free planting materials of a few important vegetatively propagated food crops such as cassava and potato have been obtained using this technique. Rapid multiplication using the apical meristem also permits the routine production of genetically identical individuals which is important in the genetic improvement of forest trees and rapid production of planting materials for reforestation purposes. This technique is also being used extensively in the production of cut flowers and potted plants.

Somatic Cell Culture

Another method of rapid propagation is by the use of somatic cells which, in theory, give rise to identical plants as in the case of meristem culture. Although several researchers maintain that there is no variation among plants derived from somatic cell culture of varieties, phenotypic changes often occur in the regenerated plants. Observations of somaclonal variation have been reported in celery in which approximately 70% of the regenerated plants were visibly normal, while 30% showed striking differences in growth rate or habit, leaf shape, color, or flowering behavior, although none of the plants exhibited characteristics considered superior to the original type.

Plants regenerated from leaf explants of *Lycopersicon esculentum* when evaluated in the field showed that several monogenic mutations segregated in the progenies of regenerated plants. The recovery of single gene mutations is evidence that tissue culture can be mutagenic. Segregation of single gene mutation in progenies of regenerated tomato plants showed variability in male sterility, jointless pedicel, tangerine fruit and lethal albino. Somatic variability has also been observed in other crops such as wheat, sugarcane and rice.

Large amount of variability could be obtained in plants regenerated from cells and tissues cultured in-vitro whether or not they are stressed with high levels of specific compounds. IRRI's current program on somatic cell culture is geared towards the production of rice varieties with increased tolerance to salt and aluminum toxicity, and resistance to *Helminthosporium* leaf spot.

One of the major constraints to increasing yield is that many soil areas suffer from chemical stresses. Mineral stresses occur in saline soils because of salt toxicity. One of the 380 million hectares of saline soils in the world, 240 million hectares are mildly saline and have crop production possibilities. About sixty million hectares of soils are affected by salinity in South and Southeast Asia. Therefore, it is crucial to breed varieties with increased tolerance to salt.

Salt-tolerant varieties usually are low-yielding and of poor plant type, but it is possible to improve their characteristics through tissue culture. For example, when NonaBokra, a salt tolerant variety was passed through in-vitro culture, a variant which had earlier maturity and 20% higher yield than the parent was regenerated. Likewise when somatic cells of high-yielding, susceptible varieties are cultured in-vitro and stressed with high concentrations of NaCl, it is possible to select some variant lines with increased tolerance.

Plants regenerated from the rice variety Taichung 65 grown in a saline culture solution containing 0.7% NaCl showed a 50% increase in survival rate compared to the parent.

Strongly acid soils cover nearly 2.6 billion hectares in humid regions and two billion hectares in the tropics. Generally they are uncultivated, but have potential for food production. Several plants regenerated from somatic cells of Taichung 65 grown in culture solution containing 30 ppm Al showed increased tolerance to aluminum toxicity.

Embryo Culture

Interspecific hybridization is an important tool for introducing valuable traits from wild species into the gene pools of cultivated plant species. A major difficulty in obtaining these hybrids appears to be, in some cases, the absence of seed development since fertilization occurs but embryos are aborted. By culturing embryos in vitro, plant breeders could rescue rare hybrids which have combined desirable traits from both parents. Embryo culture could be used to germinate those embryos that may not otherwise germinate because of a specific inhibition or prevention of key steps in pollination, pollen tube growth, fertilization, and embryo or endosperm development. Embryo culture could be used to speed up germination in certain species such as oil palms whose embryo takes up to two years to germinate under natural conditions.

Rice plant breeders are trying to find alternative gene sources of BPH resistance. *Oryza australensis*, a wild relative of the cultivated rice species, could be useful in these studies. Since embryo abortion of *O. sativa* x *O. australensis* cross occurs, it is necessary to employ embryo rescue in order to overcome such incompatibility.

Anther Culture

Anther culture technique is important in rice breeding programs because it increases the selection efficiency, accelerates the breeding cycle, and permits faster production and selection of useful mutants. As this technique involves haploid gametes (pollen grains) instead of diploid parents, the number of plants required to express the different recombinants is less than in conventional breeding. In conventional plant breeding, five to six generations of segregating populations are grown to get pure breeding lines. Anther culture eliminates these segregating populations and saves considerable labor, time, and space.

In anther culture, difficulties in the selection of mutants due to dominance are negligible, and most, if not all, genes are expressed. This allows easy isolation for recessive mutations because they are expressed immediately.

The anther culture program at IRRI has four objectives: 1) to reduce the time required to produce cold tolerant rice varieties; 2) to increase total protein and lysine in rice grains of plants regenerated from anther-derived calli, which have been stressed with amino acid analogs; 3) to increase the efficiency of callus production and green plant regeneration; and 4) to examine the variability of anther culture-derived plants under field conditions.

In South and Southeast Asia, about seven million hectares cannot be planted to modern rice varieties because of cold water and cold weather. In some areas, two cropping seasons a year are possible if cold tolerant varieties are developed.

Cold tolerance is an important plant character in areas where low temperature at seedling stage is difficult, and the progenies of many crosses of cold tolerant parents show little or no cold tolerance. Backcrossing seems to be effective although it is a very slow process. Anther culture may be another possible pathway to identify cold-tolerant materials.

Of 110 anther culture-derived lines from four crosses having at least one cold-tolerant parent, 26 plants (23.67%) have shown a high degree of tolerance at seedling stage, 30 plants (27.3%) had an intermediate degree of tolerance, and 54 plants (49.1%) were susceptible.

Rice is unique among the cereals for its superior nutritional quality. Unfortunately, it is low in several essential amino acids, including lysine, necessary for human nutrition. One way to improve the protein quality is to stress haploid cells with amino acid analogs, which will inhibit one of the enzymes necessary for the production of specific amino acids. This inhibition induces the cell to find an alternative pathway to produce the amino acid in order to survive. Once the inhibitor is eliminated, the variant cells will hopefully overproduce these specific amino acids since there will be more possible pathways for their production.

In preliminary experiments using the lysine analog S, 2-amino-ethyl-L-cysteine, a feedback inhibitor of the enzyme aspartokinase, on haploid cells of rice variety Taipei 309, we have been able to increase total protein by 9.3% and total lysine by 13.9%, over that of the parent.

We have been able to establish the optimum conditions for the induction of callus and regeneration of plants. One thousand homozygous diploid plants from nearly 30 different F1 sexual crosses and about 3000 plants from established varieties have been regenerated. Inasmuch as the plants regenerated from F1 sexual crosses showed no segregation in the preliminary field studies, yield trials were conducted.

The agronomic characteristics of regenerated plants from selected F1 crosses were compared in the field with those of their parents. Genetic variations in heading date, culm length, panicle length, panicle number, flagleaf length, yield, cold tolerance, lysine, and protein content were found.

Protoplast Culture and Fusion

Until recently, the only means of hybridization in plants was by sexual crossing. However, success in the fusion of protoplasts (somatic hybridization) from two different species by Cocking's group in 1970 gave rise to a new powerful tool for hybridization in higher plants.

Before attempting protoplast fusion, it is necessary to establish the culture and plant regeneration procedures from protoplasts of the species involved. Significant gains have recently been made in inducing cultured protoplasts directly isolated from important crop plants to divide and form colonies. Significant and stable somaclonal variation in compactness of growth, habit, maturity dates, tuber and skin color, photoperiod requirements and fruit production has been found in potato plants regenerated from isolated protoplasts of the variety Russet Burbank. These results point out that it could be simpler to selectively improve a popular variety than to create a new one.

In cereals which are acknowledged as one of the more difficult models, sustained colony development now appears possible. While much further work will be necessary before high plating efficiencies and complete reproducibility is commonplace in cereals, it is no longer assumed that this goal is unrealistic because of fundamental genetic limitations.

Somatic hybridization therefore, is an approach that should be considered in instances where sexual crosses have proven impossible or in special cases in which it is desirable to transfer cytoplasmically inherited traits. The procedure results in the merging of cytoplasm of the two parents, while in sexual crosses only the female contributes cytoplasm.

However, plants resulting from fusion between protoplasts of *Nicotiana langsdorfii* x *Nicotiana glauca* having different chloroplasts were found to be composed, with only one exception, of uniform chloroplast populations of either one of the two fusion parents. Hybrid plants resulting from fusion between protoplasts having different mitochondria contain mitochondrial genomes which, according to their mtDNA restriction patterns, were not identical to either of the parental mtDNAs. These experiments suggest that DNA recombination caused these novel restriction patterns.

The possibility of unidirectional transfer in mitochondria or chloroplast from donor to recipient plant has great promise in plant breeding. There are several known mitochondrial genome control traits like male sterility and resistance to phytotoxins which have breeding relevance. Likewise, several chloroplast genome characters are important breeding traits like pesticide resistance, although the main obstacle in utilizing cytoplasmic hybridization in plant breeding is the small number of crop plants in which protoplast culture and plant regeneration is presently possible.

Furthermore, the procedures of gene cloning in the near future will enable us to understand the genomic organization of higher plants and to isolate gene sequences important in plant productivity. The construction and insertion of complimentary DNA (cDNA) clones for the production of specific substances into a carrier could be introduced into the genetic pool of plant protoplasts. The transformed protoplast hopefully will regenerate into a plant which will express the identifiable gene function.

Monoclonal Antibodies

Antibodies, protein molecules produced by certain cells in an organism, are a basic constituent of animal and plant disease-fighting systems. Conventional serological methods for antibody production involve several extraction and purification steps, but the antibodies produced by this method are low in purity thus are usually weakly specific, low in activity, and may gradually lose their discriminatory power. Furthermore, although conventional reagents have been useful, they are difficult to maintain in sufficient supplies to meet commercial needs.

The recent discovery of the hybridoma technique has revolutionized and made possible the production of antibodies of known class, specificity and affinity. Hybridoma technique is the fusion of a carcinogenic cell (myeloma cell) with an antibody-producing cell from an immunized donor. Antibodies are formed individually by individual cells, thus this technique permits the production of monoclonal antibodies which are pure and uniform against a specific target. Since the hybridoma has the capacity to multiply rapidly and indefinitely in culture, monoclonal antibodies could be produced in commercial quantitites.

Monoclonal antibody technology is playing a growing role in animal and plant research and management. Many diseases affect the production of plant food crops for human and animal consumption, as well as the development and maintenance of pathogen-free seeds and plants. An exceptionally large number of viral, fungal and bacterial pathogens constantly threaten entire food crops and no effective methods for testing viral, bacterial or fungal infection has been developed. This problem is compounded by the fact that a number of variant forms exist in the field, with some causing severe disease. The latter types do, in fact, protect against infection those causing severe disease. The decision to destroy a field depends upon which form is present. Available serological reagents made by conventional procedures fail to distinguish one from the other or if ever they do distinguish a virus type, they could not be used to distinguish a strain variant. Monoclonal antibodies, because of their predetermined specificity, provide a means to solving this problem and could be very useful for diagnoses and epidemiological studies of plant diseases.

In rice, the technology of monoclonal antibodies could be used in solving the problems on virus diseases such as rice dwarf, grassy stunt, ragged stunt and rice tungro.

Genetic Engineering

The growth of agricultural crops is dependent on enormous supply of useful nitrogen either mechanically applied as ammonia, urea or nitrate fertilizer, or naturally produced in the soil through microbial reduction of atmospheric nitrogen. Increasing energy costs have encouraged research into novel approaches to increase the available supply of naturally reduced nitrogen. Although most organisms cannot assimilate atmospheric $N_{2,}$ limited number of prokaryotes are able to reduce N_2 directly to ammonia in a process called nitrogen fixation. One of the approaches to transfer the nitrogen fixation capability from leguminous to non-leguminous plants could be by genetic engineering.

One of the strategies to introduce nif genes could be the use of the Ti plasmid of *Agrobacterium tumefaciens*. Monocots, however, are not susceptible to the infection at the whole plant level possibly due to the lack of wound response typical in dicots, which is a drawback on the use of Ti plasmid in monocots. Alternative vectors of nif genes should therefore prove useful.

Besides the Ti plasmid, other possible candidates to be used as DNA vectors are the cauliflower mosaic (CMV) and the gemini viruses. The CMV has been extensively studied since it can infect leaves in situ and later move systematically throughout the entire plant. This system is surely limited in its host range but it may be of more practical value as a source of DNA regulatory sequences in constructions to be transferred and expressed in plants. Gemini viruses have recently attracted attention since they are pathogens of a wide variety of plants including legumes and cereals, and thus are potential candidates to introduce genes into monocotyledonous plants. However, the basic biology of the gemini virus is still being investigated. The genome of such a virus was cloned recently in *E. coli* and its sequence is being determined.

Once adequate vector(s) for monocotyledons are found, these could be useful in the insertion of monogenic characters such as those involved in pest and disease resistance. This research would lead genetic engineering scientists into the preparation of more complex plasmids which contain several genes for polygenic characters such as yield, protein quality and drought or salt tolerance. The possibility of introducing several genes could be difficult but any positive change in one of these genes could influence a total change in character which is polygenic in nature.

The euphoria caused by the genetic revolution should not however make us ignore the fact that there is no time to relax in the following two areas of agricultural production strategy: a) Deriving full benefit from the conventional tools of genetics and other scientific disciplines which can lead to valuable results and materials in the short term, and b) Helping farmers to increase food production immediately through the effective transfer of the best available technologies, since the gap between potential and actual yields is high in most farming systems even with the technologies now on the shelf.

I am glad that on the occasion of the inauguration of the National Agricultural Research Center of Pakistan, the PARC has organized this seminar on Science and Agriculture. It is only through the application of science and technology that we can achieve the goal of improving continuously the productivity and profitability of terrestrial and aquatic farming systems without detriment to the long term production potential of land and water. ▯

23

Commencement Address at the University of the Philippines

I am grateful for this opportunity to greet and congratulate you, new alumni of this University, whose hard work has received well-merited recognition at this commencement ceremony. I wish you all opportunities for professional success and fulfilment, and personal happiness. You have been fortunate to have had the opportunity to study at this great center of learning.

It is a particular privilege and honor that on this occasion, I have also become an honorary alumus of this University. I cannot find words to give adequate expression to my sense of gratitude for this kindness.

At today's commencement, the University is providing to this nation and to the world nearly 1,100 graduates. Seventy-five years ago, there were at Los Banos twelve scholars and Dr. Nora J. Cruz has described vividly and movingly how this leading agricultural, educational and research center of Asia had its humble beginning on June 14th, 1909. I quote: "In the first week, classes were held in the houses of the faculty in the town of Los Banos. Later in the week, two tents borrowed from the Bureau of Education were pitched at Camp Eldridge nearby. In the morning, with the sun beating down on the canvas, the students were introduced to the principles of scientific agriculture. More than once during those trying months, typhoons blew the tents down or turned them inside out. In the afternoon the boys hiked the three kilometers to the College farm and wrestled with fieldwork. Actually, it meant clearing the half-wild campus of weeds and underbrush."

Today, Los Banos represents a vast network of national, regional, and international institutions serving not only farmers of the Philippines but of the entire world. Even during the present decade, six institutions and four centers, together with four units of the Ferdinand E. Marcos Life Science and Agricultural Complex, have been organized. Although you know them, I will mention them to underline the wide range of disciplines and sciences they encompass: Institute of Animal Science; Farming Systems and Soil Resources Institute; Institute of Food Science and Technology; Institute of Mathematical Sciences and Physics; Institute of Chemistry; Institute of Biological Sciences; National Documentation Center for Agriculture, Forestry and Rural Development; Forestry Development Center; Gene Bank for Dairy Animals; Institute of Plant Breeding; National Institutes of Biotechnology and Applied

---

Presented at the University of the Philippines at Los Baños, April 28, 1984.

Microbiology; and National Crop Protection Center and Plant Genetic Resources Laboratory.

Success, it is often said, is due partly to inspiration and greatly to perspiration. Both vision and hard dedicated work are essential in building great institutions. Hence, an occasion such as the 75th anniversary provides an opportunity to look backward and to look forward. Looking back, we must acknowledge with gratitude the leadership of numerous visionaries starting with Dr. Edwin Bingham Copeland seventy-five years ago until the present day when UPLB has the privilege of the dynamic and far-sighted leadership of Chancellor Emil Q. Javier They all have been successful in their missions because of their ability to stimulate simplicity out of clutter, harmony out of discord, and opportunity out of difficulty. The University has been equally fortunate in having Vice-Chancellors, Deans, Directors, and faculty who are all cannitted to promoting excellence and relevance in education and research.

On behalf of the staff and scholars of the International Rice Research Institute, which has been privileged to work closely with UPLB during the last twenty-four years, I would like to express our admiration and gratitude to all the leaders who have made UPLB what it is today.

UPLB is celebrating its 75th anniversary. It will be of interest to survey what changes have taken place in Philippine agriculture during the past seventy-five years. Last week, while travelling in Central Luzon with the research workers of the Agronomy Department of IRRI and of the Maligaya Rice Research and Training Center, I met a farmer whose age is almost the same as that of UPLB. This 75-year old fanner, Mr. Francisco Carbonell, practices an integrated farming system involving crops, livestock, and fish. He does not waste even a centimeter of space and grows, in addition to rice and upland crops, a variety of horticultural plants such as jackfruit, papaya, and banana along the bunds and fences. His farm is a symbol of the green power of the Philippines and of other countries in the tropics. By green power, I mean the ability to harvest the sun throughout the year through the process of photosynthesis by green plants. I ask you to look around this commencement ground for a minute to glimpse the green power of the Philippines.

Everywhere we see beautiful trees, which Lord Buddha once said are the only living organisms capable of unilateral love, because they provide shade even to the axeman who cuts them; hills with coconuts, bananas, and medicinal plants; and not far away Laguna de Bay full of aquatic life. Charaka, an ancient physician of India, was once asked by his teacher to bring plants which were quite useless. He returned empty handed remarking that there was no such plant. I am reminded of this when I look around this lovely campus and country.

Mr. Carbonell adopts organic recycling techniques, thereby minimizing his dependence on purchased inputs. At the same time, he stresses that inputs are needed for output and therefore advocates the adoption of integrated nutrient supply and pest management procedures. He adopts modem aquaculture techniques. His aim is to maximize production per cubic volume of soil and of air. His knowledge comes not from books alone, but from spending all day in sun and rain with plants and animals. As one dependent on the sale of farm produce for his living, he is sensitive to the behavior of markets. For example, he postponed his harvest of tilapia until Good Friday so that he could get a better price. Even at this age, he bubbles with new ideas on how to improve the productivity and profitability of rice farming systems.

The famous wheat breeder and Nobel Laureate, Dr. Norman E. Borlaug once said and I quote: "Plants speak to human beings. However, they only whisper softly like a woman of great culture. They hence reveal their secrets only to those who are close to them. Those who spend their time in office rooms naturally think that plants are dumb." I asked Mr. Carbonell the single most important ingredient for his success and he said, "Science."

When I asked him the most serious problem he is facing, he said "lack of interest among youth in farming". None of his eleven children wish to follow his footsteps and make farming their life's occupation and passion. Thus, while science offers scope for imparting accelerated agricultural growth and stability, declining youth interest in farming poses a serious threat to food production and rural development. It is in this context that I am happy that so many young men and woman graduates are going out of this campus today to serve farmers and farming. The population structure of the Philippines as well as that of many other Asian countries is characterized by the predominance of youth. Unless young people take an interest in agriculture, the future of farming will not be bright. Ancient and recent history teaches us that if agriculture goes wrong, nothing goes right. Therefore we who are privileged to work for agricultural progress have a great responsibility as well as a great opportunity.

Agriculture, including crop husbandry, animal husbandry, fisheries, and forestry is the basic industry of rural areas. Agricultural advance is hence the most powerful tool for promoting rural prosperity and for preventing the migration of the rural poor to cities in search of jobs. Why then this decline in interest among young people in farming? Mahatma Gandhi said about fifty years ago that the bane of agriculture is the divorce between intellect and labor. Agriculture is still regarded by the urban elite of many developing countries as a profession which needs only brawn and no brain. Therefore, educated people tend to shy away from agriculture, leaving farming largely in the hands of semi-literate and illiterate persons. Today we know that it is only through the meaningful integration of brain, brawn, and bank that we can achieve the twin goals of food self-sufficiency and rural prosperity. You, the outgoing alumni, will have to provide the brain component of agricultural advance through your efforts in problem-solving research, inspired teaching, effective extension, and scientific farming. Graduates who have specialized in agribusiness, economics, and allied disciplines should help in harnessing the financial resources necessary to convert knowledge and technology into wealth. You should strive to generate widespread awareness of the need to reverse the current flow of brain and money from the barrio to the city.

Only yesterday at IRRI, we concluded an international seminar on biotechnology organized with the help of the National Institutes of Biotechnology and Applied Microbiology and UPLB. At this seminar, a fascinating scenario of future possibilities in agriculture emerged. Molecular genetics has opened up unusual possibilities in the manipulation and management of genetic characteristics. Recent advances in genetic engineering, microbiology, microelectronics, computer technology, and space technology have placed in our hands additional tools to make hunger and poverty largely problems of the past and to confer on every human being the right to work and to food. Will we grab this opportunity or be silent onlookers to more than 500 million going to bed hungry each day all over the world? Will we allow the fast-spreading famine of jobs to sap the vitality and self-confidence of youth? Sir Francis Bacon once said and I quote: "It would be an unsound fancy to expect that things which have never yet been done can be done except by methods which have never been tried." Therefore, it is my fervent hope and prayer that you, the 75th anniversary graduates, will have ample opportunities to use the new tools of science for enhancing hunan happiness and welfare in addition to deriving intellectual satisfaction and personal fulfilment. The ultimate goal in our work should be the happiness and well-being of all members of the human family. This can be achieved only if there is teamwork, cooperation and, above all, a sense of purpose and mission in life.

Some years ago, an eminent scholar who started compiling the ten most beautiful words in the English language, from the point of view of their ability to motivate listeners to a greater sense of purpose, placed the word *chime* at the top of his list. The quiet tinkling of a bell or chime summons in all of us a sense of peace, a sense of purpose, a sense of reverence, and a desire to serve God by being good to fellow human beings. Your alma mater represents all of these ideals, and I hope you will be blessed with abundant opportunities to draw inspiration from your life on this campus. Once again, I congratulate you and wish you well. ▯

24

Rice Research: Learning From China

We are grateful to the Chinese Academy of Agricultural Sciences (CAAS) and the China National Rice Research Institute (CNRRI) for co-sponsoring this year's International Rice Research Conference (IRRC).

China has made remarkable progress in recent years in improving the productivity and production of rice. China is not only the center of origin of considerable genetic variability both in indica and sinica (japonica) rices, but it is also the home of many innovations in rice research and improvement. The semidwarf varieties, which helped to initiate a new era in the breeding of indica rices capable of responding to good soil fertility and irrigation management, come from China. Chinese scientists also made the dream of many plant breeders for exploiting hybrid vigor on a commercial scale in a self-pollinated plant come true. The art and science of the use of biofertilizers like Azolla have been perfected by Chinese scientists and farmers. Above all, we find in this country a symbiotic linkage between research and extension, with the result that there is little gap between potential and actual yields in farmers' fields. It is therefore appropriate that the IRRC is being held in China at a time when there is concern again in the world about stability of the national food security systems of developing countries where the fate of rice production is influenced greatly by monsoon behavior.

Most parts of the South and South-east Asia are experiencing abnormal rainfall conditions during this year. In large areas of several countries – including India, Sri Lanka, Thailand, Cambodia, Vietnam and the Philippines – there is severe drought. In other areas, like Bangladesh and eastern India, there are serious floods. We do not know yet what the ultimate impact of the unfavorable monsoon conditions will be on total rice production during this year. It seems possible that there could be an eight to ten percent reduction in overall global rice production. Considering the fact that the international trade in rice covers less than four percent of total production, this degree of loss could be serious. The fragility of the food security system in many developing countries is seen from the fact that even a five percent increase or decrease in foodgrain production often makes all the difference between an uncomfortable glut and acute scarcity. Fortunately, however, there are substantial grain reserves just now in many developed and in some developing countries; hence, famine can be avoided if appropriate intervention programs are formulated and implemented.

Presented at the International Rice Research Conference, Hangzhou, China, September 21-25, 1987. Published in *Rice to Feed the World: Life and Work of H.M. Beachell*, ed. by Anwar Dil, pp. 527-30. San Diego, California/Islamabad, Pakistan: Intercultural Forum, 2001.

The unfavorable seasonal conditions, however, provide an opportunity for launching contingency plans and compensatory production programs. This is because of the availability of rice varieties which are insensitive to photoperiod and are quick-yielding. Wherever seeds are available for implementing compensatory crop production programs, we can offset to some extent the loss suffered due to drought or floods. However, for taking advantage of this opportunity, reserves of appropriate seeds are important. Just as foodgrain reserves are essential for food security, seed reserves are essential for achieving a certain measure of stability in production.

World per capita food supplies, measured as dietary energy, rose by twelve percent – from 2,340 to 2,630 kcal/day – during the eighteen-year period between 1961-63 and 1979-81. Although the gap between developed countries and developing countries still remains wide, it has narrowed during this period. The introduction of high-yielding varieties of rice and wheat in the mid-1960s has played an important role in narrowing the gap. FAO's *The Fifth World Food Survey* (FAO, 1987) indicates that in spite of all the progress made in production, the average inhabitants of the developed countries had nearly two-thirds more to eat in 1979-81 than the average inhabitants of low-income developing countries. The Fifth Survey also indicates that, with accelerating growth in per capita food supplies in the developing market economies, the proportion of the population suffering from under-nutrition has declined, even though the absolute numbers of the undernourished have increased, as a result of even more rapid population growth.

The carrying capacity of land in many developing countries is under severe stress due to increasing pressure from human and animal populations. Land is also a shrinking resource for agriculture due to competing demands for its use. It is in this context that there is a growing emphasis on sustainable food production. The issues involved have been articulated recently by the World Commission on Environment and Development (WCED). Thus, the economic viability and ecological sustainability of the production process are assuming importance from the point of view of technology development and public policy. Can rice researchers face these challenges? In this context I would like to highlight three important areas which will need increasing attention from rice scientists in the years ahead.

First, I would like to see all nations afford the highest priority to fighting what I call "the ecological fire" raging in most parts of the world. Through mechanisms of deforestation, soil erosion, desertification, water pollution, and overpopulation, developing countries are being ravaged by this fire. Developed countries suffer equally, through atmospheric pollution, acid rain, contamination of water by toxic wastes, and the spread of environmental mutagens and carcinogens. Certain phenomena – loss of biological diversity, destruction of the ozone layer, and potential climate changes such as result of carbon dioxide accumulation in the atmosphere and ocean warming – will affect us all, irrespective of the geographic origin of the problems. Today we are concerned with fighting fires in buildings and forests. We do not yet seem to be aware of the vast dimensions and potential impact of the ecological fire that is affecting our basic life-support systems of land and water, flora and fauna, and the atmosphere. IRRI is reorganizing the International Network of Soil Fertility and Fertilizer Evaluation in Rice (INSFFER) into a network which will give special attention to soil fertility monitoring and sustainability of rice production.

My second item for common action is the promotion of what I would like to christen a symphonic agricultural system, based on principles of ecological sustainability, economic viability, and equity. A symphonic research and production system based on harmony among the various links in the production-marketing-consumption chain aims to impart dynamism, coupled with stability, to agriculture. Too frequently the concept of sustainable agricultural production is used to plead for a status quo, or even a reversion to old practices in production techniques. What we need is a dynamic concept of sustainability that helps us meet the changing human needs of our expanding population at the same time that it maintains and enriches our natural resource base.

Fortunately, many developing countries have a large, untapped crop yield reservoir, even at currently available levels of technology. For example, time and again, it has been demonstrated that right now, given the optimum blend of packages of economically viable technology; of services which can enable all farmers to derive benefit from new technologies, irrespective of their innate input-mobilizing and risk-taking capacities; and of government policies on agrarian reform as well as on input supply and output pricing and marketing that are designed to stimulate both production and consumption, we can double average yields in many countries which are chronically deficit in food. This lesson we can learn from China.

Third, steps need to be taken to promote the efficiency of management of small farms. Small farms lend themselves to intensive cropping, as has been elegantly demonstrated under the Asian rice farming systems research network. Scientists can show how the available land, labor, water, and credit resources can be utilized in the most effective manner for increasing yield, income, and employment. While a small farm is not a handicap for adopting new technologies, a small farmer faces severe handicaps in having access to appropriate technologies. While scientists can develop technologies for small farms, it is only the political administration that can help small farmers to overcome their handicaps and derive economic benefit from new technologies. It is only when the small farmer is enabled to adopt new technologies and improve the efficiency of management that we find the small farm productivity reaching its potential.

Finally, we should take advantage of the emerging technologies, such as genetic engineering and other components of biotechnology, for increasing further the yield potential of the rice plant under different agro-ecological conditions. For this purpose we will have to increase the total biomass production per day while maintaining a favorable harvest index. Also, we need to transfer genes which can confer more enduring resistance to pests and diseases as well as soil stresses. Breeding varieties for adverse soil and climatic conditions deserves high priority, coupled with efforts in genetic enhancement. All other aspects of the production process, such as agronomy, crop protection and farm implements, need concurrent attention.

A fascinating panorama of new opportunities is now unfolding. It is in this context that I would like to offer our congratulations to the Chinese Ministry of Agriculture, Animal Husbandry and Fisheries for its vision in establishing CNRRI. We in IRRI feel highly privileged to have been associated with this important initiative. Dr. Robert F. Chandler, the Founder-Director of IRRI, described the history of IRRI as an "Adventure in Applied Science". I am confident that CNRRI will not only be an exciting adventure in applied science but will become a world leader in the development of symphonic rice production systems. The challenges ahead are many and complex. Achieving the desired growth rates in rice production for meeting the demands of the growing populations is not going to be an easy task. It will require both innovation and dedication on the part of scientists and total commitment to achieving national food and nutrition security on the part of governments. We have both these ingredients so strongly developed in China, and I am confident that all of us participating in this conference will return home charged with a sense of enthusiasm and hope for the future.

Before I conclude, I would like to refer briefly to the fine research collaboration we have with our Chinese colleagues. Collaboration between the Chinese Academy of Agricultural Sciences (CAAS) and IRRI is wide-ranging and far-reaching and serves as an outstanding example of the power of purposeful scientific partnership. CAAS has established one of the world's finest genetic resource conservation centers in Beijing and is actively participating in the Rice Genetic Engineering Cooperative sponsored by the Rockefeller Foundation. Last year, the Hunan Academy of Agricultural Sciences and IRRI organized an International Symposium on Hybrid Rice. This year, the Fujian Academy of Agricultural Science and IRRI organized an International Training Course in Azolla at Fuzhou. The Chinese Academy of Agricultural Mechanization Sciences (CAAMS) and IRRI plan to organize in 1988 a Monitoring Tour-cum-Workshop on Farm Machinery for Women. Academia Sinica and IRRI have useful collaboration in tissue culture and genetic research. China participates actively in the three major international research networks, namely IRTP, INSFFER and Asian Rice Farming Systems Network. With the establishment of CNRRI, the bonds of collaboration will be further strengthened. I, therefore, wish to take this occasion to thank most sincerely Minister He Kang, President Lu Liangshu, and all the scientists and officials for both the Central and Provincial Academies of Agriculture for their leadership in demonstrating the value of scientific partnership. ■

References

Food and Agriculture Organization (FAO). 1987. *The Fifth World Food Survey 1985.* Rome, Italy: Food and Agriculture Organization.

World Commission on Environment and Development. 1987. *Our Common Future*. New York: Oxford University Press.

25

Our Common Agricultural Future: First World Food Prize 1987 – Award and Acceptance Speeches

Remarks by Dr. Norman Borlaug on the occasion of the Award of the World Food Prize to Dr. M.S. Swaminathan, October 6, 1987

Tonight we celebrate the awarding of the first World Food Prize to one of the world's greatest scientists, Dr. M.S. Swaminathan, for his outstanding contributions toward increasing food production and improving the equity of its distribution. His many and diverse contributions have alleviated hunger and misery, and have contributed greatly toward improving the well-being of millions of people in the Third World.

It is especially gratifying for me to have lived to see the establishment of the World Food Prize. The road which finally led to its establishment has been a long, winding, bumpy, and often-times frustrating one. Let me share with you some of the trials and tribulations encountered along the way, which finally led to the happy event we are celebrating here tonight.

In October 1970, I was awarded the Nobel Prize for Peace for my and my many associates' contributions toward increasing agricultural production in food-deficit Third World countries. As most of you know, there is no Nobel Prize for either agriculture or food. Moreover, this was the first time an agricultural scientist had received a Nobel Prize.

While preparing my Nobel Lecture, which is a requirement for all Nobel Laureates, I tried to find out why there was no Nobel Prize for agriculture and/or food. It became apparent that Alfred B. Nobel was indeed very conscious of the importance of food, for he once wrote: "I would rather take care of the stomachs of the living than the glory of the departed in the form of monuments." Despite such clarity of interest and concern about food, when Alfred Bernhard Nobel's will and testament were drawn-up on November 27, 1885, there was no provision made for a prize for either agriculture or food. We all know how strongly and jealously the territorial domain of individual prizes, not only Nobel Prizes, but all other prestigious prizes, are guarded or protected from intrusion by apparent imposters from other fields. This reaction is equally strong in the physical, chemical, biological and medical sciences, as well as in the terrain of literature and the arts. It is, indeed, no less true of the turf of the Nobel Peace Prize, which is generally understood to be the territory of outstanding political leaders.

During the mid-1960s, both India and Pakistan were hard hit by drought and huge food shortages. Some internationally renowned scientists proclaimed that the food situation in India was hopeless and stated that millions were doomed to starvation. Widespread famine was averted only by the importation of more than fifteen million tons of grain, mainly wheat, into the two countries in 1965-66, much of it under terms of concessional sales or as outright gifts, from a number of countries. It was during this crisis that the high-yielding Mexican wheat varieties – and the accompanying package of improved agronomic practices which permitted the varieties to express their high genetic yield potential – were in the final stages of testing on thousands of farms in the two countries. Moreover, out of fear of continuing and worsening food crises, both countries by 1967 had discarded cheap food policies – which had long acted as a disincentive to production – and adopted economic policies that facilitated and encouraged the widespread use of the new wheat production technology by farmers, and a couple of years later the new high-yielding rice technology. Yields and production of wheat increased spectacularly in both countries in 1968 and 1969 and was dubbed the "Green Revolution" by Mr. William Gaud of USAID.

When the Nobel Committee of the Norwegian Storting designated me as the recipient of the Nobel Prize for Peace, it came as a great surprise to me and to the world. This was the only "window" through which work on agriculture and food production could qualify for a Nobel Prize. It was justified under Paragraph One, which laid down the conditions to be fulfilled by a recipient of the Nobel Peace Prize. It states: The award of the Prize shall be made to the person "who, during the preceding year, shall have contributed the greatest benefit to humankind".

I realized that it was not likely that a situation would again arise, within the next several decades, which would result in an agricultural or food scientist being selected as a recipient of a Nobel Peace Prize. For the next five hectic years, however, I was too busy strengthening the Green Revolution in the Indian subcontinent and spreading it to other countries to give much thought to trying to establish a deserving annual prize for agriculture and food.

By 1977, however, I wrote several letters to the Nobel Foundation inquiring about the possibility of establishing a Nobel Prize for agriculture and/or food, as had been done in establishing the Nobel Prize for Economics, (which is actually the Bank of Sweden Prize in Economics in memory of Alfred Nobel), in 1968. I was invited to present my case to the Board of Governors of the Nobel Foundation in Stockholm and to the Nobel Committee for the Nobel Prize for Peace in Oslo. Before doing so, I had done my home work well and had been assured by the three most prestigious agricultural and forestry organizations of Denmark, Norway, and Sweden that they would fund a Nobel Prize for agriculture and forestry if one could be established as had been done under the Nobel Foundation's auspices in establishing the Nobel Prize for Economics. During my presentation, the members of the Board assured me that the Nobel Foundation was well aware of the need for recognition of the importance of agriculture and food, but that since a prize for agriculture and/or food was not provided for in Alfred B. Nobel's will and endowment, it was not legally possible to establish a new prize with the use of these funds. There, my efforts to establish a new Nobel Prize for agriculture and food hit a dead end.

For the next three years, I searched ineffectively for a sponsor for a World Food Prize. In late 1981, I met the late Carleton Smith, who had been successful in establishing The Pritzker Architecture Prize, and The World Wildlife

Prize sponsored by Kerr-McGee. Smith, too, had been wondering why there was no major prize for agriculture and food, and had begun trying to do something about it.

Over the next three years, Carleton Smith and I combined our efforts in a search for sponsors. Finally, we were scheduled to meet with President James Ferguson of General Foods Corporation, to work out final details concerning the prize and hopefully its funding on the day following Memorial Day weekend in May, 1984. Unfortunately, Carleton Smith's untimely death the night before the scheduled meeting, delayed a decision.

However, in October 1985, Mr. James L. Ferguson told me General Foods Corporation had decided to sponsor and finance an annual World Food Prize. The monetary value of the Prize is the equivalent of a Nobel Prize. In the first three months of 1986, three organizational meetings were held under the leadership of Mr. A. S. Clausi, Senior Vice-President of General Foods, with participants from universities, foundations, government and industrial laboratories. The World Food Prize was initiated.

This year the Selection Committee voted unanimously to award the first annual World Food Prize to Dr. Monkombu Sambasivan Swaminathan. That brings us to this glorious evening.

Dr. Swaminathan was awarded the first World Food Prize for his outstanding contributions to expanding production and availability of food throughout the world.

During the 1960s and 1970s, I personally had the good fortune to see the important diverse roles he played in research, extension, and agricultural policy, which ushered in the Green Revolution in India. During the 1980s, I have seen this tremendous positive impact spread to many other countries.

In closing, Dr. Swaminathan, on behalf of all members of the Selection Committee of the World Food Prize, I have the very great personal pleasure to congratulate you for. your many great contributions toward the improvement of the well-being of humankind. I know you will keep up the good work. You have our very best wishes as we continue the celebration and honor of your achievements.

Dr. Swaminathan, we are confident the trail you have blazed toward increasing production, and improving the quality and availability of food for millions will be an incentive to attract some of the most talented, creative and best motivated young women and men to careers in the food system.

Acceptance Address by Dr. M.S. Swaminathan

I am grateful, both to those who conceived this prize and to those who decided to award it to me. I am aware that many in our spaceship Earth are more worthy of this recognition. I am equally aware that, although I was singled out for this recognition, I could have accomplished little in my life without the generous support of many other men and women.

My late mother and an uncle gave a sense of mission and direction to my academic and professional life after I lost my father when I was eleven. My wife, Mina, has been my principal guide and source of inspiration. She has encouraged my work, at considerable sacrifice to her own professional and personal life. My daughters Soumya, Madhura, and Nitya have helped me remain young in my thinking and bridge the generation gap in terms of values and aspirations.

My scientific colleagues have been generous in sharing their ideas, material, and suggestions – colleagues at the Central Rice Research Institute, Cuttack; the Indian Agricultural Research Institute, New Delhi; the Indian Council of Agricultural Research; the International Rice Research Institute; and in numerous universities and institutions in India and abroad. Over the years of my professional life, I have observed that progress in plant breeding was most positive when we shifted attention from individual plant characteristics to population performance. The semidwarf wheat varieties that Dr. Norman E. Borlaug sent to India in 1963 were not impressive if examined plant by plant, but were superb when examined collectively. Those varieties triggered India's wheat revolution. In other words, the total yield per hectare was a result of collective performance rather than of individual excellence.

Similarly, the contributions for which I am receiving credit are the products of teamwork and national and international collaboration. Two individual scientists to whom I am most indebted for inspiration and guidance are Dr. B.P. Pal in India, and Dr. Norman Borlaug in Mexico.

Development of improved technology alone cannot help farmers grow more food unless the diffusion of that technology is facilitated by public policy. Here again, I have fortunately had access to political visionaries: Mr. C. Subramaniam, who laid the foundation for scientific transformation of Indian agriculture at a time when the country was described by foreign observers as leading "a ship-to-mouth existence"; and the late Prime Minister of India, Indira Gandhi, whose faith in both science and food reserves is responsible for India now being able to manage severe drought and floods without importing food.

Farmers have given me guidance and encouragement that have proved extremely important in my work. I was fortunate to learn from Indian farmers rather early in life that they will respond even more to action-based communication than to lectures. For example, colleagues and I organized a seed village in Delhi State in 1964 to accelerate the multiplication of pure seeds of some of the Mexican semidwarf wheat. Farmers were rather reluctant to respond initially, but they soon became enthusiastic. When I asked a farmer why their attitude had changed, he said, "We noticed that you always visit the project on Sundays. That means you are not being paid for this job – so we know that you are here out of a conviction that what you recommend is for our welfare." Such farm education helped me to design an effective national demonstration program in the fields of small farmers, based on the principle that "seeing is believing".

I have had the good fortune of working with scientists from many nations. Numerous teachers, scholars, students, political leaders, administrators, extension workers, journalists, development communicators, and farm laborers have helped and influenced me. My brothers and sister, parents-in-law, sons-in-law, nephews, and other relatives, friends, and an invisible college of correspondents around the world have been pillars of strength. I am particularly grateful to critics, because they help me to analyze my shortcomings and to improve. The institutional support I have enjoyed during the last forty years has been a dominant factor in my scientific, organizational, and administrative work.

To all of you, I take this opportunity to say, sincerely, "Thank you."

Because I have derived immense pleasure and power from the partnerships, I have – with the support of my wife and family – decided to use the cash award associated with this honor to promote a small initiative in fostering meaningful farmer-scientist partnership in the development of an integrated approach to biological and social engineering, as applied to technology development and diffusion under small farm conditions.

May I now take a little of your time to explain my personal philosophy of agriculture.

Once, at breakfast, Martin Luther King said: "All life is interrelated. ... We are caught in an inescapable network of mutuality ... Before you finish eating breakfast this morning, you've depended on more than half the world. This is the way our universe is structured; this is its interrelated quality. We aren't going to have peace on earth until we recognize this basic fact of the interrelated structure of all reality."

In no other area of human need and endeavor is there so much global interdependence as in agriculture. Yet the urban public seldom recognizes that we live in this world as guests of green plants and of the farmers who cultivate them.

Experience shows that countries that take farmers and farming for granted come to grief sooner or later. This year's weather has been erratic – severe drought threatens some areas; floods have washed away crops in others. Its adverse impact on food production in densely populated South and South-east Asian countries is a timely warning. Complacency is creeping into the thinking of many planners and political leaders as they establish the priority accorded to the farm sector in national development plans. We have no time to relax on the food production front, as Dr. Borlaug often reminds us.

Globally speaking, reserves of foodgrains, milk powder, and butter are growing daily. Simultaneously, the number of children, women, and men who go to bed hungry is also increasing. Why have all our intellectual, technological, and financial, and spiritual resources failed to find a solution to this age-old irony? Why?

Socrates probably gave us the answers: "Nobody is qualified to become a statesman who is ignorant of the problems of wheat." If statesmen who determine national policies and priorities would all become conversant with food production and equitable distribution, hunger could be made a problem of the past sooner than otherwise would be possible. That was the hope expressed by President Franklin D. Roosevelt when he convened the 1945 conference that led to the birth of the Food and Agricultural Organization of the United Nations. Since 1945, many global congresses have been held and as many resolutions have been made calling for the elimination of hunger and its mother, poverty. But unfortunately, the situation worsens.

Sixty years ago, Mahatma Gandhi said: "It would conduce to global progress and save a great deal of time and trouble if we cultivate the habit of never supporting resolutions – either by speaking or voting for them – if we had not either the intention or the ability to carry them out." I wish that this would become the guiding principle for government representatives speaking in national and international forums, so that we do not have to listen to delegates eternally talking poor but living rich.

China, which has more than twenty percent of the world's population, has ensured basic food security for its entire people. Here is an example of how political will, coupled with appropriate development strategies, can give every child an opportunity to realize his or her potential. India, with the next largest population, has shown how agricultural production can be increased to desirable levels; but millions of people still suffer from under-nutrition due to inadequate purchasing power. India has, however, avoided famines, thanks to the policy of maintaining substantial foodgrain reserves coupled with operating a large public distribution system and rural works program.

Meanwhile, population and pressure on the supporting capacity of the land continue to grow – 4.8 billion of the expected global population of 6.1 billion will be living in developing countries in the year 2000. The famine of jobs, which is already serious in many predominantly agricultural countries, is likely to increase to quantitative and qualitative dimensions. As a consequence, inadequate purchasing power will grow in importance as the root cause of under-nutrition.

The World Commission on Environment and Development, in its report *Our Common Future*, released by Norwegian Prime Minister Gro Harlem Brundtland in April this year, has shown how ecological, economic, and ethical imperatives link the lives of all inhabitants of Mother Earth. Still, we witness increased polarization and fragmentation in terms of country, religion, language, and color.

Genetically, diversity is our greatest endowment. With the enormous diversity we see in biological populations, however, genetics shows an underlying unity, in terms of the chemical substance and mechanism of heredity. How can we reconcile these differences in human behavior and identify common action points that could bring us together?

The elimination of hunger and its real cause, poverty, should be at the top of the human agenda for common action. Unfortunately, the well-fed do not seem to be very concerned with the hunger of other people, although humanitarian responses to tragedies have time and again been manifested, as in the Ethiopian famine. Most people fear that "if others get more, I will get less." We need to accept the idea that the survival of the human species depends on a harmonious relationship with each other and with the earth on which we live.

How can we bring this change in attitude and behavior? If we are really serious about eliminating hunger, we have to overcome the resistances to a change in systems that can bring this about. Economic interests, the sociopolitical interests of small groups, and sometimes sheer ignorance or indifference combine to form a political will that makes the poor remain poor.

The roots of such attitudes are many and varied, but let me say that the taproot is fear – a fear of having to share power and resources. We therefore need to show that helping the weak to become strong solidifies the whole community. Another factor is the failure to identify ourselves as members of a global family. A Westerner may be talking about the "starving Ethiopians"; and Indian, like myself, may be talking about "the poor", "the tribal", or "the illiterate women". Each term denotes a specific group and is useful for certain purposes. But it tends to stop us from thinking about them as members of the same human family. The concern is simply not there.

How then can we generate concern for other people with economic and social handicaps? How can we help transform this concern into meaningful action? Concentrating on nutrition security for all people at all times, let me refer to three areas where people everywhere can and should work together.

First, I would like to see high priority accorded to fighting what I may call the "ecological fire" raging in most parts of the world. Developing countries are ravaged by this fire, through such mechanisms as deforestation, soil erosion, desertification, water pollution, and over-population. Developed countries suffer equally, through atmospheric pollution, acid rain, and contamination of water through toxic wastes, and environmental mutagens and carcinogens.

Certain phenomena, such as loss of biological diversity, destruction of the ozone layer, nuclear fallout, and potential climatic changes as a result of carbon dioxide accumulation in the atmosphere and ocean-warming will affect us all, regardless of the problem's geographic origin. Today, we are concerned with fighting fires in buildings and forests. Ironically, we are not fully aware of the vast dimension of potential impact of the ecological fires that affect our basic life-support systems of land, water, flora, fauna, and atmosphere. Some hopeful signs are fortunately emerging. The protocol for limiting the use of ozone-destroying chemicals signed recently in Montreal by twenty-four nations, and the development of a time-bound tropical forestry action plan, are good examples of this hopeful trend. We need to strengthen this trend. Responses to the environmental challenges we face should include measures for maintaining biological diversity and ecological processes. Planning and managing irrigation systems, promoting natural forest growth, and conserving the entire system of production from upstream forest to downstream fishery are vital for an enduring food and nutrition security. The ecological fires that can destroy the livelihood security of generations yet to be born cannot be put out in a day, or a month, or even a year. Extinguishing them needs long-term commitment and innovative approaches. Ecological relief to undertake a forestation, soil conservation, and other measures often is not easily available in seasons when there is adequate moisture in the soil. Fifteen years ago I pleaded for a good weather code for undertaking ecological rehabilitation programs in years with normal rainfall. Unfortunately, national and international aid disappears when the rains come.

Equally important is the promotion of what I would call a "symphonic agricultural system", based on the integrated principles of ecological sustainability, economic viability, and equity. The various links in the production-marketing-consumption chain of a symphonic agricultural research and development program are designed to promote growth without loss of stability. Quite frequently, the concept of a sustainable agricultural production system is used to plead for the preservation of the status quo or to revert to old production technologies. A dynamic concept of sustainability is necessary to help us meet the needs of an expanding population while maintaining and enriching the natural resource base.

In most developing countries, a vertical growth in productivity and a higher intensity of cropping are the two major pathways through which the additional food needed will have to be produced. Many of them, fortunately, have the capacity to increase both yields and cropping intensity with current technology levels. For example, given an optimum blend of three elements – viable technology, efficient services, and supportive government policies – we can double the average yields in several countries with chronic food shortages with the technologies now available. Dr. Borlaug and his colleagues are demonstrating this in several food-deficit African nations.

The president of the South Commission and former president of Tanzania, Dr. Julius Nyerere, once remarked: "Many expatriate experts come to me and say that Tanzania has a rich productive potential. I tell them I am tired of hearing about the production potential of my country. What I want is production." While external help can be useful, and is often necessary, converting untapped production potential into actual production is largely in the hands of the government and people of every country. Unfortunately, governments in developing countries tend to exhibit a strong urban bias in their resource allocation process. Rural areas get little attention as far as basic provisions and infrastructure are concerned. Mahatma Gandhi once said that the most serious form of brain drain that occurs in India is the migration of skilled and educated persons from the village to the city. Fortunately, new technologies that require both brawn and brain can provide intellectual and economic satisfaction to young farmers. They, therefore, provide an opportunity to retain educated youth in villages. Youth and the poor constitute the two genuine majorities in the Third World. Unless youth can feel excited about the opportunities now available for technological transformation of rural professions, the fate of the poor will be bleak.

The mass media term "miracle seeds" created a misleading impression that agricultural progress can be achieved through miracles. There are no short cuts to progress in agriculture or aquaculture. Above all, people should not be misled into thinking either that modern inputs are not needed for output or that biotechnology will solve all our problems. Output cannot be increased without the appropriate inputs. Tropical soils, often hungry and thirsty, support excellent crops if they are provided with water and nutrients. Therefore, we should aim to reduce not the essential inputs to the crops, but rather the cost involved. Recent research in the International Rice Research Institute and other institutes has shown that cost reduction without yield reduction can be achieved by using purchased inputs efficiently and by substituting farm-grown inputs for some of the market-purchased ones. This is an area where there is much scope for an intelligent integration of traditional and emerging technologies.

Price in international markets for primary agricultural commodities and opportunities for trade greatly influence the agricultural health of many developing countries. Some hopeful trends are now emerging that could lead to stability and equity in international trade in agricultural products. If developing countries are not given a fair deal in agricultural trade, many of them will never be able to remove the severe debt burden that is crippling them economically and politically.

John Mellor, in his lecture at the colloquium held today, stressed that the best way to improve the status of the world's poor is through the transfer of scientific resources. This is because a dynamic agricultural production program can be initiated and sustained only with the help of a dynamic national research system. The International Agricultural Research Centers (IARCs) supported by the Consultative Group on International Agricultural Research (CGIAR), have helped many nations buy time by providing the needed backstopping, in both research and training. The stronger a national research system, the greater is the benefit derived from an IARC. Unfortunately, many developing countries do not realize the pivotal role of science and technology in converting the natural endowment of a country into wealth meaningful to its people. Research gets low priority in resource allocation, and research institutions are ill equipped and inadequately funded. Consequently, good scientists go to the better-endowed institutions in developed countries. Apart from better working conditions, the growing gap in pay, privileges, and perquisites between scientists employed in UN organizations, IARCs, and advanced institutions in developed countries and scientists in national research systems of developing nations is becoming a major cause of brain drain. A low-cost mechanism on the Peace Corps model will have to be developed to fill critical gaps in technological expertise, where needed.

The trend toward total privatization of plant breeding research in developed countries and the ever-widening scope of patent rights also cause concern in developing countries. From such a concern arises the recent move in FAO to promote, in developing countries rich in crop-genetic resources, the concept of farmers' rights to compensate for the breeders' rights prevalent in developed nations. Seeds, which from the dawn of agriculture have known no political frontiers until now, may soon be required to have, in addition to phytosanitary certificates, "passports" to be issued only upon substantial payment. Thus technology, as well as trade, is becoming a source of dispute and discord. It is sad that agriculture, which could be a major instrument of international cooperation, is becoming another source of polarization. It is in this context that the strengthening of a nonpolitical and noncommercial scientific organization like the CGIAR assumes urgency.

A third area that requires attention is an analysis of the reasons for our inability to achieve freedom from hunger. The problem needs understanding both at the production and consumption levels.

Taking production first, political leaders often tend to take a short-term view of essentially long-range problems. They are therefore interested in "crash agricultural production programs" that are launched with a good deal of political fanfare. Such programs generally ignore the human beings whose toil is vital to their success. It is not surprising that many of them collapse – after having wasted scarce resources.

I have come across many agricultural plans developed by agricultural departments in Asia and Africa where the word "farmer" is not mentioned, even by mistake. And these are documents related to raising production! All the other ingredients are there – credit, seeds, irrigation, fertilizers, pesticides, and farm implements – but not the human being who will use all these inputs. And when the term "farmer" is finally used, it is mostly taken to refer to men, and seldom to women farmers and farm labor. Yet we suspect women played a leading part in the domestication of plants more than twelve thousand years ago. In most developing countries, they continue to perform key functions in seed selection, food production, post harvest conservation, and organic recycling. Also, the poorer the household the greater the need for women to have access to independent income.

Yield targets are frequently set in most government plans, but seldom do we see a minimum income target for farming families. Public and private sector employees are always concerned with their net take-home pay, continuously adjusted against inflation. But the same officials, when formulating policies for the farm sector, do not think that small farmers in the self-employed sector deserve a minimum take-home pay. A minimum income target for families engaged in rural occupations also ensures adequate resource flow to the villages and helps correct imbalance in the terms of trade between the farm and non-farm sectors. We need a new deal for the rural self-employed in the Third World whose livelihood security depends mainly on land- and water-based occupations.

In this context, I would like to pay tribute to President Corazon C. Aquino of the Philippines, whose government has decided that the aim of agricultural progress is to increase in real terms the income of Filipino farmers – from about $60 a month at present to $100 per month by 1992 (in US dollars). Unless we can reorient the thinking of governments along these lines, we will always have politicians and professionals talking about the untapped potential of their countries but not doing anything to realize that potential.

The livelihood security of the poor in rural areas depends upon diversified opportunities for employment both on the farm and in off-farm sectors. To undertake this task on a scale that matches needs, more investment is essential in rural areas. FAO has estimated that an investment of $1,500 billion (in US dollars, 1980 prices) will be needed up to the year 2000, if developing countries are to achieve the necessary growth rates in agriculture. The Brundtland Commission has dealt with the interrelationships among ecological health, food security, and arms expenditure. The global expense for arms is now estimated at $2 million per minute, or nearly $20 billion a week. In spite of such expenditure, international surpluses in terms of cash, commodities, and commercial technology are abundant in developing countries. A recent study by a group of economists at the World Institute for Development Economics Research (WIDER) points out that during the next five years, Japan alone could easily transfer resources worth $125 billion (in US dollars) to developing nations. Resources can thus become available if there is a political will to help the poor. Father Byron rightly stressed that political will by itself will have little impact unless there is also a political way to accomplish the desired goal.

We should constantly remind ourselves that hunger robs millions of individuals of the opportunity to lead fuller lives. And today, although we know that there is no valid excuse for hunger to exist, we still do not know how to remove this stigma for our civilization.

May I suggest three major lines of action to supplement present efforts?

First, we need to make people aware of the fact that we can create a truly joyful world, where not a single person spends the night hungry. How can we do this on a scale that can have an impact? Television is obviously the communication medium of choice. If about two billion people can watch the Olympic Games or the World Cup football tournament on TV around the world, can we not let them see a "Freedom from Hunger Olympics", where countries and organizations demonstrate the methods they have used to provide physical and economic access to food to everyone? May I appeal to the General Foods Fund to take the lead in organizing this World Without Hunger Olympics? A global event like this might be organized every two years, with the help of national and international organizations committed to promoting ecological and nutritional security. Among them are the Better World Society, the Hunger Project, the Worldwide Fund for Nature, the International Union for the Conservation of Nature and Natural Resources, the International Council of Scientific Unions, the UN/FAO World Food Programme, UNICEF, and many others.

A second suggestion is the development of a mechanism through which scientists and farmers from different countries can share experience and know-how. Currently, there are "peasant to peasant" organizations. Similarly, scientists have many opportunities to meet and exchange ideas. The communication revolution we are witnessing now will increasingly facilitate continuous intellectual interaction across continents and reduce the need for formal conferences. What is missing is a mechanism for promoting joint scientist-farmer undertakings. If agricultural research becomes a joint-sector activity involving scientists and farmers, the growing gap between knowing and doing can be bridged. Also, the pace of development and diffusion of location-specific technologies can be accelerated.

Finally, we need intervention programs tailored to meet the specific needs of countries where hunger is prevalent. Professor Amartya Sen has shown in several of his books and articles how well designed public intervention measures that enhance entitlements can reduce or eliminate hunger and expand life expectancy. Agrarian reform and employment guarantee schemes are important in this respect.

The challenge lies in developing strategies at the national level for utilizing food and commodity aid in a way that will help end the need for such aid, not perpetuate dependence and erode self-reliance. Considerable experience in this field already exists within international organizations such as the World Food

Programme and UNICEF as well as within many bilateral and national organizations. Based on an analysis of this experience, an action code for initiating appropriate "Food for Self-Reliance" and employment guarantee programs can be prepared.

To sum up, I plead for coordinated non-governmental initiatives at first, generating awareness by taking advantage of the powerful mass media tools now available; second, analysis of field problems and developing solutions from them jointly by farmers and scientists; and third, action in utilizing the growing global grain surpluses for enabling all those in need of help to earn their daily bread.

We live in an age of unparalleled opportunity for promoting sustainable nutrition security. This prospect for a world without hunger is a glorious legacy given to our contemporary world by scientists and technologists; communicators and social scientists; administrators and industrialists; and workers in the factories, fields, forests, pastures, rivers, and oceans.

As we depart for dinner this evening, what could be a more satisfying and joyful feeling than knowing that every other member of the human family will also go to bed after a nourishing meal? Until such a wholly attainable world becomes a reality, our task remains unfinished. ▯

60. At World Food Prize, Washington DC, 1987

61. Speaking at World Food Prize Convocation, IRRI, 1987

26

Population, Environment and Food Security

Introduction

All available experience indicates that demographic transition leading to low birth and death rates takes place when an enabling environment where children are born by choice, prevails. The major components of such an enabling environment are also now widely recognized, such as female literacy and the status of women; infant and maternal mortality rates; improved health, nutrition, livelihood and ecological security for all; and shared decision-making by couples. According to several estimates, the world population could increase to ten billion by 2050. About ninety-seven percent of that increase will occur in the developing countries and two-thirds of this growth is expected to take place in the cities. Hence, the urban poor will need special attention.

The UN Conference on Environment and Development held at Rio de Janeiro in June 1992 was unique in having united the countries of the world through global conventions on biodiversity and climate. The Desertification Convention, for which the foundation was laid at Rio, is likely to come into force soon. Above all, a plan of action for promoting environmentally sustainable development, known as Agenda-21, was also adopted.

Agenda-21 of Rio de Janeiro, designed for implementation during the remaining years of this century and during the first decade of the 21st century, is a blueprint for economic development without environmental destruction. The growing number of ethnic and economic conflicts indicates that development, even if environmentally benign, will not be socially sustainable if it ignores equity. In other words, a better common present is essential for a better common future.

Food Security

Food and drinking water are first among the hierarchical needs of human beings. This is why Thomas Robert Malthus in his essay on population published in 1798, wrote, "the period when the number of men surpass their means of subsistence has long since arrived". Now, the global population is nearly six billion and yet famines of the kind Malthus predicted have been averted in recent decades except in regions affected by ethnic or civic strife and/or prolonged drought or other natural calamities. Globally, the problem of hunger is now more related to what Professor Amartya Sen refers to as 'entitlements' and not to food production lagging behind population growth.

Lecture at the International Conference on Population and Development, Cairo, Egypt, September 9, 1994.

My colleagues and I at the Center for Research on Sustainable Agricultural and Rural Development at Madras are currently engaged in designing a Hunger-free District Program, starting with hunger-free villages and towns. This program aims at freedom from hunger at the level of the individual, since intra-household variability in food intake exists, often to the disadvantage of women and girl children. To eliminate endemic hunger at the level of the individual, it will be necessary to pay integrated and concurrent attention to the following: a) Sensitize and mobilize public opinion through mass media leading to appropriate political action; b) Achieve agricultural intensification and diversification, so that the income and employment potential of small farm agriculture is enhanced through economically and ecologically sustainable farm and off-farm enterprises; c) Ensure access to food at affordable prices both by maintaining food security reserves and by operating an efficient public distribution system; d) Ensure universal and compulsory school education; e) Develop an integrated health security system, involving attention to sanitation, hygiene, preventive and curative health measures, reproductive health and access to safe and acceptable contraceptive services; f) Initiate public action for providing: i) protective social security for the assetless and vulnerable sections of the population, such as employment guarantee and food for nutrition programs; ii) promotional social security involving access to information, technological empowerment through training in new skills, credit and remunerative self-employment and marketing opportunities; g) Mobilize the voluntary sector and private sector industry for promoting literacy, ensuring health security and fostering a job-led economic growth strategy.

Such a program for achieving freedom from endemic hunger will succeed only where there is public commitment to political (i.e., democracy and independent judiciary and mass media), social (such as group action) and technological empowerment of all citizens, and more particularly of the economically and socially disadvantaged sections of the population. The early achievement of this goal will be possible, if in addition to the needed public action, private sector industry will enter into a social contract with the economically and socially disadvantaged sections of the population in order to improve the livelihood security of the poor.

Demographic Trends and Famines

Recent predictions of the growing imbalance between our ability to produce food from diminishing land and water resources and expanding biotic and abiotic stresses point towards the incidence of serious famines if efforts in the area of population stabilization do not bear fruit. Such predictions have been made from time to time during recent decades. For example, in 1966, Paul Ehrlich predicted as follows: "Some time between 1970 and 1985 the world will undergo vast famines – hundreds of millions of people are going to starve to death. That is, they will starve to death unless plague, thermonuclear war, or some other agent kills them first. The United States should announce that it will no longer ship food to countries such as India where dispassionate analysis indicates that the unbalance between food and population is hopeless."

The above prophecy of Ehrlich would have proved to be correct, had it been made a century earlier, i.e., in 1866. Between 1870 and 1900, nearly thirty million famine deaths occurred in India. The population of undivided India was 281 million from 1891 and ten million starvation deaths took place between 1893-94. The last big famine in the Indian sub-continent occurred in 1942-43, when over two million died out of starvation. The famine prevention strategy of independent India, comprising steps such as bringing more area under assured irrigation, intensification of agricultural research, education and development, land reform, assurance of minimum prices, maintenance of food security reserves, operation of an extensive public distribution system, integrated rural development and employment generation both for unskilled and skilled workers, has been very effective in avoiding calamities of the kind predicted by Ehrlich and others. Endemic hunger, however, persists largely due to inadequate purchasing power among those living below the poverty line. The hunger problem is thus better described in terms of million person-years of jobs than in tons of foodgrains.

The packages of technology, services and public policies which helped to prove the prophets of doom wrong, are described in the case of wheat, in a book titled *The Wheat Revolution* (Swaminathan, 1993). It is estimated that the total food demand in developing countries will increase by approximately one hundred fifty percent by 2025. The continuing population increase will result in a decline of available cultivable land per capita world wide from 0.3 ha in 1988 to 0.17 ha in 2050, with only 0.11 ha per capita in developing countries. About 8 percent of the increase in food production in the last half century has come from the expansion of the cultivated area and ninety-two percent came from higher yields per hectare. The main inputs have been new crop varieties, increased use of fertilizers and biocides and a large increase in irrigated area, at about three percent per annum from 1950 and 1980. The total world irrigated area has increased from 80 million ha in 1950 to 237 million ha in 1993. Today, 46.5 percent of all grain is produced under irrigation. The practical limit for full intensification under irrigated conditions is about 8.5 tons per ha for a single crop. In the Punjab part of India and in Egypt, a yield of six tons per ha is being obtained by farmers under conditions of good water and soil fertility management. The challenge now is to both sustain such yields in favorable areas and expand good land, water and crop management to more areas and farming systems, particularly in areas without irrigation.

Thus, there is no time to relax. The Roman philosopher Seneca said long ago: "A hungry person listens not to reason nor cares for justice, nor is bent by any prayers." On the eve of the Cairo Conference, several dire predictions have again been made on the impending food crisis in developing countries and the consequent potential for widespread famines and starvation deaths. Lester R. Brown and Hal Kane (1994) have calculated that China and India will face serious food shortages (Table 1). Such forecasts are based on the fear that the human carrying capacity of supporting ecosystems is being exceeded particularly in developing countries where most of the population increase is taking place. There is of course scientific basis for such apprehensions. In per capita terms, land, water and forest availability is shrinking (Table 2).

Table 1. Grain Production, Consumption, and Net Trade in the "Big Four" Countries, 1950 and 1990 with projections to 2030

Countries	Years	Grain Production (millions of tons)	Grain Consumption (millions of tons)	Net Trade
United States	1950	133	121	+12
	1990	290	214	+76
	2030	377	295	+82
China	1950	109	109	0
	1990	329	335	-6
	2030	263	479	-216
India	1950	57	55	+2
	1990	158	158	0
	2030	222	267	-45
Former Soviet Union	1950	79	80	-1
	1990	182	219	-37
	2030	237	262	-25

Source: Data from US Department of Agriculture cited by Brown and Kane (1994).

Table 2: Population Size and Availability of Renewable Resources, 1990, with Projections to 2010

	1990 (million)	2010 (million)	Per Capita Change
Population	5,290	7,030	–
Fish Catch	85	102	-10
Irrigated Land (hectares)	237	277	-12
Crop Land (hectares)	1,444	1,516	-21
Forests	3,413	3,165	-30

Source: Food and Agriculture Organization of the United Nations (1991).

Table 3. Human-induced Land Degradation, 1945 to 1992

Regions	Total Degraded Area (millions of hectares)	Degraded Area as Percentage of Total Vegetated Land
Asia	746	20
Africa	494	22
South America	244	14
Europe	220	23
North and Central America	158	8
Oceania	103	13

Sources: World Resource Institute (1992) and Pimentel (1993).

The Green Revolution technologies based on chemical inputs gave us a breathing spell for arriving at a balance between human population and the food production potential of the available arable land and water resources. But they have also led to adverse environmental repercussions. This is why Agenda-21 of the UN Conference on Environment and Development has placed much stress on the adoption of ecologically desirable technologies.

Sustainable Food Security

In my view, there can be enough food for feeding a global population of about ten billion if the six-pronged strategy indicated below is widely adopted. However, I must add that even if food is available on the market, the poor will have no access to it unless they have opportunities for remunerative employment. Hence sustainable food production and consumption need integrated attention in national and international food policies.

a) *Bridging the gap between potential and actual yields with the technologies currently available on the shelf.* In most farming systems in developing countries, there is a large gap between potential and actual yields even with currently available technologies. For example, in India, the average yield of rice from about forty million hectares is only a little over two tons per hectare. This can at least be doubled. Therefore, through a combination of location-specific technologies, timely delivery of appropriate inputs and public policies in land reform, input-output pricing and marketing, it will be possible to tap the untapped yield reservoir currently existing in most farming systems.

b) *Globally, over fifteen percent of good farmland is now degraded due to a variety of human induced causes (Table 3).* The highest priority will have to go for upgrading the biological potential of such wasted lands. In the population-rich but land-hungry countries of South and South East Asia, nearly twenty percent of farm land remains unproductive due to a variety of causes, such as salinity, alkalinity, water logging and loss of top soil due to erosion. The Desertification Convention can provide an opportunity for launching a mass movement for protecting and improving soil health.

c) *Ecologically sound practices will have to be introduced both in agriculture and capture and culture fisheries.* Here, Mahatma Gandhi's dictum "Nature provides for everybody's need but not for everybody's greed" will have to be kept in view while exploiting the resources of land, inland rivers and reservoirs and oceans. Programs which are environmentally destructive and socially disruptive should be avoided.

d) *Promotion of group cooperation among families with small holdings so as to empower them with the economic and ecological advantages of scale in farm operations.* This will be particularly important in water harvesting and efficient on-farm management of water, inland and coastal aquaculture, integrated pest management and improved post-harvest technology. This will also enable them to benefit from producer oriented marketing arrangements so that a fair share of the consumer's money goes to the producer.

e) *A new trade ethic which leads to the industrialized nations buying at a fair price on a long-term basis agricultural commodities from the developing world.* Otherwise, several features of the World Trade Agreement particularly those relating to Trade-related Intellectual Property Rights, will work against the interests of the poor. Today, industrialized nations have capital and technology-driven advantages both in agriculture and in industry. They also enjoy leadership in the services sector. This is why the rich-poor divide is continuously increasing year after year. Developing countries, which are predominantly agricultural both in economic and livelihood terms, should improve their agriculture and not depend on imports. Only then will rural employment and prosperity improve.

f) *A new deal for the self-employed through credit, technology, training, techno-infrastructure and trade.* This will call for appropriate public and private actions resulting in harmony between public good and private profit. Non-governmental organizations and private sector industry can play a pivotal role in this area.

At present, the number of people living on one ton of foodgrain per year varies from one to six. If efficiently used and if healthy foods are encouraged, about four individual can live on one ton of foodgrains per year. This number will be three per ton per year, where the plant-animal-man food chain is widely prevalent. Therefore, to feed a population of about ten billion, we need about 3,500 million tons of foodgrains, including provisions for seeds and spoilage.

Producing this should be possible if we step up R&D efforts on sustainable intensification and diversification of farming systems. Diversification of food habits will further improve the nutritional status of children and adults.

Local Level Socio-Demographic Charters

The draft Population Policy Statement submitted to the Government of India by an Expert Group chaired by me in May 1994, indicates an approach to promoting development based on concurrent attention to the principles of ecology, economics, equity and employment. The Expert Group's major recommendation is on promoting an environment, which will enable couples to decide for themselves the size of their family. Such planning will have to come from an awareness of the obligations of parents to their children as well as to society. To implement a population policy based on the methodology, "think, plan and act locally and support nationally and internationally", the Expert Group recommended that elected grassroot level democratic institutions should prepare a socio-demographic charter for their respective village or town.

The principle aim of such socio-demographic charters is to stimulate action on meeting the unmet needs of the local population in relation to minimum needs. This will also include the provision of quality contraceptive services based on informed choice. Another major purpose of the socio-demographic charters is to create greater awareness of the need to achieve quality of life improvement, while living within the carrying capacity of supporting ecosystems. The Expert Group stressed that "it is high time the limits to the human carrying capacity of the supporting ecosystems are recognized." This is becoming increasingly clear with reference to water, good quality farmland and forests (Table 2).

The blueprint for such action is the local level socio-demographic charter prepared by the people. Participatory rural appraisal studies have amply demonstrated that, whether literate or illiterate, rural families can articulate their priority needs in social and human development. The major components of a socio-demographic charter are the following: a) Gender equity and problems relating to women and children including sex ratio, crèches and child care facilities; b) Habitat, including housing, environmental hygiene, garbage and sewage recycling and equitable management of common property resources; c) Education for all at school level; d) Health security including reproductive health of women and availability of quality contraceptive services; e) Environment with particular attention to the human carrying capacity of land and water and conservation of flora and fauna; f) Access to balanced diets and safe drinking water at the level of each individual in a household; g) Opportunities for skilled employment in the primary, secondary and tertiary sectors of economic activity both in rural and urban areas. People living in urban slums need urgent attention, since the coexistence of luxury and misery in cities and towns ignites social unrest and conflicts.

Such a charter prepared by the principal stake-holders, based on appropriate training modules and manuals, will help priority setting in meeting their unmet needs. The socio-demographic charter will help to place population goals in the context of health, livelihood, and ecological security. It will help to articulate the kinds of support state and national governments and bilateral and multilateral donors should extend for achieving low birth and death rates and gender equity. The Expert Group has suggested the establishment of a National Population and Social Development Commission and a National Population and Social Development Fund to respond to the needs and priorities articulated in the socio-demographic charters. This will help to achieve a proper match between response and the generic and specific needs of villages and towns. This suggestion made in the context of emerging grassroot level democratic institutions in India is worthy of being examined by all countries with a system of decentralized and democratic decision making.

The example of Kerala where the birth rate fell from forty-four per thousand in the 1950s to eighteen by 1991 has been the subject of many studies. All the studies have shown that where coercive disincentives are substituted with an enabling environment such as emphasis on universal education, primary health care and environmental hygiene, population growth gets limited on its own. The population problem is serious in many regions not because of the proportion between the natural increase of population and food, but because of environmental reasons. This is why the socio-demographic charter places emphasis on local communities studying the relationship between human and animal populations and the supporting capacity of the ecosystems in which they live.

Population Policies for Human Happiness

The above agenda for giving our children and the children yet to be born not only existence but also happiness will need substantial financial, managerial and infrastructure resources. It will need for its efficient implementation, the active involvement of non-governmental organizations and private and public sector industry.

I pointed out earlier that so far we have been able to ensure that the predictions of Malthus and Ehrlich do not come true. The continuation of a situation where the rate of growth in food production remains above the population growth rate will depend much on our ability to implement the action plan I have outlined. An approximate estimate of the additional funds required for implementing some of the suggestions relating to ecological security is given in Table 4. The total is less than a trillion dollars. Where will this new money come from?

A world in which twenty percent of the population enjoys eighty-four percent of the annual income, while another twenty percent struggles for survival on a mere 1.4 percent of the world's income, as stressed in UNDP's *1994 Human Development Report*, can never provide a secure and sustainable way of life for humankind. Mahatma Gandhi urged that those who possess and enjoy affluence should hold it in trust with a responsibility to use it to build a better world for all. A one-time contribution of $1,000 per capita by the one billion people living in luxury in all countries of the world would help to form a Trillion Dollar Trust Fund to create a world with an enabling environment for achieving the desired demographic goals. I am citing these figures just to illustrate the vast scope for voluntary action in ensuring that both unsustainable lifestyles and unacceptable poverty become anachronisms of the past. We should foster a new human ethic which sensitizes us to the need on the one hand, for population not exceeding the human carrying capacity of supporting ecosystems, and on the other, for enhancing the human caring and sharing capacity.

The peace dividend arising from an easing of mega-tensions is now regarded as another emerging opportunity for redeployment of resources. UNDP in its *1994 Human Development Report* calculates that the actual peace dividend between_1987 to 1994 was $935 billion and that a further $460 billion may be saved from military expenditure during 1995-2000. However, for such redeployment to really occur on a sustained basis, we must have greater understanding, tolerance and love of diversity and pluralism, whether in terms of religion, language, color, culture, ethnicity or political belief. This has to be a major task for 1995, which has been declared as the International Year for Tolerance. While we should be tolerant of diversity in human societies,

Table 4. Estimates of Additional Expenditures Required to Achieve Sustainable Development between 1990 and 2000

Items	1990	1994	1998	2000	Total
	(billions of US dollars)				
Protecting the Topsoil on Cropland	4	24	24	24	154
Reforesting the Earth	2	6	7	7	32
Slowing Population Growth	13	28	32	33	159
Raising Energy Efficiency	5	25	45	55	180
Developing Renewable Energy	2	123	24	30	94
Retiring Debts of Developing Countries	20	50	10	-	130
Total	46	145	142	149	729

Source: MacNeill (1989)

we must be intolerant of inequity either in terms of gender or economic well-being.

In the ultimate analysis, the purpose of this Conference should not be just how to limit population growth or promote food and environmental security, but it should be to achieve what the French mathematician-cum-social scientist Condorcet emphasized in his paper of 1795, namely, "Population growth can be limited if people have a duty towards those who are not yet born, that duty is not to give them existence but to give them happiness." For making Condorcet's aspirations come true, we need a global movement for integrated population and social development. We need a Global Trust Fund for Population and Social Development which will help all women, men and children, wherever they are, to meet their minimum needs in food, education, health care, employment and housing. Only then we will witness the desired global demographic transition as well as harmony between humankind and nature. If urgent steps are not taken to curb the unsustainable lifestyles and consumption on the part of the rich billion and to mitigate the poverty and deprivation experienced by another billion children, women and men, we will be living in a world of social, economic and technological apartheid far worse in its human implications than the skin color based apartheid which has just ended. This should be the message from Cairo, a city that represents the synthesis of the best in human civilizations and cultures.▯

References

Brown, L.R. and H. Kane. 1994. *Full House: Reassessing the Earth's Population Carrying Capacity*. New York: W.W. Norton and Company.

Condorcet. 1975. *Esquisse d' un Tableau Historique des Pogres de 1' Esprit Humain.*

Ehrlich, P. 1968. *The Population Bomb*. New York: Ballantine Books.

Malthus, T.R. 1987. *Essay on the Principle of Population.* Harmondsworth, UK: Penguin Books.

Sen, Amartya. 1994. Population and reasoning. Paper presented at the United Nations, New York, April 18, 1994.

Swaminathan, M.S. (Ed.). 1993. *Wheat Revolution: A Dialogue*. Madras: Macmillan India.

-----. 1994. *Uncommon Opportunities: An Agenda for Peace and Equitable Development.* Report of the International Commission on Peace and Food. London: Zed Books.

27

Agriculture, Food Security and Employment: Changing Times, Uncommon Opportunities

The production of more food to meet the needs of a burgeoning population has been one of the outstanding global achievements of the last half-century. It has demonstrated the extraordinary capacity of humankind to meet a collective challenge of monumental proportions. It has harnessed the power of science and technology for the most cherished of purposes – preserving human lives. It has released the initiative and pride of long suppressed peoples to achieve self-sufficiency and determine their own destinies. It has finally lifted the suffocating blanket of pessimistic economic determinism that has stifled human hopes and aspirations for nearly two hundred years, since Thomas Malthus first set forth the thesis that human population will always outrun the growth of food production, thus ensuring the perpetual poverty of humankind.

Food has become a symbol of our collective human endeavor to create a better world for all. However, the victory has been partial, and neither the challenges nor the opportunities that food presents have been fully addressed. It is of crucial importance, not only to the poor but also to the peace and stability of global society, that we complete the task of banishing famine and hunger once and for all. Hunger anywhere threatens peace everywhere. Hunger leads to political instability, social unrest, massive migrations, rebellions, civil war, crime, and violence. Prosperity, which eliminates hunger, also tends to eliminate violence. The last decade has seen growing recognition that the problem of food is inextricably intertwined with the problems of peace, political and social stability, and employment, and that no comprehensive solution to one is possible without substantial progress on the others.

This basic tenet was among those that suffused the work of the International Commission on Peace and Food (ICPF), which presented its findings in 1994 in a report entitled *Uncommon Opportunities* (ICPF, 1994). The present article draws on some of the deliberations of the Commission, particularly on the role of agriculture as a powerful engine for employment generation and economic growth in the developing world. A central philosophy is that a great deal can be accomplished, if only we shift the emphasis from preoccupation with cataloguing the problems to focusing on the opportunities available to eliminate them.

Increasing Agricultural Production

Contrary to the fears raised in earlier centuries and revived in recent decades, today, the world does possess the capacity to feed everyone. This is so, even at

---

First published in *Nature and Resources* (UNESCO) 31(1): 2-15 (1995).

current levels of food production – subject as they are to disincentives, quotas, restrictions, and trade barriers that food surplus countries impose to curtail food production, while other nations remain in perpetual deficit. In spite of a doubling of population in the developing countries since 1960, their average food supplies have increased from 1,950 to 2,475 calories per capita per day, or from an average of ninety percent to one hundred and seven percent of the minimum caloric requirement. During a recent twelve-year period, per capita food production in developing countries has risen by fifteen percent. Overall, the proportion of people in developing countries suffering from hunger and malnutrition has dropped dramatically, both in relative and absolute terms, from 941 million people, constituting thirty-six percent of the population of these nations in 1970, to 781 million, constituting less than twenty percent of the population in 1990.

Current projections indicate that the growth rate in world agricultural output will continue to exceed population growth over the next two decades. By 2010, foodgrain production is expected to reach four times the level in 1960. Increased production of other food crops is expected to raise per capita availability in the developing world to 2,700 calories per day. Per capita meat production is expected to grow by sixty percent and milk production by twenty percent. During the same period, malnutrition is projected to decline to 640 million persons, constituting 11.6 percent of the population of developing countries, a little over half the level in 1990.

Yet, the achievements have not been uniform throughout the developing world. Per capita food production has actually declined in more than half of the developing countries over the past fifteen years. In eighteen countries with high rates of population growth, primarily in Africa, it has been deteriorating for the past three decades. While the Near East, North Africa, East Asia, and Latin America and the Caribbean regions are expected to achieve average food supplies of 3,000 calories per day by 2010, two hundred million people will still be malnourished in South Asia and three hundred million, or one-third of the population of sub-Saharan Africa, will suffer from severe food deficits. Due to rapidly expanding population, agricultural production in sub-Saharan Africa would have to grow at four percent annually in order to significantly improve per capita availability of food. Thus far, only East Asia has sustained such high growth rates in agriculture, averaging 6.1 percent from 1970-1990.

The projection of future prospects, based on past trends, is especially questionable with regard to food. Not even some of the most prominent agricultural economists expected the gains that have occurred in food production in the past several decades. An international team of experts, visiting India in 1963, projected a mere ten percent growth in foodgrain production by 1970, whereas growth actually achieved during this period was fifty percent. Even with technological, financial, and natural resources constraints, there are no insurmountable obstacles to achieving dramatically more over the next fifteen years than is indicated by past trends or current projections.

The idea that hunger cannot be conquered because we are running out of land to support rapidly burgeoning populations is contradicted by facts. Globally, there is no correlation between population density and hunger. China, with only half as much arable land per capita as India, produces thirteen percent more foodgrains per capita. Taiwan and the Republic of Korea have only half the farmland per capita of Bangladesh, yet they produce forty percent more food per capita. Tiny Netherlands, with the highest population density in the world, produces more than sufficient food to feed itself and remain a large net food exporter. Currently, eleven percent of the world's land surface is used for agricultural crops, just four percent more than in 1960. A comprehensive theoretical study of soils, climate, vegetation, and topography, conducted in 1975, indicated that both land and water utilized for agriculture could be doubled, if necessary, and that the earth could support thirty-six times the 1975 level (eighteen times the 1990 level) of cereal production using the same share of cultivated land. There would be severe practical obstacles to such a vast expansion of croplands, but these findings suggest that physical limitations to food production are not the primary constraints. A more commonly accepted estimate indicates that the world's land and water used for agriculture could more than double.

The time is now ripe for imaginative national and international action for providing every individual with economic and physical access to balanced diets, safe drinking water, primary health care, education and a clean environment. Freedom from hunger and political freedom go hand in hand. Just as subsistence agriculture and periodic famine were the economic foundations of monarchy and feudalism, the generation of agricultural surpluses that stimulated commercialism, and later industrialization, have formed the basis for the rise of democratic institutions. Greater freedom for individual action and ownership both stimulate, and are supported by greater productivity in agriculture. Authoritarian government is frequently either the result or the cause of food shortages – its use of force, justified, on the one hand, to meet a crisis situation or necessitated, on the other, to restore order and initiate emergency measures. Only under democracy is government compelled to pay attention to the needs of people at the lowest levels of society.

The dual pressures of a free press and electoral system have helped a free and democratic India to avoid famine for nearly fifty years, despite the recurring incidence of widespread famine in previous centuries, up to as recently as four years before the country gained independence. Even in China, which might appear to be a contradiction of this thesis, it was the liberalization of the agricultural sector and greater freedom given to the peasant community that were responsible for the remarkable increase in food production. For countries still at an early stage of political development, where government lacks even the force of authority to govern, the first essential steps may necessarily be towards greater centralized authority and control, a common stage in the political evolution of the nation state. But that central authority cannot release the full initiative of its farmers, or tap the full potentials of agriculture, without first instituting broader democratic measures. Recent events in Eastern Europe demonstrate that where states attempt to use authority as the lever for agricultural development, the achievements are likely to be limited and short-lived. The inability of the authoritarian system to produce enough food for its people was one of the major factors contributing to its downfall. Democracy is the most potent fertilizer to ensure food security at the household level.

Agriculture's Dual Role

Achieving food security necessitates increasing food production and employment opportunities. Agriculture plays a dual role in the abolition of hunger – it produces food and it can also produce a great many of the jobs needed by households to buy food in developing countries. Since agriculture is the world's single largest employer, raising production and productivity in this sector can immediately place additional purchasing power in the hands of the rural poor, who will, in turn, utilize the added income for purchasing more food, clothing, and other basic consumer goods that will create more jobs and higher incomes for countless others. The increased agricultural produce becomes raw material for a wide range of agro-based industries and services that stimulate formation of new enterprises and create downstream jobs, as well as products for further processing, domestic sale, or export.

Promoting job creation in agriculture appears, at first glance, to contradict the global trend of the past hundred years, in which employment in this sector has declined steadily from historic highs of more than seventy percent in the industrial nations, to the point where, today, only two percent of the work force in the United Kingdom, three percent in the United States and four percent in Germany, are directly engaged in farm operations, although a much larger percentage of the work force in these countries work in businesses and industries linked to agriculture.

Nevertheless, this strategy follows a natural course of development that has occurred in many countries which now have a relatively small portion of their work force in agriculture. Agricultural surpluses and rising farm incomes are preconditions and stimuli to economic growth and industrialization. For most developing countries, with the vast majority of people still residing in rural areas, the most cost effective and practical strategy to generate more jobs and raise personal incomes is through agriculture. Today, more than 1.1 billion people in developing countries, constituting fifty-eight percent of the economically active population, work in agriculture. The decline in proportion of the work force in this sector must necessarily be gradual. An employment strategy that generates a large number of new jobs in the non-farm rural sector could contribute substantially to diversification of rural employment opportunities. The biovillage model of job-led economic growth of the M.S. Swaminathan Research Foundation is based on this concept.

A similar strategy has proved highly effective as an engine for growth in a number of East Asian countries that employed crop-intensive and labor-intensive technologies to achieve increasing levels of employment and productivity in agriculture. Empirical evidence from these countries confirms that wherever agriculture becomes prosperous, labor becomes scarce.

Between 1952 and 1968, land reform in Taiwan increased the number of cultivators fivefold, leading to dramatic increases in output and productivity, a shift from foodgrains to higher value- added fruit and vegetable crops, and the creation of more than one hundred thousand jobs in post-harvest and processing activities. These changes in employment led to enhanced rural incomes and purchasing power, growing domestic demand for goods and services, including manufactured goods, and further job growth. Land reform in the Republic of Korea, during the early 1950s, increased the number of owner cultivators from fifty percent to ninety percent and led to a 4.7 percent annual growth in labor productivity per hectare over a fifteen-year period. Then, as agricultural technology improved and industrialization gained momentum, the proportion of the Republic of Korea's work force engaged in agriculture fell from fifty-five percent to under sixteen percent over the following three decades. Thailand, which has had the fastest growth of the East Asian economies in recent years and still employs seventy percent of its work force in agriculture, has also attained high rates of production and employment in the rural sector through diversification in agriculture, from traditional cultivation of rice and rubber, to high-value crops and agro-based industries.

Increasing the Demand for Agricultural Products

There are powerful social forces active in the world today that can stimulate significantly greater growth in both food demand and food production: liberalization of world trade, especially trade in agricultural products; emphasis on aggressive strategies to expand employment opportunities in developing and developed countries; advances in technology for agricultural production, food processing and dissemination of information; rising levels of education, which spur rising expectations; and the energizing impact of democratization.

The gap between the availability of food in industrial nations and in developing countries remains large. Food supplies per capita for all developing countries, measured in terms of total calories available, are only seventy-two percent of the levels in the industrial countries. The availability of protein for consumption is forty percent lower in developing countries than in industrial nations. Viewed from another perspective, this gap represents a huge potential for the growth of agriculture, and through it for more rapid industrialization and job creation in developing countries, with rising exports from industrial nations.

Rising incomes are accompanied by a diversification in diet which generates greater demand for wheat, meat, dairy products, fish, vegetables, fruits, and processed foods. Projections for the year 2010 anticipate a sixty percent increase in cereal consumption in developing countries, a fifty-two percent increase in meat consumption, and a sixty-nine percent increase in milk consumption. Meat consumption in China has tripled since 1978 and is expected to more than double again over the next two decades. Sugar consumption in developing countries is currently less than half the average of industrial nations. In India, sugar consumption is projected to rise from 13 kg to more than 25 kg, generating a demand for a hundred percent increase in sugarcane production, the establishment of 300 to 400 new sugar mills in the country, and the creation of three to four million new jobs in this industry alone. Worldwide, raising the average level of sugar consumption in developing countries could generate tens of millions of additional jobs in agriculture, industry, and services.

Viewing the future demand for food from this perspective reveals a tremendous opportunity. Vast sections of humanity now aspire to the higher quantity, wider variety, and greater nutritional content of food once consumed only by the wealthy. An effort to raise nutritional standards in the developing countries nearer to the levels prevalent among industrial countries will not only eradicate hunger but also dramatically spur economic growth, employment, and purchasing power among the rural poor, stimulating growth of production and jobs in industry and services, as well as increasing exports and imports. Raising the entire world population to the level of the prosperous nations would require a seventy-two percent increase in total world food production, measured in terms of calories. In order to meet the people's nutritional requirements for fruit and vegetables, India's production of these crops needs to double within the next decade. Achieving that high level will generate more than six million new jobs in the country.

Broadening the scope to include other agricultural products, particularly textiles, the potential for accelerating global economic growth by an agriculture-led strategy is even greater. Cotton is a crop with a very high income and employment multiplier effect. Per capita cotton consumption in poorer developing countries, such as India, is less than half the level in China and less than one-third the level in industrial nations.

Challenges in Agriculture

The ratio of growth of foodgrain production to population growth has entered into a period of decline over the past decade – increasing at just one percent per year, compared to three percent during the previous two decades. There is a similar trend for other staple crops and meat. For fish, there has actually been a net decline of seven percent in per capita world production. To a large extent, this slow-down is the result of slower growth in demand, especially in industrial countries, underlining the fact that food security is predominantly an economic problem rather than a technological one. Yet, the decline in the growth of global agricultural productivity over the past decade is also attributable to a variety of political, economic, and environmental factors.

Political priority. After the Second World War, developing countries that had suffered from recurring famine in early periods struggled to increase food production as an urgent national priority to keep up with surging population growth and spiralling demand for food. India's Green Revolution was spurred by the imminent threat of severe famine in the mid-1960s. China was compelled to produce more food by the loss of perhaps as many as thirty or forty million lives to famine in the late 1950s.

The very motive which stimulated these achievements moderated them as well. For, while these governments took concerted steps to meet the minimum needs of the population for food, they tended to overlook the equally great potential to utilize agriculture as an engine for economic growth and job creation. Once minimum food needs were met, attention was diverted to other sectors of the economy.

Politically, the elimination of the famine threat, arising from food shortages, is one of the reasons for slower growth in agricultural productivity in recent years. This has resulted in a marked decline in investment by developing countries in government-sponsored agricultural research and rural techno-infrastructure. This tendency has been compounded by the false notion that in order to achieve economic growth, new jobs must be created in industry and services.

Economic policy. The resurgence of faith in the free market and liberalized trade has placed developing countries under increasing pressure from donors and international financial institutions to abandon special types of economic assistance to the agricultural sector (e.g., subsidies for supply of critical inputs, such as fertilizer and seeds, and price supports for marketable crops). This pressure has recently been aggravated by the impact of structural adjustment programs that usually involve drastic reduction in government subsidies. In Ghana, for instance, the cost of fertilizer as a percentage of total cultivation cost has risen tenfold in recent years. The impact of macro-economic reforms on agriculture has been most severe in the countries of Eastern and Central Europe, where production in this sector has fallen drastically over the past few years.

Technology. Slower growth in productivity can be attributed to a slowdown in the development of improved hybrid varieties, shortages of quality seeds and fertilizers, and the absence of techno-infrastructure facilities needed for storage and processing, expansion of markets, transport and distribution.

Of considerable contemporary importance is the role of technology and, particularly, biotechnology. The advent of biotechnology has opened up new opportunities for crop, tree, animal, and fish improvement, by rendering novel genetic combinations feasible. It has generated new opportunities for skilled employment in villages, through biological software industries, such as the manufacture of biofertilizers and biopesticides and the establishment of biomass refineries. At the same time, the advent of biotechnology has created a potentially dangerous situation in the pattern of international cooperation in science and technology.

The feedstock for the biotechnology industry is biological diversity. The Convention on Biological Diversity recognizes in Article 16 the importance of both access to, and transfer of, biotechnology in achieving the objectives of the Convention in the areas of genetic conservation, evaluation, and utilization.

The World Trade Agreement makes patenting, or some appropriate form of varietal protection, mandatory. The systems of plant variety protection so far in operation in industrialised countries, under the Union for the Protection of New Varieties of Crop Plants (UPOV) Convention, recognize and reward only the breeders. In the forum of the Food and Agriculture Organization of the United Nations (FAO), the concept of Farmers' Rights has been developed during the last decade to stress the need for also recognizing and rewarding the contributions of rural and tribal women and men who have, over the millennia, not only conserved plant and animal genetic resources but have also practised selection for resistance to biotic and abiotic stresses, and for medicinal and other economic properties. Hence the current interest in mechanisms for integrating equity and ethical considerations in the sharing of profits from plant-breeding research, involving both conventional and molecular approaches.

Environment. A number of factors are posing serious obstacles to productivity growth and threaten even current levels of production. Eighty percent of the poor in Latin America, sixty percent in Asia, and fifty percent in Africa live on marginal land of low productivity and high susceptibility to environmental degradation. Raising productivity on these lands can be extremely difficult.

Quality farm lands are being lost at an astonishing rate to diversion for non-farm uses, desertification, deforestation, soil erosion, and the depletion and pollution of water resources. These factors have resulted in the degradation of nearly one billion hectares worldwide since the Second World War. In addition, climatic changes resulting from the increased emission of carbon dioxide into the atmosphere are expected to exert additional negative impact over the long-term, the extent of which is difficult to assess.

Opportunities in Agriculture

Rapid and sustained expansion of food and agricultural production cannot be achieved without concerted efforts to address these factors. At the same time, there are a number of positive factors that offer opportunities to increase productivity significantly in the near to middle term.

Closing the productivity gap. A survey, comparing the productivity for major crops between countries, reveals a wide disparity between proven technological potentials and actual field results for every crop (Swaminathan and Nair, 1994). Cereal yields in Western Europe and North America average 4.5 tons less per hectare, compared to three tons in Asia, 2.3 tons in Eastern Europe and the former USSR, two tons in Latin America, and slightly more than one ton in Africa. FAO projects a forty percent increase in wheat yields in ninety-two developing countries, excluding China, by 2010.

In rice, countries with widely differing climatic conditions, such as Australia, Egypt, Italy, Japan, the Republic of Korea, and Spain, achieve average yields nearly double the world average and two to three times the yields achieved in countries such as India, Malaysia, the Philippines, and Thailand. Total rice yields in ninety-two developing countries are expected to rise to thirty-seven percent by 2010. Maize yields in Austria, Chile, Germany, Greece, and Italy average more than double the world average, and five to six times the averages achieved in large maize producing countries, such as India. Even comparing areas under irrigated maize, average yields range from 1.5 to 8.4 tons per hectare. New varieties of maize that yield thirty percent more grain in a drought season than conventional varieties are under development. They could be especially effective in the lowland tropics, including much of Africa. Potato yields in Belgium and the Netherlands are roughly three times the world average.

This productivity gap, between what is routinely achieved by different countries, represents a vast immediate potential for improving yields that does not require significant additional investment of time or money in R&D. The world already possesses the knowledge, technology, and organizational capacities to considerably raise average world yields in major crops.

Emphasizing cultivation practices. The phenomenal increases in productivity, achieved over the past three decades through the development of improved and hybrid varieties of seeds, have overshadowed and distracted attention from the equally great potential of raising yields through improvements in methods of cultivation. This is especially true since much of the slowdown in productivity growth is attributable to depletion of soil and water resources that can be partially rectified or offset by improved practices.

Growing environmental concerns in the West have generated pressure for reducing fertilizer consumption in developing countries when, in most instances, fertilizer application levels are lower than those in Western Europe by a factor of five or ten times. Increased, rather than reduced use of chemical fertilizers, will be essential to the expansion of agriculture in these countries and can considerably raise crop yields. Total fertilizer consumption by developing countries is expected to double by 2010.

Furthermore, replacing macronutrients lost during intensive cultivation does not compensate for the depletion of more than a dozen micronutrients that also determine the quantity and quality of crops. Even in such agriculturally diverse countries as India and the United States, it has been demonstrated that better management of micronutrients can raise productivity substantially – in some intensively cultivated areas by as much as fifty percent to one hundred percent or more, in a wide variety of crops – without significant changes in the structure or method of cultivation.

Greater attention is needed to conserving and applying organic sources of manure and raising nitrogen-fixing crops, as well as to the use of biofertilizers. Unlike the creation of new irrigated lands, the application of such environmentally sustainable field practices is neither very costly nor technology-intensive, and can be widely propagated through improvements in agricultural training and extension. Such approaches are an integral part of the biovillage program in India.

Increasing exports. Over the next six years, the agreements reached during the Uruguay Round of the General Agreement on Trade and Tariffs (GATT) negotiations are expected to reduce the tariffs levied by the industrial nations on agricultural products by thirty-seven percent and on tropical food exports by forty-three percent. Developing countries will cut agricultural tariffs by an average of twenty-four percent. In addition, North American and Western European countries have agreed to cut agricultural subsidies by one-third.

Together, these cuts are expected to result in a twenty percent rise in exports of agricultural and processed foods by 2005. The share of developing countries in this increase has been estimated at US $20 billion to US $60 billion annually. Increased trade can act as a strong stimulus to the growth of agricultural, agro-industrial and rural employment, provided that these countries respond dynamically to the opportunity by raising productivity and strengthening rural infrastructure for storage, processing, and distribution. However, rising demand within developing countries themselves will limit the overall growth of exports and increasing exports will push up domestic food prices, unless much greater priority is given to matching production with opportunities for home and external trade.

Increased private investment. The modernization of the farm sector greatly needs the investment, technology, professional management, and marketing expertise that private firms can bring. Wherever land reforms have resulted in the division of farms into small parcels, farmers need to be supported by well-organized services, particularly for post-harvest handling, provided either by companies or cooperatives.

Economic liberalization offers expanded opportunities for private enterprises to work with small producers in a wide range of agro-industries, to combine the social benefits of small holdings with the economies and marketing expertise of corporate management. The contract system of agricultural production, introduced in backward regions of Thailand, has enabled small farmers to work closely with private businesses to produce labor-intensive, value-added crops, with technology, training, and marketing provided by the companies, resulting in a rapid rise of farm incomes and job opportunities.

Developing countries possess significant advantages in terms of climate, year-round production, and low labor costs that are conducive to foreign investment. With private capital flows on the rise, there is an opportunity for developing countries to upgrade technology and obtain direct access to foreign markets, by providing a conducive atmosphere for foreign investments and collaboration in agribusiness ventures, such as hybrid seed production, flowers, processed fruit and vegetables, fresh and salt-water aquaculture, and sericulture.

Shift from commodity-based to resource-based planning. The responsibility of governments to achieve dramatic increases in foodgrain production, to keep pace with population growth, naturally led to a commodity-based approach to agricultural development in many developing countries. Governments projected demand, set production targets for specific crops, and instituted programs to help farmers produce the food needed to meet the minimum needs of the population.

But, once shortages are eliminated, the focus on specific commodities becomes a barrier to further agricultural development, because it retards diversification into more profitable commercial crops. Contrary to free market theory, farmers in many countries have tended to continue production of traditional crops, even when prices have fallen in response to increased supplies.

There is still a need for government to educate and encourage farmers to adopt resource-based planning, oriented to national and international market potentials and based on principles of economics, employment, and ecology. Resource-based planning examines how the available land and water resources can best be utilized to achieve maximum and sustainable economic return to the farmer. Such a shift can lead to diversification into commercial crops, like fruit and vegetables that generate significant increases in on-farm employment and incomes and act as a stimulus to downstream agro-industries.

From minimum yields to maximum potentials. The factors that have perpetuated commodity-based planning in agriculture have also fostered an emphasis on minimum targets rather than maximum goals. Often, government has perceived that its responsibility was to ensure sufficient food to prevent famine, not maximum output and profitability.

Government machinery has proved effective in forcing through radical measures to disseminate new technology, seed, and production methods in the face of crisis, but it usually lacks the driving impetus to work for the highest benefit of individual farmers. The shift needed is for planners to study the potentials of the agricultural sector to serve as an engine for job creation and higher incomes, and as a stimulus to industrialization and exports, and then to formulate national goals to maximize exploitation of these potentials.

Integration of agriculture, marketing, and processing. Conscious efforts can be made to foster the natural linkage between agricultural and industrial development by placing emphasis on crops that have the greatest potential for stimulating the growth of agro-industries, services, and exports. Filling in the missing links in the chain of processing and distribution, such as pre-cooling for fruit and vegetables, can enable small farmers to produce for national and export markets; it also results in higher at-farm prices and creates alternative forms of rural employment.

Empowerment of women. The vast majority of women living in rural areas are engaged in agriculture. Therefore, upgrading productivity, skills, and incomes in this sector is the single most effective means of improving the livelihood of women in developing countries. Skill development programs, in areas such as hybrid seed production, floriculture, inland aquaculture, vegetables and poultry, can be particularly beneficial. Promotion of micro-level credit institutions and savings programs can generate capital for the establishment of small rural enterprises by women.

International Agenda for Food Security

The crucial importance of food security to world peace and economic development demands that the international community take collective responsibility and initiative to eradicate hunger and famine on a global basis, as a complement to the initiatives of individual countries to deal with the problem domestically. Conditions vary too widely, from country to country, for detailed recommendations to be broadly applied. However, a number of strategies are relevant to the majority of developing countries. Some of the components of an Action Plan, designed to achieve this goal, are given in the following paragraphs.

Elimination of agricultural trade barriers. The agreements reached for liberalizing agricultural trade in the Uruguay Round of GATT are an important step forward, but they will not release the full dynamism of the agricultural sector that is so critical for more rapid growth of incomes and employment in both developing and developed countries. A rapid reduction, leading to a complete elimination of agricultural subsidies and other trade barriers by industrial nations, could generate a positive multiplier effect.

It has been estimated that complete liberalization of trade in agricultural commodities would yield an annual gain (in 1992 dollars) of about US $25 billion for OECD countries, and US $22 billion for developing and formerly centrally planned countries. Actual gains could be very much higher. These subsidies cost Western consumers several hundred billion dollars annually. In addition, they lead to dumping of surplus sugar, cereals, milk, and beef in developing countries, often pushing down the prices that farmers in these countries receive for their produce to far below their production cost.

The benefits of eliminating these subsidies would be multiple. For instance, eliminating subsidies to sugarbeet growers and dairy farmers in Europe would stimulate greater demand, not only for imported sugar from developing countries, but also for cocoa which, combined with sugar and milk, is the basis for the chocolate industry. Measures of this type will result in substantially increased demand for agricultural products from developing countries – the first step in an upward spiral of global economic growth that will stimulate industrial exports and job growth in both the East and the West.

Global environmentally sustainable Green Revolution. The term "Green Revolution" has been praised for its positive impact on crop productivity, and criticized for its sometimes adverse effects on the environment and social equity. It is often overlooked that one of the most beneficial consequences of the Green Revolution has been its "forest-saving" nature. If agricultural production had not been increased through higher productivity, more forest land would have been diverted to annual cropping.

The Green Revolution was originally applied to improving productivity of wheat and rice in high potential areas. After the first rounds of phenomenal success, the effort to extend it to other agricultural regions and other crops lost momentum because the urgent necessity had been eliminated. Recognizing the importance of agricultural surplus to rural employment, industrialization, and exports, renewed efforts are called for to apply the comprehensive, integrated approach of the Green Revolution to high potential commercial crops and to all regions.

With the knowledge gained over the past thirty years, greater emphasis can be placed on ensuring that the productivity improvements are not only economically viable but also ecologically sustainable. Region-specific strategies should be adopted for mountain areas, high rainfall tropics, uplands, and irrigated plains. The crops covered should include foodgrains and oil seeds, cotton and jute, sugarcane, fruit, vegetables, dairy, meat, medicinal plants, spices, agroforestry for fodder, fuel, and industrial raw materials.

Curbing the "greed revolution" and promoting an "evergreen revolution". Many of the negative consequences of Green Revolution (i.e., yield enhancing) technologies arise from the excessive and unscientific use of pesticides, mineral fertilizers, and other chemical inputs. The adverse consequences of aquaculture techniques also arise from trying to achieve very high levels of production.

An evergreen revolution on the farm is essential for producing more food and other agricultural commodities. It is, however, important that the fast-spreading "greed revolution", leading to the unsustainable exploitation of the environmental capital stocks, is curbed. We must make a distinction between green and greed revolutions.

UN development force for food deficit regions. The complex task of planning, managing, and executing nationwide programs to eradicate food shortages may be beyond the political, administrative, and management capacities of some governments, hampered by political instability or social strife, or the absence of training and experience. The slowest member country retards the progress of the whole world. The interests of global peace, political stability, and basic human rights justify, and may necessitate, external assistance to help countries establish viable food security systems.

In countries that are unable either to produce sufficient food or to initiate coordinated programs to overcome present deficits, the international development force, proposed by the ICPF, can act in a trusteeship role to assist in designing and implementing integrated programs to upgrade food production and distribution. Emphasis should be placed on sustainable production and distribution – not just relief operations – through the introduction of effective systems for planning, administration, education, demonstration, and implementation.

Model districts. The potential benefits of applying a resource-based, location-specific approach to agriculture can best be demonstrated by the establishment of model programs in different agroclimatic regions. The models should be large enough, geographically, to serve as a viable index of national-level potentials, and of economic and ecological sustainability. Although the concept may vary from place to place, the central approach is to examine in detail the current usage of both land and water resources, and then formulate a district-level plan for utilizing available physical, technological, managerial, and financial resources in a sustainable manner. The underlying motivation would be to optimize, over time, production, productivity, farm incomes and employment, non-farm occupations, self-employment opportunities, agro-industrial development, exports, and expansion of the service sector.

This analysis is likely to identify a wide gap between present and potential achievements, which can form the basis for creating an alternative district development plan. Government can seek the assistance of farmers' organizations, private enterprise, educational and research institutes, and voluntary agencies for gathering information, analysis, and plan formulation. Implementation of the plan should be primarily through activities that facilitate more efficient operation of market forces, such as assisting farmers to identify and transfer improved technology, establishing commercial organizations of small farmers and linkages between farms and industry, and disseminating information on markets and technology.

World food model. The actual structure and dynamics of the development of global agriculture and its relationship to industrialization, employment, and trade, needs to be fully understood in order to plan for, and achieve, a world free of hunger. Although national and international institutions maintain a variety of databases and models to track the impact of production, trade, and prices in agriculture, these models are too limited in scope and detail to construct a working global model of the agricultural sector. Such a model could serve as a valuable tool in projecting the medium- and long-term interactions of changes in food production, consumption, employment, trade, productivity, technology, and environmental factors. A world food model that forms the essential basis for evolving a global strategic plan, can help eliminate both food deficits and surpluses.

On-farm training. The massive demonstration programs, conducted in farmers' own fields, by countries such as India to propagate the Green Revolution technology, proved that even uneducated, traditional peasant farmers in developing countries would rush to adopt new technology when its application and benefits are clearly demonstrated. The slow dissemination of new technologies that still hampers agriculture in these countries is largely due to the ineffectiveness of more traditional types of extension and farmer education, conducted on a relatively limited scale, by agricultural colleges, technical institutes, and research stations.

This gap in the agricultural education system can be closed by expanding the agricultural training network to the village level. The establishment of farm schools in villages, on land temporarily leased from local farmers, can combine the advantages of both formal training and demonstration on farmers' fields, with emphasis on economic viability and ecological sustainability. This is a low-cost strategy that requires prior training of a large number of village-level instructors, but little investment in infrastructure. The schools can demonstrate the economic benefits of new and improved crops and cultivation practices, engaging local farmers as both students and staff, and covering most of the cost of training in the form of sale proceeds from the farm schools' production.

Water conservation. The inefficient and negligent use of water in agriculture is one of the most serious barriers to a sustainable expansion of agricultural production. Public policy regarding the cost of water supplied by major irrigation projects and low-cost or free distribution of power for pumping underground water aggravate the problem. Technologically, the solutions to water depletion are largely in hand. What is needed is a massive effort in public education and demonstration, coupled with incentives to encourage their adoption.

In monsoon regions, it has been demonstrated that recharging underground aquifers by reverse pumping during rain-surplus seasons can restore ten years of water depletion in one rainy season. Proven methods are available and have been commercially applied wiith great success, as the achievements of Israel so dramatically demonstrate. Water consumption can be reduced radically, by as much as five-to-tenfold, at the same time as significantly increasing crop yields. The most productive of these methods, involving the construction of greenhouses, is too capital-intensive for immediate widespread adoption in developing countries. Drip and sprinkler irrigation are far more moderate in cost and still result in enormous savings of water. The cost of adopting these methods can be further reduced by more widely popularizing them, so as to achieve increasing economies of scale in their manufacture.

Low-input sustainable agriculture. While the case can be argued against pressure on food-deficit developing countries to reduce their comparatively low-level usage of mineral fertilizers and chemical pesticides, empirical studies, sponsored by the National Academy of Sciences and the Department of Agriculture in the United States, found strong evidence, long resisted by scientists and commercial farmers in the West, that chemical-free farming can be as productive and cost-effective as chemical-based methods. Industrial countries that offer subsidies to encourage chemical usage should modify their policies to provide active incentives for the use of chemical-free methods instead.

During the last ten years, Indonesia has succeeded in bringing down the consumption of chemical pesticides, without adversely affecting grain production, through the nationwide adoption of integrated pest management systems for protection of food crops. India has launched a massive national training program to popularize biological methods of pest control that can reduce the consumption and cost of chemical pesticides by fifty to seventy-five percent.

Inland aquaculture. Depletion of deep-sea fishery resources has resulted in a significant decline in world harvests from the sea. Proven technology is available for substantially replacing sea-grown fish with fresh or salt-water pond-grown varieties. The new intensive and semi-intensive methods of cultivation can generate yields ten to one hundred times higher than traditional extensive methods. Due to the warmer climate and lower labor costs of the developing world, inland aquaculture can become a major stimulus to rural employment, incomes, processing industries, and exports.

Micro-level indicators of food security. Development initiatives have been severely limited by the absence of accurate and sensitive measures to assess the impact of national-level policies and programs on food security at community and household level. New indices, such as the Sustainable Livelihood Security Index, now under development in India, need to be evolved and applied; they should combine measures of income and employment; food and nutritional status, including the availability of safe drinking water and sanitation facilities; ownership of productive assets; and education and productive skills.

Summing Up

Three points from the executive summary of the ICPF bear repeating. First, nowhere are we in greater need of fresh thinking than when it comes to the issue of employment. Full employment would seem more appealing if it were not so widely believed to be impossible. The western world has come to accept the myth that technology is inevitably eliminating jobs. Again, it is intellectual limitations that stand in the way of progress, more than material constraints. Although it is certainly true that technology eliminates jobs, at the same time, it creates them and, on balance, it creates many more than it destroys.

Second, the creation of one billion jobs in developing countries, over the next decade, is achievable through a comprehensive strategy, based on the promotion of commercial agriculture, agro-industries and agro-exports, improved marketing, expansion of rural enterprises and the service sector, dissemination of commercial information, extending basic education and upgrading skills. Achievement of this goal requires that the industrial countries adopt agricultural trade policies designed to enhance the export opportunities of developing nations.

A third related point is that neither the industrial, nor the developing countries, can resolve the problem of unemployment in isolation. The industrial nations require a significant increase in demand that only the faster-growing developing countries can provide. The latter require greater investment and access to markets, especially for agricultural products and textiles. A global employment program of the sort discussed at the World Summit for Social Development, Copenhagen, in March 1995, should set forth a plan to expand job creation dramatically, worldwide, during the rest of the decade. The plan should focus on elimination of protectionist trade policies, debt rescheduling for the poorest debtor nations, accelerated transfer and dissemination of technology, and international cooperation to encourage labor-friendly tax policies.

These proposals reiterate what UNESCO Director General, Federico Mayor points out in his Foreword to the ICPF report: "... The uncommon opportunities that have arisen as the world shifts into new alignments ... include the chance to tackle the problems of hunger and injustice at the root of so many of the global challenges confronting us – not the least, the overriding challenge of nurturing a worldwide culture of peace."

Biovillages and Sustainable Agriculture

The biovillage program was initiated in 1991 by the M.S. Swaminathan Research Foundation, in three villages (Kizhur, Sivaranthakam, and Pillayarkuppam) of the Union Territory of Pondicherry, with a primary focus on the economically and socially disadvantaged in rural areas. The program seeks to explore and test a new development paradigm that addresses concurrently the twin concerns of rural poverty and resource base degradation – the former through technological empowerment of the resource poor and the latter through integrated resource management systems, including recycling and value addition. The biovillage approach involves the development and popularization of ecotechnologies, based on the integration of the ecological prudence and technologies of the local population with breakthrough technologies, such as biotechnology, information technology, renewable energy technologies, and space technology.

The biovillage program aims at the full utilization of the production potential of both biophysical and human resources. It is pro-nature, pro-poor, and pro-women in its basic development orientation. The program focuses on the rural poor, especially the rural women, by developing ways and means of overcoming generic constraints to widespread adoption of small-scale income-generating activities on a site-specific basis. Remunerative activities are tailored to the specific circumstances of resource-poor households. The resource-poor are enabled to benefit from the products and processes of science and technology by translating acquired knowledge and skills into sustainable income- and employment-generating activities that add value to their principal assets, labor, and time.

Technological empowerment of the resource-poor is an integrated process, involving the identification of technologies that are environmentally compatible, economically viable, and socially equitable; testing and adapting them to the specific socioeconomic conditions of the participants; internalizing the technology through on-production-line training into income- and employment-generating production activities, by introducing the participants to sources of knowledge and know-how and to credit institutions; and devolving innovative institutional mechanisms to facilitate the sustainability of enterprises, by providing infrastructure support for ensuring the supply of inputs and the marketing of outputs. Operationally, a mix of enterprises appropriate to the resource endowments of the area are identified, and the resource-poor are enabled to translate them into income- and employment-generating activities.

Enterprises, for landless women, include production of oyster mushroom from paddy straw, vermicompost from used straw waste, goat rearing based on biomass from fodder plantations on wastelands, aquaculture in community ponds, dairying based on fodder from fodder banks, broiler production based on local feed resources, and production of hybrid vegetable seeds. For marginal and small farm households, the focus is on small-scale production of vegetables and flowers, flower nurseries, fodder banks, and mulberry banks. Integrated resource management is fostered through flower-vegetable-fish culture and fodder-dairy-biogas enterprises. By fostering group action, landless people are encouraged to undertake resource-dependent activities, such as dairying and sericulture, by linking with fodder and mulberry banks established by small farmers and by accessing to common property resources, such as fodder plantations on wastelands and aquaculture in community ponds.

In sum, the biovillage approach is based on the premise of sustainable job-led growth that will become possible only if the development programs are people-centered, if they are environmentally sound, if they are participatory, and if they build local and national capacity for self-reliance.

On Ethics, Biodiversity Patenting, and Economic Opportunities

Over seventy years ago, J.B.S. Haldane, in his book *Daedalus, or Science and the Future* (1923), urged that unless our ethical outlook evolves to keep pace with development of science and technology, social chaos and disaster would result. In this context, some recent developments in biotechnology and biodiversity patenting are disturbing. Examples include patenting by commercial companies in industrialized countries, of Margosan-O developed from the neem tree; an extract from the African soapberry plant Endod, used for killing zebra mussels; an anti-cancer extract from the Indian medicinal plant *Camptothescin*; and, to cap it all, a broad patent that could cover all genetically-engineered cotton.

These events are leading some countries to take pre-emptive measures that will prevent others from filing patent claims. For example, Queensland (in Australia) has passed legislation giving it intellectual property rights over genetic information embodied in all the plants and animals found in the State. Similar initiatives are yet to be taken by developing countries, although such pre-emptive strikes are likely to be taken soon by other countries to prevent a few cashing in on the silent work, over centuries, of nature and human families.

Action is urgent, particularly in the field of medicinal and aromatic plants, given their importance in many developing regions and the strong likelihood of considerable growth in the trade of these products in the next decade. Two examples from recent research in India provide glimpses of the untapped potential. Firstly, the compound piperine, extracted from the fruits of *Piper longum*, helps in the absorption of antibiotics, such as rifampicine; tuberculosis patients can get cured with a lower dose of rifampicine when it is taken in combination with piperine. Secondly, baccosides, extracted from the brahmi plant *Baccopa miniori*, help to improve memory among the aged; this compound has been patented by the Central Drug Research Institute in Lucknow (India). There are numerous opportunities of this kind that the biotechnologists of the region concerned should seize and thus help break the present situation where genetic richness and economic impoverishment co-exist. ▯

References

International Commission on Peace and Food (ICPF). 1994. *Uncommon Opportunities. An Agenda for Peace and Equitable Development*. London: Zed Books.

Haldane, J. B. S. 1923. *Daedalus, or Science and the Future*. London: Kegan Paul.

Swaminathan, M. S. and K.N.N.S. Nair. 1994. *Agenda for a Food- Secure Asia-Pacific Region by 2010 A.D.* Madras: MSSRF and UNDP.

28

Toward a Food-Secure World

Introduction

Since the advent of the industrial and technological revolutions, economic indicators have been used as the principal criteria for measuring sustainability. Population expansion, rapid industrialization, commercialization of agriculture, and quantum jumps in economic activity have been some of the results of the development paradigm adopted after World War II. Technological progress in several areas such as space, information technologies, biotechnology, energy, and new materials has been impressive. At the same time there is a growing understanding of the ecological and social costs of such progress.

The UN Conference on the Human Environment held at Stockholm in 1972 and the UN Conference on Environment and Development held at Rio de Janeiro in 1992 helped to articulate the serious environmental repercussions of contemporary development pathways. The UN Conference on Social Development held at Copenhagen in March 1995 warned that development which is not socially equitable will lead to social disintegration and conflicts. Jobless economic growth and feminization of poverty are the other consequences of the current pattern of development. Thus, the concept of sustainable development has now to be viewed in terms of ecology, social and gender equity, employment, and economics. How to achieve such a synthesis in developmental thinking, planning, and implementation, thereby enabling humankind to take to the pathway of green productivity, is the task facing us today.

In its report titled "Changing Course" the International Business Council pointed out that where there is a will there is a way (Schmidheiny, 1992). If technology has so far been a major cause of ecological damage, it can be a leader in finding methods to ensure that development is sustainable. In a recent study, Repetto et al (1996) of the World Resources Institute have shown that environmental protection not only need not reduce productivity growth but can in fact stimulate growth without accompanying ecological damage.

Presented at the Symposium Honoring the Tenth Anniversary of the World Food Prize, 1996. First published in *Food Security – New Solutions for the Twenty-first Century*, ed. by Amani E. El Obeid et al, pp. 107-28. Ames, Iowa: Iowa State University Press, 1999.

Ecotechnology and Sustainable Livelihoods

The Twentieth Century: A Balance Sheet. As we approach a new century we can look back and draw a balance sheet of our achievements and failures (Table 1). Spectacular progress in science and technology ranks first among our major accomplishments. Recent advances in biotechnology and genetic engineering, space technology, information technology, and new materials have opened up uncommon opportunities for a world where every individual can lead a healthy and productive life. The spread of democratic systems of governance, the breakdown of skin color-based apartheid and the advent of the information age have created the sociopolitical substrate essential for integrating the principles of intra- and intergenerational equity in public policy. The power of a right blend of technology and public policy is strikingly evident from the progress made in recent decades to keep the growth rate in food production above the rate of growth in population, thereby ensuring that the Malthusian prediction of population overtaking our ability to produce adequate food does not come true.

While the positive achievements are many and make us proud of the power of the human intellect, we will be entering the new millennium with some of the greatest social and scientific challenges humankind has ever faced. Several of these challenges have been articulated with great clarity in the "Human Development Report" series published by the United Nations Development Program (UNDP) in recent years.

Environmental degradation and increasing economic and gender inequality are among the most serious problems we face today. The rich/poor divide is increasing at an alarming rate. The pattern of development adopted by rich societies is leading to jobless economic growth, pollution, and potential changes in climate. Unsustainable lifestyle on the part of the rich billion and unacceptable poverty on the part of another billion co-exist. The absence of an educational and health environment that is conducive to every child achieving his or her innate genetic potential for physical and mental development leads to the spread of poverty in capability. UNDP has proposed indicators for measuring both human development and human capability.

The US National Academy of Sciences, the Royal Society of London, the Indian National Science Academy, and fifty-five other scientific bodies in a statement made in 1993 pointed out that stress on the environment is caused by four interacting factors; population growth, consumption habits, technology, and social organization (Graham-Smith, 1994). Concurrent attention is needed on all these four factors to promote sustainable development and sustainable societies. The report "Sustainable America" indicates what an affluent society should do. In poor nations, the social sustainability of the development process is as important as ecological and economic sustainability (The President's Council on Sustainable Development, 1996). Also, if the current pace of damage to the ecological foundations essential for sustainable advances in biological productivity – land, water, flora, fauna, forests, oceans, and the atmosphere – continues, sustainable food and nutrition security cannot be achieved. Therefore, as we

Table 1. Contemporary development: a 50-year balance sheet (1945-1995)

Achievements	Concerns
Impressive progress in science and technology	Scientific dimensions of sustainable development
Information Age: Global Village	Era of unfulfilled expectations
Unprecedented economic growth	Damage to life support systems and potential adverse changes in climate and sea level
Uncommon opportunities for food, health, literacy, and jobs for all	Population growth, jobless economic growth, and social disruption; gross economic and gender inequity
Peace Dividend: from defense to development	Growing violence in the human heart: from AIDS to ethnic strife
End of skin color-based apartheid: spread of democratic tradition and renewed faith in UN and other multilateral institutions	Debt burden and emerging technological and economic apartheid

approach the new millennium, we need a broader concept of sustainability which encompasses environmental, economic, and social parameters. Among social factors, gross economic and gender inequity need priority attention. If such a paradigm shift in developmental thinking and pathways does not occur, the successes achieved in the twentieth century in abolishing skin color-based apartheid, in conquering space, and in splicing genes will be overshadowed by the spread of technological and economic apartheid. If these forms of apartheid are allowed to grow and spread, social disintegration and ecological genocide will be the result.

Ecotechnology: The Emerging Solution. Technologies rooted in the principles of ecology, economics, and equity are now referred to as ecotechnologies. The United Nations Educational, Scientific, and Cultural Organization (UNESCO) and the Cousteau Foundation established by Jacques Cousteau are promoting ecotechnology networks in different parts of the world. The M.S. Swaminathan Research Foundation (MSSRF) at Madras is the coordinating center for the Asian Ecotechnology Network. A major purpose of this network is the creation of ecojobs, which are economically viable, environmentally benign, and socially equitable. A multimedia database on opportunities for ecojobs is being developed, because the dissemination of information on ecojobs is essential for creating opportunities for sustainable livelihoods in rural and urban areas.

The most serious manifestation of poverty is hunger. It is now recognized that endemic hunger is largely the result of inadequate livelihood opportunities which in turn lead to inadequate purchasing power. "Hidden hunger" results from both micronutrient deficiencies and poor environmental hygiene, which impair the biological absorption and retention of food.

A Science Academies Summit on uncommon opportunities for a food-secure world held at MSSRF in July 1996 stressed that national policies for sustainable food and nutrition security should ensure two things: first, that every individual have the physical, economic, social, and environmental access to a balanced diet that includes the necessary macro- and micronutrients, safe drinking water, sanitation, environmental hygiene, primary health care, and education necessary to lead a healthy and productive life; second, that food originate from efficient and environmentally benign production technologies that conserve and enhance the natural resource base of crops, animal husbandry, forestry, and inland and marine fisheries.

During the next three decades, population is expected to increase by another 2.5 billion people. Food requirements will grow due to increases in both population and per capita purchasing power. World grain production has grown from 631 million tons (metric tons) in 1960 to nearly 1,900 million tons in 1995. Such a phenomenal growth has had its environmental cost in terms of soil degradation, aquifer depletion, genetic erosion, and pesticide pollution. This is why we have to produce more in the coming decades but produce it differently. To achieve such a shift, the following ground rules must be followed in technology development and dissemination as well as in public policy.

First, production advances must be based on linking the ecological security of an area with the livelihood security of the people in a symbiotic manner.

Second, steps must be taken to create widespread awareness of the human and animal population-supporting capacity of different ecosystems. Sustainable systems of management of soil, water, biodiversity, and forests should be internalized in rural societies.

Third, since the poor remain poor because they have no productive assets and there is no value to their time, asset creation and value addition to time should receive high priority in poverty alleviation programs. Women belonging to the economically underprivileged sections of the society, in particular, are often overworked and underpaid. What they need is a reduction in the number of hours of work and an increase in the economic value of each hour of work. This will call for massive efforts in information and skill empowerment of the poor, particularly women. The emerging technologies are largely knowledge intensive and hence the transfer of knowledge and market-driven skills can become the most powerful instrument. Modern information technology affords opportunities for reaching the unreached and thereby achieving a learning revolution within a short span of time.

Fourth, equal attention is needed to the problems of the rural and urban poor. Lack of livelihood opportunities in rural areas leads to the proliferation of urban slums. Damage to common property resources in villages results in the growth of environmental refugees. In many developing countries, agriculture, including crop and animal husbandry, forestry, fisheries, and agro-processing, provides most of the jobs and income in rural areas. Therefore the triple challenge of producing more food, income, and jobs from diminishing per capita land, water, and non-renewable energy sources can be met only through agricultural intensification, diversification, and value addition. Integrated, intensive farming systems that are ecologically sustainable are needed for this purpose.

Fifth, microcredit and microenterprises should be linked in a symbiotic manner, as is being done successfully in Bangladesh through the Grameen Bank and other innovative efforts. Finally, an "Evergreen Revolution" of the kind described above can be given a self-propelling and self-replicating momentum only if it is based on the self-mobilization of the people. In all externally funded and introduced development projects, there should be a built-in withdrawal strategy, so that the program does not collapse when the external inputs are withdrawn.

Meeting the Multidimensional Challenges:
Response of the M. S. Swaminathan Research Foundation

I will describe briefly each of the responses being developed and field tested by MSSRF to identify implementable approaches at the micro and policy levels to meet the challenges outlined earlier.

Linking the ecological security of an area with the livelihood security of the local community: Creating an economic stake in conservation. The community biodiversity program of MSSRF illustrates how such mutually beneficial linkages can be fostered in biodiversity-rich areas. It is a sad fact that the tribal and rural families who have conserved and enhanced biodiversity remain poor, while those who are utilizing the products of their efforts become rich. When the conservers have no social or economic stake in conservation, denudation of natural ecosystems becomes more rapid. The MSSRF has adopted a three-pronged strategy for creating an economic stake in biodiversity conservation.

First, a transparent and implementable methodology has been developed for incorporating in sui generis systems of plant variety protection procedures for recognizing and rewarding informal innovations in genetic resources conservation and enhancement.

Second, a symbiotic social contract between commercial companies and tribal and rural families is being fostered for the purpose of promoting the cultivation by local communities of genetic material of interest to the companies on the basis of buy-back arrangements. Such a linkage will prevent the primary material being unsustainably exploited.

Third, local women and men are trained in the compilation of biodiversity inventories and in biomonitoring, so that they themselves become custodians of their intellectual property. Such trained women and men constitute an agrobiodiversity conservation corps and will be able to help their respective communities to deal with issues such as "prior informed consent" in the use of genetic resources.

For assisting the community biodiversity movement, MSSRF has established a technical resource center for the implementation of the equity provisions of the Convention on Biological Diversity. This is the first technical resource center of its kind in the world and contains six major components:

a) Chronicling the contributions of tribal and rural families to the conservation and enhancement of agrobiodiversity through primary data collection in the states of Tamil Nadu, Kerala, Andhra Pradesh, and Orissa, as well as in the Lakshadweep and Great Nicobar group of islands.

b) Organization of an agrobiodiversity conservation corps of young tribal and rural women and men who have a social stake in living in their respective villages and who, with appropriate training, can undertake tasks such as compilation of local biodiversity inventories, revitalization of the in situ genetic conservation traditions of their respective communities, monitoring of ecosystem health with the help of appropriate bio-indicators, and restoration of degraded sacred groves. The members of the corps will be able to assist their respective communities in dealing with the prior informed consent provision of the Convention on Biological Diversity in the use of genetic resources.

c) Development of multimedia databases documenting the contributions of tribal and rural families in the conservation and improvement of agrobiodiversity, for the purpose of enabling them to secure their entitlements from the national and global community gene funds.

d) Maintenance of a community gene bank and herbarium: A community gene bank with facilities for medium-term storage has been established to conserve farmer-preserved and developed seeds from the tribal areas of South India. The material will be catalogued and linked to the technical resource center database. The herbarium serves as a reference center for the identification of landraces, traditional cultivars, and medicinal plants conserved by tribal and rural families.

e) Revitalization of genetic conservation traditions of tribal and rural families through social recognition of their contributions and the creation of an economic stake in conservation: For this purpose, replicable models of private sector engagement in contract cultivation of commercially valuable plants by tribal and rural families are being developed.

f) Establishment of a Legal Advice Cell to make available to tribal and rural families appropriate legal advice relating to intellectual property rights and plant variety protection.

The population-supporting capacity of ecosystems: Local level socio-demographic charter. In order to help internalize an understanding of the vital need to restrict population growth within the supporting capacity of land, water, forests, and the other components of the ecosystem, training modules have been developed to enable the women and men members of village-level democratic institutions to prepare sociodemographic charters for their respective villages. A gender code is an important component of the charter. Such sociodemographic charters will help local communities to view population issues in the context of social development and to ensure that children are born for happiness and not just for existence.

Information and skill empowerment. For this purpose, the concept of information villages has been developed. Trained rural women and men will operate information shops where generic information on the meteorological, management, and marketing factors relevant to rural livelihoods will be converted into location-specific information. Trained farm women and men themselves will become trainers. The computerized extension system adopted in the information shops also helps to sensitize local families on their entitlements from government and other programs. Information technologies provide numerous opportunities for value-added jobs in rural areas. While new technologies are important, folk media are often even more effective in reaching the unreached. Hence, folk plays, folk arts, and theater are fully mobilized for achieving information empowerment. For ensuring the success of information empowerment programs, the information disseminated should be demand driven and locale specific.

Agricultural intensification, diversification, and value addition. These approaches are achieved through participatory research with farm families. Eco-technologies such as integrated pest management and integrated nutrient supply are used. Ecotechnology development involves blending the best in frontier technologies with traditional wisdom and practices. Modern science and the ecological prudence of the past can thus be combined.

Ecotechnologies are also practiced in aquaculture. Integrated agriculture and aquaculture techniques enhance both farm income and the nutrition security of the household. Whole villages are being enabled to adopt such Integrated, Intensive Farming Systems (IIFS). This approach is essential for meeting the triple goals of more food, income, and jobs from the available land and water resources. There are seven basic principles guiding the IIFS movement.

a) Soil health care: This is fundamental to sustainable intensification. The IIFs approach fosters the inclusion of stem-modulating legumes like *Sesbania rostrata*; the incorporation of Azolla, blue-green algae, and other sources of symbiotic and nonsymbiotic nitrogen fixation; and the promotion of cereal-legume rotation in the farming system. In addition, vermiculture composting and organic recycling constitute essential IIF8 components. Farmers using IIFs are trained to maintain a soil health card to monitor the impact of farming systems on the physical, chemical, and microbiological components of soil fertility.

b) Water harvesting and management: The IIFS farm families include in their agronomic practices measures to harvest and conserve rainwater so that it can be used in a conjunctive manner with other sources of water. Where water is the major constraint, technologies that can help to optimize income and jobs from every liter of water are chosen and adopted. Maximum emphasis is placed on on-farm water use efficiency and on the use of techniques such as drip irrigation that help to optimize the benefits from the available water.

c) Crop and pest management: Integrated Nutrient Supply (INS) and Integrated Pest Management (IPM) systems form important components of IIFS. The precise composition of the INS and IPM systems depends on the components of a farming system as well as on the agro-ecological and soil conditions of the area. Computer-aided extension systems will provide farm families with timely and precise information on all aspects of land, water, pest, and post-harvest management.

d) Energy management: Energy is an important and essential input. Besides the energy efficient systems of land, water, and pest management described earlier, every effort will be made to harness biogas, biomass, solar, and wind energies to the maximum extent possible. Solar and wind energy will be used in hybrid combinations with biogas for farm activities like pumping water and drying grains and other agricultural produce.

e) Post-harvest management: Farmers using IIFs will not only adopt the best available threshing, storage, and processing measures, but will also try to produce value-added products from every part of the plant or animal. Post-harvest technology assumes particular importance in the case of perishable commodities like fruit, vegetables, milk, meat, eggs, fish, and other animal products and processed foods. A mismatch between production and post-harvest technologies adversely affects both producers and consumers. Growing urbanization leads to a diversification of food habits. Therefore, there will be increasing demand for animal products like milk, cheese, eggs, and processed food. Agro-processing industries can be justified on the basis of an assessment of consumer demand and should be promoted in villages to increase employment opportunities for rural youth. In addition, these industries can help to mitigate micronutrient deficiencies in the diet. Investment in sanitary and phytosanitary measures is important for providing quality food, both for domestic consumers and for export. To assist the spread of IIFS, government should make major investments in storage, roads, and transportation, as well as in sanitary and phytosanitary measures.

f) Choice of the crop and animal components of farming systems: In IIFS, it is important to give very careful consideration to the composition of the farming system. Soil conditions, water availability, agro-climatic features, home needs, and above all, marketing opportunities will have to determine the choice of crops, farm animals, and aquaculture systems. Small and large ruminants will have a particular advantage among farm animals, because they can live largely on crop biomass. Backyard poultry farming can help to provide supplementary income and nutrition.

g) Information, skill, organization, management, and marketing empowerment: The IIFS is based on the principle of precision farming. Hence, for its success the IIFS needs a meaningful and effective information and skill empowerment system. Decentralized production systems will have to be supported by a few key, centralized services, such as the supply of credit, seeds, biopesticides, and animal disease diagnostics. Ideally, an information shop will have to be set up by trained local youth to give farm families timely information on their entitlements as well as on meteorological, management, and marketing factors. Organization and management are key elements, and depending on the area and farming system, steps will have to be taken to provide small producers with the advantages of scale in processing and marketing.

The IIFS is best developed through participatory research between scientists and farm families. This will help to ensure economic viability, environmental sustainability, and social and gender equity in IIFS villages. The starting point is to learn from families who have already developed successful IIFS procedures.

It should be emphasized that IIFS will succeed only if it is a human-centered rather than a mere technology-driven program. The essence of IIFS is the symbiotic partnership between farm families and their natural resource endowments of land, water, forests, flora, fauna, and sunlight. Without appropriate public policy support in areas like land reform, security of tenure, credit supply, rural infrastructure, input and output pricing, and marketing, small farm families will find it difficult to adopt IIFS.

Increasing farm and non-farm employment. The biovillage program addresses three key areas: preventing resource degradation, improvement of crop and animal productivity, and alleviation of poverty. The biovillage program in progress in villages in the Pondicherry region of India places equal emphasis on off-farm livelihood opportunities and on-farm jobs. This program avoids a patronage approach to poverty alleviation. It regards the poor as producers and innovators and helps to build their assets through value addition to time and labor. The basic approach is on asset building and sustainable human development leading to the growth of entrepreneurship.

The programs are designed on a pro-nature, pro-poor, and pro-women foundation. By placing emphasis on strengthening the livelihood security of the poor, the biovillage model of sustainable development revolves around the welfare of the economically and socially underprivileged. It is thus a human-centered pattern of development. The enterprises chosen are based on marketing opportunities. The technological and skill empowerment of the poor is the major approach. Because of the market-driven nature of the enterprises, the economic viability of the biovillage approach is assured. Production and post-harvest technologies and farm and nonfarm occupations are brought together in a manner that benefits both producers and consumers.

Biovillages around biosphere reserves would help provide alternative sources for meeting the day-to-day needs for food, fuel, fodder, and other commodities of the families living near such biodiversity-rich areas. Also, biovillages near urban areas would help to link the rural producer and the urban consumer in a mutually beneficial partnership.

Producing the processed and semiprocessed food products required in urban areas in the villages around towns and cities minimizes the need for the rural poor to migrate to urban centers for livelihood opportunities. Also, food processing can be used as a method of providing the needed micronutrients by including millets and grain legumes in the food.

Toward an Evergreen Revolution in Agriculture

The term "Green Revolution," coined by Dr. William Gaud of the US Department of Agriculture in 1968, has come to be associated not only with higher production through enhanced productivity, but also with several negative ecological and social consequences. There is also frequent reference to the "fatigue of the Green Revolution", due to stagnation in yield levels and to a larger quantity of nutrients required to produce the same yield as in the early 1970s. Experts such as Lester Brown of the Worldwatch Institute have been warning about an impending global food crisis due to increasing population, increasing purchasing power leading to the consumption of more animal products, increasing damage to the ecological foundations of agriculture, declining per capita availability of land and water, and the absence of technologies that can further enhance the yield potential of major food crops.

Should we therefore assume that as we enter a new millennium we will not have the benefit of new technologies which can help farmers to produce more food and other agricultural commodities from less land and water?

I believe we are now in a position to launch an Evergreen Revolution that can help to increase yield, income, and livelihoods per unit of land and water, if we bring about a paradigm shift in our agricultural research and development strategies. The Green Revolution was triggered by the genetic manipulation of yields in crops such as rice, wheat, and maize. The Evergreen Revolution will be triggered by farming systems that can help to produce more from the available land, water, and labor resources without either ecological or social harm. Thus, progress can be achieved if we shift our mind-set from a commodity-centered approach to an entire cropping or farming system. This does not mean that we should decelerate our efforts in crop improvement research. But such research should be tailored to enhancing the performance and productivity of entire production systems. The transition from the fatigue of the Green Revolution to an Evergreen Revolution involves a shift from a crop-centered to a systems-based approach to technology development and dissemination.

Let us take for example the prospects for "super-rice" capable of yielding over ten tons of rice per hectare. Such a rice plant will need a minimum of 200 kg of nitrogen per hectare, together with other major- and micro-nutrients. Addition of such nutrients solely through mineral fertilizers will lead to serious environmental problems; hence, the introduction of legumes in the rotation becomes important.

Scientists now have unique opportunities for designing farming systems to achieve the triple goals of more food, more income, and more livelihoods per hectare of land provided we harness the tools of ecotechnologies resulting from a blend of traditional knowledge with frontier technologies. Such tools include biotechnology, Geographic Information Systems (GIS) mapping, space technology, renewable energy technologies (solar, wind, biomass, and biogas), and management and marketing technologies.

We can enter a millennium of hope only if we abandon the old concept of a crop-centered Green Revolution and replace it with farming systems- and frontier technologies-centered Evergreen Revolution. Ecotechnologies should have pride of place among frontier technologies.

Industrial countries are responsible for many global environmental problems such as potential changes in temperature, precipitation, sea level, and incidence of ultraviolet â radiation. While further agricultural intensification in industrialized countries will be ecologically disastrous, the failure to achieve agricultural intensification and diversification in developing countries where farming provides most of the jobs will be socially disastrous. This is because agriculture, including crop and animal husbandry, forestry and agro-forestry, fisheries, and agro-industries, provides livelihood to over seventy percent of India's population. The smaller the farm, the greater is the need for higher marketable surplus to increase income. Eleven million new livelihoods will have to be created every year in India and these must come largely from the farm and rural industries sectors. Importing food and other agricultural commodities will thus have the same impact as importing unemployment. What we need now is an environmentally sustainable and socially equitable Green Revolution or what may be termed an Evergreen Revolution.

Those who advocate going back to the old methods of farming ignore the fact that only a century ago when the population of undivided India was 281 million, famines claimed thirty million lives between 1870 and 1900. The famine eradication strategy comprising the following steps is perhaps the most important achievement of post-independence India: enhanced production and productivity; better distribution through the public distribution system; adequate grain reserves; purchasing power enhancement through various employment generation and guarantee schemes; and special intervention programs for children, pregnant and nursing mothers, and old and infirm persons.

While famines have been prevented, widespread under-nutrition prevails among the economically underprivileged. Since nonfood factors such as health care, environmental hygiene, and literacy play an important role in promoting sustainable food security at the level of the individual, we should redefine security and the steps for achieving sustainable food security as follows. First, sustained physical access to food will involve a transition from chemical- and machinery-intensive to ecological farming technologies. Second, the emphasis on economic access underlines the need for promoting sustainable livelihoods through multiple income-earning opportunities. Third, environmental access involves on the one hand attention to soil health care, water harvesting management, and the conservation of forests and biodiversity, and on the other to sanitation, environmental hygiene, primary health care, and primary education.

If the political vision to implement this mission is forthcoming, population stabilization can be more readily achieved. The prediction of the French philosopher. Marquis de Condorcet, made in 1795, that population will stabilize itself if children are born for happiness and not just for existence will then come true.

The emphasis on the individual is important, since the household often is not a homogeneous unit. Women and girls tend to suffer more from under-nutrition than men and boys. The UNDP's 1995 "Human Development Report" (UNDP, 1995) contains distressing data on the growing feminization of poverty. To give operational content to such a concept of food security, we should initiate a Hunger-Free Area Program (HFAP) with the following objectives.

Ensure sustainable availability of food by maintaining the growth in food production over population growth through the development and dissemination of ecotechnologies, supported by appropriate packages of services and public policies. Ecotechnology involves blending the ecological prudence and technologies of the past with the best in frontier technologies, particularly biotechnology, information technology, space technology, renewable energy technology, and management technology. Without ecotechnological empowerment, farmers will not be able to produce more food and other agricultural commodities on an environmentally sustainable basis from fewer land, water, and energy resources.

Sustain the productivity of the natural resource base by conserving and improving the ecological foundations essential for continuous advances in crop and animal productivity.

Ensure adequate household incomes through promotional social security, such as accessing assets, employment, and organizational and marketing empowerment. Agricultural programs should aim concurrently at more food, more jobs, and more income. Integrated attention to farm and nonfarm employment and value-addition to primary agricultural commodities will be necessary to enhance income and rural livelihood security.

Provide entitlement to food for the vulnerable groups through protective social security measures such as employment guarantees and food for nutrition programs.

Introduce a National Food and Livelihood Security Act with the concurrence of the National Development Council for the purpose of paying integrated attention to important issues. These include conserving land, water, forests, and biodiversity, and protecting the atmosphere; enhancing productivity through ecotechnologies; improving distribution to eliminate endemic hunger; maintaining adequate food security reserves; strengthening the techno-infrastructure for better post-harvest technology; expanding the coverage of sanitary and phytosanitary measures; and developing efficient research, education, extension, and marketing systems to take full advantage of emerging opportunities in international trade and to ensure that research and extension designed to promote public good receive adequate support.

Potential Future Challenges: Climate Change

In 1995, the Intergovernmental Panel on Climate Change (IPCC) published a second assessment of climate change and its impact. The study confirmed the potential for climate change due to the injection of greenhouse gases such as carbon dioxide (CO_2), methane, nitrous oxide, and others. Further refinement of general circulation models (GCM) and coupling of them with ocean models and negative feedback from aerosols led IPCC to project an increase in global mean surface temperature of 2°C, which is lower than that projected in 1990. The low value of climate change projects a change of 1°C by 2100. However, changes at the regional level continue to remain difficult to project. A mean change of 2°C, which is the "best estimate", could result in changes of from less than 0°C to more than 4°C in different regions. It is these changes that are important for biota, particularly the biodiversity and agriculture production system. The projected changes for this part of the world (India) show a change of 1°C with increased cloudiness and approximately five percent change in rainfall.

Sinha and Swaminathan (1991) showed that in India a temperature change of 1.7°C in a large grid caused a change from -0.4 to 4.5°C from coastal to interior areas. Recent studies in the United States also show that an increase of 0.4°C has resulted in changes from 0 to 2°C or more.

These changes would have an effect on major crops, especially grain crops. An increase of 1 to 2°C combined with reduced radiation causes enhanced sterility of spikelets in rice. Fortunately, there is already evidence of genetic variability in rice for tolerance to both higher temperature and reduced radiation. Crops such as pigeon pea promote vegetative growth with enhanced humidity and reduced radiation. We do not know the response of most vegetable and fruit crops.

The general view has been that enhanced CO_2 concentration would benefit crops through an improved photosynthesis rate. At the doubling of CO_2 equilibrium, the actual concentration of CO_2 could be only around 450 parts per million (ppm) in the atmosphere. Therefore, we should be cautious in extrapolating data from a 600 ppm or higher concentration of CO_2. We need more studies on the interaction of CO_2, temperature, and radiation for projecting the impacts of climate change on individual crop and agricultural systems.

The projected changes in climate, particularly temperature, humidity, and radiation could have a profound effect on diseases, pests, and micro-organisms. Incorporation of disease and pest resistance at higher temperature levels should be an important objective in crop improvement programs. It is known that a change of 1°C changes the virulence of some races of rush infecting wheat. The whole lifecycle of insects can be affected by a small change in temperature.

The impact of change in precipitation and temperature on crops has received much less attention than it deserves. Records show that the rainfall pattern has changed during the last seventy-five to eighty years. In some areas there has been either an overall decrease in rainfall or the total number of rainy days has decreased while maintaining the same total rainfall. This causes increased intensity of rain and results in greater soil erosion and runoff. Thus new crops and varieties will have to adapt to a greater degree of soil and water stress. In this respect traditional agricultural systems such as mixed cropping may prove advantageous. Consequently, crop improvement for intercropping/mixed cropping should possibly receive a greater emphasis.

The Final Milestone: A Hunger-Free World

In his budget speech delivered on July 17, 1996, the Chief Minister of Tamil Nadu, India, made the following announcement: "Feed the people who are hungry, Educate the people to uplift the World." To fulfill this dream of Mahakavi Bharatiyar (a Tamil poet who fought for India's independence), this Government will launch a new "Hunger-Free Area Program" with an aim to eradicate poverty-induced hunger. A number of schemes are already under implementation to alleviate poverty and to cater to the nutrition requirements of different groups of the population. Gaps in this coverage will be identified which can then be specifically targeted under the Hunger-Free Area Program. Provision has been made in the Budget for preparing a detailed strategy to implement this program in association with Dr. M. S. Swaminathan. (Karunanidhi, 1996)

Studies at MSSRF have shown that by adding a horizontal dimension to numerous vertically structured programs and by promoting a coalition of all concerned with ending hunger and deprivation, it is now possible to provide opportunities for a healthy and productive life for all.

Designing a Hunger-Free Area Program

The problem of food and nutrition security at the level of the individual has to be viewed in three dimensions. First, inadequate purchasing power leads to calorie-protein under-nutrition. Second, the lack of the needed quantity and variety of micronutrients and vitamins in the diet leads to several nutritional disorders, including blindness caused by vitamin A deficiency. This kind of problem is referred to as hidden hunger, which affects more than two billion people in the world today. Third, lack of environmental hygiene and sanitation leads to a low biological absorption and retention of food due to intestinal infection and diarrhea. Thus, both food and nonfood factors assume importance in determining the nutrition security of an individual. Concurrent action at all levels is necessary in a Hunger-Free Area Program. The steps involved in this process are the following: a) Identify the basic underlying reasons for chronic under- and malnutrition; b) Collate information on available programs and opportunities for the sustainable end of hunger; c) Articulate the steps needed to provide the "missing elements" in achieving the end of hunger; d) Mobilize local level community action and commitment for achieving the end of endemic hunger by the end of the Ninth Five Year Plan and enlist mass media support for this purpose; e) Assess the extent of financial resources required, technical resources necessary for technological empowerment, and managerial and organizational resources needed; f) Foster the organization of a grassroots-level coalition for ending hunger in each HFAP region. Each coalition would include delegates elected by the people as well as representatives from government agencies, nongovernmental organizations, voluntary agencies, service organizations (Rotary clubs and Lions clubs), academia (research, training institutions, and universities), the corporate sector, financial institutions, and the mass media.

Global Action for Sustainable Food Security

The commitment of the international community to ensure food and nutrition security to every child, woman, and man on our planet has been reiterated at several international conferences held during this decade.

1990 World Summit for Children, New York
1992 International Conference on Nutrition, Rome
1992 UN Conference on Environment and Development, Rio de Janeiro
1993 World Conference on Human Rights, Vienna
1994 International Conference on Population and Development, Cairo
1995 UN Conference on Social Development, Copenhagen
1995 Fourth World Conference on Women, Beijing
1996 World Food Summit, Rome

Despite such high-level political reaffirmation of the need to achieve rapidly the goal of "Food for All", over eight hundred million human beings now suffer from hunger and malnutrition. Hidden hunger, arising from the deficiency of essential micronutrients, is widespread. Endemic hunger is now due more to the lack of adequate purchasing power at the household level than to the lack of availability of food in the market. Both poverty and nonfood factors such as environmental hygiene, sanitation, and lack of safe drinking water are becoming major contributors of food insecurity at the level of communities and individuals. A majority of those who are food insecure live in South and South-east Asia.

There is also increasing concern about the earth's capacity to produce adequate food for the growing population. The increase in human population, now confined mostly to the developing countries, coupled with enhanced purchasing power leading to greater capability to buy food as well as increasing urbanization will lead to a higher growth rate in demand for food as well as for more diversified food products in the coming millennium. Thus, on the one hand, there will be a need for intensification and diversification of agriculture, particularly in the developing countries. On the other hand, the ecological foundations essential for sustainable advances in crop and animal productivity are being eroded. This situation has led to the ringing of alarm bells by experts concerning the earth's capacity to produce adequate food to meet growing needs (Brown, 1995). Lester Brown has concluded that the growth in world food production has been hindered by constraints imposed by the earth's natural systems, the environmental degradation of land and water resources, and the diminishing backlog of yield-raising agricultural technologies. This raises questions about the earth's population-carrying capacity. The overriding environmental issue for the twenty-first century is the earth's capacity to produce enough food to satisfy expanding demand (Brown and Kane, 1994).

The year 1998 will mark the bicentenary of Malthus's essay on population. The onward march of science and technology, supported by appropriate public policies, has so far helped humankind to keep Malthusian forebodings on the population-food supply equation at bay. It is, however, becoming increasingly clear that without concerted national and global action to arrest environmental degradation, promote economic and gender equity, and check population growth, the Malthusian specter may still come true in the next century. The Food and Agriculture Organization of the United Nations (FAO), therefore, organized a World Food Summit in Rome from November 13 to 17, 1996. At this summit, a Rome Declaration on World Food Security and a World Food Summit Plan of Action were adopted. Following are highlights of these resolutions.

Rome Declaration on World Food Security. The Declaration makes a commitment to reduce the number of under-nourished people to one-half their present level no later than 2015. It also states: "We will pursue participatory and sustainable food, agriculture, fisheries, forestry and rural development policies and practices in high and low potential areas which are essential to adequate and reliable food supplies. We will strive to ensure that food, agricultural trade, and overall trade policies are conducive to fostering food security for all through a fair and market-oriented world trade system" (FAO, 1996: 3).

World Food Summit Plan of Action. The Plan of Action includes seven commitments: a) Create an enabling political, social, and economic environment designed to create the best conditions for the eradication of poverty, and for durable peace; b) Implement policies aimed at eradicating poverty and inequality and improving physical and economic access to sufficient, nutritionally adequate, safe food and its effective utilization; c) Pursue participatory and sustainable food, agriculture, fisheries, forestry, and rural development policies and practices in high- and low-potential areas; d) Ensure that food, agricultural trade, and overall trade policies are conducive to fostering food security for all; e) Prevent and be prepared for national disasters and man-made emergencies; f) Promote optimal allocation and use of public and private investments to foster human resources, sustainable food, agriculture, fisheries and forestry systems, and rural development, in high- and low-potential areas; g) Implement, monitor, and follow up the Rome Plan of Action at all levels in cooperation with the international community.

The above plan of action integrates three principles: conservation of the ecological foundations essential for sustainable advances in agricultural productivity and production; sustainable production through ecotechnologies; and equity in access to food through an appropriate blend of national and international action in the area of entitlements and employment. Food and livelihood security must be viewed in an inter-linked manner.

Conclusion

We are facing a battle against time in safeguarding our natural resources. In his book, *The Diversity of Life*, Edward O. Wilson has warned that Homo sapiens are in imminent danger of precipitating a biological disaster of a greater magnitude than anything we have witnessed so far in our evolutionary history (Wilson, 1992). There is no time to relax if we are to ensure that the Malthusian prophecy of famine and pestilence does not come true in the coming millennium. Legal, educational, and participatory measures of program implementation and benefit-sharing will all be needed for promoting a peoples' movement for conservation.

It is clear that the concept of sustainable development should be broad based to incorporate considerations of ecology, equity, employment, and energy, in addition to those of economics. This will call for a systems approach in project design and implementation. Both unsustainable lifestyles and unacceptable poverty have to be eliminated. Factors that influence climate and sea level have to be addressed with the seriousness they deserve and need. Sustainable development will become a reality if we keep in mind that the greatest responsibility of our generation, as suggested by Dr. Jonas Salk is to be good ancestors.

The first and foremost responsibility of governments, the corporate sector, and civil society is to give every child an opportunity for the full expression of her or his innate genetic potential for physical and mental development. This is an achievable goal provided the rich regard themselves as trustees of their financial and intellectual wealth and not just as owners. That there is adequate surplus wealth in the world is clear from twenty percent of the human population earning over eighty percent of the global annual income. According to the UNDP's 1996 "Human Development Report", the world's 358 billionaires are worth more than the combined income of the poorest 2.5 billion people (UNDP, 1996). If the concept of trusteeship of financial and intellectual wealth becomes the twenty-first century human ethic, we can lay the foundation for a socially, ecologically, and economically sustainable future for humankind. This is the task as well as the hope for the new millennium.

It is appropriate to conclude with a statement from a document by the US President's Council on Sustainable Development titled "Sustainable America": "Prosperity, fairness, and a healthy environment are interrelated elements of the human dream of a better future. Sustainable development is a way to pursue that dream through choice and policy." (The President's Council, 1996: 4)

References

Brown, Lester R. 1995. *Who Will Feed China? Wake-up Call for a Small Planet*. New York: W.W. Norton.

—— and Hal Kane. 1994. *Full House: Reassessing the Earth's Population Carrying Capacity*. New York: W. W. Norton.

Food and Agriculture Organization of the United Nations (FAO). 1996. Rome Declaration on World Food Security and World Food Summit Plan of Action. A report from the World Food Summit held in Rome, November 1996.

Graham-Smith, F. (Ed.) 1994. *Population – The Complex Reality: A Report of the Population Summit of the World's Scientific Academies*. London: The Royal Society.

Karunanidhi, M. 1996. Budget 1996-97. Government of Tamil Nadu, July 17, 1996.

The President's Council on Sustainable Development. 1996. *Sustainable America: A New Consensus for Prosperity, Opportunity, and a Healthy Environment for the Future*. Washington DC: The President's Council.

Repetto, Robert and the World Resources Institute. 1996. *Has Environmental Protection Really Reduced Productivity Growth? We Need Unbiased Measures*. New York: World Resources Institute.

Schmidheiny, Stephan. 1992. *Changing Course: A Global Business Perspective on Development and the Environment*. Cambridge, Massachusetts: MIT Press.

Sinha, S.K. and M.S. Swaminathan. 1991. Deforestation, climate change and sustainable nutrition security: A case study for India. *Climate Change* 19:201-09.

United Nations Development Program (UNDP). 1995. *Human Development Report*. Delhi: Oxford University Press.

-----. 1996. *Human Development Report*. Delhi: Oxford University Press.

Wilson, Edward 0. 1992. *The Diversity of Life*. New York: W.W. Norton.

Background References

International Commission on Peace and Food. 1994. *Uncommon Opportunities: An Agenda for Peace and Equitable Development*. London: Zed Books.

M.S. Swaminathan Research Foundation. 1996. *Uncommon Opportunities for Achieving Sustainable Food and Nutrition Security. Conference Proceedings of the Science Academies Summit*. Madras: M.S. Swaminathan Research Foundation.

Myers, Norman and Jennifer Kent. 1995. *Environmental Exodus – An Emergent Crisis in the Global Arena*. Washington DC: Climate Institute.

Swaminathan, M.S. (Ed.) 1993. *Reaching the Unreached: Information Technology – A Dialogue*. Bangalore: Macmillan India.

-----. (Ed.) 1994. *Reaching the Unreached: Ecotechnology and Rural Employment – A Dialogue*. Bangalore: Macmillan India.

-----. (Ed.) 1996. *Agrobiodiversity and Farmers' Rights*. Madras: M.S. Swaminathan Research Foundation.

29

Ethics and Equity in the Collection and Use of Plant Genetic Resources: Some Issues and Approaches

Sharing of benefits and the concomitant increase in the recognized value of the resources are the most effective pathways to foster conservation and the continued availability of plant genetic resources. In general, the issue of benefit-sharing has not received the detailed attention it deserves among those formulating legal measures for operationalizing the Convention on Biodiversity (CBD). The Conference of Parties also is yet to come to grips with this vital issue.The only concrete step taken so far is the inauguration by Ms. Elizabeth Dowdeswell, Executive Director, UNEP of a Technical Resource Centre for the Implementation of Equity Provisions of CBD at M.S. Swaminathan Research Foundation (MSSRF), Madras, India, in July 1996.

Starting with the establishment of an International Commission on Plant Genetic Resources by the FAO Council at its meeting held in November 1983, under my chairmanship, the concept that informal innovation by tribal and rural women and men in the field of genetic resources conservation and improvement needs recognition and reward has found universal acceptance. Women in particular have played a key role in seed and plant selection and preservation. The term "Farmers' Rights" was developed in the forum of the International Commission on Plant Genetic Resources of FAO to denote "the rights arising from the past, present, and future contributions of farmers in conserving, improving, and making available plant genetic resources, particularly those in the centers of origin/diversity."

At a dialogue organized by MSSRF in January 1990 at Madras, in collaboration with the Keystone Center in the United States, the participants stated, "we agree on the concept of Farmers' Rights and we agree that the contributions to a fund in recognition of these rights should not be voluntary." Practically speaking, a voluntary fund is a fund without resources. Thus, there should be a compulsory funding mechanism. This would insure that Farmers' Rights are recognized in a "real way". The concept of Farmers' Rights includes recognition of the fact that farm women and men have developed and continue to help develop and maintain genetic diversity. In many cases, farmers engage in conscious and creative practices as they select and breed their crops.

Overview Lecture presented at the Workshop on Ethics and Equity in the Management of Biodiversity held under the auspices of the Consultative Group on International Agricultural Research (CGIAR) at Foz do Iguacu, Brazil, April 22-25, 1997. First published in the *Plant Genetic Resources Newsletter* No. 112, pp. 7-18, 1997. Reprinted in *A Century of Hope: Harmony with Nature and Freedom from Hunger,* by M.S. Swaminathan, pp. 99-120. Chennai, India: EastWest Books (Madras), 1999.

Ethnobotanists around the world have chronicled the invaluable contributions of tribal and rural women and men in the conservation and enhancement of genetic diversity in plants (Swaminathan and Kochhar, 1989).

The Plant Variety Protection Acts introduced in developed countries generally conform to the provisions of the conventions agreed to from time to time under the forum of the International Union for the Protection of NewVarieties of Plants (UPOV). They are designed to promote invention and investment in plant breeding. However, they do not accord recognition to the conservation by farm families of genetic diversity, which is the basic feedstock for the plant breeding and biotechnology industries. By introducing provisions for Farmers' Rights, Plant Variety Protection Acts will concurrently foster innovation and conservation. They will thus help to strike a balance between homogeneity and heterogeneity in the genetic make-up of new cultivars.

At two dialogues on methods of recognizing and rewarding informal innovations in the area of genetic resources conservation organized by MSSRF in 1994 and 1996, a detailed draft legislation was developed for converting the "know-how" relating to Farmers' Rights into "do-how" (Swaminathan, 1995, 1996). From these drafts, the Government of India has developed a draft legislation for a Plant Variety Protection and Farmers' Rights Act for adoption by the Indian parliament.

Scenario after the Coming into Force of the Convention on Biological Diversity

The legally binding Convention on Biological Diversity (CBD) which came into force on December 29, 1993, and which has since been ratified by nearly 170 countries, recognizes "the close and traditional dependence of many indigenous and local communities, embodying traditional lifestyles, on biological resources, and the desirability of sharing equitably benefits arising from the use of traditional knowledge, innovations, and practices relevant to the conservation of biological diversity and the sustainable use of its components."

The CBD also recognizes the pivotal role of women in the conservation of biodiversity in the following words: "... Recognizing also the vital role that women play in the conservation and sustainable use of biological diversity and affirming the need for the full participation of women at all levels of policy making and implementation for biological diversity conservation." In spite of such an affirmation, the gender dimension of biodiversity management is yet to receive serious consideration.

In CBD, the fair and equitable sharing of benefits arising out of the utilization of genetic resources has been emphasized in Articles 8, 9, 10, 15, 16 and 20. The challenge now lies in developing transparent and implementable methodologies for giving effect to these provisions. The internalization of the gender dimension in the implementation mechanisms is equally important. I would like to indicate some methods of achieving these goals.

Benefit-sharing through Access to Genetic Resources

The post-CBD era, marked also by the coming into force of the World Trade Agreement, is witnessing the emergence of a new order of business, predicated on international negotiations and multilateral and bilateral arrangements. Access to genetic resources is cited in the Convention as a means of benefit-sharing because the use of genetic resources leads to the creation of new genetic traits and combinations that in turn will be useful to breeders using both Mendelian and molecular methods of crop improvement. As a result, countries will have to be selective about what they access, what arrangements they negotiate, and how the benefits are shared. In such negotiations, it would be important to differentiate between access for research and commercial purposes.

Regulating access to genetic resources does not mean prohibiting access. Article 15 (4) of the Convention requires that access to genetic resources, where granted, should be on "mutually agreed terms". Article 15 (5) of the Convention clearly states that such access shall be subject to prior informed consent (PIC). Developing legally-binding PIC procedures and standardization of material transfer agreement (MTA) procedures need priority attention in countries wishing to regulate access to their genetic resources and ensure equity in sharing benefit-sharing. The initiative taken by CGIAR institutions in this respect could serve as a model.

The Convention provides countries with the opportunity to assert sovereignty over their genetic resources, and to ensure that the benefits arising from their utilization are shared fairly and equitably. Countries must move fast, however, to take advantage of this opportunity by establishing national legislation on genetic resources access and, just as importantly, by developing procedures and institutional capacities to implement that legislation.

Appropriate Instruments

If the benefits of genetic resources utilization are to be shared fairly and equitably, Governments will need to design specific mechanisms to ensure that those benefits actually reach intended beneficiaries, be they local communities, national research institutions, or government conservation agencies. The nature of the intended beneficiaries will influence the appropriateness of different kinds of benefits. Financial reward and the organization of support services which can help to strengthen their on-farm conservation traditions might be an appropriate way of recognizing the role of communities. Training and technological empowerment will enable them to improve their economic condition, while sticking to their conservation ethics.

Intellectual Property Rights (IPR) and Benefit-sharing

In India, a program is underway to develop a system of community registers of local informal innovations related to genetic resources, and natural resources management generally. This is intended, inter alia, to revitalize traditional knowledge, skills, and techniques; protect local rights of tenure over resources by providing proof of customary use; and protect local knowledge from exploitation by commercial users, including protection against imposition of IPR by outsiders and by providing proof of prior use. Such community biodiversity chronicling and documentation will also help them to take decisions on PIC with reference to the use of indigenous knowledge.

Another proposed route for solving the lack of IPR protection for informal innovations related to Plant Genetic Resources (PGR) is enactment by governments of some form of community IPR law. Such a law would redefine IPRs to encompass the collective, incremental innovations, practices, and knowledge of local communities, and vest those communities with enforceable IPRs in those innovations (Mugabe et al, 1996).

One simple and quite feasible change that would bring IPRs more in line with the benefit-sharing provisions of the Convention would be for governments, primarily in developed countries, to amend their IPR systems to include a requirement that patent applications disclose information relating to the country of origin or sources of the genetic material used in the research leading to the innovation for which they are seeking IPR protection. The confidentiality of such information could be protected and their use restricted solely for dealing with issues related to benefit-sharing.

Systems of Biodiversity Management

Biodiversity management systems are now broadly grouped into the following three categories: in situ, in situ/on-farm, and ex situ. The systems in place and the persons involved in operating them in India are depicted in Figure 1. The precise methods of benefit-sharing may have to be tailored to suit the practices and procedures adopted in their conservation and utilization. The area which has received the maximum attention relates to the conservation and use of genetic material in plant and animal breeding and biotechnological enterprises. This category of diversity which is of economic value is now referred to as agrobiodiversity. It would be useful to discuss this area first, since much more work has been done in relation to the sharing of benefits with the conservers of agrobiodiversity. The Indian legislation relating to Plant Variety Protection includes a provision for the recognition and reward of the contributions of tribal and rural women and men to the conservation and enhancement of agro-biodiversity. Enhancement takes place through selection for tolerance/resistance to biotic and abiotic stresses as well as through information on the useful properties contained in the genotypes which have been conserved by local communities.

Most decisions affecting the relationship between agriculture and biodiversity are made by individual farmers, not by national planners. The incentive structure under which farmers and other local resource managers make decisions about conversion of natural habitats for alternative uses is influenced by macro-economic policies relating to farm and non-farm enterprises and development programs. The resulting agricultural practices will affect both on-site and off-site biodiversity. Changes in the society through alterations in the level of services provided by biodiversity influence loss or gain in extractable products and valuable ecosystem services. An important subset change in their level will affect agricultural production. Changes in agricultural production and services influenced by biodiversity may also lead to changes in societal benefits, but these have been largely ignored. We have yet to study carefully the socio-economic and food security implications of the general eclipse of in situ on-farm conservation practices.

Table 1. Extent of genetic uniformity in selected crops

Crop	Country	Number of varieties
Rice	Sri Lanka	From 2,000 varieties in 1959 to less that 100 today; 75% descend from an common stock
	Bangladesh	62% of varieties descend from a common stock
	Indonesia	74% of varieties descend from a common
Wheat	USA	50% of crop in 9 varieties
Potato	USA	75% of crop in 4 varieties
Soybean	USA	50% of crop in 6 varieties

Source: WCMC 1992.

Table 2. Past crop failures due to genetic uniformity

Date	Location	Crop	Effects
1846	Ireland	Potato	Potato famine
1800s	Sri Lanka	Coffee	Farms destroyed
1940s	USA	US Crops	Crop loss to insects doubled
1943	India	Rice	Great Bengal famine
1960s	USA	Wheat	Rust epidemic
1970	USA	Maize	US $1 billion loss
1970	Philippines, Indonesia	Rice	Tungo virus epidemic
1974	Indonesia	Rice	3 million tons destroyed
1984	USA(Florida)	Citrus	18 million trees destroyed

Source: WCMC 1992.

Table 3. Donors for various traits being used for developing the new plant type in rice (Tropical Japonicas)

Trait	Donors	Country of Origin
Short stature	MD2, Sheng-Nung 89-366	Madagascar, China
Low tillering	Merim, Goak, Gendjah, Gempol, Gendjah Wangkal	Indonesia
Large panicles	Daringan, Djawa Serang, Ketan Gubat	Indonesia
Thick stems	Sengkeu, Sipapak, Sirah Bareh	Indonesia
Grain Quality	Jhum Paddy, WRC4, Azucena, Turpan4	India, Philippines, Thailand
Donor for resistance		
Bacterial blight	Ketan Lumbu, Laos Gedjah, Tulak Bala	Indonesia
Blast	Moroberekan, Pring, Ketan Aram, Mauni	Indonesia
Tungro	Gundil Kuning, Djawa Serut, Jimburg, Lembang	Indonesia
Green leafhopper	Pulut Cenrana, Pulut Senteus, Tua Dikin	Indonesia

Source: WCMC 1992.

The need for the revitalization of the in situ on-farm conservation traditions of rural and tribal families is very clear from the growing extent of genetic uniformity in many crops. Such homogeneity enhances genetic vulnerability to pests and diseases (Table 2).

In the coming millennium, most developing countries have no option except to produce more food and agricultural commodities from diminishing per capita arable land and irrigation water resources. In other words, they will have to achieve higher production through a vertical growth in productivity than through a horizontal expansion of area. The further enhancement of the yield potential of major food crops can be achieved only through new and novel genetic combinations employing, where necessary, recombinant-DNA technologies. The position of rice is an example. The gradual improvement in yield potential in rice has been achieved through techniques like crosses between indica and japonica, exploitation of hybrid vigor using a male sterile strain originally identified in the Hainan Island of China, and, more recently, through distant hybridization among species of Oryza. To achieve a new plant type yielding over ten tons of rice, there is need for germplasm from a wide range of countries and agro-ecological regions (Table 3).

Today, the poverty of the rural conservers stands in contrast to the prosperity of the breeders and genetic engineers working in commercial enterprises. Thus, there is urgent need for developing an implementable and transparent methodology for sharing benefits with the communities which have long practiced conservation ethics. It would be useful to consider this issue at the following three levels: in situ on-farm conservation of agrobiodiversity; in situ conservation of habitats and ecosystems; and, ex situ conservation. The methods which can be adopted in these cases are discussed below.

1) *In situ on-farm conservation of agrobiodiversity*:

a) Implementing the concept of farmers' rights: Reward and recognition have to be at two levels. First, the contributions of an individual farm woman or man for the development of a new strain can be recognized in the same manner as in the case of a breeder working in a scientific organization or a company. This is particularly true of farmers who identify spontaneous mutants or make selections in field crops, tree species, and in fruit, vegetable, and ornamental crops. They will, however, have to be helped to undertake the necessary paperwork that will be needed to file applications for plant variety protection.

The more difficult task is the identification of communities whose collective work in the area of genetic resources conservation and enhancement deserves recognition and reward. The Madras draft Act for Plant Variety Protection and Farmers' Rights (Swaminathan, 1996) calls for the submission of pedigrees of new varieties indicating the source from where the critical genes responsible for the desirable character came from. For example, the pedigree of IR 66, a high-yielding rice variety bred at the International Rice Research Institute, provides information on the wide range of landraces which have gone into the making of the variety.

Pending the development of an internationally agreed method, different institutions have been following their own procedures. The procedure adopted in the Indian draft Act for identifying the communities and providing them with group recognition can be the starting point of integrating equity considerations in benefit-sharing. Under this draft Act, a Community Gene Fund will be created by imposing a one percent cess (levy) on all agricultural produce. This amount will be used for rewarding communities that have contributed valuable germplasm. The money available for the community will be utilized for the purpose of strengthening the in situ on-farm conservation traditions of these communities. In addition, they can use the funds for training in seed technology and for developing an infrastructure for seed drying and storage and other similar requirements. For example, a farmer who maintains a genetically diverse crop that provides a lower return than a genetically uniform high-yielding variety, but who serves a valuable function as the conserver of genetic diversity, is meeting an opportunity cost. How should such costs be compensated? It is clear that the process is not an easy one. Refinement in benefit-sharing methodologies can come about only through experience. This is why no further time should be lost in initiating the process of recognition of farmers' rights. In such work, it is equally important that a gender perspective is internalized in the procedures. Otherwise, women's specific knowledge, priorities, and efforts will tend to get ignored. Unfortunately, most of the existing literature on Farmers' Rights is not gender sensitive. A recent workshop held at MSSRF on "Gender Dimensions in Biodiversity Management in India" has clearly brought out the need for removing this deficiency.

Farmers do not exist in vacuum, but live and work in a setting where social rules and norms of their communities have an influence on their decisions. They are also sensitive to options which can enhance their marketable surplus and thereby their income. Following are some forms and examples where conservation and benefit-sharing can be made mutually reinforcing. Insecure land tenure can affect, in several ways, the initiatives of farmers to conserve biodiversity in several ways.

b) Farmer incentives: i) Insecure land tenure can affect farmer incentives to conserve biodiversity in several ways. Insecure tenure reduces farmer incentives to consider long-term productivity effects on their land, including any long-term effects from damage to biodiversity, and reduces farmer incentives to minimize on-site damage. Insecure tenure reduces incentives to intensify agricultural production. Intensification often requires investment in land and infrastructure improvement that farmers may be unwilling to make if their tenure is insecure. Agricultural growth may therefore be extensive rather than intensive, which is much more harmful to biodiversity. Some relatively biodiversity-friendly land uses, such as agroforestry, tend to be discouraged by insecure tenure since they require fairly long periods to provide a return on investment. In some cases, tenure security can be established or increased by clearing land. This has notably been the case in the Brazilian Amazon, where it has resulted in substantial conversion of natural ecosystems. ii) Tenure and usufruct rights in communal areas. Communal areas (including community forests, pastures, and wastelands) are generally regarded as open-access resources that anyone can exploit (an image popularized by Hardin's *Tragedy of the Commons* (1968). Because habitats in these areas are often less modified than areas used more intensively for agriculture, they often tend to have higher levels of biodiversity. If resources in communal areas are misused, damage to biodiversity could be substantial. Hence, local communities need to be empowered in the sustainable management of common property resources.

c) Recognition of farmers' seed systems: The formal breeding system often ignores the critical role farm families play in breeding and seed selection. National Seed Laws should recognize and support the farmer-centered seed system, since this contributes considerably to the maintenance of intraspecific genetic diversity. Formal gene banks should support farmers' efforts in breeding and selection. Community gene banks also need to be established. Participatory plant breeding procedures involving the formal and informai systems need to become internalized in national agricultural research systems.

2) *In situ conservation*: As shown in Figure 1, in situ conservation of habitats and ecosystems is being carried out under the auspices of national governments through biosphere reserves, national parks, and protected areas. Some of the unique natural biodiversity areas are also being designated World Heritage Sites under UNESCO's Convention on Cultural and Natural World Heritage Sites. The primary responsibility for the conservation of protected areas vests with governments. Of late, joint forest management procedures involving forest-dependent communities and the concerned government departments are evolving.

In the case of in situ conservation, the issues needing attention are: fostering an economic stake in conservation among communities depending on forests for their livelihoods, by linking conservation and commercialization in a symbiotic and mutually-reinforcing manner; recognizing and rewarding the contributions of tribal and forest dependent communities in the field of conservation as well as in the generation of knowledge and information (commonly referred to as the IPR Rights of indigenous people).

In most of its operational clauses, the CBD leaves the final decision to bilateral negotiations. The Convention has stipulated that access to genetic resources shall be subject to the prior informed consent of the germplasm conservers and providers. As pointed out by Lesser (1997), real equity can be achieved only with regard to value-based systems, but it is extremely difficult to gain international agreement over their provisions. "Charity begins at home" and therefore it will be better if biodiversity-rich developing countries start standardizing and introducing in their respective countries methods of recognizing and rewarding informal innovations in genetic resources conservation. In the draft biodiversity legislation being prepared in India, a National Biodiversity Fund has been proposed for operationalizing the concept of equity in benefit-sharing. One source of the National Biodiversity Fund will be bioprospecting. Vandana Shiva et al (1997) have termed current patterns of

Figure 1. Biodiversity management systems in place and the persons involved in operating them in India

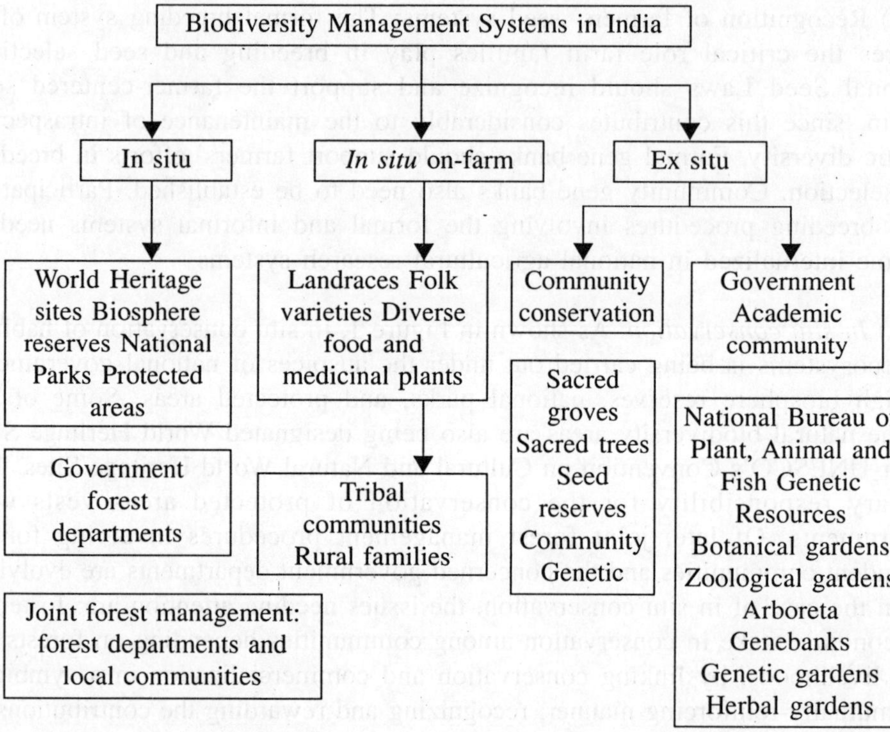

bioprospecting as "sophisticated biopiracy", since they often lead to the patenting of indigenous knowledge and biodiversity. According to them, "biodiversity-based traditional knowledge systems of the forest dwellers, farmers, and healers are fast becoming the private property of the TNCs."

Bioprospecting can help to increase the income of forest-dependent communities, if carried out in an ethical manner. It is also important that technical resource centers (TRC), which can help tribal and rural women and men to derive economic benefits from their contributions to biodiversity conservation, are established. As I mentioned earlier, the first such TRC established by MSSRF was dedicated to indigenous and rural communities by Ms. Elizabeth Dowdeswell, Executive Director, UNEP, in July 1996.

It was 125 years ago that the first national park was established by the US Congress at Yellowstone. The establishment of national parks was inspired by the conviction that the integrity of such landscapes rich in wildlife and other forms of biodiversity should be protected forever. Today, the UN's list of national parks and protected areas has as many as 10,000 sites larger than 1,000 hectares. Unfortunately, many national parks have become "hot spots" of biodiversity (MSSRF, and Rajiv Gandhi Foundation, 1995). It is clear that such critical ecosystems can be saved and conserved only with community participation and understanding.

The multifaceted strategy for conserving critical ecosystems through the establishment of biosphere reserves, developed at the Seville Conference organized by UNESCO in 1995, shows the way to a more sustainable future for protected areas. The Seville vision emphasizes, "rather than forming islands in a world increasingly affected by severe human impacts, biosphere reserves can become theatres for reconciling people and nature; they can bring knowledge of the past to the needs of the future." They can thus contribute to the needs of the society as a whole by showing the way to a more sustainable future.

Benefit-sharing in the field of in situ conservation could take the following forms. A joint forest management procedure similar to the joint forest management methods now in operation in several countries. Under this system, the local families could derive benefits both wood and non-wood forest products based upon sustainable utilization principles; equitable sharing of the economic benefits from bioprospecting; cultivation on contract of medicinal and other economically useful plants with commercial companies on the basis of remunerative buy-back arrangements; provision of additional public support in terms of livelihood opportunities to the families around protected areas in order to give explicit recognition to the role of conservation.

3) *Ex situ conservation*: The CGIAR collections are also value-added since they have been classified for their useful genetic traits. This collection of nearly 600,000 accessions has now been placed under the overall umbrella of FAO. The CGIAR has also developed a detailed material transfer agreement format for the purpose of sharing the genetic material with users. The CGIAR is also committed to providing a complete set of the germplasm collected from different countries to the designated agency in that country. By and large, benefit-sharing in the field of ex situ conservation has to be based on the principles earlier enunciated for agrobiodiversity.

The mission of IARCs (International Agricultural Research Centers) is to work on behalf of small-scale farmers to increase world food security. The IARCs have benefited by full and free exchange of plant genetic resources worldwide. Until 1993 (pre-CBD), IARC's seed distribution policies were guided by an international approach to conservation and exchange of genetic resources. The situation in the post-CBD era has changed and thus came into existence the need to sign a Material Transfer Agreement with every transfer.

With the GATT-TRIPS regime and changing Intellectual Property (IP) initiatives by several countries, the role of IARCs will change, especially in activities relating to access to genetic resources and benefit-sharing. To overcome some of these critical issues, CGIAR established a Genetic Resources Policy Committee to monitor the changing scenario and recommend appropriate steps in accordance with changing international scenarios and national legislation.

The Next Steps

It is now nearly four years since CBD came into force. Yet, there has been no serious dialogue or effort to give effect to the equity provisions of CBD. In contrast, items of interest to the biotechnology industry, like biosafety, have received abundant attention. This is not surprising since the farmers' rights concept of FAO also remains mainly on paper. What then should be the next steps in efforts to promote benefit-sharing with reference to the use of Plant Genetic Resources? What can be the contribution of CGIAR in this field?

Most nations, particularly developing ones, are currently in the process of enacting two kinds of legislation – one relating to CBD and the other to TRIPS of the World Trade Agreement. This is hence an appropriate time for developing and implementing provisions in national legislation which can help countries to implement the equity provisions of CBD. An important step in this direction is to give explicit recognition to in situ on-farm conservation by tribal and rural families as being equally important for food and nutrition security as in situ and ex situ conservation. Further, it is important to recognize in national legislation the gender roles in the conservation of plant genetic resources. Women play a particularly important role in the conservation and improvement of crops of importance in strengthening household food and nutrition security. Methods of benefit-sharing can be dealt with at the following three levels: local communities; national level; global level. The following are some suggestions about the action that would be appropriate at these three levels.

Local level. Tribal and farm women and men are conserving agro-biodiversity, often at considerable economic sacrifice. The formal breeding system derives immense benefit from landraces and folk varieties conserved in farm households. Biodiversity also constitutes the feedstock for the biotechnology industry. Since farmers are conserving biodiversity as a "public good" resource, their work must be supported and rewarded from public funds. Both the Biodiversity legislation and the Plant Variety Protection Act of countries should make provision for recognizing and rewarding in situ on-farm conservation from National Biodiversity and Community Gene Funds (Swaminathan, 1996).

Further, in the conventional in situ conservation procedures such as biosphere reserves, national parks, and protected areas, provision should be made for joint forest management involving the forest departments and local communities. This will enable tribal and farm families to have access to non-wood forest products and to develop a stake in the conservation of protected areas.

A third step which can be taken in fostering linkages between commercial companies and tribal and rural families, relates to arranging the sourcing of the material in which the commercial firms are interested. This can be done in the case of medicinal plants and nutritious grains. Such economic empowerment based on buy-back arrangements will lead to a symbiotic social contract between the conservers and utilizers. The absence of such symbiotic partnerships is leading to the collection of material from the wild and to many economic species facing extinction.

A recent example of this situation is the report by IUCN of the threat to the tree species *Aniba duckei*, occurring in the rainforests of north Brazil, which contains an essential oil of great demand in the perfume industry. Over 3,000 trees of *Aniba duckei* are felled each year to meet the demand for about 30 tons of essential oil. Instead of cutting the trees from natural forests – thereby bringing them under the Red Data Book Category, it will be appropriate if the perfume industry promotes tree farming by local communities on the basis of an agreed buy-back arrangement. Many examples of this kind can be cited in each country to emphasize the scope for harmonizing the goals of conservation and commercialization.

Another method of assisting local communities is the inclusion of farmers' Seed Systems within the purview of the pre-breeding work of NARS. Suitable participatory breeding procedures should be developed for this purpose.

National level. At present, the plant variety protection legislations of countries adhering to the UPOV convention are primarily designed to protect the interests of commercial breeding and seed industries. The interests of the farmers' conservation and seed systems urgently need similar protection. In sui generis systems of plant variety protection, as required under the TRIPS component of the World Trade Agreement, provision should be made to recognize and reward the contributions of farmer conservers and breeders. An examination of the pedigrees of most modern cultivars will reveal that the critical genes responsible for resistance/tolerance to biotic and abiotic stresses came from landraces/folk varieties selected and conserved by tribal and farm households. Since Farmers' Rights will often be community rights and not of particular individuals, there is an urgent need for UNEP to convene a meeting of legal experts to develop an internationally-agreed procedure for recognizing and rewarding community contributions to genetic resources conservation and enhancement. While patent and plant variety protection acts reward individuals, there has been no similar effort to accord recognition to the collective contributions of tribal and farm communities.

National biodiversity legislation should make provision for legalizing the Prior Informed Consent clause of CBD. Tribal and rural families should be empowered to prepare local biodiversity inventories and to take decisions on PIC. The step taken by MSSRF to organize a local level Agrobiodiversity Conservation Corps could be used as a model. National Biodiversity and Community Gene Funds should be used to strengthen, revitalize and reward the in situ on-farm conservation traditions of tribal and rural communities.

International level. The only substantive discussion at the international level on issues relating to gender and social equity and ethics has been in the forum of the International Commission on Genetic Resources of FAO. Even here, no concrete ideas have emerged on giving practical meaning and content to the concept of Farmers' Rights.

In a recent publication titled *Business and Biodiversity* published by the World Business Council for Sustainable Development and IUCN (Stone et al, 1997), the need for sharing benefits has been reiterated. However, the only concrete recommendation is the following: "Company managements should ensure that local communities are closely involved with decisions affecting the use of natural resources. Of particular importance in some areas is the indigenous peoples' knowledge and the specific issues related to their use of, and relationship to, the land and biological resources. This relationship needs to be harnessed in a positive way, both for the company and the peoples, by developing agreements that promote the equitable sharing of resources."

Thus, the pathway suggested is bilateral negotiations between companies and local communities. The ethical and equity principles underlying some benefit-sharing models adopted to date have aroused much debate and divergence of viewpoints (Bunders et al, 1996).

We also have information on both megabiodiversity areas and "hot spot" locations with reference to threats to biodiversity (Heywood, 1995). Hence, extending support to the communities conserving and improving agro-biodiversity is not difficult, if there is a will to do so. UNEP, FAO, UNESCO, and IPGRI of CGIAR could jointly develop guidelines for according support from a global biodiversity fund.

CGIAR's Role

The extensive breeding work done in the CGIAR has given broad indications of the sources of valuable genetic material. CGIAR Institutes also have over 600,000 accessories preserved in Gene banks. CGIAR's ex situ collection is a value-added one, since the collections have been studied for their useful traits.

The CGIAR is in a unique position to take a proactive role in matters relating to the ethical and equity aspects of plant genetic resources. CGIAR has excellent relationships with the national agricultural research systems of developing countries and, now, also has a NGO committee to help in linking science and society. I hope this workshop will help IARCs to become role models in helping to end the present unethical situation where the primary conservers suffer from poverty, while those who utilize their knowledge and the products of their in situ on-farm conservation traditions are prosperous. ▯

References

Bunders, J., B. Haverkort and W. Hiemstra. (Ed.) 1996. *Biotechnology: Building on Farmers' Knowledge*. London: Macmillan Education Ltd.

Heywood, V.H. (Ed.) 1995. *Global Biodiversity Assessment. United Nations Environment Programme*. Cambridge: Cambridge University Press.

Lesser, W. 1997. The Role of Intellectual Property Rights in Biotechnology Transfer Under the Convention on Biological Diversity. *ISAAA Briefs No-3*. Ithaca, New York: ISAAA.

Mugabe, J., C.V. Barber, G. Henne, L. Glowka and A.L. Vina. 1996. *Managing Access to Genetic Resources - Towards Strategies/or Benefit*. Nairobi, Kenya.

Rajiv Gandhi Institute for Contemporary Studies and M.S. Swaminathan Research Foundation. 1995. *Protecting Endangered National Parks. RGICS Project No. 7*. New Delhi/Madras: Rajiv Gandhi Institute for Contemporary Studies and M.S. Swaminathan Research Foundation.

Shiva. Vandana, A.H. Jafri, G. Bedi and R. Holla-Bhar. 1997. *The Enclosure and Recovery of the Commons: Biodiversity, Indigenous Knowledge and Intellectual Property Rights*. New Delhi: Research Foundation for Science, Technology and Ecology.

Stone, D., K. Ringwood and F.Vorhies. 1997. *Business and Biodiversity - A Guide for the Private Sector*. Switzerland: World Business Council for Sustainable Development (WBCSD) and The World Conservation Union (IUCN)

30

Science in Response to Basic Human Needs

The 20th century will soon end with an impressive array of accomplishments in nearly every field of science and technology. The last part of this century has been particularly rich in innovations in the fields of biotechnology and information and space technologies. Genomics and molecular genetics have conferred on human beings such levels of control over the very blueprints of life, that member nations of UNESCO have adopted a Universal Declaration on the Human Genome and Human Rights. The whole global population was only nine hundred and forty million in 1798 when Thomas Malthus expressed his apprehensions about human capacity to achieve a balance between food production and population. Human numbers will reach the six billion mark in 1999. Yet today the food security challenge is not one of availability of food in the market but one of economic access to it. Science-based technologies supported by appropriate public policies are responsible for food famines becoming rare. The famine of food at the household level today is mostly due to inadequate purchasing power arising from the famine of jobs or livelihood opportunities.

Progress in preventive and curative medicine leading to the control of many communicable diseases has been an important factor in determining demographic trends. From 1804, when the world population reached a billion, it took one hundred twenty-three years to reach two billion in 1927, thirty-three years to attain three billion in 1960, fourteen years to reach four billion in 1974, thirteen years to attain five billion in 1987 and twelve years to reach six billion in 1999. The medium fertility projection, which is usually considered as "most likely" by the Population Division of the United Nations indicates that world population will reach 8.9 billion in 2050. Most of the additional population will be in developing countries. India and China together now account for thirty-eight percent of the world's population.

Basic Human Needs: A 20th Century Balance Sheet

Poverty and social and gender inequity are increasing globally and nationally. According to the World Bank, 1.2 billion people lived on less than $1 per day in 1987. This number increased to 1.3 billion in 1993. About three billion people worldwide lived on less than $2 per day in 1993. Nearly 1.5 billion of the world population of six billion will live in severe poverty at the dawn of the new millennium. The gap between the rich and the poor is increasing day by day, with disastrous consequences to achieving the UN goal of reducing poverty by half by

---

Keynote Address presented at the World Conference on Science, Budapest, Hungary, June 26, 1999. First published in *Current Science* 77(3): 341-53 (August 10, 1999).

2015. Achieving FAO's World Food Summit goal of reducing the number of children, women and men going to bed hungry by half from the present number of over eight hundred million by 2015 may also not be achieved, since hunger today is essentially poverty induced. In addition to energy-protein under- and malnutrition, over two billion suffer from micronutrient deficiencies.

Global per capita water supplies are declining and are now thirty percent lower than they were twenty-five years ago. By 2050, as much as forty-two percent of the world's population will live in countries with insufficient freshwater stocks to meet the combined needs of agriculture, industry and domestic use. In addition, water is needed for the maintenance of ecosystems. Water conflicts are likely to grow. Hence, a Committee chaired by me has recommended the establishment of an International Center for Cooperation in Water Management at Valencia in Spain, in order to initiate proactive action in resolving potential water conflicts.

Contaminated drinking water causes diseases that account for ten percent of the total burden of disease in developing countries. Also in 1996, about 1.4 billion low-income and over 400 million middle-income people lacked access to sanitary facilities. At the present rate of progress, over nine hundred million persons will lack adequate sanitation in the year 2015. In one out of four major cities surveyed by UN Habitat, fewer than ten percent of households had sewerage connections.

Gro Harlem Brundtland, Director General of the World Health Organization, stressed in her address delivered at the 52nd World Health Assembly on May 18, 1999, that while the health gains of the 20th century count as one of the biggest social transformations of our times, more than a billion fellow human beings have been left behind in the health revolution. Whereas ninety percent of the disease burden is in the developing countries, these countries have only got access to ten percent of the resources going to health. The following initiatives of WHO, if accorded the needed political, public and financial support will make a large difference to the health of the poor: Roll back malaria; Tobacco free initiative; Fight against HIV/AIDS and tuberculosis. The International Conference on Health Research for Development planned for the year 2000 should accord priority to developing an implementable strategy for reaching the unreached in the health sector.

Disparities are also growing in access to technology. For example, 97% of all Internet hosts are in developed nations, home to 16% of the world's population. The rapid expansion of proprietary science covered by intellectual property rights is leading to the emergence of "technological apartheid" and to "orphans remaining orphans" in relation to the choice of research areas for priority attention. Support from public funds for research aimed at public good is tending to decline.

The coming century is being referred to as a "knowledge century", a century of innovation, enterprise, eco-entrepreneurship and genetic enhancement. Yet, UNESCO's goal of literacy for all is still a far cry in many developing nations. Enrollments by female students are catching up with those of boys in some regions but continue to lag in others. Intensive efforts have, however, significantly increased primary school enrollments during the nineties. In India, an innovative Education Guarantee Scheme was first introduced a few years ago in Madhya Pradesh. This approach is now being extended to the entire country in order to accelerate progress in achieving the goal of total literacy. However, in many countries of the Middle East, North Africa, South Asia and sub-Saharan Africa, girls' enrollments continue to lag behind boys' and illiteracy rates are still high.

It is not only in opportunities for education that children of many developing countries remain handicapped, but even more alarmingly, in opportunities for the full expression of their innate genetic potential for physical and mental development. For example, twenty-five to fifty percent of children born in several developing countries are characterized by low birth weight (LBW), caused by maternal and fetal under- and malnutrition. The UN Commission on Nutrition in its recent report (1999) has warned about the serious consequences of LBW for both brain developments in the child as well as to the health of the child in his/her later life (Figure 1).

The rich-poor economic and technological divide is not only causing inequity at the level of the present generation, but is also enhancing inter-generational inequity. For example, Panayotou et al (1999) point out that the affluent economies of the temperate zone are likely to impose severe net costs on the tropical regions because of their excessive consumption of fossil fuels leading to a large release of greenhouse gases. Since the temperate-zone economies are rich and the tropical zone economies tend to be poor, global climate change represents a burden imposed on the poorer countries by the richer nations.

Panayotou et al (1999) also point out that much of the damage caused by a continuous rise in CO_2 concentrations in the atmosphere will occur in tropical countries, especially Africa and India. India and countries in Africa are predicted to suffer adverse agricultural consequences, possibly severe vector-borne diseases, increased risks of severe tropical storms, vulnerability to rising ocean levels, and other stresses from an increase in temperature from already high levels. The temperate zone countries are either little affected in general or even benefited on an average by the prospects of global climate change. Agricultural productivity is predicted to rise in the high latitudes, through a combination of longer growing seasons plus CO_2 fertilization of crops.

Thus, the affluence as well as poverty-induced environmental damage leads to multiple forms of inequity, namely, inequity at birth in mental capacity, inequity in opportunities for a productive and healthy adult life, and inter-generational inequity.

Among the more serious dangers to sustainable human development is the increasing damage to our basic life support systems comprising land, water, flora, fauna, forests, the oceans and the atmosphere. The number of environmental refugees is increasing day by day due to damage to environmental capital stocks resulting in the loss of rural livelihoods (Myers and Kent, 1995).

The decade of the nineties has highlighted our dismal record in the area of building sustainable food, health and livelihood security systems. Noteworthy among the recent UN Conferences in relation to basic human needs are: Child Summit, New York, 1990; Earth Summit, Rio de Janeiro, 1992; Human Rights, Vienna, 1993; Population and Development, Cairo, 1994; Poverty and Social

Development, Copenhagen, 1995; Gender Justice and Equity (the World Conference on Women), Beijing, 1995; Food Summit, Rome, 1996; Urban Quality of Life, Istanbul, 1997.

These Conferences have led to global plans of action, which if implemented, could lead to a better life for all. The Conventions on Climate, Bio-diversity and Desertification, the Law of the Sea and Agenda-21 of UNCED provide a framework for the sustainable and equitable management of nature and natural resources.

In my view, the World Food Programme (WFP), FAO, UNICEF, WHO and UNDP should immediately launch in collaboration with interested national governments a global program to fight maternal and fetal under-nutrition, thereby helping to minimize the frequency of children with low birth weight. The program may be titled "Global Movement for Children for Happiness". A similar program involving horticultural and direct intervention approaches needs to be launched to eliminate hidden hunger caused by the deficiency of micronutrients in the diet. Without these two steps, the foundation for achieving the goals of the UN Conferences listed above cannot be laid. Millions of children will continue to be born for mere existence and not for happiness if this area of nutrition continues to receive inadequate priority. This is why the proposed WFP-FAO-UNICEF-WHO-UNDP initiative for fighting the incidence of LBW children is urgently called for.

A review of the recommendations of these international conferences reveals that apart from political, social and gender rights, the goal of poverty eradication should receive the highest priority from both national governments and bilateral and international organizations. Poverty is the root cause of hunger, lack of shelter and access to clean drinking water, illiteracy, ill health and other forms of human deprivation. The pathway to meeting the basic needs of every human being is poverty eradication.

Science, Technology and Economic and Social Inequity

The contributions of India, China and Greece to scientific discovery and knowledge since pre-historic times are now widely acknowledged. In contrast, modern science in the European tradition is only about five hundred years old, dating from the time of Copernicus. Maddox (1998) has urged that a clear distinction should be drawn between modern science and its precursors. The interplay between observation and explanation was formerly less important than it is now. A theory qualifies as an explanation only if it can be and has been tested by experiment or observation, employing when necessary measurements more sensitive than the human senses can yield. Similarly, a distinction should be drawn between science and technology. Science helps to advance the frontiers of knowledge, while science-based technologies help to advance the frontiers of economic wealth. Maddox has also emphasized that while we are right to marvel at what has been accomplished in the past one hundred years, we should not forget that there would have been the same sense of achievement at the end of each of the preceding centuries. Thus, we should celebrate many different visions and technologies, both historical and contemporary.

Virginia Postrel (1998) in her book *The Future and its Enemies* points out that knowledge, like experience, is in many ways cumulative. Dynamism and innovation are about creating the future. However, we should not ignore the past from which the future evolves. It is this concept which led Federico Mayor of UNESCO and the late Commandant Jacques Costeau to propose the concept of ecotechnology. Ecotechnology is the blend of the ecological prudence of the past and modern science. It involves an inter-disciplinary approach and a proactive and continuous interaction among social scientists and technologists working in the frontier areas like space and information technologies, biotechnology, nuclear science and renewable energy technologies (see Von Weizsacker et al, 1983 for approaches to technology blending). Eco-technologies are more knowledge than capital intensive and provide opportunities for decentralized production supported by a few key centralized services.

Virginia Postrel also points out that "loss of rural employment and migration from the countryside to the cities causes a fundamental and irreversible shift. It has contributed throughout the world to the destabilization of rural society and to the growth of vast urban concentrations. In the urban slums congregate uprooted individuals whose families have been splintered and whose cultural traditions have been extinguished." The proliferation of urban slums as a result of the loss of rural livelihoods presents a grave threat to achieving the goal of ensuring every human being access to basic human needs like safe water, sanitation, balanced diets, health care and education.

To develop an effective strategy for poverty eradication, it is important to understand how poverty arose first and why it is affecting developing countries more. Jared Diamond (1997) points out that until the end of the last Ice Age, around 11,000 BC, all peoples on all continents were hunter-gatherers. Different rates of development of different continents occurred between 11,000 BC and AD 1,500. While aboriginal Australians and many native Americans remained hunter-gatherers, most of Eurasia and much of the Americas and sub-Saharan Africa gradually developed agriculture, herding, metallurgy and complex political organization.

One of the most interesting questions in comparative economic history is the gap in levels of real income between developed and developing countries. David Landes (1990) pointed out that "Western Europe was already rich before the industrial revolution because of substantial technological progress, not only in the production of material goods, but in the organization and financing of their exchange and distribution. The appropriation of extra-European resources and labor further increased the wealth of Western Europe." Angus Maddison (1983) generally supports this view. The industrial revolution in Europe led to a widening of the prosperity gap between industrialized and developing nations. The transition to the technological and industrialized age also marked the transition to a world with increasing economic inequity.

Roger Farmer (1998) has stressed that although differences in rates of growth in the standard of living seem like small numbers, they can have a very big impact on real standard of living because of the increase each year getting compounded. Today, developing countries face additional handicaps including severe debt and debt servicing burdens, an unfair trade regime where trade is becoming free but not fair, and an expansion in social exclusion through patents and other forms of intellectual property rights.

Fortunately, there are no human differences in intelligence that parallel human differences in technology. Hence, it should be possible to embark upon a dynamic program for the technological empowerment of the poor. Jared Diamond (1997) points out that people who until recently were technologically primitive – such as aboriginal Australians – speedily master industrial technologies when given opportunities to do so. We have seen in India, especially in the Punjab and Haryana, barely literate farmers acquiring sophisticated electrical and mechanical engineering skills. Therefore, UNESCO, UNDP, WTO, ICSU and other interested multilateral and bilateral donors should develop a plan of action for mobilizing technology, training, techno-infrastructure and trade for poverty eradication. Trade policies should be formulated in such a manner that they will strengthen and not erode the livelihood security of the women and men living in poverty. Developing countries should ensure that their import and export policies are based on a livelihood impact analysis.

The pervasive poverty we witness today is the most serious indictment of contemporary developmental pathways. The poor are poor because they have no productive assets – no land, no education and no technical skill. They earn their livelihood largely through unskilled work. They have been bypassed by modern technological advances. They suffer from a sense of exclusion from the exciting scientific adventures we are witnessing today. Their sense of frustration is enhanced when they see on television the benefits of technological progress enjoyed by a section of humankind. Reaching the unreached and including the excluded have to be important components of the science and technology policy and strategy for the new century, if the huge stock of scientific and technological knowledge and innovations with which we will be entering the next century is to become a blessing for humankind as a whole. A priority task for the Inter-Academy Center, proposed by Bruce Alberts, President of the National Academy of Sciences, USA, should be the closing of the vast knowledge and skill gap prevailing between rich and poor nations on the one hand and between the rich and the poor within all nations. New information technologies provide a unique opportunity for the knowledge and skill empowerment of the poor. The women and men now living in poverty can be helped to experience a better quality of life only by increasing the economic value of their time and labor. A transition from unskilled to skilled work resulting in value addition to labor and time will be needed for enabling the poor to experience a productive and healthy life. Opportunities exist today for achieving such a transition speedily.

Gender Dimensions of Poverty: Women in Science and Technology and Science and Technology for Women

UNDP's Human Development Report (1995) states "human development, if not engendered, is endangered". The report further states that the revolution towards gender equality must be propelled by a concrete strategy for accelerating progress. The Beijing Platform for Action adopted by the Fourth World Conference on Women on September 15, 1995, says: "In the past decade, the number of women living in poverty has increase disproportionately to the number of men, particularly in the developing countries. The feminization of poverty has also recently become a significant problem in the countries with economies in transition as a short-term consequence of the process of political, economic and social transformation. In addition to economic factors, the rigidity of socially ascribed gender roles and women's limited access to power, education, training and productive resources as well as other emerging factors that may lead to insecurity for families are also responsible. The failure to adequately mainstream a gender perspective in all economic analysis and planning and to address the structural causes of poverty is also a contributing factor."

"Science curricula in particular are biased. Science textbooks do not relate to women and girls' daily experience and fail to give recognition to women scientists. Girls are often deprived of basic education in mathematics and science and technical training, which provide knowledge they could apply to improve their daily lives and enhance their employment opportunities. Advanced study in science and technology prepares women to take an active role in the technological and industrial development of their countries, thus necessitating a diverse approach to vocational and technical training. Technology is rapidly changing the world and has also affected the developing countries. It is essential that women not only benefit from technology, but also participate in the process from the design to the application, monitoring and evaluating stages."

The Beijing Conference recommendations concerning the technological empowerment of women were considered at a meeting of women scientists and technologists in Asia and the Pacific region at the M.S. Swaminathan Research Foundation (MSSRF), Chennai, in December 1996 under the auspices of UNDP and UNIFEM. The meeting recommended several institutional and policy devices to enlarge the role of women in science and technology development and dissemination. Recommendations were also made to mainstream gender considerations in ongoing S&T programs. A major requirement for women entrepreneurs is flexi-time and flexi-duration of work. Also, women engaged in micro-enterprises supported by micro-credit will need institutional structures, which will provide them with the power of scale particularly in marketing. An example of the kind of support needed is provided by the Biotechnology Park for Women coming up near Chennai in India. This Park will provide centralized services like information, training and electronic marketing facilities for many women entrepreneurs seeking to find avenues for remunerative self-employment. There is need for a global movement for the knowledge, skill and technological empowerment of women living in poverty. Because of the multiple burden on their time, they are over-worked and under-paid. Hence, the aim of scientific research designed for helping resource-poor women should be to add economic value to each hour of work and reduce the total number of hours of work. The Biovillage program of MSSRF aims to achieve these twin goals. Successful examples of women's technological empowerment need to be replicated worldwide. Opportunities for assured and remunerative marketing will determine the success of the micro-credit supported micro-enterprises now being advocated to end the feminization of poverty. It would be useful if UNIFEM, ILO, the

World Women's Banking and WTO undertake a careful study of the public policy support needed at the national and global levels to make women's enterprises supported by micro-credit economically viable.

Science and Food and Water Security

The 20th century began with the rediscovery of Mendel's laws of inheritance. It ends with moving specific genes across sexual barriers with the help of molecular mapping and recombinant DNA technology. The impact of science and technology in every field of crop and animal husbandry, inland and marine fisheries and forestry has been profound. Let me illustrate this, taking the improvement of wheat production in India as an example.

Wheat cultivation started in the Indian subcontinent over four thousand years ago. Wheat kernels have been found in the Mohenjodaro excavations dated 2,000 BC. From that period up to August 1947, when the colonial rule ended, Indian farm men and women developed the capacity to produce seven million tons of wheat per year. Between 1964 and 1968, when semi-dwarf strains containing the Norin 10 genes for dwarfing were introduced in irrigated areas, wheat production rose from ten to seventeen million tons per year. In other words, four thousand years of progress was repeated in four years (Swaminathan, 1993). During 1998-1999, wheat production in India exceeded seventy million tons, i.e., a ten-fold increase in about fifty years.

Similar progress has been made in improving the production and productivity of rice, maize, soybean, potato and several other crops as well as in farm animals in many developing countries around the world. New technologies supported by appropriate services and public policies have helped to prove doomsday predictions wrong and have led to the agricultural revolution (the Green Revolution) becoming one of the most significant of the scientific and socially meaningful revolutions of this century. A world without hunger is now within our reach. A hunger-free world will be possible if every nation pays concurrent attention to improving food availability through ecologically sustainable methods of production, to enhancing economic access to food by promoting a job-led economic growth strategy, and to ensuring the biological absorption of food in the body through the availability of safe drinking water and environmental hygiene. Steps should also be taken to enlarge the base of the food security basket by revitalizing the earlier tradition of cultivating a wide range of food crops (National Research Council, 1989, 1996; M.S. Swaminathan Research Foundation, 1999).

Emerging farming technologies will be based on precision farming methods leading to plant scale rather than field scale husbandry. Farming will be knowledge intensive, using information from remote sensing, Geographical Information System (GIS), Global Positioning Systems (GPS), and information and computer technologies. Farmers in industrialized countries are already using satellite imagery and GPS for early detection of diseases and pests, and to target the application of pesticides, fertilizers and water to those parts of their fields that need them urgently. Among other recent tools, the GIS methodology is an effective one for solving complex planning, management and priority setting

Figure 1. Consequences of low birth weight

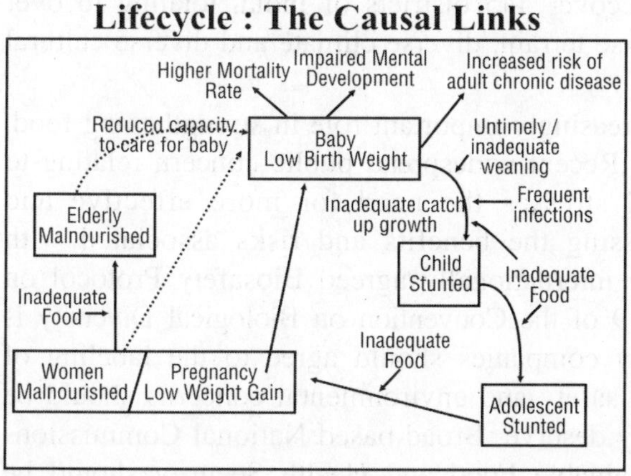

Figure 2. The Gulf of Mannar region in southern India, indicating the pressure of landless poor in the area

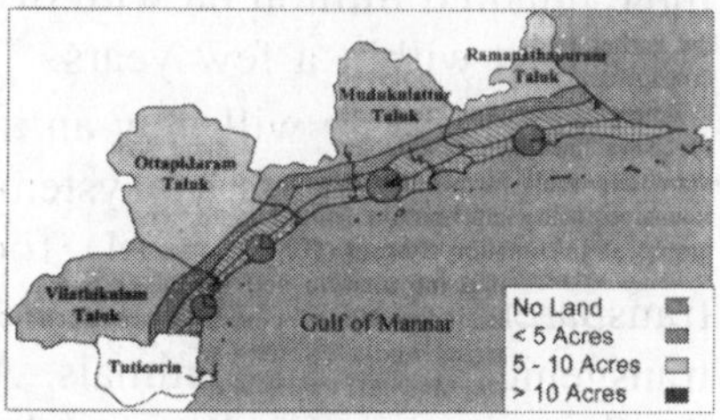

problems. Similarly, remote sensing technology can be mobilized in programs designed to ensure drinking water security. GIS can be applied for a variety of planning and management activities like: Environment and natural resources management; Forestry and wild life protection; Biodiversity conservation; Groundwater assessment; Land use and thematic mapping; Urban and town planning; Health care; Pollution management.

Let me cite two examples of their value from our recent work. First, GIS was applied for determining priorities in a program designed to launch a total attack on hunger in the Dharmapuri district in Tamil Nadu, India. Socio-economic data like the percentage of poor population, percentage of unemployment, literacy rate, and infant and maternal mortality rates were mapped in GIS. The layers were prepared for each factor and registered together. Different levels were given to classify each factor. They were overlaid to get a profile map showing the poorest villages which need to be accorded priority in the hunger-free area program. Second, GIS proved to be an invaluable tool in developing strategies for the conservation and sustainable and equitable use of biodiversity. The Gulf of Mannar region in South India (Figure 2) is a biological paradise. Unfortunately, anthropogenic pressures and the unsustainable use of coral reefs, sea grass beds and mangroves are causing serious damage to this priceless heritage. With financial support from the Global Environment Facility (GEF) a Gulf of Mannar Biosphere Reserve Trust is being created by the Government of Tamil Nadu. The aim is to make all stakeholders regard themselves as Trustees of this area. This evolution of the Gulf of Mannar Biosphere Reserve into a Biosphere Trust held in trust for posterity is an example of UNESCO's vision of Biosphere Reserves for the 21st century articulated at Seville becoming a reality. "Rather than forming islands in a world increasingly affected by severe human impacts, biosphere reserves can become theatres for reconciling people and nature. They can bring the knowledge of the past to the needs of the future."

Under a program of the Government of India designed to provide drinking water to all, ground water surveys through satellite remote sensing data were used for hydrogeomorphological mapping. Based on statistics of over 170,000 bore wells dug with these maps, it was found that the success rate of finding water was as high as ninety-two percent. Without such data becoming available, it would have been impossible to cover 445 districts of India, totaling to over three hundred million ha with diverse terrain, diverse climate and diverse cultural conditions within a few years.

Biotechnology will play an increasingly important role in strengthening food, water and health security systems. Recent widespread public concern relating to genetically modified (GM) food stresses the need for more effective and transparent mechanisms for assessing the benefits and risks associated with transgenic plants and animals. An internationally agreed Biosafety Protocol on the lines recommended in Article 19 of the Convention on Biological Diversity is an urgent necessity. Biotechnology companies should agree to the labeling of GM foods in the market. All food safety and environmental concerns should be addressed with the seriousness they deserve. Broad-based National Commissions on Genetic Modification for Sustainable Food and Health Security should be set up, consisting of independent professionals, environmentalists, representatives of civil society, farmers' and women's organizations, mass media and the concerned Government regulatory authorities. This will help to assure both farmers and consumers that the precautionary principle has been applied, while approving the release of GM crops. The Government of the United Kingdom has set an example by setting up two multi-stakeholder commissions to take a systems approach to all aspects of genetic engineering and GM foods (*Nature*, 1999: 339, 287). These are: Human Genetics Commission, and Agricultural and Environment Biotechnology Commission

The US Government has taken a similar step by setting up a Biotechnology Advisory Committee made up of scientists, farmers, industry groups, professionals, environmentalists, regulatory agencies and members of the public (*Nature*, 1999: 339, 508). Such broad-based consultative and policy guidance bodies can help Governments to take decisions on issues like the application of "Genetic Use Restriction Technologies" (GURTs) or what are popularly referred to as "terminator" and "verminator" technologies which cause the non-viability of seeds. The other aspect of biotechnology which is causing serious concern, relates to the commercial exploitation of biodiversity. The commercial importance of biodiversity will be evident from the following data on the size of world markets in 1997 for a few important items of commerce (Kate and Laird, 1999): Genetic resources: $500-800 million; Petrochemicals: $500 million; Computer market: $800 million.

Biodiversity-rich but biotechnology-poor countries are adversely affected by the prevailing non-adherence to the ethical and equity principles in benefit-sharing contained in Articles 8 and 15 of CBD. The primary conservers, largely tribal and rural women and men, live in poverty, while those who use their knowledge and material for producing commercial products become prosperous. The invaluable contributions of tribal and rural families to genetic resources conservation and enhancement have been recognized in the Convention on Biological Diversity (Figure 3). Yet, the political will to implement the equitable benefit-sharing provisions of CBD is lacking. We need urgent steps to recognize

Figure 3. Equity principle in the conservation of biological diversity; the plant in the picture is *Bacopa monnieri* (Brahmi)

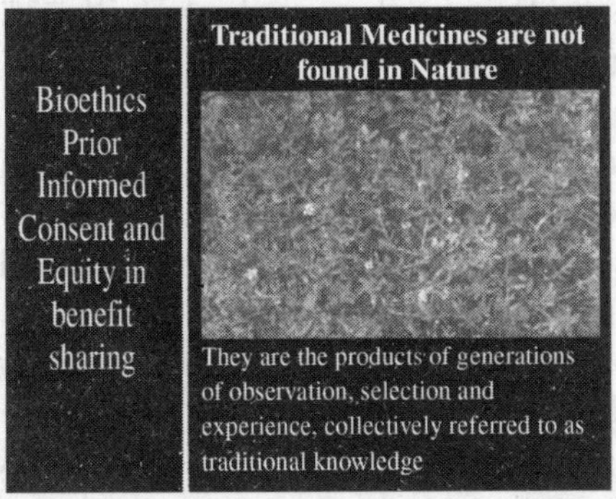

Table 1. Aging-related plant remedies

Country	Plant	Property
India	*Trichopus zeylanicus*	Helps to remove fatigue
India	*Bacopa monnieri*	Helps to improve memory
Tropical Africa	*Prunus africana*	Treatment for benign prostatic hyperplasis

and reward their contributions to providing material of great importance to global food and health security. Three validated findings (Table 1) will be adequate to stress the significance of traditional knowledge and conservation efforts to help mitigate handicaps caused by aging in human beings.

Article 27(b) of the Trade Related Intellectual Property Rights (TRIPS) component of the World Trade Agreement will come up for review later this year. All nations should agree to incorporate in this clause the ethics and equity principles enshrined in Articles 8(j) and 15 of CBD. The World Intellectual Property Rights Organization (WIPO) which has launched a study of the need to recognize the intellectual property rights of the holders of traditional knowledge, should complete this study soon and help to make the principles of ethics and equity the foundation of IPR.

UNESCO, FAO and WHO should formulate a Universal Declaration on the Plant Genome and Farmers' Rights on lines similar to the Declaration on the Human Genome and Human Rights. If such steps are taken, biopiracy will give way to symbiotic biopartnerships. Conservation and commercialization will then become mutually reinforcing. Such a step will not only help to strengthen biodiversity conservation but will make an important contribution to poverty alleviation. The life sciences industry based on the modification of living organisms to create new products and services will then have a sustainable future.

The biotechnology industry should lose no further time in giving a helping hand to evolve internationally agreed protocols for biosafety, bioethics and biosurveillance. If this is not done, biotechnology and other life science industries, which can be invaluable allies in a global science and technology for basic human needs movement, will remain clouded in controversies, suspicion and rich-poor conflicts.

Emerging Scientific Revolutions and an Ecology of Hope

Fortunately, as we approach the new century we are experiencing three major revolutions in science and technology, which will influence agriculture and industry in a fundamental manner. It will therefore be appropriate to make a brief reference to them:

1) The gene revolution – which provides a molecular understanding of the genetic basis of living organisms, as well as the ability to use this understanding to develop new processes and products for agriculture, industry, the environment, and for human and animal health.

2) The ecotechnology revolution – which promotes the blending of the best in traditional knowledge and technology with frontier technologies such as biotechnology, space and information technologies, renewable energy and new material.

3) The information and communication revolution – which allows a very rapid growth in the systematic assimilation and dissemination of relevant and timely information, as well as a dramatically improved ability to access the universe of knowledge and communicate through low cost electronic networks.

In principle, these three types of advances – when coupled with improvements in the management and governance – greatly increase the power of a scientific approach to genetic improvement, the integrated management of natural resources and ecosystems, and the management of local and regional development strategies. Eco-technologies enable the adoption of ISO 9000 and ISO 14000 standards of environmental management. These scientific revolutions seem to be proceeding at an ever-increasing pace, with most of the action occurring in industrialized nations. Also, new discoveries of great relevance to sustainable food and health security are coming under the purview of proprietary science, since they are covered by Intellectual Property Rights. How then can we mobilize recent advances in science and technology for meeting the basic needs of the economically and socially underprivileged sections of the human family?

The Gene Revolution. The past ten years have seen dramatic advances in our understanding of how biological organisms function at the molecular level, as well as in our ability to analyze, understand, and manipulate DNA molecules, the biological material from which the genes in all organisms are made. The entire process has been accelerated by the Human Genome Project, which has poured substantial resources into the development of new technologies for working with human genes. The same technologies are directly applicable to all other organisms, including plants. Thus, a new scientific discipline of genomics has arisen. This discipline has contributed to powerful new approaches in agriculture and medicine and has helped to promote the biotechnology industry.

Several large corporations in Europe and the United States have made major investments in adapting these technologies to produce new plant varieties of agricultural importance for large-scale commercial agriculture. The same technologies have equally important potential applications for addressing food security in the developing world.

The key technological developments in this area are: a) Genomics: The molecular characterization of species; b) Bioinformatics: Data banks and data processing for genomic analysis; c) Transformation: Introduction of individual genes conferring potentially useful traits into plants, trees, livestock, and fish species; d) Molecular breeding: Identification and evaluation of useful traits by use of marker-assisted selection, which greatly speeds up traditional breeding processes; e) Diagnostics: Identification of pathogens by molecular characterization; f) Vaccine technology: Use of modern immunology to develop recombinant DNA vaccines for improved control against lethal diseases of animals and fish.

Let me cite one example from the work of MSSRF scientists to illustrate the value of the new tools. As a part of the anticipatory research program to meet the consequences of sea level rise arising from global climate change, genes responsible for conferring the ability to withstand seawater intrusion were identified in a few mangrove species through molecular mapping. They have been transferred to annual economic plants through recombinant DNA technology.

The sequencing of the genome of rice (*Oryza sativa* L. cv. Nipponbare) by an international consortium supported by the Rockefeller Foundation and the International Rice Research Institute will permit allele mining for all genes of rice and possibly for other cereals. Thus, altogether unforeseen opportunities for creating novel genetic combinations have been opened up.

As mentioned earlier, there are widespread public concerns about the potential adverse impact of genetically modified organisms (GMOs) on human health, biodiversity and the environment. Several of these concerns are genuine. In order to take advantage of recombinant DNA technologies without associated harm to human or ecological health, it is important that every country has in place suitable institutional structures and regulations for biosafety, bioethics and biosurveillance. At the same time, there is need for greater investment of public funds for public good research, the results of which can reach the unreached. For example, in food and agriculture, there is need to strengthen both National Agricultural Research Systems and the International Agricultural Research Centers supported by the Consultative Group on International Agricultural Research (CGIAR).

The Ecotechnology Revolution. Knowledge is a continuum. There is much to learn from the past in terms of the ecological and social sustainability of technologies. At the same time, new developments have opened up new opportunities for developing technologies which can lead to higher productivity without adverse impact on the natural resources base. Blending traditional and frontier technologies leads to the birth of ecotechnologies with combined strengths in the following areas: Economics, Ecology, Equity, Employment, Energy.

For example, in the area of water harvesting and sustainable use, there are many lessons to be learned from the past. In the desert area of Rajasthan, India, drinking water is available even in areas with 100mm, annual rainfall, largely because women are continuing to harvest water in simple structures called kunds. In contrast, drinking water is scarce during summer months in some parts of North-east India with an annual rainfall of 15,000 mm. Thus, there is need to conserve traditional wisdom and practices, which are often tending to become extinct (Agarwal and Narain, 1997). The decision of the WIPO to explore the intellectual property needs, rights and expectations of holders of traditional knowledge, innovations, and culture is hence an important step in widening the concept of intellectual property. FAO has been a pioneer in the recognition of the contributions of farm families in genetic resources conservation and enhancement by promoting the concept of "Farmers' Rights". Like WIPO, Union for the Protection of New Varieties of Crops (UPOV) should also undertake the task of preparing an integrated concept of breeders' and farmers' rights. UPOV itself should be restructured to become a Union for the Protection of Farmers' and Breeders' Rights.

The Information Technology Revolution. New communication and computing technologies are already influencing life on our planet in a profound manner: a) Access to the Internet will soon be universal, and it can provide unrestricted low-cost access to information, as well as highly interactive distance learning. The Internet will not only facilitate interactions among researchers, but also greatly improve their ability to communicate effectively with the potential users of their research knowledge; b) Computing makes it possible to process large-capacity databases (libraries, remote sensing and GIS data, gene banks) and to construct simulation models with possible applications in ecosystem modeling, preparation of contingency plans to suit different weather probabilities and market variables; c) The software industry is continuously providing new tools that increase research productivity and create new opportunities for understanding complex agroeco systems; d) Remote sensing and other space satellite outputs are providing detailed geographic information useful for land and natural resources management.

The promotion of ecotechnology development and dissemination, the effective adoption of integrated systems of gene and natural resources management and the effective harnessing of information technologies should become essential elements of the "science and technology for basic human needs" movement.

Launching a Knowledge System for Sustainable Food and Livelihood Security

The explosive progress in science and communication has led to the christening of the communities who will live on our planet in the coming century as Knowledge Societies. Information technology is beginning to play a pivotal role in stimulating and sustaining such Knowledge Societies. Let me describe in some detail an innovative experiment we are carrying out at MSSRF with a view to using information technology for the benefit of the rural poor. The experiment aims to provide the information people need and can use in their daily lives. The whole program is need-based and demand-driven. As we cannot hope to provide telephones, computers and other tools of technology to each household for a long time to come, we have chosen to reach the unreached through community access.

Opportunities for a Learning Revolution. Results from our work indicate that computer-aided knowledge dissemination mechanisms can help to reach the unreached and foster new voices and new leaders. For success, a user-controlled and demand-driven system is essential. Use of the local language as the medium of communication has ensured ease of access to the entire target population.

Figure 4.

Computer aided *Virtual College* linking scientists and women and men living in poverty

The MSSRF Knowledge System deals with the following three components of food security in an integrated manner.

a) Availability: The first requisite is the adequate availability of food for home consumption and for markets. In the coming century, more food and other agricultural commodities will have to be produced from less per capita arable land and irrigation water, first, to meet the needs of the growing population; second, to combat the under-nutrition and malnutrition currently prevailing among 800 million children, women and men; and, third, to meet the need for a larger volume of feed grains because of greater consumption of animal products in urban areas.

b) Access: Even if foodgrains, fruits, vegetables and animal products are available in abundance in the market, inadequate purchasing power inhibits access to balanced diets among the economically underprivileged sections of society. Women and children are particularly vulnerable. Economic access to food depends very much on opportunities for remunerative employment and on multiple livelihood sources in the case of the landless poor. An important aspect of food security thus relates to the creation of opportunities for sustainable livelihoods. Fortunately, ecological farming is both knowledge and skill intensive. There is need for spreading information on the opportunities for eco-jobs in the farm and non-farm sectors. Urban and peri-urban agriculture helps to link the producers, food processing industry and consumers in a symbiotic chain.

c) Absorption: Even if food is consumed in adequate quantities, the biological absorption and utilization of food in the body depends upon the availability of clean drinking water, environmental hygiene, sanitation and facilities for primary health care. Diarrheal diseases and liver ailments are common among infants and children living under insanitary conditions.

Thus, the Knowledge System for Sustainable Food Security has as its goal the empowerment of rural women, men and children with information relating to ecological agriculture, sustainable livelihoods and the biological absorption and utilization of food in the body. The Knowledge System is managed by local women at the Village Knowledge Center, from where the computer-aided information system is operated. The Knowledge Centers can be coupled with training centers in order to create a cadre of eco-entrepreneurs.

Producing more but producing differently. The Knowledge System provides information on soil health care, water, pest and energy management, post-harvest technology, farming systems design and marketing. Since the conservation of soil fertility is fundamental to sustained advances in productivity, farm families are trained to maintain Soil Health Cards. Similarly, those possessing livestock are helped to maintain Animal Health Cards. Since the system aims to convert generic information on meteorological, management and marketing factors into location-specific details, a high degree of interactive learning is built into the process.

The ecological farming database provides information on the environmentally sound management of pests as well as on eco-friendly methods of soil fertility and irrigation water management. The aim of the Knowledge System is to help in spreading an evergreen revolution in small farms rooted in the principles of ecology, economics, gender and social equity and employment generation. Precision farming methods help to lower the cost of production and enhance both income and ecological sustainability.

In the field of agriculture, knowledge dissemination and supply of the inputs essential for deriving benefit from new knowledge have to be synchronized in space and time. It will hence be desirable to link the Knowledge Center not only with a Training Center but also with a single window system of input supply.

Information empowerment and freedom from hunger. The Village Knowledge Center enables farming families not only to produce more farm products without associated ecological harm, but helps everyone in the village to create a hunger-free area. The Action Plan for the elimination of hunger and malnutrition consists of the following seven steps.

First, the villagers themselves identify who the hungry amidst them are. They are generally very poor without land or livestock or fish pond or any other productive asset. They also tend to be illiterate.

Second, a Household Entitlements Database provides information on all the government, bilateral and international schemes available to rural families. The information is disaggregated according to age, gender and occupation. The families covered by the Knowledge Center are given Household Entitlement Cards. They are encouraged to bring these cards to the Knowledge Center and fill them up by themselves with the help of the lady operating the computer. This helps them to know what their entitlements are and how to access them.

Third, a database on elimination of under-nutrition or calorie deprivation indicates the schemes available for overcoming protein-calorie under- and mal-nutrition. Information on projects run by civil society organizations is also provided.

Fourth, a Hidden Hunger Elimination Database deals with the micronutrient deficiencies prevailing in the village. Usually, a pediatric survey is conducted to identify the extent of prevalence of deficiencies of Vitamin A, iron, iodine and other micronutrients and information is provided on the programs available for the total elimination of micronutrient deficiencies. The Knowledge Center also organizes interactions between the affected families and those who provide remedies such as the use of iron and iodine fortified salt and spirulina and the cultivation of appropriate vegetables and fruits in home gardens.

Fifth, a plugging the leaky pot database provides information on schemes available for clean drinking water, environmental hygiene and sanitation, latrines, primary health care and other methods of improving the biological absorption and retention of food in the body. Information is also provided on oral dehydration and other methods of managing diarroheal diseases, particularly among children. These steps will help to ensure that food is used by the body for the purpose for which it is consumed.

Sixth, the Eco-jobs Database provides information on the opportunities for economically, socially and environmentally sustainable self-employment. The jobs suggested are chosen on the basis of credit and market availability. The Knowledge Center thus performs the role of a Rural Placement Center for the Remunerative self-employment. A wide range of on-farm and off-farm employment opportunities are included in the database. Only those goods for which there is a reliable market are recommended for production. The Knowledge Center also becomes a place for interaction between credit and extension agencies and rural families, thereby triggering a New Deal for the Self-employed.

Finally, a Women and Children Database provides information on the projects available to preschool children, pregnant and nursing mothers and old and infirm persons. Special focus is on the nutritional health of pregnant women and on children in the 0 to 2 age group. Information on reproductive health care and on other special programs for women is included in the database. Community managed support services, such as day-care centers, for working mothers are included in the database. This is essential because of the multiple burdens shouldered by women, as a consequence of which they tend to remain over-worked but underpaid or even unpaid.

Once a Knowledge Center gets established, the subsequent development of data bases can be done by the rural families themselves. Unless rural families have a sense of ownership and are willing to provide space and meet the salary of the computer operators, who are drawn from the village itself, the Knowledge Center is likely to become a supply driven show piece. The Knowledge Center is also designed to promote the monitoring of the impact of information empowerment. The following indicators are used: Infant mortality rate, maternal mortality rate, low-birth weight children, sex ratio, incidence of adult malnutrition, percentage of workers performing unskilled operations and incidence of micronutrient deficiency-induced ailments. Depending upon the interests of rural families, other criteria like the population supporting capacity of the ecosystem, per capita income and average life span can be included. The growth of the Knowledge Center must be an evolutionary process, with information empowerment in areas relevant to meeting the basic human needs receiving over-riding priority.

Bruce Alberts, President of the US National Academy of Sciences, in his anniversary address delivered in Washington DC on April 26, 1999 referred to the significance of the MSSRF Knowledge System for rural families in the following words: "Connecting scientists to each other is only the first step. Scientists everywhere must use these initial connections as a tool for spreading their knowledge, skills, and values throughout their own nations, including their local communities. By taking full advantage of new information technologies, the scientific community has an unprecedented opportunity to close the vast 'knowledge gap' between all peoples. How might this be possible?

"I want to highlight a wonderful example that points the way forward. The M.S. Swaminathan Research Foundation has established an experimental network in India that will soon connect more than twenty isolated rural villages to a wireless Internet service. About half of the population in most of these villages has a total family income of less than $25 per month. The project is designed to provide knowledge on demand to meet local needs using the World Wide Web, and it does so through a bottom-up process. The process starts with volunteer teams that help poll the villagers to find out what knowledge they want. Particularly popular thus far are women's health information, advice on growing local crops and protecting them from diseases, the daily market prices for these crops, local weather forecasts, and clear information about the bewildering array of programs that are provided by the Indian government to aid poor families. To participate, each village must provide a public room for the computer system, as well as the salaries for a set of trained operators. In return, the village receives the needed hardware and maintenance for the communication system, especially designed Web sites in the local language that convey the requested information, and training programs for those villagers who have been selected to run their local knowledge system.

"Drawing on this concept, I envision a global electronic network that connects scientists to people at all levels - farmers' organizations and village women, for example. The network will allow them to easily access the scientific and technical knowledge that they need to solve local problems and enhance the quality of their lives, as well as to communicate their own insights and needs back to scientists."

Thus, we now have an opportunity to develop a new form of computer-aided Virtual College, linking scientists and resource-poor women and men living at a per capita income of one to two dollars per day. Such a Virtual College should be truly the product of partnership between the poor and those who have the requisite technological know-how and do-how. At the global level also there could be a similar Virtual College for basic human needs operated jointly by UNESCO and ICSU.

Harnessing Resources for a
"Science for Basic Human Needs" Movement

Since the onset of the Industrial Revolution in Europe, technology has been a major source of economic inequity among nations and among communities within nations. If technology has been a cause of economic and social inequity in the past, today we have an opportunity for making technology an ally in the movement of social, gender, and economic equity. Modern information technology provides this opportunity. Knowledge and skill empowerment can now be achieved at a fast pace. However, the technological and skill empowerment of the poor cannot be achieved through programs designed on the basis of a patronizing and a top-down approach. The information provided should be demand- and need-driven and the knowledge centers should preferably be managed by women belonging to the socially and economically underprivileged sections of the society. Our aim in the early part of the coming century should be the initiation and spread of a Knowledge Revolution for ending economic and gender inequity.

The accomplishment of the tasks I have outlined so far requires considerable technical, managerial and financial resources. Scientists of the International Peace Research Institute, Oslo, have studied the causes of armed conflicts during the last ten years. They found that violent conflicts in most cases could be traced to economic rather than ideological differences (de Soysa and Gleditsch, 1999). They have hence suggested that investing in agriculture, which helps to promote food and livelihood security in many nations is an effective strategy for preventing future wars, eradicating poverty, preventing environmental destruction and reducing violence. Unfortunately, even now, far too high a proportion of national GDP is being spent on arms and military equipment as compared to programs designed for poverty eradication and meeting the basic needs of the underprivileged sections of humankind. The so-called peace dividend still remains only in the realm of possibility (Swaminathan, 1994).

The year 2000 has been appropriately designated the International Year for the Culture of Peace. Without peace and human security, it will not be possible to ensure the basic human needs to every child, woman and man. It will be appropriate to recall on this occasion what Dwight D. Eisenhower, a great war leader who subsequently became the President of the United States, stated on August 16, 1953: "Every gun that is made, every warship launched, every rocket signifies in the final sense a theft from those who are hungry and are not fed, from those who are cold and are not clothed. This world in arms is not spending money alone. It is spending the sweat of its laborers, the genius of its scientists, the hopes of its children." (Eisenhower, 1953)

Harnessing science and technology for fulfilling the basic minimum needs of every child, woman and man living on our planet will be possible only if this message becomes central to the ethos of human culture.

Conclusion

To sum up, we are ending this century with a huge stockpile of scientific discoveries and technological innovations. This stockpile is more than adequate to help all nations to provide every adult human being an opportunity for a healthy and productive life and every newborn child a happy future. It is, therefore a sad commentary on our political, social and spiritual value system that the number of children, women and men living in poverty today exceeds the entire human population of our planet at the beginning of this century. Unsustainable life styles and degrading poverty co-exist everywhere. This is the greatest failure of the developmental pathways and strategies adopted during this century. Can we lay the foundation at this Conference for the emergence of a new political, social and scientific commitment to end the irony of widespread human misery and deprivation prevailing in the midst of uncommon opportunities for a better common present and future for all? In my view we can, provided every one of the nearly two billion persons who are enjoying a healthy and productive life today will keep the following advice of Mahatma Gandhi as the guiding principle in his/her day-to-day life and work: "Recall the face of the poorest and the weakest man whom you have seen, and ask yourself, if the steps you contemplate are going to be of any use to him. Will he gain anything by it? Will it restore to him control over his own life and destiny?"

In other words, the formidable power of science and technology can benefit humankind only if we know how to temper it with humanism. Let us hope the new millennium will bring forth a new crop of leaders of science who will be able to usher in an era of humanistic science. It is the duty of scientific establishments and science academies to nurture and foster the growth of young men and women research leaders capable of initiating and managing change in goals and strategies in the coming century.

References

Agarwal, Anil and Sunita Narain. 1997. *Dying Wisdom: Rise, Fall and Potential of India's Traditional Water Harvesting Systems.* Faridabad: Thomson Press (India).

De Soysa, Indra and Nils P. Gleditsch. 1999. *To Cultivate Peace – Agriculture in a World of Conflict.* Oslo, Norway: International Peace Research Institute (PRIO).

Diamond, Jared. 1997. *Guns, Germs and Steel: The Fate of Human Societies.* New York: W.W. Norton and Company.

Eisenhower, D. 1953. Talk delivered before the American Society of Newspaper Editors.

Farmer, R. A. 1998. *Macroeconomics.* Cincinnati, Ohio: South-Western College Publishing Company.

Kate, K.T. and Sarah A. Laird. 1999. *The Commercial Use of the Biodiversity.* London: Earthscan Publications.

Landes, David S. 1990. *Unbound Prometheus.* Cambridge, UK: Cambridge University Press.

Myers, Norman and Jennifer Kent. 1995. *Environmental Exodus: An Emergent Crisis in the Global Arena.* Washington DC: Climate Institute.

Maddox, John. 1998. *What Remains to be Discovered.* London: Macmillan.

Maddison, Angus. 1983. *Journal of Economic History.*

M.S. Swaminathan Research Foundation. 1999. *Proceedings of the International Consultation organized by the Genetic Resources Policy Committee of the CGIAR.* Madras: MSSRF.

National Research Council, USA. 1989. *Lost Crops of the Incas.* Washington DC: National Academy Press.

National Research Council. 1996. *Lost Crops of Africa.* Washington DC: National Academy Press.

Panayotou, Theodore, Jeffrey Sachs and Alix Peterson. 1999. *Developing Countries and the Control of Climate Change: Empirical Evidence.* Cambridge, Massachusetts: Harvard University Press.

Postrel, Virginia. 1998. *The Future and its Enemies: The Growing Conflict Over Creativity, Enterprise and Progress*. New York: The Free Press.

Swaminathan, M.S. (Ed.) 1993. *Wheat Revolution: A Dialogue*. Madras: Macmillan India.

-----. 1994. (Ed.) *Uncommon Opportunities: An Agenda for Peace and Equitable Development.* Report of the International Commission on Peace and Food. London: Zed Books.

-----. 1998. Farmers' Rights and Plant Genetic Resources. *Biotechnology and Development Monitor No. 36* (September-December).

UN Commission on Nutrition. 1999. *Ending Malnutrition by 2020: An Agenda for Change in the Millennium.*

UNDP. 1995. *Human Development Report*. Delhi: Oxford University Press.

Von Weizacker, U. Ernst, M.S. Swaminathan and Aklilu Lemma. 1983. *New Frontiers in Technology Application: Integration of Emerging and Traditional Technologies.* Dublin: Tycooly International Publishing.

31

Government-Industry-Civil Society Partnerships in Integrated Gene Management

At the outset, I would like to express my deep sense of gratitude to the Volvo Environment Prize Foundation for choosing me for this prestigious award. I accept it with humility on behalf of my numerous colleagues, students as well as the rural and tribal women and men with whom I work. Through the institution of this Prize, the founders of the Volvo Environment Prize Foundation have shown their deep commitment to fostering a better care of the Earth's ecosystems and their management. I wish to pay my tribute to all those connected with the management of this Prize for their dedication to the cause of a better today and tomorrow for all members of the human family.

We all live on this planet as guests of the green plants which help to convert sunlight into various agricultural commodities, and of the farm families who toil in sun and rain to cultivate them. I wish to dedicate this lecture to the millions of rural and tribal families who not only conserve the world's rich bio-resources but also serve as the backbone of the global food and health security systems.

On the eve of the UN Conference on Environment and Development held at Rio de Janeiro in June 1992, the Union of Concerned Scientists published an Open Letter titled, "World Scientists' Warning to Humanity", which stated that "human beings and the natural world are on a collision course". The letter stated further, "if not checked, many of our current practices put at serious risk the future that we wish for human society and the plant and animal kingdoms, and may so alter the living world that it will be unable to sustain life in the manner that we know." This warning was signed by over sixteen hundred scientists from leading scientific academies in seventy countries. The list included one hundred and four Nobel Laureates.

Colborn, Dumanoski and Myers in their book *Our Stolen Future* (1996) and James Morgan in his book *The Last Generation* (1999) also provide a picture of the grim future that awaits the generations yet to be born, if we lose further time in restoring harmony between humankind and nature.

A Chinese proverb warns, "If you do not change direction, you will end up where you are headed." Since we do not want to reach where we are presently headed, what change of course should we bring about?

Presented on the occasion of receiving the Volvo Environment Prize at Columbia University, New York, October 26, 1999. First published in *Current Science* 78 (5): 555-62 (March 2000).

I wish to take biodiversity, one of the key components of our basic life support systems, as an example to illustrate what changes are needed in the management of our biological resources. It is now widely realized that the genes, species, ecosystems and traditional knowledge and wisdom that are being lost at an increasingly accelerated pace limit our options for adapting to local and global changes, including potential changes in climate and sea level. *The Global Biodiversity Assessment* published in 1995 by the United Nations Environment Program estimates that about thirteen to fourteen million species may exist on our planet. Of this, less than two million species have so far been scientifically described. Invertebrates and micro-organisms are yet to be studied in detail. In particular, our knowledge of soil micro-organisms is still poor. Also, biosystematics as a scientific discipline is tending to attract very few scholars among the younger generation.

Another important paradigm shift witnessed in recent decades in the area of management of natural resources is a change in the concept of "common heritage". In the past, atmosphere, oceans and biodiversity used to be referred to as the common heritage of humankind. However, recent global conventions have led to an alteration in this concept in legal terms. Biodiversity is now the sovereign property of the nation in whose political frontiers it occurs. Under the UN Convention on the Law of the Sea, nations with coastal areas have access to a two hundred mile Exclusive Economic Zone (EEZ). For example, the ocean surface available to India under the EEZ provision is equal to two-thirds of the land surface available to the country. The Climate Convention and the Kyoto Protocol provide for both common and differentiated responsibilities to countries. Thus, the global commons can be managed in a sustainable and equitable manner only through committed individual and collective action among nations.

Biodiversity and Human Security

While we have some knowledge of variability at the eco-system and species levels, our knowledge of intra-specific variability is poor, except in the case of plants of importance to human food and health security. *The Global Biodiversity Assessment* warns, "unless actions are taken to protect biodiversity, we will lose forever the opportunity of reaping its full potential benefit to humankind." What kind of action will help us to ensure not only the conservation of biodiversity, but also its sustainable use? I would like to discuss this issue with reference to the following three major threats to biodiversity in general, and agrobiodiversity in particular.

Habitat Destruction

UNEP's *Global Environment Outlook-2000* states: "Reduced or degraded habitats threaten biodiversity at gene, species and ecosystems level, hampering the provision of key products and services. The widespread introduction of exotic species is a further major cause of biodiversity loss. Most of the threatened species are land-based, with more than half occurring in forests. Freshwater and marine habitats, especially coral reefs are also very vulnerable." How do we arrest this trend and prevent further genetic erosion? I would like to summarize briefly the approach adopted in India as well as at our Research

Center in Chennai (Madras) to foster an Integrated Gene Management strategy in the country. We use the term management in the context of natural resources to include conservation, sustainable use and equitable sharing of benefits. It is only such a concept of management that can result in the conservation as well as enhancement of natural resources.

The Integrated Gene Management system includes in situ, ex situ and community conservation methods (Figure 1). The traditional in situ conservation measures comprising a national grid of National Parks and protected areas are generally under the control of government environment, forest and wildlife departments. The exclusive control of such areas by government departments has often led to conflicts between forest dwellers and forest-dependent communities, and forest officials. The non-involvement of local communities in the past in the sustainable management of forests has resulted in a severe depletion of the forest resources in India. It has become clear that sole government control alone will not be able to protect prime forests or regenerate degraded forests.

Participatory Forest Management. The Participatory Forest Management (PFM) procedure initiated first at Arabari in the state of West Bengal in India in 1972, largely at the initiative of a young forest officer, Ajit Kumar Banerjee, and later extended to the other states in the country, became a significant turning point in the history of forest management in India as well as other Asian countries (Singh et al, 1997; Poffenberger et al, 1997). The essential feature of this system is that the state and community become partners in the management of the forest resource. The state continues to own the resource but the benefits are shared. Access to non-timber forest products becomes an important avenue of sustainable livelihoods to the forest-dependent communities. Thus, the community develops an economic stake in the preservation of forests, leading to conservation and sustainable use becoming mutually reinforcing components of a Forest Management Policy. The experience gained in India during the last twenty-five years shows that the process of natural forest degradation can be reversed through FMP and that forests can provide to the local community non-wood forest products on a continuous or seasonal basis, if there is a more widespread understanding of their regenerating capacity. Since forests are the home for a large proportion of naturally occurring biodiversity, saving forests results in saving genes.

Community Gene Management. Both in situ on-farm conservation of infra-specific variability, particularly in plants of food and medicinal value and ex situ on-farm conservation through sacred groves have been part of the cultural traditions of rural and tribal families in India. In the *Old Testament* also, there are several references to sacred groves. Among the important trees usually preserved in Indian Sacred Groves are: *Ficus religiosa, Saraca asoca, Shorea robusta, Alstonia scholaris* and many other species of ecological, economic and spiritual value. Gadgil and Vartak (1975) defined sacred groves as tracts of forest that

Figure 1. Integrated gene management system

Government-community partnership in integrated gene management

Government		Community	
In situ conservation	Ex situ conservation	In situ on-farm	Ex situ on-farm
• Protected areas • National parks • Biosphere reserve	• Gene banks • Botanical gardens • Zoological gardens • Tissue culture and DNA libraries	• Land races • Intra-specific diversity	• Sacred groves • Arboreta

have been completely immune from human interference on the basis of religious beliefs. Unfortunately, several of these traditions are now tending to wither away. It is only by giving explicit recognition to the pivotal role of community conservation in strengthening ecological, food and health security systems that we can succeed in the revitalization of these traditions. In national integrated gene management systems, in situ, ex situ and community conservation methods should receive adequate and concurrent attention. A recognition and reward system, based on FAO's concept of Farmers' Rights and CBDs provisions for ethics and equity in benefit-sharing, should become an integral part of national legislation in the fields of biodiversity conservation and plant variety protection.

Biosphere Management. Methods of fostering the sustainable management of Biosphere Reserves were discussed at a conference convened by UNESCO in 1995 at Seville in Spain. The Seville vision states: "Rather than forming islands in a world increasingly affected by severe human impacts, biosphere reserves can become theatres for reconciling people and nature. They can bring the knowledge of the past to the needs of the future."

To convert the above vision into field level reality, we have been working for a shift in the nature of management of Biosphere Reserves in India. A brief description of the approach adopted is given below.

India has currently eighty-five National Parks covering 3.6 million sq km and four hundred and forty-eight wildlife sanctuaries covering about 120,000 km^2 in the major biogeographic zones. The total extent of protected areas includes five designated as World Heritage sites, nine biosphere reserves and six Ramsar sites. In 1994, the M.S. Swaminathan Research Foundation (MSSRF) conducted a detailed study of the threats to the Gulf of Mannar Biosphere Reserve in Tamil Nadu. The study showed that unless the livelihood security of the families without assets living in that area can be strengthened, unsustainable exploitation of biological resources will continue (Swaminathan, 1999).

During 1996-1998, MSSRF initiated a proposal for bringing the Gulf of Mannar Biosphere Reserve, under a participatory management mode. The project proposal was considered by the Global Environment Facility (GEF) Council at its meeting held in May 1999. The GEF Council approved the proposal and commended it as a model project which deserves to be widely emulated by those preparing similar projects. The Scientific and Technical Advisory Panel (STAP) of GEF endorsed the project for approval with the following remarks: "The project addresses a major challenge, namely the conservation of coastal biodiversity of the highest ecological value in a large area subject to considerable pressure from poor populations upon the sole resources that appear to be at their disposal. To meet this challenge, the project follows the only framework which can succeed, namely to combine the necessary protection of the threatened ecosystem and ecological processes with economic and social benefits which will meet the essential needs of local people, through providing appropriate institutional, financial and managerial arrangements." This proposal was submitted by the Governments of Tamil Nadu and India to GEF through UNDP for support.

The management structure through which people and nature will get united in the Gulf of Mannar area is through a Gulf of Mannar Biosphere Trust. Representatives of fishermen and rural communities as well as all the principal civil society stakeholders will, together with government representatives, form members of the Trust. The members of the Trust will hold this unique biological treasure in trust for present and future generations. Considerations of intra- and inter-generational equity will guide the work of the Trust, which will provide a horizontal dimension to the numerous vertically structured activities undertaken in this area both by Central and State governments and private sector organizations. It is hoped that the Gulf of Mannar Biosphere Trust, whose long-term sustainability will be ensured with the help of a Trust Fund, will show the way for promoting a management by partnership system of governance in all the other Biosphere Reserves in the country.

Project Tiger launched in 1972 with support from WWF-International was the first step in institutionalized efforts in the conservation of habitats of endangered animals. An all-India tiger census conducted in 1972 revealed only 1,827 tigers in the country as against an estimated 40,000 at the beginning of the 20th century. The 1993 census placed the tiger population at 3,750.

One of the tiger reserves is the Similipal Biosphere Reserve in Orissa. This reserve was set up in June 1994 and covers an area of 2,750 km^2. Of late, threats to this reserve have increased due to both population pressure and commercial greed. Experience with participatory or joint forest management programs in India has shown that wherever the local communities are fully involved as partners in saving forests, the results have been very encouraging.

It is proposed to develop a similar plan for the Similipal Biosphere Reserve in collaboration with the Government of Orissa. The outcome of this program will be the establishment of a Similipal Biosphere Trust managed jointly by the major stakeholders.

A major need in such programs is the strengthening of the livelihoods of the poor families living in the vicinity of the Biosphere Reserve. For this purpose, the Biovillage model of livelihood security will be introduced in the villages around the reserves. The Biovillage concept of human-centered rural development aims to address concurrently the challenges of natural resources conservation and poverty eradication (Swaminathan, 1999). Market-driven livelihood opportunities will be identified and local families will be assisted with the help of institutional credit. The steps involved are: Micro-level planning with

GIS maps; Micro-enterprises identified on the basis of market needs; Micro-credit; Enlisting the support of industry for contract cultivation by small farmers on the basis of buy-back agreements.

Genetic Resources Enhancement and Sustainable Use. The main approaches adopted are summarized in Figure 2. Some methods of promoting symbiotic biopartnerships between conservers of genetic resources and holders of traditional knowledge and the utilizers of knowledge and material are indicated in Figure 3. An example of the Community Gene Management system being implemented in the state of Orissa is given next, in order to indicate how such concepts can be operationalized at the village level.

Community Gene Management. Tribal and rural farming communities have a long tradition of serving as custodians of genetic wealth, particularly landraces often carrying rare and valuable genes for traits like resistance to biotic and abiotic stresses, adaptability and nutritional quality. Several land types that carry valuable genes are preserved by farmers for religious functions (Table 1) and they constitute valuable material for conservation and sustainable use. Local landraces are still being maintained largely by the tribal poor. They thus serve public good at personal cost.

Such poverty-ridden custodians of genetic wealth are increasingly confronted with severe socio-economic problems which are rendering the maintenance of their traditional conservation ethics difficult. Steps are urgently needed to link their conservation efforts to the strengthening of their livelihoods.

It is now widely recognized that the conservation continuum consists of the following three links:

In situ →	In situ →	Ex situ
Conservation of habitats	On-farm conservation by rural and tribal communities	Conservation through botanical and zoological gardens and cryogenic gene banks

While the two ends of this conservation chain (namely in situ and ex situ) receive support from public funds, in situ on-farm conservation by rural and tribal women and men largely remains unrecognized and unrewarded. Yet, this link in the chain is responsible for the conservation of valuable intra-specific variability.

MSSRF's partnership with local communities and government agencies is designed to strengthen this neglected component in the conservation chain. For this purpose, a Community Gene Management System (CGMS) is being developed, initially in three biodiversity-rich districts of the State of Orissa in India.

Structure of CGMS. CGMS comprises the following three steps: a) Field Gene Bank (FGB) at the village level – In situ on-farm conservation by rural and tribal families; Participatory breeding (selection); b) Area Seed Bank (ASB) at the level of a cluster of villages – Conservation of representative samples of seeds in seed stores; Insurance against total loss of seeds during drought years;

Figure 2. Approaches to genetic resources management

Integrated gene management

Enhancement and sustainable use

- Location-specific selection and seed multiplication (mostly carried out by women)
- Participatory plant breeding (Farm family-plant breeder collaboration)
- Pre-breeding (Generation of novel genetic combinations for use by grassroot plant breeders)

Figure 3. Promoting Symbiotic biopartnerships

Integrated gene management

Equitable sharing of benefits (Guidelines for biopartnerships)

Individual benefits
- Sui generis – System of varietal protection
- Recognition and reward for individual plant selectors

Code of conduct
- Prior informed consent
- Maintenance of community biodiversity register

Community benefits
- Community gene fund
- Assistance to strengthening in situ on-farm conservation traditions
- Contract cultivation of local strains of food and medicinal plants and enlisting commercialization as an ally in conservation

Table 1. Some valuable land types preserved by Orissa tribal farmers for use in their religious functions

Rice variety	Predominant quality	Festival	Time of maturity (month)
Kalakrishna	Scented	All festivals	January
Tulsi (*Ocinmum sanctum*)	Scented	Chaitra Parva	April
Haldichudi	White slender long grains, good taste	Shakti Puja	December
Machhakanta	White slender long grains, good taste	Manabasa and Laxmi Puja	November
Mer	Black grains with medicinal properties	Annual ceremony of forefathers	November
Deulbhoga	Bold and short grains, reddish tinge on cooking with mild scent preferred during worship at temples	Temple deities	December

c) Community Gene Bank and Herbarium (CGB) – Cryogenic preservation; Evidence for getting tribal families reward and recognition under the proposed Plant Variety Protection and Farmers' Right Act.

FGBs are basically in situ on-farm centers of conservation. Landraces and location-specific plant genetic resources (PGR), usually identified on a participatory basis with the local families, are conserved in FGBs. The local landraces are periodically grown in their native habitats for seed renewal. They serve as effectively decentralized and highly cost-effective arms of a CGMS.

Those FGBs which are situated at higher elevations are likely to have cooler and conducive climates for storing seeds without loss of viability. No expensive infrastructure is needed to set up FGBs. The FGBs would ensure in situ on-farm maintenance of landraces and preservation of the cultural and cropping practices under which PGR acquired their distinctive traits. The communities in natural resource areas would be stakeholders in the FGBs. They can then provide active support not only to prevent genetic erosion in those areas, but can add new genetic stocks as and when they discover them. FGBs serve the cause of both conservation and continuous evolution through selection, recombination and mutation.

Several FGBs can be linked to an ASB taking into account factors like distance, communicability, conservation space and the like. There could be two to three seed banks in a district. The ASB will help to strengthen coping mechanisms for facing the problem of seed scarcity caused by drought-induced crop failures. The ASB would be managed by a committee of the stakeholders. Logically this committee can monitor IPR issues such as prior informed consent and benefit-sharing. One or more ASBs will be linked to the CGB. The CGB would hold ex situ the seed stocks of landraces, etc., along with herbarium sheets and other information needed to get the primary conservers reward and recognition under the proposed Indian Act for Plant Variety Protection and Farmers' Rights. The CGB could deposit duplicate sets of accessions in the National Gene Bank (NGB) of the National Bureau of Plant Genetic Resources at New Delhi.

The operational efficiency, low transaction costs and conservation potential of such a network will encourage their replication and implementation at the state and national levels. Thus, an effective integrated national gene management system can be built up.

The CGMS is being introduced in three predominantly tribal districts of Orissa, namely Koraput, Khandamal and Kalahandi. Koraput district is a primary center of origin (particularly of the aus ecotype) of rice. Though genetic variability in rice has been well chronicled, there is a need to purify the seeds of landraces, multiply and evaluate them as they are direct and indirect sources of valuable genes.

Among the millets, ringer millet (*Eleusine coracana*) has been popular as a tribal crop for use as staple. Sorghum is another crop generally grown by the tribal families in their backyards and the grains are used as "pop" sorghum. The rich variability in "pop" sorghum needs to be conserved and used sustainably. Among the pulses (i.e., grain legumes), perennial types of *Cajanus cajan* are widely cultivated. They are well adapted to rainfed condition and poor soils.

Khandamal district is located in central Kalahandi in the south-western and Koraput in the southern region of Orissa. These tribal districts have been chosen for fostering the growth of the first community-centered Integrated Gene Management System in India.

Science and Society - A New Social Contract. The primary partners in this project are the economically and socially underprivileged tribal women and men. In addition to the revitalization of the traditional conservation ethic of local communities, such a system will help to implement the conservation and equitable sharing provisions of the Convention on Biological Diversity (CBD, particularly articles 8 and 15). The CGMS provides an opportunity for fostering symbiotic partnerships between rural/tribal women and men and scientists in areas like participatory breeding and the development of new varieties adapted to local conditions using novel genetic combinations provided by genetic enhancement centers.

Alien Species Invasion

Alien invasive species are posing a serious threat to biodiversity conservation worldwide. Such species include introduced weeds, invertebrate and vertebrate pests and plant and animal diseases. The growing volume of trade and new trade routes are increasing the prospects for the unconscious introduction of alien invasive species. The problem is particularly serious in developing countries due to inadequate quarantine arrangements. The existing sanitary and phytosanitary measures are proving to be insufficient to eliminate the risk of introduction of invasive species through foodgrain shipments by "Aid agencies", as well as through trade in horticultural and agricultural products and animal pets.

Article 8(h) of CBD calls upon the parties to the Convention to "prevent the introduction of, control or eradicate those alien species which threaten ecosystems, habitats or species."

The water hyacinth, which is a native of the Amazonian region, and which has now spread worldwide, is a good example of an aggressive alien invasive species. In many countries, thick mats of water hyacinth impede navigation, fisheries, displace native aquatic flora and fauna, and interfere with irrigation, dams and hydroelectric power generation.

The north-eastern region of India is classified as a mega-biodiversity area. Unfortunately, this region has also become a "hot spot" area with reference to threats to biodiversity. The region is characterized by the continued prevalence of shifting cultivation leading to the depletion of forest resources. Alien invasive species like *Eupatorium*, *Mekenia* and *Lantana* have spread fast in the post-shifting cultivation areas. Unique mammals like *mithun* are now finding it difficult to get adequate food, since the alien species are non-edible. This is one of the causes for the dwindling population of *mithun*.

Every country, as proposed under Article 8(h) of CBD should develop a strategy for the prevention, control and ultimate eradication of alien invasive species. Biological control and, where feasible, finding economic uses for such species will help in managing the menace posed by such species.

This is also an area where the opportunities provided by recombinant DNA technologies may help. The possibility of causing seed abortion through approaches popularly christened as "terminator mechanism" could be explored in the case of species such as *Parthenium* which also cause human health problems like dermatitis. Since such research may not be of interest to private sector R&D institutions, research on the control of alien invasive species through genetic modification techniques will have to be supported from public funds.

Genetic Homogeneity

Genetic homogeneity associated with mono-cropping and modern agriculture leads to the replacement of large numbers of local varieties with a few high yielding strains. The dangerous consequences of covering large contiguous areas with one or two genetic strains are now well known. Varietal diversification and crop rotations involving crops with non-overlapping pest sensitivity are important for sustainable agriculture. The transition from "Green Revolution" to an "evergreen revolution" involves the substitution of a commodity-centered approach to a farming systems approach. The farming systems approach involves the adoption of mixed farming (crop-livestock-fish) methodologies, based on an integrated natural resources conservation strategy. Participatory breeding including the use of novel genetic combinations arising from genomics and molecular breeding will help to combine advances in yield and quality with location-specific adaptation.

In most developing countries, particularly in India and China which have fifty percent of the global farming population, continuous advances in farm productivity per units of land, water and energy are essential for sustainable food security. Hence, there is need for developing and disseminating ecotechnologies, based on appropriate blends of traditional wisdom and technologies with bio-information, space and renewable energy technologies. The ecological prudence of the past and the fruits of contemporary innovation can then be combined in a symbiotic manner.

With the growing privatization of plant breeding and expansion of proprietary science, it is important that an ecologically, economically and socially sustainable farming systems policy is developed for each agro-ecosystem. Such a policy will have to be developed jointly by farming families, official extension agencies and private sector companies. Unilateral introduction in large areas of one or two genetically modified strains of important food crops could cause irreparable harm in a few years' time both to food security and technological credibility. The pathway to an evergreen revolution on the farm is the adoption of integrated natural resource and gene management strategies.

Summary and Conclusions

Biodiversity Management, Transition from Social Exclusion to Inclusion

The threats to biodiversity arising from habitat destruction, alien species invasion and genetic homogeneity are increasing worldwide. While on the one hand, species extinction is taking place unabated, the realization that national and global food, health, livelihood and environmental security systems depend upon the conservation of biodiversity is also growing. Genomics and molecular breeding have opened up uncommon opportunities for creating novel genetic combinations which can help humankind to face new challenges arising from population increase and potential adverse changes in temperature, precipitation, ultraviolet-â radiation and sea level. A time has come when further analyses of the causes of biodiversity loss alone are not enough. Effective and implementable methods of stopping further genetic erosion and fostering the rehabilitation of degraded eco-systems in megabiodiversity regions are the need of the hour. The remedial action which needs to be taken, will have to be different in the following three methodologies of conservation.

In situ conservation through a national grid of protected areas. In this system of conservation, the following paradigm shift in management systems will be needed to ensure the achievement of goals for which the protected area networks exist.

Participatory management of forests and habitats rich in biodiversity: Participatory systems of forest management involving both the forest dwellers and forest-dependent communities, and the government forest and wildlife departments will have to be fostered. Unless the livelihood security of local communities and the ecological security of protected areas are linked symbiotically, conservation will be a lost cause under conditions of poverty and deprivation. Local communities should have access to non-wood forest products, so that they develop an economic stake in the conservation of the forests.

Biosphere trusts: Unique biosphere areas should be managed in a manner that all the stakeholders hold the biosphere in trust for the generations yet to be born. Management of the biosphere in a trusteeship mode will result in every group of stakeholders understanding their rights and responsibilities.

Harmonizing, conservation and commercialization: Commercialization can become an important instrument of conservation, if commercial companies, such as pharmaceutical firms, will help rural and tribal families to cultivate on contract rare medicinal plants. This will help them to source the raw material they need, without directly exploiting plants growing in forest canopies. Many important medicinal plants are now in the Red Data Books of several developing countries due to the unsustainable exploitation of such plants from the wild. Domestication of economically valuable Red Data Book species will help the cause of conservation, while at the same time meeting the needs of the commercial user.

Ex situ conservation. The management of Gene Banks, Botanical and Zoological Gardens and other ex situ conservation centers should be restructured so as to include the equity and ethics principles enshrined in the CBD. Provisions for prior informed consent and equity in benefit-sharing should be incorporated in voluntary codes of conduct. Gene Bank Managers should ensure that the accessions maintained by them for public good research are not used in a monopolistic manner. Material and Knowledge Transfer Agreements should form the basis for the supply of the material and knowledge gathered from the primary conservers of genetic resources and holders of traditional wisdom.

Community conservation. Urgent steps are needed, particularly in countries in which centers of diversity/origin of economic plants occur, for the revitalization of the in situ on-farm and ex situ on farm conservation traditions of tribal and rural families. No further time should be lost in introducing a recognition and reward system for giving social prestige to the conservation ethics of farming and rural families. Since such recognition will be for entire communities and not individuals, community systems of recognition and reward should be developed. These should be accorded legal status through sui generis systems of Plant Variety Protection and Farmers' Rights. Community Gene Funds based on a cess on all agricultural produce can be established for rewarding communities who, through in situ on-farm conservation, have helped to maintain rich intra-specific variability in economic plants. Also, a CGMS comprising FGBs, ASBs and CGB could be organized.

Alien Species Invasion

With increasing liberalization of trade in foodgrains and other agricultural commodities, the threat posed by alien species invasion to native biodiversity is increasing. A global strategy for the control of such invasion comprising the following collaborative steps is urgently needed: a) Improving sanitary and phytosanitary measures; b) Introducing effective methods of biological control; c) Finding economic uses for the invasive species, so that local communities can participate in their removal; d) Mobilizing the new tools of genetic engineering to render such aggressive weeds sterile and thereby incapable of sexual propagation. An integrated strategy for managing the problems created by alien species will help eliminate the threats to native terrestrial and aquatic biodiversity.

Genetic engineering research to introduce infertility in aggressive invasive species like *Parthenium*, *Mekenia, Eupatorium*, and *Lantana* will have to be supported from public funds, since such research will not attract the attention of commercial companies. If research priorities are entirely dictated by market opportunities, orphans will continue to remain orphans. Scientific progress will then have little meaning for the poor and the public.

Avoiding Genetic Homogeneity Informing Systems

Since the time modern plant breeding started with the rediscovery of Mendel's laws of inheritance in 1900, it has been known that genetic homogeneity in crop plants enhances genetic vulnerability to pests and diseases. There is therefore apprehension among scientists about the potential adverse impact of growing GM varieties of crops like rice, wheat, maize, potato and soybean over large contiguous areas. Commercial biotechnology companies recommend a refuge area in conjunction with pest resistant GM crops, in order to promote a resistance management strategy which will help to prevent the breakdown of pest resistance in the GM variety. Nevertheless, proprietary control of new strains may lead to a situation where the same genetic strain is widely grown.

Sustainable agriculture needs for its success a wide range of locally adapted varieties. Under the integrated gene management strategy, participatory breeding with farm families should be encouraged. Also, institutions working for public good should create novel genetic combinations under a well-designed pre-breeding program and make such material available to grassroot farmer breeders who could incorporate the novel genetic traits for resistance to biotic and abiotic stresses in locally adapted varieties. Thus, advanced molecular breeding techniques can be brought to the service of resource-poor farm families. In this manner, the techniques of prebreeding and participatory breeding can be blended in a manner that the requirements of ecological, economic and social sustainability are concurrently satisfied.

Gender Dimensions of Biodiversity Management

It is important that in all the above aspects of a community-centered integrated gene management strategy, gender roles in all aspects of genetic resources conservation, sustainable use and equitable sharing of benefits are given attention. Women in many developing countries are the primary seed selectors and savers. Their contributions to the evolution of a biodiversity conservation ethic should be fully recognized in any system which is designed to operationalize the equity in benefit-sharing provisions of CBD.

If we promote worldwide a community-centered integrated gene management strategy, we will soon stop hearing about vanishing species and vanishing landscapes and habitats. None of the policies and procedures I have suggested is difficult to implement. What is needed is as much interest in preventing species extinction and biodiversity loss, as in the chronicling of threats and preparation of Red Data Books.

Today, Governments and the scientific community often play the dominant role in organizing and managing structured in situ and ex situ conservation measures. By recognizing that conservation efforts represent a continuum, with rural and tribal families performing a vital function in preserving precious genetic variability in important plants and farm animals, we will be able to attend to all the links in the conservation chain. Community conservation is a value-added link in the conservation system, since local families not only conserve but also add value to the conserved material through selection and information.

Let me conclude with the hope that if the 20th century was a period of understanding and chronicling threats to biodiversity and bioresources both in land and water, the 21st century will be one where the threats are terminated and benefits harnessed for a better common present and future for humankind. ■

References

Colborn, T., D. Dumanoski and J.P. Myers. 1996. *Our Stolen Future.* New York: Dutton.

Gadgil, M. and V.D. Vartak. 1975. *Bombay Natural History Society.* Mumbai: Natural History Society.

Poffenberger, M., P. Walpole, E. D'Silva, K. Lawrence and A. Khare. 1997. *Asia Forest Network - Research Network Report, No. 9.*

Singh, Samar, A. Avenash Datta, Arvind Bakshi, A. Khare, Sushil Saigal and Navin Kapoor. 1997. *Participatory Forest Management in West Bengal.* New Delhi: World Wide Fund for Nature - India.

Swaminathan, M.S. 1999. Science in Response to Basic Human Needs. *Current Science* 77: 341-53.

-----. 1999. A *Century of Hope: Towards an Era of Harmony with Nature and Freedom from Hunger.* Chènnai, India: EastWest Books (Madras).

32

Acceptance of the Franklin Delano Roosevelt Four Freedoms Medal

Your Majesty, Your Excellencies, Ladies and Gentlemen:

My wife Mina and I feel highly privileged to be here at the beautiful and historic Abbey. It is a particular privilege to receive the Roosevelt Four Freedoms Medal in the gracious presence of Her Majesty the Queen.

Your Majesty, your father Prince Bernard has been a pioneer in the areas of environment protection and sustainable development. As the founder President of the World Wildlife Fund (WWF), he has been a source of great inspiration to all of us in our struggle to safeguard and improve our life support systems.

Your Majesty, while your father has been a leader in the field of environmentally sustainable development, your husband Prince Klaus has been a leader in socially sustainable development. He has been a strong advocate of mainstreaming ethical and equity considerations in all areas of human endeavor. Thus, your father and husband have laid the foundations for environmentally and socially sustainable development, which are fundamental pre-requisites for achieving freedom from hunger and poverty.

My sense of happiness at being here is enhanced by two more factors. First, the Netherlands is the first country outside my own, where I had the opportunity of benefiting from higher education. During 1949-50, I studied at the Landbouw Hogeschool at Wageningen, where I learnt not only the science and art of crop improvement, but also that the term "impossible" exists primarily in the minds of human beings, since the Dutch showed that seemingly impossible tasks like recovering vast areas of land from the ocean can be achieved through clear vision supported by appropriate action.

Secondly, I recall as a University student during World War II, my great admiration for President Franklin Delano Roosevelt. Like most young students of those days in India, I was an ardent admirer and follower of Mahatma Gandhi, about whom Albert Einstein once remarked: "It is difficult to believe that such a person ever walked on our earth in flesh and blood." Gandhi's principles of *antyodaya* (concern for the poorest person) and *sarvodaya* (harmony between the interests of individuals and the community) have guided my life and work. This is why our Foundation established at Chennai (Madras) ten years ago is devoted to imparting a pro-nature, pro-poor, pro-women and pro-livelihood orientation to technology development and dissemination.

Acceptance speech at Franklin and Eleanor Roosevelt Institute and the Roosevelt Stichting, Middelburg, The Netherlands, May 27, 2000.

Commending President Roosevelt for his efforts to bring World War II to a speedy conclusion, Gandhi wrote to him on July 1, 1942, and I quote: "Many of my countrymen have received and are still receiving higher education in America. I know too that several have taken shelter there. I have profited greatly by the writings of Thoreau and Emerson. I say this to tell you how much I am connected with your country."

Seeking President Roosevelt's intervention in getting independence from British rule, Gandhi said in the same letter, "I would like to turn into goodwill the ill-will which exists in India towards Great Britain and thus enable the millions of India to play their part in the present war." Roosevelt's constructive role in the emergence of Independent India is widely recognized. What is, however, not so well-known is the pivotal role Roosevelt played in bringing into existence a global institution dedicated to the cause of ending hunger, namely, the Food and Agriculture Organization of the United Nations (FAO).

Early in 1943, Mrs. Eleanor Roosevelt convened a small group, which included Frank L. McDougall of Australia, to discuss future cooperation among nations at the end of World War II. It was agreed at that meeting that food, as a problem common to all, should form the base for cooperation. This led to President Roosevelt convening a Conference on Food and Agriculture in Hot Springs, Virginia, from May 18 to June 3, 1943.

This Conference was attended by forty-four nations including The Netherlands and India. The Conference concluded: "Meeting in the midst of the greatest war ever waged and in full confidence of victory, it declares its belief that the goal of freedom from want of food, suitable and adequate for the health and strength of all peoples, can be achieved." The Hot Springs Conference led to the establishment of FAO on October 16, 1945 at Quebec in Canada.

Thus, FAO owes its birth to the initiative of Eleanor and Franklin Roosevelt. Their vision of a hunger-free world is however yet to be achieved, since at least eight hundred million children, women and men will go to bed partially hungry even tonight, more than fifty-five years later. Millions of newborn children are characterized by low birth weight due to maternal and fetal under-nutrition, endangering the realization of their full genetic potential for mental and physical development. Economic, ecological, technological and social inequity is increasing in the world.

My work during the last forty-five years relating to a hunger-free India and world has been greatly influenced by President Roosevelt's concept: "The test of our progress is not whether we add more to the abundance of those who have much; it is whether we provide enough for those who have little." Translated into action, this has meant concentration on improving the productivity of small farm holdings and the well-being of the economically and socially under-privileged women and men who toil day and night and in sun and rain to produce food for their fellow human beings. I am firmly convinced that hunger and deprivation can be eliminated sooner than most people consider feasible, provided there is synergy among technology, public policy and social action. This is the major lesson we can learn from the farm revolution that transformed India's agricultural destiny during the last part of the last century.

Differences in education and technological empowerment have played an important role in the skewed distribution of wealth and power we now see within and among nations. I have observed in very poor parts of India as well as in many countries of Asia and Africa that the poor take to new technologies like fish to water, provided they are enabled to do so through opportunities for "learning by doing" and for producer-oriented marketing. Our work in rural India has demonstrated that technology can become a powerful ally in the movement for gender and social equity, if programs for knowledge and skill empowerment are demand-driven, user-controlled and economically meaningful. There are uncommon opportunities now for every woman and man to achieve freedom from want, since the growing rich-poor divide is not due to innate differences among the people themselves but more due to human-made differences in their environments, which can therefore be overcome by human effort. Striving to create a political, social, economic and technological climate, which will help to create opportunities for every human being to lead a productive and healthy life and ensure that children are "born for happiness and not for mere existence", is the best tribute we can pay to Franklin and Eleanor Roosevelt.

Gandhi advised, "Be the change you want to bring about". By internalizing in our thought and action Roosevelt's Four Freedoms, we can contribute to making the knowledge and skills we posses, be they little or much, instruments for promoting harmony within humankind and between humankind and nature. Modern information and communication technologies have made it possible to reach the unreached and include the excluded in terms of knowledge and skill empowerment. At the dawn of a new century and millennium, we recognize that a better common present is essential for a happy common future. This will call for concerted efforts to end both unsustainable lifestyles and unacceptable poverty. This is the pathway to achieving for all and forever freedom from want and fear.

I once again thank you for giving me this opportunity to rededicate myself to the cause of ending food insecurity at the level of every child, woman and man on our planet.

62. With the Queen of Netherlands after receiving Franklin D. Roosevelt Four Freedoms Medal, 2000

33

A Century of Mendelian Breeding: Impact on Wheat

Introduction

The rediscovery of Mendel's laws of genetics in 1900 opened up a new era in crop breeding in general, and wheat breeding in particular. Although the art of plant breeding is as old as the beginning of agriculture nearly 12,000 years ago, systematic research in the areas of genetics and cytogenetics, which commenced in the early part of the 20th century, created uncommon opportunities for improving the productivity, profitability, stability and sustainability of wheat production. Even before Mendel (1822-1884), plant hybridizers like Kolreuter, Knight, Gartner and Burbank were able to produce improved varieties of crops through careful observation and selection. Even the concept of sustainability to which we now attach great importance was recognized long ago as essential for sustained agricultural progress. For example, the Roman farmer, Varro, explained the basic principles underlying sustainable agriculture in the following words over 2,000 years ago: "Agriculture is a science which teaches us what crops should be planted in each kind of soil, and what operations are to be carried out, in order that the land may produce the highest yields in perpetuity."

During the past one hundred years, Mendelian genetics has helped not only to exploit naturally occurring genetic variability but has also accelerated the process of generation, manipulation and combination of new variability. For example, H.J. Muller, the pioneer in induced mutagenesis, wrote in 1935: "Organisms are found to be far more plastic in their hereditary basis than has been believed, and we may confidently look forward to a future in which the surface of the earth will be overlaid with luxurious crops, at once easy to raise and to gather, resistant to natural enemies and climate, and readily useful in all their parts."

Keynote Address at the 6th International Wheat Conference, Budapest, Hungary, June 6, 2000. First published in *A Century of Mendelian Breeding: Impact on Wheat in a Global Environment,* ed. by Z. Beda and L. Lang, pp. 5-21. Netherlands: Kluwer Academy Publications, 2001.

The author is very grateful to Dr. Prem P. Jauhar, Professor of Cytogenetics, State University Station, Fargo, North Dakota, USA and to Dr. S. Nagarajan, Project Director, All India Wheat Project, Karnal, India, for their kind help in assembling data relating to the progress in the improvement of wheat yield and production in USA and India respectively.

Muller's prediction has since come true. The term "Green Revolution" coined in 1968 by Dr. William Gaud was triggered by the quantum jump in wheat productivity and production brought about by the semi-dwarf wheat varieties originating from the wheat breeding program of the International Maize and Wheat Improvement Center (CIMMYT) in Mexico, headed by the Nobel Laureate Norman Borlaug. In India, a special stamp titled "Wheat Revolution" was released in July 1968, by the then Prime Minister of India, Mrs. Indira Gandhi (Figure 1). This stamp helped to raise public awareness of the role of science in bringing about revolutionary progress in wheat production. In India, while there was a decline in per acre yield during the period 1905-1950 (Figure 11), a linear positive increase in wheat yield has been observed between 1955-99 (Figure 12).

We are now in a state of transition from Mendelian to molecular breeding. However, it is clear that for success in crop improvement, we will always depend upon an appropriate blend of Mendelian and molecular breeding. Breeder's eye for selection and for spotting the winner will continue to play an important part in successful plant breeding.

During the period 1900 to 2000, the following significant developments took place with reference to the application of Mendelian genetics to wheat improvement: Selection from naturally occurring genetic diversity; Collection, conservation, evaluation and utilization of wheat genetic resources; Introduction of new varieties from other countries; Hybridization and selection for a wide range of biotic and abiotic stresses and for grain quality; Induced mutagenesis and mutation breeding; Monosomic and aneuploid analysis; Wide hybridization and alien gene transfer; Exploitation of Heterosis, using genetic and chemical methods of male sterility; Apomixis and fixation of heterosis; DNA marker assisted selection and breeding; Functional genomics and molecular breeding.

It should be recorded that, in the early part of the 20th century, the first major success in achieving a yield revolution in crops was through the exploitation of hybrid vigor in maize. It is said that hybrid maize helped not only to revolutionize maize production, but also helped to improve the efficiency of farm management in the United States of America. This is because of the fact that farmers who learnt the value of good agronomic management of soil health and water use in maize also began to adopt good farm management practices in other crops.

This is also true in India. The areas where the wheat revolution began, such as the Punjab, Haryana and Western Uttar Pradesh, also became the centers of origin of the rice, potato, milk and other forms of agricultural revolution. The radiating impact of the wheat revolution, which began with the breeding of semi-dwarf strains, has been immense, since unproved genetic strains triggered a farm management revolution.

To quote Dr. Norman Borlaug, "there is no time to relax". It is projected that global demand for wheat will increase by forty percent by the year 2020. Also, sixty-seven percent of the world's wheat consumption will be in developing countries. Between 1961 and 1990, yield increases accounted for ninety-two percent of the additional cereal production in developing countries. In the years ahead, there is no option except to produce more from less per capita land and water resources. Can we sustain the yield revolution in wheat? It will be useful to consider this issue in the context of the genetic pathways which led to the wheat revolution of the 20th century.

Progress in Yield Improvement

Wheat is a crop of great antiquity. The ancestry of the cultivated bread wheat, *Triticum aestivum* (2n = 6x = 42) has been studied in detail. Synthetic hexaploids have also been produced by crossing *Triticum turgidum* (2n = 4x = 28) and *Triticum tauschii* (2n = 2x = 14) (Figure 2). Wheat is the first major cereal where extensive cytogenetic analysis by several scientists like E.R. Sears, Ralph Riley, Hitoshi Kihara, Otto Frankel and James McKey led to important advances in crop improvement. The development of a complete set of monosomics in the wheat variety, Chinese Spring, by E.R. Sears and his colleagues, helped breeders to plan their work based on the location of genes in different chromosomes. Analysis of induced mutations further helped in understanding genetic relationships among *Triticum* species.

We can identify at least four major phases in the evolution of wheat breeding during the 20th century.

Phase I (1900 -1930): Early Days of Mendelian Genetics. Soon after the re-discovery of Mendel's laws of genetics in 1900, systematic work on the genetics of resistance to stem, leaf and stripe rusts started. Selection from naturally occurring genetic variability also began. For example, in India an early improved wheat variety, NP 4, originated through simple selection. Soon, hybridization work started involving the use of landraces. The CIMMYT publication titled, *World Wheat Facts and Trends* (1995) gives data on the extensive use of land races in the breeding programs of all wheat growing countries. In recent years, the contributions of landraces to the pedigrees of successful varieties has increased (Figure 3).

According to CIMMYT (1995), the more widely used landraces are: Yaroslav Emmer (an emmer wheat from the former Soviet Union), Turkey (a landrace grown in the Crimea by Turkish farmers), Fife (believed to have originated in Poland), Daruma (source of dwarfing genes and believed to have originated in Korea), Rieti (an early maturing Italian landrace) and Zeeuwse White (a Dutch landrace). Some of the frequently used landraces from developing countries are: Hard Red Calcutta, Etawah, Indian G (all from India); and Alfredo Chaves and Polyssu (from Brazil).

During the early part of the 20th century, the major breeding challenges were in the area of resistance to rusts and grain quality improvement. A study of the yield improvement achieved between 1900 to 1930 in USA shows only limited progress (Figure 4). The emphasis was more on stability of production through disease resistance than on achieving quantum jumps in yield.

Phase II (1930 -1960): Enlarging the base of theory and its application. This period was marked by the introduction of cytogenetic knowledge and tools in wheat improvement. The search for genes for yield improvement started. The first breakthrough came when Japanese scientists developed the Norin 10 semi-dwarf wheat, with Daruma as the donor of the semi-dwarf character (Figure 5). The Norin semi-dwarf wheats brought into the United States from Japan by Dr. S.D. Salmon led to the breeding of the semi-dwarf winter wheat variety, Gaines, by Dr. Orville Vogel in the late fifties. The same material was used by Dr. Norman Borlaug in Mexico for developing semi-dwarf spring wheat (Borlaug, 1989). Other important concepts introduced during this period were shuttle breeding and international testing nurseries. The shuttle breeding program developed by Dr N.E. Borlaug in Mexico involved growing alternate generations under two diverse environments. The locations were: a) The Yaqui Valley located at 27° N latitude and forty-nine meters elevation located in the state of Sonora in north-west Mexico; b) Toluca, located at 18° N latitude and twenty-six hundred meters elevation in the state of Mexico.

These locations differed in soil type, temperature, rainfall and photoperiod. The shuttle breeding procedure led to the selection of strains possessing relative insensitivity to photoperiod as well as a broad spectrum resistance to stem rust. In addition, it helped to reduce considerably the time needed to breed a new variety.

Another important feature of this era was accelerated breeding and testing for resistance to rusts and other diseases. As early as in 1898, Farrer in Australia recognized the importance of general resistance to rust in wheat. With the discovery of physiological specialization in rust by Stakman et al (1962) and the clarification of the genetic basis of resistance (Biffen, 1905) the hypersensitive type of resistance was incorporated in many wheat varieties. Van der Plank defined clearly the theoretical basis of genetic resistance to rusts. Thus, this phase of wheat improvement was characterized by widening the gene pool used by breeders, incorporation of genes for the semi-dwarf plant type, shuttle breeding and breeding to meet the challenge of physiologic specialization in pathogens.

Phase III (1960-1980): The Green Revolution Phase. This phase is also generally referred to as the Green Revolution era. It was characterized by revolutionary progress in improving wheat production and productivity in several developing countries like India and Pakistan (Figure 6). The introduction of the semi-dwarf plant type enabled the wheat plant to yield well under conditions of good soil fertility and irrigation water management (Figure 7). Farmers who were used to harvesting one to two tons of wheat per hectare started harvesting over five tons per ha. The various steps involved in this revolution in India have been chronicled (Swaminathan, 1993). In view of the widespread interest in this remarkable transformation in India's agricultural destiny, it will be useful to summarize some of the highlights.

The Food and Agriculture Organization of the United Nations (FAO) was established even before the end of World War II, since combating food insecurity was recognized as a major challenge of the post World War II era. During 1942-1943, the Indian sub-continent witnessed a severe famine in Bengal resulting in the death of nearly three million children, women and men. This prompted Jawaharlal Nehru, the first Prime Minister of Independent India to remark in 1948 that "everything else can wait, but not agriculture". Priority to agriculture was reflected in the form of land reform, expansion of irrigation facilities and greater support to research, extension and the production of inputs, particularly seeds and fertilizer. Fertilizer trials with varieties of wheat and rice in the fifties revealed that the then cultivated varieties lodged even when about 20 kg of nitrogen were applied per hectare.

The quest for the breeding of crop varieties capable of responding to higher levels of plant nutrition started in 1952 when, at the instance of the late Dr. K. Ramiah, a program for incorporating genes for fertilizer response from japonica rice varieties into indica strains was initiated at the Central Rice Research Institute (CRRI), Cuttack, under the sponsorship of FAO and the Indian Council of Agricultural Research (ICAR). The major aim of this project was to select from segregating populations of indica × japonica crosses, lines which showed the ability to utilize effectively about 100 kg of N per hectare. With this quantity of nutrient supply, about five tons of rice per ha can be produced. This program led to varieties like ADT 27 in Tamil Nadu and Mashuri in Malaysia. Several genetic problems like semi-sterility arose, rendering the speedy selection of high-yielding rice varieties from indica × japonica crosses difficult. With the advent, in the early sixties, of the semi-dwarf, non-lodging, relatively photo-insensitive varieties based on the Dee-gee-woo-gen dwarfing gene identified in China, interest in transferring genes for fertilizer response from japonica varieties waned. Semi-dwarf indica rices like Taichung Native-1, IR 8 and Jaya provided the initial material for the breeding of a wide range of high yielding rice varieties.

When I joined the Indian Agricultural Research Institute (IARI), New Delhi in late 1954, I started a research program for developing non-lodging and fertilizer responsive varieties of wheat, on the lines of the indica-japonica hybridization program in which I was earlier involved at CRRI, Cuttack. At that time, the research strategy adopted had three components. First, crosses were made between cultivated bread wheat varieties and the semi-dwarf, stiff straw compactum and sphaerococcum sub-species of *T.aestivum* as well as with the naturally occurring dwarf, Tom Thumb. Second, attempts were made to induce erectoides mutants in commercial wheat varieties through the use of radiations and chemical mutagens. Third, studies on the potential for increasing straw stiffness through different chemical treatments were initiated. Unfortunately, in all these three approaches, short and stiff straw was always associated with short panicles with fewer grains. The reason why straw stiffness became such an essential prerequisite for favorable response to water and fertilizer was the tendency, among the then cultivated tall wheat varieties, to lodge when mineral fertilizer was applied. Also, such lodging made it difficult to give irrigation during the grain development phase, when the crop will benefit much from water availability. Thus, with the earlier tall varieties, it was difficult to get economic response to the application of mineral fertilizers and adequate irrigation water. Average yield stagnated at less than one ton per ha. This is why the breeding of non-lodging varieties was accorded such high priority during the fifties, when India had taken to the path of expanding the area under irrigation and manufacturing mineral fertilizers (Swaminathan, 1993).

During the late fifties, scientific publications on the work done under the leadership of Dr. Orville Vogel in Washington State of the United States of America, on the transfer of dwarfing genes from the Norin-10 wheat variety to North American winter wheats, started appearing. When requested, Dr. Vogel was kind enough to send seeds of Gaines, a semi-dwarf winter wheat variety with red grains. He further indicated that Dr. N.E. Borlaug in Mexico had semi-dwarf varieties in a spring wheat background, which will grow better under the short day conditions prevailing in India during winter months. I hence wrote to Dr. Borlaug seeking his help.

In March 1962, a few dwarf spring wheat strains entered by Dr. Borlaug in the International Wheat Rust Nursery and sent by the US Department of Agriculture were grown in the fields of the Indian Agricultural Research Institute (IARI). Their phenotype was most impressive. They had reduced height and long panicles, unlike the earlier hybrids between *aestivum* and *compactum* and *sphaerococcum* and the induced erectoides mutants which had short height and small panicles. In 1963, Dr. Borlaug visited India and sent a wide range of semi-dwarf material which provided the initial material for achieving an accelerated advance in wheat productivity and production.

In 1964, a National Demonstration Program was started in farmers' fields, both to verify the results obtained in research plots and to introduce farmers to the new opportunities opened up by semi-dwarf varieties for improving very considerably the productivity of wheat. When small farmers, who with the help of scientists organized the National Demonstration Program, harvested over five tons of wheat per hectare, its impact on the minds of other farmers was electric. The clamor for seeds began and the area under high yielding varieties of wheat rose from four ha in 1963-64 to over four million ha in 1971-72. A small Government program became a mass movement. The rest of the history is recorded in a book on the Wheat Revolution (Swaminathan, 1993).

I anticipated in 1967 a tendency among farmers in north-west India with relatively large holdings to use high doses of fertilizers and to grow in large and contiguous areas a single genetic strain. In my Presidential Address to the Agricultural Sciences Section of the fifty-fifth Science Congress held at Varanasi in January 1968, I stressed in the following words the need for adding the dimension of ecological sustainability in programs designed to achieve high yields per ha. "Intensive cultivation of land without conservation of soil fertility and soil structure would lead, ultimately, to the springing up of deserts. Irrigation without arrangements for drainage would result in soils getting alkaline or saline. Indiscriminate use of pesticides, fungicides and herbicides could cause adverse changes in biological balance as well as lead to an increase in the incidence of cancer and other diseases, through the toxic residues present in the grains or other edible parts. Unscientific tapping of underground water will lead to the rapid exhaustion of this wonderful capital resource left to us through ages of natural farming. The rapid replacement of numerous locally adapted varieties with one or two high-yielding strains in large contiguous areas would result in the spread of serious diseases capable of wiping out entire crops, as happened prior to the Irish potato famine of 1854 and the Bengal rice famine of 1942. Therefore, the initiation of exploitative agriculture without a proper understanding of the various consequences of every one of the changes introduced into traditional agriculture, and without first building up a proper scientific and training base to sustain it, may only lead us, in the long run, into an era of agricultural disaster rather than one of agricultural prosperity." (Swaminathan, 1968)

The above analysis led to the intensification of research on integrated pest management and integrated nutrient supply and to the development of a gene deployment strategy in wheat with reference to resistance to stem, leaf and stripe rusts.

The remarkable speed with which the high-yielding varieties were identified from the initial Mexican material and later developed within the country was the result of the multi-location testing and inter-disciplinary research organized under the All India Coordinated Wheat Research Project of the Indian Council of Agricultural Research. Wheat production in India rose from ten million tons in 1964 to seventeen million tons in 1968. In 1999, Indian farmers harvested about seventy-two million tons of wheat, taking India to the second position in the world in wheat production.

Rajaram has chronicled the various steps taken at CIMMYT during the Green Revolution period to internationalize wheat breeding. Some of the significant steps were: a) Exploitation of the spring × winter gene pool through a cooperative venture between CIMMYT and Oregon State University in the USA; b) Septoria leafspot resistance in semi-dwarf wheats through a cooperative venture between CIMMYT, Tel Aviv University in Israel and IPO, Wageningen, The Netherlands; c) Slow rusting genes to leaf rust were identified, quantified, and bred with initial guidance from Dr. Caldwell at Purdue University, USA; d) Industrial quality characters were emphasized; e) Breeding for resistance to aluminum toxicity was initiated through a cooperative venture between CIMMYT and several Brazilian Agricultural Research Institutes; f) Breeding programs for *durum* wheat and triticale were initiated; g) Germplasm dissemination through formalized International Nurseries for bread wheat, *durum* wheat, and triticale were established.

Emphasis on productivity improvement was coupled with attention to stability of production. Breeding for slow rusting received particular attention (Figure 8). Also, the sources of resistance were kept very diverse. This helped in developing effective gene deployment strategies to ensure that widespread epidemics do not occur. For example, a high degree of stability of performance in relation to leaf rust (*Puccinia recondita f. sp. tritici*) was achieved through introduction of genes from the Brazilian variety, Frontana.

In the case of stem rust (*Puccinia graminis f.sp tritici*), the variety Hope has been the mainstay. The Sr-2 gene complex actually consists of Sr-2 plus eight to ten genes pyramided in three to four gene combinations. Sr-2, referred to as the backbone gene in stem rust resistance breeding is a slow ruster and has conferred in wheat varieties an impressive level of durability of resistance.

For resistance to stripe rust *(Puccinia striliformis)*, more than eight genes have been identified and gene pyramiding has been the major pathway in resistance breeding. During the Green Revolution era, Karnal bunt caused by *Tilletia indica* assumed importance. Several winter × spring wheat crosses showed wide adaptation and stability. A major factor for such wide adaptation was the 1 BL/1 RS translocation. The 1B/IR translocation carried several resistance genes, such as Lr 26, Sr 31, Yr 9 and Pm 8 (McIntosh, 1983).

During this period, crosses with *Secale cereale*, *Triticum dicoccoides*, *Triticum tauschi* and other alien species transfers were intensified. CIMMYT rendered very valuable service in spreading diverse genetic material through international screening nurseries and trials (Rajaram and Hettel, 1995).

Thus during 1900 to 1980, the Mendelian methods of selection from segregating populations, recombination through inter-varietal and inter-specific crosses and induced mutations helped to breed both durum and aestivum wheat varieties, adapted to different photoperiods and climatic conditions. Wheat productivity and production went up all over the world. The progress achieved in USA and India is depicted in Figures 4 and 6.

Greater interdisciplinary collaboration among breeders, plant pathologists, agronomists, physiologists, soil scientists, entomologists, nemotalogists, economists and other social scientists, climatologists and policy makers was the principal factor responsible for the success of the Green Revolution. The Green Revolution era can also be termed "the golden age in interdisciplinary and international collaboration" in wheat improvement for sustainable food security. The concept of shuttle breeding transcended continental boundaries and a global college of wheat scientists emerged. Above all, the Green Revolution showed how to generate synergy between technology and public policy.

Phase IV (1980-2000):Transition from Mendelian to Molecular Breeding. The last twenty years have witnessed great progress in using sophisticated approaches to wheat breeding. Hybrid wheat is reaching the possibility of large-scale commercial cultivation. The use of genetic-cytoplasmic male sterility and of chemical hybridizing agents (CBA) are responsible for progress in the commercial exploitation of hybrid wheat. Different management practices such as lower seed rate, raised bed planting, split nitrogen application and different row width are being tried to enhance the expression of hybrid superiority. The cultivation of hybrid wheat is slowly gaining in momentum in South Africa, Australia (New South Wales), China, Argentina and France. The use of wild relatives in genetic engineering is growing (Khush and Baenziger, 1998). The global average yield of wheat is 2.5 t/ha; the low average yield of wheat is because of large areas of wheat being under rainfed conditions. Progress in improving yield is however steady, as is clear from the data of yield and area in Hungary (Figure 9). The factors governing wheat yields in Australia have been summarized in Figure 10. Some of the critical issues in yield improvement in wheat have recently been discussed in a book edited by E.H. Satorre and G.A. Slafer (1999). The explosive progress being witnessed in the areas of functional genomics and molecular manipulation will influence very much future trends in wheat breeding in the 21st century. Briggs (1998), McCouch (1998), and Qualset et al (1999) have stressed the value of expressed sequence tags (ESTs). Several thousand RFLP markers, in addition to AFLP, SSR, morphological and isozyme markers have been mapped. Close to one hundred QTLs have been identified. This should help to accelerate the pace of breeding for yield. So far, advances in yield improvement have been associated with increases in harvest index (i.e., grain-straw ratio). Further advances will depend upon greater biomass production and not merely on partitioning the phytosynthates.

Challenges Ahead

At the dawn of the 21st century, we can look back with pride and satisfaction on the revolution which farm men and women have brought about in our agricultural history during the 20th century. Writing about the role of Indian farm families in initiating the Wheat Revolution, I wrote in the *Illustrated Weekly of India* in 1969 and I quote: "Brimming with enthusiasm, hard-working, skilled and determined, the Punjab farmer has been the backbone of the revolution. Revolutions are usually associated with the young, but in this revolution, age has been no obstacle to participation. Farmers, young and old, educated and uneducated, have easily taken to the new agronomy. It has been heart-warming to see young college graduates, retired officials, ex-armymen, illiterate peasants and small farmers queuing up to get the new seeds. At least in the Punjab, the divorce between intellect and labor, which has been the bane of Indian agriculture, is vanishing."

While we can and should rejoice about the past achievements of farmers, scientists, extension workers and policy makers, there is no room for complacency. We will face several new problems, of which the following are important.

First, increasing population leads to increased demand for food and reduced per capita availability of arable land and irrigation water.

Second, improved purchasing power and increased urbanization lead to higher per capita foodgrain requirements due to an increased consumption of animal products.

Third, marine fish production is tending to become stagnant and coastal aquaculture has resulted in ecological and social problems.

Fourth, there is increasing damage to the ecological foundations of agriculture, such as land, water, forests, biodiversity and the atmosphere and there are distinct possibilities for adverse changes in climate and sea level.

Fifth, while dramatic new technological developments are taking place, particularly in the field of biotechnology, their environmental, food safety and social implications are yet to be fully understood.

Finally, gross capital formation in agriculture is tending to decline in both public and private sectors during the present decade. The rate of growth in rural non-farm employment has been poor.

Since land and water will be shrinking resources for agriculture, there is no option in the future except to produce more food and other agricultural commodities from less per capita arable land and irrigation water. In other words, the need for more food has to be met through higher yields per units of land, water, energy and time. It would, therefore, be useful to examine how science can be mobilized for raising further the ceiling to biological productivity without associated ecological harm. It will be appropriate to refer to the emerging scientific progress on the farms as an "evergreen revolution", to emphasize that the productivity advance is sustainable overtime since it is rooted in the principles of ecology, economics, social and gender equity and employment generation.

The Green Revolution based on Mendelian genetics has so far helped to keep the rate of growth in food production above population growth rate. The Green Revolution was however the result of public good research, supported by public funds. The technologies of the emerging gene revolution based on molecular genetics in contrast are spearheaded by proprietary science and can come under monopolistic control. How can we take the fruits of the gene revolution to the unreached? This is a challenge which this Congress should address.

I would like to list five major challenges which will confront the wheat breeders of the early part of this century.

a) Ecology: Ecological sustainability of high productivity will be an important determinant in relation to the choice of technologies. For example, if hybrid wheat can enable us to produce 8 to 10 t/ha, over 300 kg of nitrogen will be needed by the crop. It is obvious that if the nutrient needs of hybrid or other high-yielding wheat varieties are to be met entirely through mineral fertilizers, there will be serious environmental problems including nitrate pollution of ground water. Hence, success in achieving high productivity on a sustained basis will depend upon our ability to develop new methods of feeding the plant. Research on breeding and feeding should be carried out concurrently by a team of breeders, physiologists, agronomists and soil scientists.

b) Equity: The Convention on Biological Diversity (CBD) stipulates that plant exploration, collection and introduction should be based on the principles of prior informed consent and equity in benefit-sharing. Therefore, exchange of wheat genetic resources in the future will be possible only on the basis of Material and Knowledge Transfer Agreements.

c) Concerns Relating to Genetically Modified Organisms (GMOs): There are growing public and political concerns relating to GMOs. The concerns relate to food and environmental safety and bioethics. It is essential that these concerns are carefully addressed through a mechanism for risk-benefit analysis which inspires public confidence. An integrated disease management strategy should be developed to ensure that GMOs with novel genetic combinations for disease resistance do not break down due to the emergence of new physiological strains of pathogens.

d) Expansion of Proprietary Science: The world is witnessing an expansion of proprietary science governed by Intellectual Property Rights (IPR). Public good research supported from public funds, in contrast, is shrinking. What will be the impact of such a situation on international varietal or other trials organized by CIMMYT? Is the golden age of cooperative research coming to an end? How can we find a balance between public good and private profit?

e) Climate Change and Safeguarding Genetic Diversity: Will molecular breeding resulting in "super wheats" lead to a high degree of genetic homogeneity in farmers' fields? We know that genetic homogeneity will enhance genetic vulnerability to biotic and abiotic stresses. Hence, we should foster an integrated program of pre-breeding and participatory breeding. Pre-breeding will help to generate novel genetic combinations, while participatory breeding with farm families will help to combine genetic efficiency with genetic diversity. Numerous location specific varieties can be developed in this manner. This will be the most effective in meeting challenges arising from potential changes in temperature, precipitation and sea level as a result of global warming arising from the growing imbalance between carbon emissions and absorption.

To conclude, wheat geneticists, breeders, scientists and policy makers should continue to work together and share ideas and material, if the challenge of sustainable food security is to be faced successfully. This Conference, being held at the beginning of a new century and millennium, should demonstrate to the world the power of partnership. Let us increase the genes for cooperation both in wheat varieties and among wheat researchers.

Figure 1.

Figure 2. Schematic showing the production of synthetic hexaploids derived by crossing *Triticum turgdum* × *T. tauschii* and their utilization

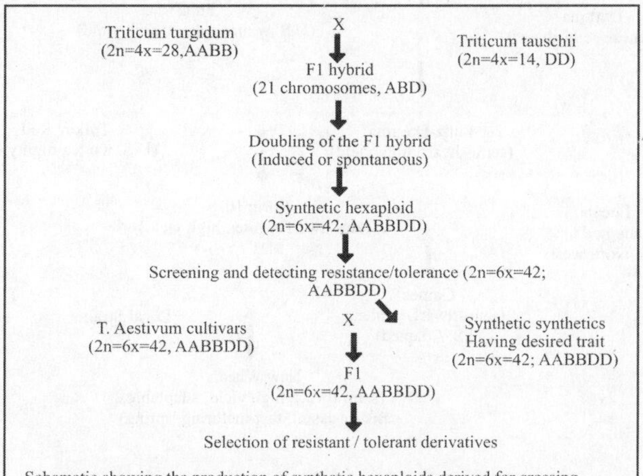

Schematic showing the production of synthetic hexaploids derived for crossing Triticum turgdum x T. Tauschii and their utilization

Source : Biodiversity in Trust

Figure 3.

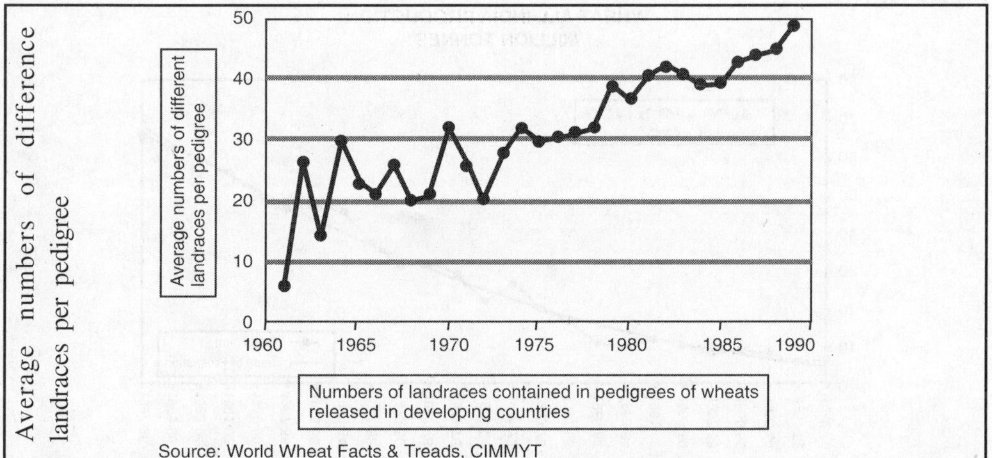

Numbers of landraces contained in pedigrees of wheats released indeveloping countries

Figure 4.

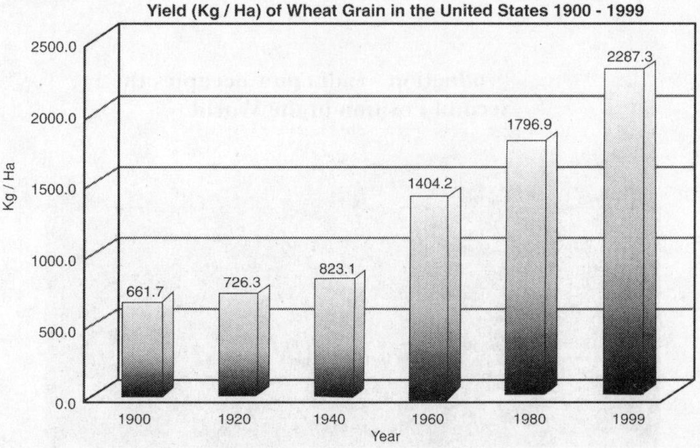

Figure 5.

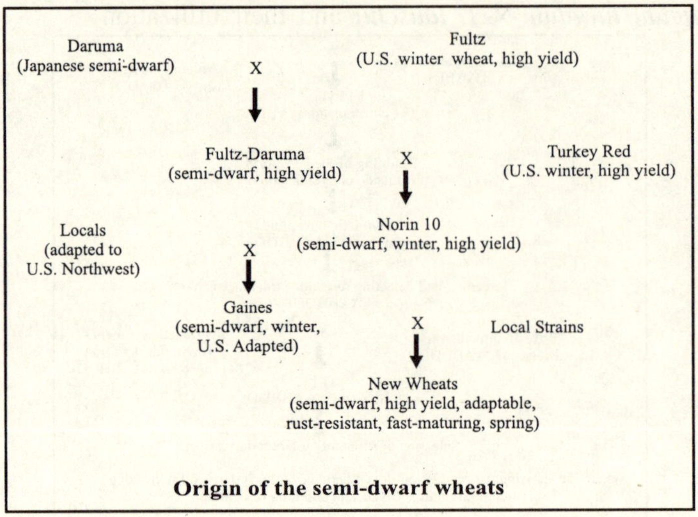

Figure 6.

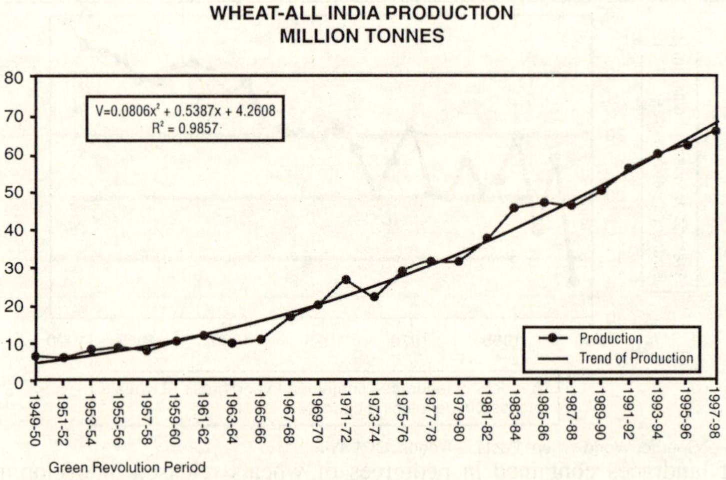

Source: Agricultural Research Data Book

Figure 7.

Figure 8. Wheat Breeding at CIMMYT (1994)

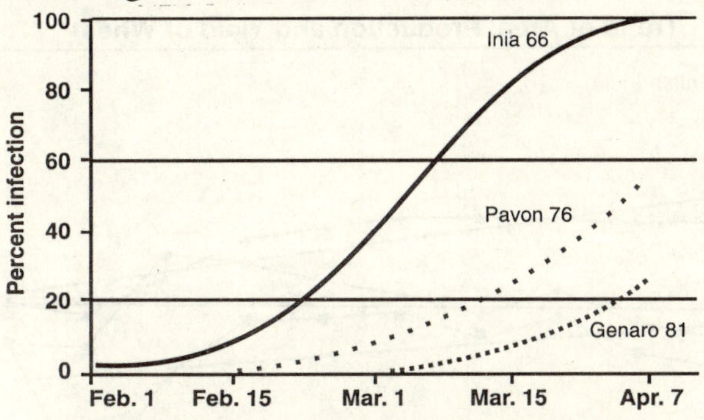

Slow resting resistance of Genaro 81 and Pavon 76 to leaf rust when compared to Inia 66 (Cd. Obregon 1984-85)

Source : Wheat Breeding at CIMMYT (1994)

Figure 9. The Hungarian Agriculture and Food Industry in figures

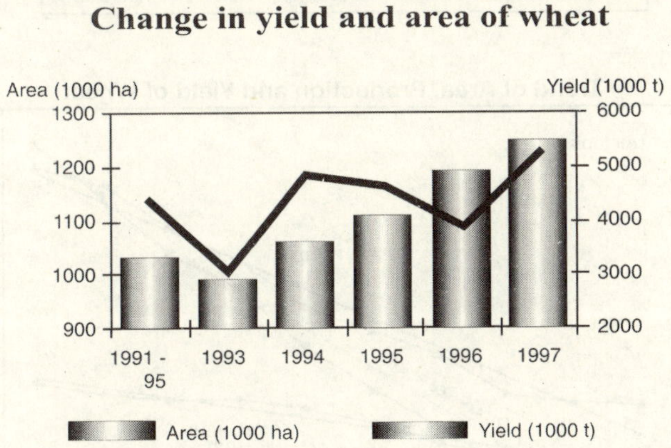

Source: The Hungarian Agriculture and Food Industry in Figures

Figure 10. Wheat Ecology and Physiology of Yield Determination

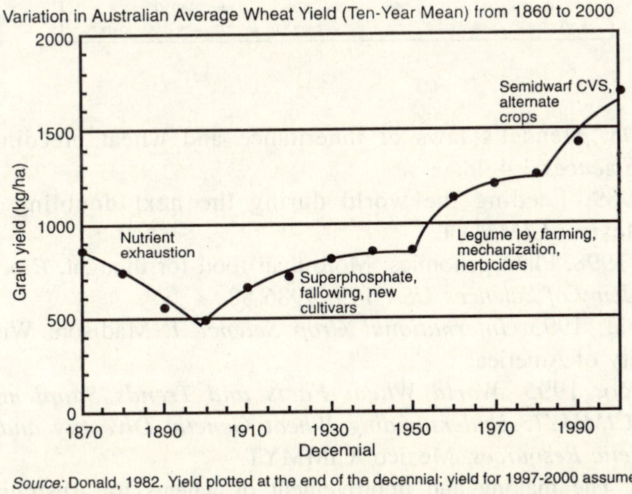

Source: Donald, 1982. Yield plotted at the end of the decennial; yield for 1997-2000 assumed to average 1.71 t/ha.

Source : Wheat Ecology and Physiology of Yield Determination

Figure 11.

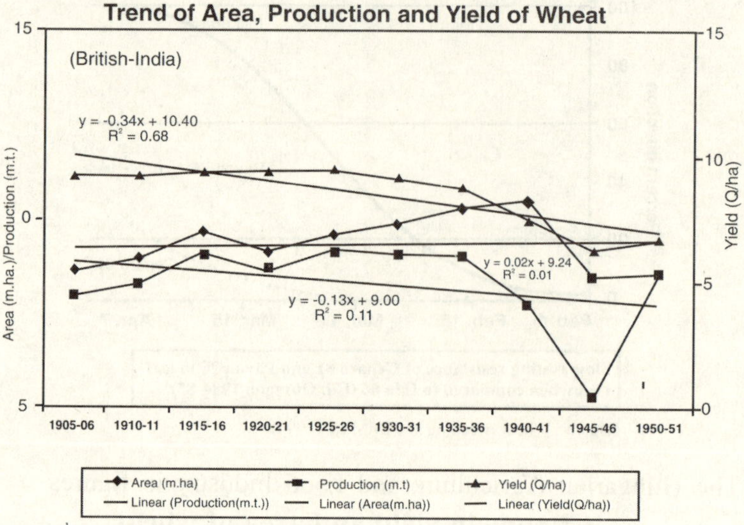

Figure 12.

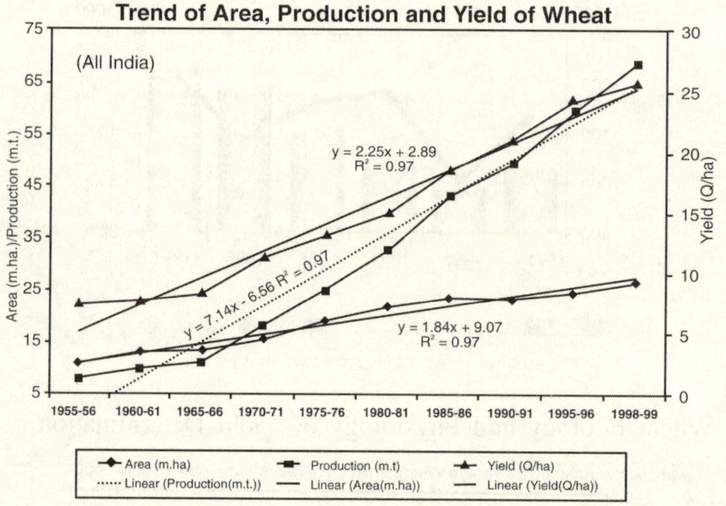

References

Biffen, R.H. 1905. Mendel's laws of inheritance and wheat breeding. *Journal of Agricultural Sciences* 1:4-48.

Borlaug N.E. 1989. Feeding the world during the next doubling of the world population. Mexico: CIMMYT.

Briggs, Steven P. 1998. Plant genomics: More than food for thought. *Proceedings of the National Academy of Sciences USA* 195: 1986-88.

Buxton, D.R. et al. 1993. *International Crop Science I.* Madison, Wisconsin: Crop Science Society of America.

CIMMYT, Mexico. 1995. *World Wheat Facts and Trends Supplement. Ongoing Research at CIMMYT: Understanding Wheat Genetic Diversity and International Flows of Genetic Resources.* Mexico: CIMMYT.

Farrer, W. 1898. The making and improvement of wheats for Australian conditions. *Agricultural Gazzette (New Series) Wales* 9:131-68.

Fuccillo, Domini et al. 1997. *Biodiversity in Trust Conservation and Use of Plant Genetic Resources in CGIAR Centers*. Cambridge, UK: Cambridge University Press.

IARI. 1999. *Agricultural Research Data Book*. New Delhi: IARI.

Khush, Gurdev S. and P. Stephen Baenziger. 1998. Crop improvement: emerging trends in rice and wheat. *Crop Productivity and Sustainability – Shaping the Future (1998). Proceedings of the 2nd International Crop Science Congress*, ed. by V.L. Chopra, R.B. Singh, Anupam Varma, pp. 113-25. In *Plants and Society* (1989), ed. by M.S. Swaminathan and S.L. Kochhar, pp. 523-55. London: Macmillan Publishers.

McCouch, Susan. 1998. Toward a plant genomics initiative: Thoughts on the value of cross-species and cross-genera comparisons in the grasses. *Proceedings of the National Academy of Sciences USA* 95:1983-85.

McIntosh, R.W. 1983. A catalogue of gene symbols. *Proceeding of the 6th IWGS*, pp. 1197-1255. Kyoto, Japan.

Ministry of Agriculture and Regional Development, Hungary. 1998. *The Hungarian Agriculture and Food Industry in Figures*.

Muller, H.J. 1935. *Out of the Night*. New York: Vanguard Press.

Qualset C.O. et al. 2000. *The Structure and Function of the Expressed Portion of the Wheat Genomes*.

Rajaram, S. and G.P. Hettel. 1994. *Wheat Breeding at CIMMYT: Commemorating 50 Years of Research in Mexico for Global Wheat Improvement*. Ciudad Obregon, Sonora, Mexico: CIMMYT.

Reeves, Timothy G. et al. 1999. *New Wheats for a Secure, Sustainable Future*. Mexico: CIMMYT.

Satorre, Emilio H. and Gustavo A. Slafer. 1999. *Wheat: Ecology and Physiology of Yield Determination*. New York: The Howorth Press.

Stakman, E.G., D.M. Stewart and W.Q. Loegering. 1962. *Identification of physiologic races of Puccinia gramims var. tritici. USDA-ARS E-617* (Rev.).

Swaminathan, M.S. 1966. Mutational analysis of the hexaploid Triticum Complex. *Proceedings of the 2nd International Wheat Genetics Symposium, Lund 1963, Hereditas* 2: 418-38

-----. 1968. The Age of Algeny, Genetic Destruction of Yield Barriers and Agricultural Transformation. Presidential Address, Agricultural Science Section, Fifty-fifth Indian Science Congress. Published in *Proceedings Indian Science Congress, Varanasi.*

-----. (Ed.) 1993. *Wheat Revolution: A Dialogue*. Madras: Macmillan India.

34

The Past, Present and Future Contributions of Farmers to the Conservation and Development of Genetic Diversity

Introduction

The significant contributions of rural and tribal families, particularly women, to the conservation and enhancement of agrobiodiversity, remained unrecognized until a few decades ago. Studies by Darwin, de Candolle, Vavilov, Harlan, Wellhausen and many others have however shown a strong positive correlation between cultural diversity and genetic diversity, thereby demonstrating the impact of human communities on the conservation and use of biodiversity. The component of biodiversity, now referred to as agrobiodiversity, is largely the result of human-nature interaction.

Although ethnobiologists and social anthropologists had long recognized the role of rural and tribal families in the domestication and improvement of economic plants, it is only during the last twenty-five years that serious efforts have begun on standardizing methods of recognizing and rewarding the invaluable contributions of farm families to the conservation and development of genetic diversity. A serious debate on this issue started in the forum of FAO in 1981 when I was Independent Chairman of the FAO Council. Later, in November 1983, the FAO Council meeting under my chairmanship set up an International Commission on Plant Genetic Resources (now renamed as FAO Commission on Genetic Resources for Food and Agriculture, so as to include within its scope animal genetic resources also). Soon debate started in meetings of this Commission on the concept of Farmers' Rights. Although there has been a near consensus for nearly two decades on the need for recognizing and rewarding the contributions of farm families to genetic resources conservation, no internationally agreed methodology exists even now for this purpose.

The Convention on Biological Diversity (CBD), which came into force in December 1993, has for the first time in a legally binding document given explicit recognition to the role of indigenous communities to both conservation and accumulation of knowledge and information (Article 8j). The prior informed consent and benefit-sharing provisions of CBD have also not become operational so far.

Conservation Methods and Economic Stakes

Sharing of benefits and the concomitant increase in the recognized value of

---

Presented at the International Conference on Science and Technology for Managing Plant Genetic Diversity in the 21st Century, Kuala Lumpur, Malaysia, June 12-16, 2000.

the resources are the most effective pathways to foster conservation and to ensure the continued availability of plant genetic resources. Today, there is a growing economic stake in the exploitation of the habitats rich in plant genetic resources. That this will be disastrous is evident from the loss of nearly 3.5 percent of world's forests since UNCED in Rio de Janeiro in 1992. There is therefore an urgent need to create an economic stake in conservation. In general, the issue of benefit-sharing is yet to receive the detailed attention it deserves among those formulating legal measures for operationalizing the Convention on Biodiversity. If FAO's revised International Understanding on PGR incorporating provisions for Farmers' Rights becomes a protocol of the legally binding CBD, follow-up action will become mandatory. The adoption of the Cartagena Protocol on Biosafety is a sign of hope.

The different systems of conservation in vogue today are: In situ conservation, Ex situ preservation, and In situ on-farm conservation.

Unlike cryogenic preservation in Gene Banks, conservation involves both preservation and evolution. Therefore, ex situ preservation alone cannot compensate for the lasting benefits that accrue from the conservation of habitats and ecosystems rich in biodiversity. The different systems of conservation are shown in Figure 1, which also indicates the agencies actively involved in the different methods of conservation/preservation.

There is need to standardize methods of ending the prevailing dichotomy where the primary conservers remain poor, while those using their material and / or information become rich. In the case of in situ and ex situ conservation, public funds are provided to establish Biosphere Reserves/Protected Areas/Botanic Gardens/Gene Banks, etc. However, in the case of in situ on-farm conservation of agro-biodiversity, rural and tribal women and men conserve land races for folk varieties for public good at personal cost. Equity demands that their work must be recognized and compensated. Thus, we should develop a framework for promoting benefit-sharing under the different conditions of conservation depicted in Figure l.

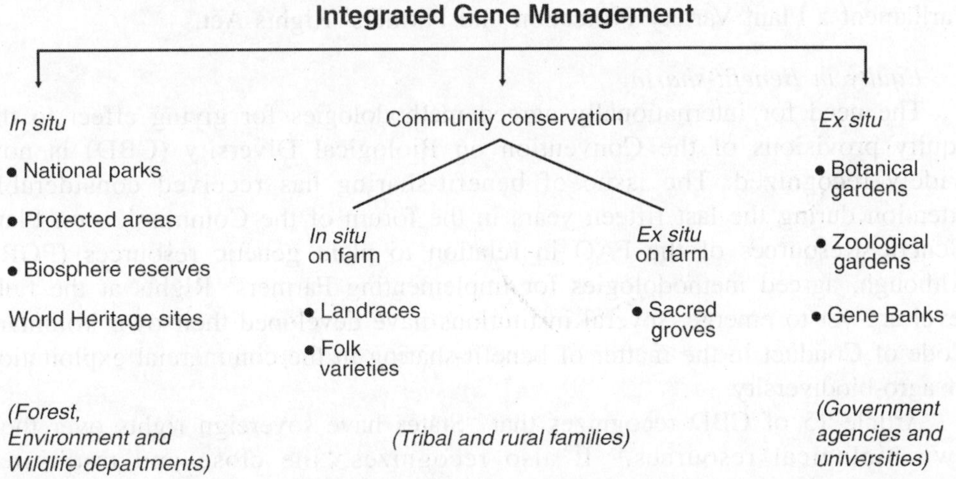

Role of community conservation in integrated gene management.

At a dialogue organized in January 1990 at Madras by MSSRF in collaboration with the Keystone Center in the United States, the participants stated, "we agree on the concept of Farmers' Rights and we agree that the contributions to a fund in recognition of these rights should not be voluntary. Practically speaking, a voluntary fund is a fund without resources. Thus, there should be a compulsory funding mechanism. This would insure that Farmers' Rights are recognized in a real way." The concept of Farmers' Rights includes recognition of the fact that farm women and men have developed and continue to help develop and maintain genetic diversity; in many cases, farmers engage in conscious and creative practices as they select and breed their crops. Ethnobotanists around the world have chronicled the invaluable contributions of tribal and rural women and men in the conservation and enhancement of genetic diversity in plants (Swaminathan and Kochhar, 1989).

The Plant Variety Protection Acts in vogue in developed countries generally conform to the provisions of the Conventions agreed to from time to time under the forum of the International Union for the Protection of New Varieties of Plants (UPOV). They are designed to promote invention and investment in plant breeding. They however do not accord recognition to the conservation and enhancement by farm families of genetic diversity, which is the basic feedstock for the plant breeding and biotechnology enterprises. By introducing provisions for Farmers' Rights, Plant Variety Protection Acts will concurrently foster innovation and conservation. They will thus help to strike a balance between homogeneity and heterogeneity in the genetic make-up of new cultivars. Such a balance is essential for sustainable agriculture. Breeders and farmers are allies in our struggle for a hunger-free world. Hence, their rights should be mutually reinforcing and not projected as antagonistic.

At two dialogues on methods of recognizing and rewarding informal innovations in the area of genetic resources conservation organized by MSSRF in 1994 and 1996, a detailed draft legislation was developed for converting the know-how relating to Farmers' Rights into field level do-how (Swaminathan, 1995; 1996). Based on these drafts, the Government of India has introduced in Parliament a Plant Variety Protection and Farmers' Rights Act.

Equity in Benefit-sharing

The need for internationally agreed methodologies for giving effect to the equity provisions of the Convention on Biological Diversity (CBD) is now widely recognized. The issue of benefit-sharing has received considerable attention during the last fifteen years in the forum of the Commission on Plant Genetic Resources of the FAO in relation to plant genetic resources (PGR). Although, agreed methodologies for implementing Farmers' Rights at the field level are yet to emerge, several institutions have developed their own Voluntary Code of Conduct in the matter of benefit-sharing in the commercial exploitation of agro-biodiversity.

Article 15 of CBD recognizes that "States have sovereign rights over their own biological resources." It also recognizes "the close and traditional dependence of many indigenous and local communities embodying traditional lifestyles on biological resources, and the desirability of sharing equitable benefits arising from the use of traditional knowledge, innovations and practices relevant to the conservation of biological diversity and the sustainable use of its components." Implementation of the above provisions of CBD will require both material and information transfer agreements within and among nations, which incorporate provisions for prior informed consent and equity in benefit-sharing between the primary conservers and those utilizing their knowledge and material in plant breeding, genetic engineering and pharmaceutical and other biological enterprises. The CGIAR has developed well-defined procedures for material and knowledge transfer.

The Trade Related Intellectual Property Rights provisions (TRIPS) of the World Trade Agreement call for either a patenting or a sui generis system of plant variety protection. This is for the purpose of providing incentives for innovation and investment in the use of plant genetic resources. Such Plant Variety Protection Acts have been in existence in several industrialized countries for over fifty years. Since 1961, the International Union for the Protection of New Varieties of Plants (UPOV) has been prescribing guidelines for harmonizing the provisions of such legislation in different countries. The UPOV convention, however, provides for recognizing and rewarding only the contributions of breeders and not of those who have conserved the basic raw material used by breeders. The time has come to end the sad irony of the poverty of the conservers in contrast to the prosperity of those utilizing the fruits of their knowledge and conservation ethics. Several developing countries in South America, Africa and Asia are currently in the process of enacting legislation incorporating provisions for benefit-sharing. Individual institutions and commercial companies have also been practicing their own procedures for sharing benefits with the communities which provided the genetic material. The IUCN Environment Law Center (1996) has prepared a detailed paper on legislative and institutional considerations relating to benefit-sharing. These different steps illumine the path towards an internationally agreed protocol on benefit-sharing under the auspices of FAO or CBD.

Equity in benefit-sharing is fundamental to the retention and revitalization of the in situ on-farm conservation traditions of rural and tribal families. Material and Information Transfer agreements should safeguard the interests of those providing the concerned material/information. CGIAR institutions are already adopting a Material Transfer Agreement procedure which will prevent the monopolistic exploitation of public funded research for commercial profit. Benefit-sharing procedures will have to be developed at the individual and community levels. At the level of an individual farmer conserver/innovator the same procedures for seeking recognition and reward as those available to professional breeders can be used. Only they may have to be helped in obtaining patents/plant variety protection in accordance with the prescribed national legislations. The problem is more complex in the case of benefit-sharing with entire communities. Procedures are available for identifying the area from which critical genes responsible for the commercial success of a new variety came. Thanks to molecular techniques, this possibility also extends to genes controlling quantitative traits like yield and quality. Therefore, appropriate reward can be given from the Community Biodiversity and Gene Funds proposed to be established under Biodiversity and Plant Variety Protection Acts in several developing countries. Breeders will have to be requested to disclose the full pedigrees of their new varieties and indicate to the extent possible the area from where the critical genes, including QTLs (Quantitative Trait Loci) came. The communities concerned can decide how to use the funds provided. Obviously, they should be used for community benefits, including the funds needed for strengthening on-farm conservation of land races and seed technology.

The benefit-sharing methods appropriate to different systems of genetic resources conservation are summarized below.

In situ Conservation: This largely takes the form of Biosphere Reserves, National Parks and Protected Areas. Here, many opportunities are available for benefit-sharing, such as the following: Participatory Forest Management, involving benefit-sharing with reference to wood and non-wood forest products; Bioprospecting based on equity in sharing commercial profit; Ecotourism which strengthens the livelihood security of local communities; Symbiotic social contracts between tribal and rural families and the corporate sector with reference to sourcing of material based on assured buy-back arrangements.

Ex situ preservation: Here, methods of sharing royalty with the providers of the critical information and material can be developed. Community Gene Funds can be established to provide recognition and reward to entire communities based on the analysis of the pedigrees of successful varieties.

In situ on-farm conservation: This is an area of vital importance to global food and health security. Community Biodiversity and Gene Funds could be used to compensate local communities for the yield and income loss they may incur by continuing to conserve land races and folk varieties of economic plants, instead of abandoning them in favor of high yielding varieties; in addition, both individual and community benefits can be conferred on the basis of transparent procedures. Technical Resource Centers for the implementation of the equity provisions of CBD, like the one existing at MSSRF, will be of help in this process. Thus, benefits can be in cash and kind. Social recognition and material benefits are both important. There are several informal models of such recognition and reward developed and operated by non-governmental organizations, academia, business and industry. Through a combination of political will, professional skill and peoples' participation, fair and equitable methods of benefit-sharing can now be introduced in all systems of conservation and use of plant genetic resources.

The FAO Global Plan of Action developed at Leipzig provides an excellent blueprint for plant genetic resources conservation. In the ultimate analysis, the success of conservation efforts will depend upon peoples' participation. Equity in benefit-sharing is the trigger for fostering a people-centered conservation movement. Implementation of benefit-sharing procedures, rooted in the principles of ecology, social and gender equity and economics will be an important step in achieving a better common future for humankind. This is why industrialized countries and biotechnology industries, for whom biodiversity constitutes the basic feed stock, should contribute to a Global Community Gene Fund designed to reward the tribal and rural families who are today protecting public good at personal cost.

The three fundamental concepts of Article 15 of CBD namely, sovereign rights over genetic resources (article 15.1), access subject to prior informed consent (article 15.5) and access subject to mutually agreed terms (article 15.4), provide a framework for a new partnership between holders and providers of knowledge and genetic material and users, based on equity and ethics in sharing benefits.

Unilateral exploitation of biodiversity for commercial purposes is now widely regarded as biopiracy. This method of exploitation should give way to biopartnership based on procedures like co-patenting and equitable benefit-sharing. Such biopartnerships will help to foster symbiotic relationships between biodiversity providers and users and will foster the causes of conservation and sustainable use.

The recent initiative of the World Intellectual Property Rights Organization (WIPO) to consider questions relating to according recognition to traditional knowledge systems and informal innovations is a welcome first step in enlarging the concept of IPR. The World Trade Organization (WTO) should also initiate steps to enlarge the concept of Trade Related Intellectual Property Rights (TRIPS), so as to evolve a new TRIPS Plus paradigm, where the "plus" refers to equity and ethics in the sharing of benefits between traditional knowledge holders and conservers and commercial companies. Finally, UPOV should also restructure itselt so as to become an International Union for the Protection of Breeders and Farmers Rights. Such action at the international level and appropriate legislative and non-legislative measures at the national level will help to give meaning and content to the concept of equitable benefit-sharing.

Global Plan of Action

At an International Technical Conference on Plant Genetic Resources held at Leipzig, Germany, from June17 to 23, 1996, under the auspices of FAO, a global plan of action was developed for both in situ and ex situ conservation of PGR. Although at this Conference, benefit-sharing methodologies were not specifically discussed, the following areas which have a bearing on benefit-sharing were given consideration: Development and commercialization of under-utilized crops and species; Supporting seed production and distribution; Developing new markets for local varieties and "Diversity-rich" products; Expanding and improving education and training.

The Leipzig declaration called "for a new and more productive partnership between scientists and farmers to build upon the ongoing efforts of farmers to manage and improve their plant genetic resources, especially in marginal areas." Such partnership will help to address the twin needs for: Revitalization of the in situ on farm conservation of agrobiodiversity by rural families; Creating an economic incentive system which provides for both compensation for loss of yield and reward for contributions to genetic resources conservation and enhancement.

FAO Global Plan of Action represents the most comprehensive strategy for the conservation and sustainable use of PGR.

In a recent publication titled *Business and Biodiversity,* published by the World Business Council for Sustainable Development and IUCN (Ringwood and Vorhies, 1997), the need for sharing benefits has been reiterated in the following statement: "Company management should ensure that local communities are closely involved with decisions affecting the use of natural resources. Of particular importance in some areas is the indigenous peoples' knowledge and the specific issues related to their use of, and relationship to, the land and biological resources. This relationship needs to be harnessed in a positive way, both for the company and the peoples, by developing agreements which promote the equitable sharing of resources." Thus, the pathway suggested is bilateral negotiations between companies and local communities.

We also have information on both megabiodiversity areas and "hot spot" locations with reference to threats to biodiversity (see Heywood (ed). *Global Biodiversity Assessment*, 1995). Many of these areas are also characterized by high population density (Cincotta et al, 2000). Hence, extending support to the communities conserving and improving agro-biodiversity is not difficult, if there is a will do to so. UNEP, FAO, UNESCO and IPGRI of CGIAR could jointly help to develop internationally acceptable guidelines for according support from a Global Biodiversity Fund.

Community Food and Water Security System

It is gratifying that several developing countries in South America, Africa and Asia are in the process of institutionalizing benefit-sharing procedures in legal measures relating to CBD. Experience in operating such legislation will lead to the refinement of benefit-sharing procedures by the beginning of the new millennium. This will then provide the basis for the adoption of an agreed protocol on benefit-sharing by the Conference of Parties to CBD. Inaction is not the answer to imperfections in currently available benefit-sharing methodologies. Learning by doing is the pathway to perfection.

In future, the contributions of farm women and men to in situ on-farm and ex situ on-farm conservation will depend much upon the social recognition and economic incentives provided to the primary conservers. While the older generations conserved as a part of tradition and ethics, the younger generation is likely to continue the earlier traditions only if a recognition and reward system is put in place.

In addition to implementing the ethics and equity principles enshrined in CBD, it will also be important to integrate the revitalization of the in situ on-farm conservation traditions with a Community Food and Water Security System of the type described below.

There is need for a community centered food security system based on attention to all the links in the conservation-cultivation-consumption chain. One male and one female member of Panchayats should be trained to serve as members of a Food and Water Security Corps. Such a system will include the following four major components: a) Field Gene Bank: This involves the in situ on-farm conservation of land races and local varieties of crops, through the revitalization of the conservation traditions of rural and tribal families, particularly women; b) Village Seed Bank: The rural families often lose their seed stocks due to drought, flood and other natural calamities. Therefore, in each village a Community Seed Bank will be established through a Seed Security Self-help Group, supported by microcredit; c) Village Water Bank: Conservation of rainwater, sustainable management of ground water and the conjunctive use of surface, ground and recycled water are important components of the Village Water Bank; d) Block Grain Bank: It is important to maintain grain reserves of local staples to meet emergencies like drought and natural calamities. For this purpose, a Community Grain Bank will be established at a suitable location, each to serve about 25,000 families. The Grain Bank will be operated by a Self-help Group supported by a revolving fund.

Figure 2. Components of a community food and water security system

Community Food and Water Security System

↓

Community Food and Water Security Committee

↓

Community Food and Water Security Corps

↓

Field Gene Bank | Village Seed Bank | Village Water Bank | Area Grain Bank

Thus the Community Food and Water Security System will foster a sustainable people-centered and people-controlled method of ending food and drinking water insecurity at the level of each individual. It will help to ensure both food security and genetic resources conservation.

Contributions of Farming Families:
A Case Study in Kolli Hills, Tamil Nadu

Past – Subsistence mode of production: The local tribal community is known as Malayali. For ages, subsistence farming has been the predominant pattern and continued to be so until two decades ago. It was not just for survival but it was a way of life for them. Their traditional management practices enabled them to have a sustainable food system. Their traditional cultivated food crops include minor millets such as Ragi (*Eleusine coracana*), Samai (*Panicum miliare*), Thinai (*Sefaria italica*), Thirivaragu (*Paspalum scrobiculatum*), Panivaragu (*Panicum miliaceum*), Rice (*Oryza sativa*), both wetland as well as dryland varieties, and pulses.

Almost all the food crops with one or two exceptions had a minimum of three to four varieties with different agronomic characters. Based on their forefathers' knowledge and their own experience, the Malayalis were motivated to adopt different crop varieties to suit local environmental, socio-economic and cultural conditions.

They adopted varieties based on a wide range of criteria. The Malayali farmers, both men and women, pointed out that different food crops and different varieties formed their agricultural system, because of their high yield, cooking qualities, desirable maturity period, ease of harvest. Further, some of the varieties provided good fodder and thatching material for their house roofs. Another significant aspect is the compatibility of different varieties with climatic and edaphic factors, resources, and cropping patterns (e.g., mixed cropping) of the Malayali farming system.

Many varieties were adopted for different reasons, which may be explained by giving a few examples.

Sadan samai, a medium duration variety gives good yield. Straw serves as good fodder material to their cattle, but it is difficult to process.

Malliya samai, matures in ninety days and gives good yield, but straw does not have fodder value. The straw is good thatching material. While the above two varieties are cultivated in terraced uplands, the long duration variety of *Perun samai* is cultivated in the sloppy terraces because it fits seasonal characteristics better.

There are cultural reasons for cultivation of particular food crops. A rice variety known as *Kam Nellu*, even today is grown by the households worshipping goddess Kongayi Amman, a local deity, for ritual offering. Similarly, Thinai flour is also offered as ritual offerings to local female deities. The delicacies prepared out of this grain are considered very tasty, and the gruel prepared out of it is given to women after childbirth. Thinai is also considered *soodu* (hot) by the locals. So the food prepared out of it is taken once a week particularly during winter season. Incidentally, pounding this grain is quite easy when compared to other millets. The concept of "hot" and "cold" foods is also inherent in tribal families, the "hot" food being taken during winter months and the "cold" food being consumed during summer.

Ragi was cultivated on a large-scale until recently as it is considered very healthy and nutritious and can be consumed year around. Ragi is followed by samai, *Puzhudikar Nellu* (a dryland rice variety) and other crops.

Malayalis practice mixed cropping (mainly ragi based) where a minimum of six varieties of crops are sown simultaneously and cultivated in a single plot. It is a risk minimizing device as well as a means of extending the cropping season.

Present – Transformation from food production to commodity production: In recent times, i.e., in the late 1970s and early 1980s, a trend towards market oriented economy set in which altered their food and production system. This transformation took place as a result of various developmental activities such as improving accessibility through construction of roads, setting up cooperative societies and schools. Entry of outsiders and penetration of market economy gradually replaced subsistence economy with a market oriented one.

Subsistence based food production system has largely been replaced by the cultivation of cassava in the terraced uplands and pineapple in the sloppy and rocky lands. Between 1970 and 1984, area under cassava and pineapple increased more than fifty times.

Exposure to urban life brought changes in socio-economic and cultural values. Monetary considerations favored a shift from subsistence economy to market oriented one. Changes in agricultural practices and access to education forced people to reduce their livestock.

Another crucial factor is increase in population. It was approximately 19,000 in 1951 and it increased to 32,130 persons in 1991.

Their food security is gradually declining, making them dependent on the market. Most importantly, agro-biodiversity of Kolli Hills is at great risk. Another crucial factor which has a direct bearing on Kolli Hills agro-biodiversity is land alienation which has been taking place at an alarming rate.

Disturbances in the exchange relation may affect their food supply. To avoid any such disturbances, an alternate system of the kind described earlier that will ensure their food security is a must.

Future – Revitalization of traditional agricultural and food system: If the present trend continues there is more scope for greater commodification in future. Those who continue with farming will concentrate on the commercial crops such as tapioca, pineapple and plantation crops that have economic value. And those who are away due to education or because of new opportunities drifted towards non-farm activities. Such people and households would be completely diverted from their traditional agricultural activity. Unless they derive economic benefit from their conservation traditions, many of the local crops and varieties will soon become "lost crops".

If this scenario has to be changed it needs to have a positive and effective intervention with peoples' involvement and participation.

A two-pronged strategy can help to mitigate the problem. One deals with cultural aspects and the other with the economic aspect. There should be a cultural motivation for the people whereby they will feel proud about their traditional crops. A value should be created in such a way that producing their traditional food crops is prestige in itself. People should be made to rediscover the cultural value of raising their traditional millets and other crops, by according social prestige to such traditions.

Value addition in terms of monetary gains is another crucial factor. Value addition both in terms of cultural as well as monetary returns would help to reinforce conservation of traditional millet crops and agro-biodiversity of Kolli Hills. The implementation of the concept of Farmers' Rights will also help. The fast changing scenario in Kolli Hills is depicted in Figure 3.

Gender Dimensions of Biodiversity Management

Most of the studies on contributions of farming families are not gender disaggregated. MSSRF in collaboration with the FAO Regional Office in Bangkok has initiated a series of studies on the gender dimensions of biodiversity management (Swaminathan, 1998). More recently, detailed studies have been undertaken by Dr. Virendra Kumar and Dr. (Ms.) Hemal Kanvinde of MSSRF on gender roles in biodiversity conservation and management in Sri Lanka and Maldives. The results so far gathered stress the need to recognize the specific role of women in genetic resources conservation and enhancement. This is particularly important in the context of the equitable benefit-sharing provisions of CBD.

Conclusion

The books *Lost Crops of the Incas* (1989) and *Lost Crops of Africa* (1996) bring out very clearly how ancient civilizations depended upon a wide range of food and medicinal plants. Also, the book *Biodiversity in Trust* (Fuccillo et al, 1997) brings out clearly the enormous range of intra-specific diversity maintained in the Gene Banks of IARCs belonging to the CGIAR system. We owe such rich variability to the conservation ethos of farm and tribal families. Often the conservation habits were not due to any commitment to conservation per se, but because of the understanding that genetic variability diminishes vulnerability to biotic and abiotic stresses. Farming families in the past were motivated by a desire to minimize risks rather than to maximize profits. Risk aversion agronomy involved growing mixtures of crops and crop varieties.

Modern agriculture involves monoculture and the cultivation of the same genetic strain over large areas. There is a fear that when genetically modified varieties covered by proprietary science and released by commercial companies become popular, genetic homogeneity will increase in farmers' fields. The challenge lies in integrating genetic efficiency with genetic diversity. The procedure adopted by MSSRF scientists for this purpose is the integration of pre-breeding with participatory breeding (Figure 3). By adopting such procedures, we can conserve agrobiodiversity and at the same time enhance it through participatory breeding with farm families. This will help to select location specific varieties adapted to the local agro-ecological and socio-cultural conditions. If we adopt such scientific procedures and at the same time introduce legal measures for social recognition and economic reward, farm families will continue to conserve, nurture and enhance agrobiodiversity.

Figure 3.

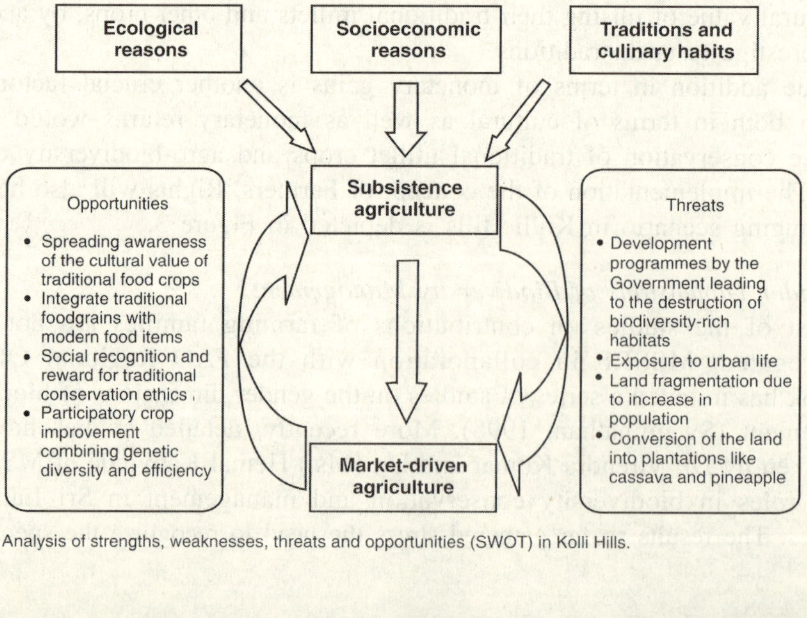

Analysis of strengths, weaknesses, threats and opportunities (SWOT) in Kolli Hills.

References

Cincotta, Richard P., Jennifer Wisnewski and Robert Engleman. 2000. Human population in the biodiversity hotspots. *Nature* 404: 27, 990-92. (April)

Fuccillo, Domini, et al. 1997. *Biodiversity in Trust: Conservation and Use of Plant Genetic Resources in CGIAR Centers*. UK: Cambridge University Press.

Heywood, V.H. (Ed.) 1995. *Global Biodiversity Assessment*. United Nations Environment Program. Cambridge, UK: Cambridge University Press.

National Research Council, U.S.A. 1996. *Lost Crops of Africa*. Washington, DC: National Academy Press.

National Research Council, U.S.A. 1989. *Lost Crops of Incas*. Washington, DC: National Academy Press.

Ringwood, Stone D. and K.F. Vorhies. 1997. *Business and Biodiversity - A Guide for the Private Sector*. WBCSE - World Business Council for Sustainable Development, IUCN. Switzerland: The World Conservation Union.

Swaminathan, M.S. and S.L. Kochhar. (Eds.) 1989. *Plants and Society*. London: Macmillan Publishers.

Swaminathan, M.S. (Ed.) 1995. *Farmers' Rights and Plant Genetic Resources, Recognition and Reward: A Dialogue*. Madras, India: Macmillan India.

-----. (Ed.) 1996. *Agrobiodiversity and Farmers' Rights*. Delhi: Konark Publishers.

-----. (Ed.) 1998. Gender Dimensions in Biodiversity Management. Delhi: Konark Publishers.

35

Higher Education for Social and Economic Progress

I feel very privileged to be here today to deliver the commencement address of this unique University, which is leading the academic world in showing how higher education can become a powerful instrument for furthering the social and economic well-being of scholars from under-privileged families. The efforts of this University in helping first generation learners to overcome the handicaps of their childhood years are truly path breaking.

As one who has spent most of his life in harnessing science and education for overcoming gender, genetic and economic inequities, I am grateful to the University for enrolling me as an honorary alumnus. On behalf of the distinguished men and women who have been conferred honorary degrees today, and on my own behalf, I express our sincere gratitude to the Board of Trustees of the University of Massachusetts for the honor done to us.

Today is an occasion for all graduating from this University to honor the past, to celebrate the joy of living in the present, and to make plans for shaping the future. In your individual and collective aspirations to shape the future, I commend to you the example of a small Himalayan nation, Bhutan, which has chosen the concept of Gross National Happiness, to measure progress in human well-being. How can we measure happiness? The indices used by Bhutan are: care for nature, economic progress rooted in the principles of ecology and gender and social equity, good governance as reflected in freedom from corruption, and promotion of cultural and spiritual values. An important component of cultural and spiritual values are love of diversity and pluralism in human societies. We have a global convention on biodiversity to promote the conservation and sustainable and equitable use of diversity in flora and fauna, including micro-flora and micro-fauna. Unfortunately, such commitment to the conservation of biodiversity does not seem to extend to love of human diversity. There are still widespread conflicts in different parts of the world based on religion, language, color, class and political belief. While we can be proud of the end of skin color-based apartheid towards the end of the last century, various other forms of apartheid based on religion, gender and economic status continue to prevail. In my view, it is the first and foremost duty of educated individuals to end all forms of intolerance of diversity and pluralism in human societies.

Convocation address at the University of Massachusetts, Boston, June 2, 2001.

The spectacular progress made in the study of the human genome in recent years has further underlined the unity in the genetic architecture of all members of the human family. Genomic, proteomics, information technology and nanotechnology are opening up uncommon opportunities for a better common present for human kind. However, it is becoming equally clear that unless the technological push is matched by an ethical pull, the products of our intellect may become a curse rather than a blessing, as Albert Einstein warned many years ago.

We must ensure that the widespread rich-poor divide now existing is not further enlarged as a result of digital, genetic and other forms of technological divides. Franklin D. Roosevelt once said that "the test of our progress is not whether we add more to the abundance of those who have much; it is whether we provide enough for those who have little." How can we translate this goal into action and achievement in our day-to-day life? Mahatma Gandhi of my country suggested two basic guidelines to bridge technological, economic and social divides. The first is the principle of *antyodaya* or attention to the poorest of the poor in all development programs. For this, he advised, "Recall the face of the poorest and the weakest person whom you have seen, and ask yourself, if the steps you contemplate are going to be of any use to him or her. Will it restore to the poor control over their life and destiny?" Gandhi advised all committed to including the excluded in deriving benefit from technological and economic progress to consider themselves as trustees and not owners of their financial and intellectual endowments.

The United States is today the richest nation in the world, possessing nearly a third of global financial wealth. It is also a technology rich country. Fortunately, there is a trend towards an expansion of private philanthropy in this country. Even those with moderate wealth have begun to care and share. Unfortunately, in the field of intellectual wealth, the opposite is happening. Intellectual property rights are covering findings of great importance to food and health security. The recent controversy about the availability of drugs, which can help to cure, at affordable prices, AIDS victims in Africa, is a case in point. Also, if the market becomes the major determinant of priorities in investment in scientific research, "orphans will remain orphans" and the health and nutritional problems affecting the poor nations will continue to remain unattended. While there are many benefits from competitive private sector R&D investments, deceleration of investment in university and public sector research will cause incalculable harm to the goal of rendering both unsustainable consumption and unacceptable poverty problems of the past. There must be a balance between public and private sector research, and universities should structure their priorities in a manner that there is synergy between the two.

Rabindranath Tagore, Nobel Laureate in Literature, illustrated the source of true happiness in a lovely poem, which I would like to quote on this occasion.

> Child, how happy you are sitting in the dust
> Playing with a broken twig all the morning;

I smile at your play with that little bit of broken twig
I am busy with my accounts adding up figures by the hours
Perhaps you glance at me and think what a stupid game to spoil
your morning with,
Child, I have forgotten the art of being absorbed in
sticks and mudpiles
I seek out costly playthings and gather lumps of gold and silver
With whatever you find, you create glad games;
I spend both my time and strength over things I can never obtain;
In my frail canoe I struggle to cross the sea of desire;
And forget that I too am playing a game ...

Here in Boston, Prof. Edward O. Wilson has been drawing attention to the path of ecological suicide humankind is taking. The ecological footprint, which is the average amount of land and sea appropriated by each person for food, water, housing, energy, transportation, commerce and waste recycling, is roughly one hectare in developing countries, and ten hectares in the United States. If every citizen of this world aspires to reach the consumption level of an average US citizen, we will need at least four more planet earths to fulfill such a desire. This is why, to quote Gandhi again, "nature provides for everybody's need, but not for everyone's greed".

Remaining silent spectators of the expansion of intra- and inter-generational inequity is the greatest tragedy of our times. Environmental degradation, economic and gender inequity, and jobless economic growth are growing concurrently with unprecedented progress in biotechnology, information and space technologies as well as the production of biological software for sustainable development. Universities should become leaders in promoting job-led economic growth and not joyless growth. Everyone taking his or her degrees today at this visionary University should strive to qualify for being included in the special category of human beings described by Maria Ranier Rilke in the following words:

Again and again in history
Some special people wake up
They have no ground in the crowd
They move to broader laws
They carry strange customs with them
And demand room for bold audacious actions
The future speaks ruthlessly through them
They change the world.

I hope you will all be blessed with opportunities to become such special persons. I wish you all great personal happiness and professional achievement. ▯

36

Towards Freedom from Hunger: A Global Food for Sustainable Development Initiative

I am most grateful for the privilege of delivering this year's Gresham Special Lecture. Sir Thomas Gresham's passion was in fostering a culture of "new learning". I have therefore chosen a theme which is as old as human civilization and yet even more relevant today, and shall try to explore what an attitude of new learning can contribute to finding a lasting solution to the human quest for freedom from hunger. In developing my contribution to the solution of this age-old problem, I shall follow Marcel Proust's exhortation: "The real voyage of discovery does not consist of seeking new landscapes, but in having new eyes."

Looking Back

From ancient times until today the relationship between peace and food has been stressed from time to time. The International Peace Research Institute at Oslo, in a carefully documented report titled *To Cultivate Peace - Agriculture in a World of Conflict* by de Soysa and Gleditsch (1999), found that violent conflicts in most cases could be traced to economic rather than ideological differences. The International Commission on Peace and Food which I chaired, in its report titled *Uncommon Opportunities: An Agenda for Peace and Equitable Development* (1994), has suggested that investing in agriculture which helps to promote food and livelihood security in many nations is an effective strategy for preventing future wars, eradicating poverty, preventing environmental destruction and reducing violence. Unfortunately, even now, far too high a proportion of national GDP is being spent on arms and military equipment as compared to programs designed for poverty eradication and meeting the basic needs of the underprivileged sections of humankind. Even a war hero like Dwight D. Eisenhower had this to say when he became President of the United States: "Every gun that is made, every warship launched, every rocket, signified in the final sense a theft from those who are hungry and are not fed, from those who are cold and are not clothed. The world in arms is not spending money alone. It is spending the sweat of its laborers, the genius of its scientists, the hopes of its children."

'Right to Food' has been recognized as a fundamental right in numerous UN meetings. Mary Robinson, UN High Commissioner for Human Rights, in a recent lecture in Rome, stressed the need for a 'Right to Food' approach

---

The Gresham Special Lecture delivered at Mansion House, London, June 24, 2002.

to agricultural trade in the framework of the WTO's Agreement on Agriculture. I hope this approach will lead to a Food Security or Livelihood Security Box in the revised WTO Agreement on Agriculture, the negotiation for which will start in Cancun next year. Some twenty countries have so far adopted constitutions that more or less explicitly refer to the right to food. One of the most explicit references is in the South African Constitution that states in section 27: "Everyone has the right to have access to sufficient food and water." Norway is a world leader in this field. The Norwegian Parliament adopted a Human Rights Act on May 21, 1999, which adopts a rights-based approach to agricultural policy.

Many industrialized countries including the United Kingdom are now shifting their approach from food aid to enabling countries to build sustainable food security systems. This is because of the growing realization that agriculture is not just a food-producing machine, but is the very backbone of the livelihood and ecological security system in most developing countries. Hunger is the extreme manifestation of poverty, since the poor spend over seventy percent of their daily earnings on food. Food aid can also become a powerful engine of sustainable economic development, as the experience of the World Food Programme in designing Food for Work programs has shown. In spite of the growing awareness of the need to accord the highest priority to ending poverty-induced hunger, progress in achieving internationally agreed targets has been deplorable.

The World Food Summit (WFS) convened by FAO in Rome in 1996 resolved to bring down the number of hungry and undernourished people by twenty-two million every year. A meeting of Heads of States and Governments and Ministers for Agriculture was convened by FAO in Rome from June 10-13, 2002, to review the progress made since 1996 in achieving the targets set at the WFS. The survey at the WFS – Five Years Later meeting in Rome indicated that the actual achievement in hunger reduction was only six million per year. Over 200 million persons or twenty-eight percent of the entire population of Africa were found to be chronically hungry. Twenty-four thousand children, women and men die every day due to hunger related causes. The world population will be over seven billion by 2015 and hence the estimates of WFS 1996 will have to be suitably adjusted upward, taking into account the increase in human numbers. The task before the WFS – Five Years Later meeting was to understand why the very modest target set in 1996 has not been achieved.

Both in 1996 and now, civil society organizations (CSO) and non-governmental organizations (NGOs) held parallel conferences and issued separate declarations. In 1996, the CSO declaration was titled "Profit for Few or Food for All". The 2002 CSO Forum declaration stressed that food sovereignty, right to food and agro-ecological models for agriculture are the key elements for any strategy towards ending hunger and malnutrition. This declaration further pointed out that "genetically modified organisms (GMOs) represent a threat to family farmers, other food producers, the integrity of genetic resources and human and environmental health. It will affect particularly the rural poor, who cannot afford this costly alternative."

The official WFS Plus Five declaration titled "International Alliance Against Hunger" urged concerted action to fulfill the 1996 commitments and stated: "We are committed to study, share and facilitate the responsible use of biotechnology in addressing development needs." The official participants in the WFS Plus Five meeting were largely Ministers of Food and Agriculture and senior government officials. The participating Heads of State or Government were largely from countries in Africa. OECD countries were mostly represented at the level of Agriculture Ministers or senior officials, excepting Spain as well as Italy whose Prime Minister served as the Chair of the Conference.

In addition to chronic protein-energy malnutrition caused by poverty, two billion people in the developing countries, many of them women and children, suffer from hidden hunger caused by one or more micronutrient deficiencies, like lack of iron, iodine or Vitamin A. Currently thirty-four countries are experiencing severe food shortages, the most seriously affected being countries in Southern Africa as well as Afghanistan and North Korea. Developing countries are likely to spend over 3.3 billion dollars this year for import of cereals, largely from rich nations.

Women and children are the worst sufferers. A special session dealing with this issue, stressed the need for implementing in letter and spirit the provisions of Article 14 of the Convention on the Elimination of All Forms of Discrimination Against Women (CEDAW) adopted in 1979 by the UN General Assembly. This will involve providing equal access to and control of natural and productive resources and the full participation of rural women in policy making at all levels and throughout development activities.

The African Heads of State or Government identified civil strife and ethnic wars, debt servicing burden, trade barriers resulting in lack of market access, incidence of HIV/AIDS, and drought as the principal causes for the widespread persistence of hunger. African leaders resolved to strengthen the recently formed The New Partnership for Africa's Development (NEPAD) in order to promote a Marshall Plan kind of revitalization of African economies. How far their hope for substantial additional development assistance from rich nations will be forthcoming, remains to be seen. Between 1900 and 2000, concessional assistance from developed countries and loans from the international financing institutions fell by fifty percent for agriculture. At the same time, the OECD countries increased farm subsidies to more than three hundred billion US dollars, amounting to 12,000 dollars per farmer per year. The assistance given by OECD countries to farmers in developing countries works out to six dollars per farmer per year. The recently approved US Farm Bill envisages a further substantial increase in the already high subsidies given to US farmers. Thus, agriculture in the rich nations is supported by heavy inputs of subsidy, capital and technology. There is hence no level playing field in global agricultural trade.

The President of the European Commission, Mr. Romano Prodi, announced at the WFS Plus Five meeting that by the end of 2006, EU official development aid will be raised to an overall rate of 0.39% of GDP. This is in contrast to the target of 0.7% set by the UN nearly two decades ago. He also announced that the EU will try to improve market access to the farm products of developing countries. If trade does not become fair, in addition to being free, small-scale farmers' farming will be eclipsed by the factory farming enterprises of developed countries.

FAO has prepared a comprehensive Anti-Hunger Program involving an additional investment of twenty-four billion dollars annually. This amount is equivalent to 2.5% of the subsidies paid to farmers in OECD countries. The FAO program aims to address both hunger today as well as sustained progress in agricultural and rural development. FAO also presented an analysis of the progress made by different developing countries in achieving the WFS goals. China led the countries that had achieved the largest reduction in the population of under-nourished between 1990-92 and 1997-99. Others in the category of achievers included Thailand, Vietnam, Ghana, Mozambique, Kuwait and Peru. The worst performing nations include Iraq, Guatemala, Somalia, Venezuela, Tanzania, Cuba, Congo and North Korea. Out of the one hundred and sixteen million children, women and men saved from the hunger trap between 1990-92 and 1997-99, seventy-six million are from China. Unfortunately, the number of undernourished increased by eleven million during the same period in India, in spite of substantial grain reserves with the Government of India. This should be a cause for alarm and serious introspection.

The policies adopted by China, in achieving speedily the goal of food for all and forever, involved priority to rural and agricultural development and to on-farm and non-farm employment. China's policy of a rural and agricultural development based economic growth strategy has obviously paid rich dividends in terms of human nutrition and well-being. The contribution of agriculture to GDP is going down in many developing countries, while the onus of providing jobs still remains with the farm sector. This explains why rural poverty is more acute in many poor nations, as compared to urban poverty.

Looking Forward

Since the establishment of FAO in 1945, billions of pages of analysis on hunger have been written. Every conference organized at a cost of millions of dollars ends with a plea for converting words into action. If we heed the advice given 2,000 years ago by the Roman farmer Varro and the Roman philosopher Seneca, then we will not be discussing hunger but will be living in a hunger-free world.

> "Agriculture is a science which teaches us what crops should be planted in each kind of soil, and what operations are to be carried out, in order that the land may produce the highest yields in perpetuity." – Varro
>
> "A hungry person listens neither to reason nor cares for justice, nor is bent by any prayers." – Seneca

As I mentioned earlier, in most developing countries, agriculture, including crop and animal husbandry, fisheries, forestry and agro-processing, constitutes the backbone of the livelihood and ecological security systems. Household food security in such countries is best defined in terms of person-years of jobs or livelihoods, rather than in terms of tons of foodgrains. A famine of jobs or of purchasing power leads to a famine of food at the level of consumption.

Agricultural progress serves as the best safety net against poverty, hunger and unemployment. Therefore, sustainable agriculture is the foundation for sustainable human security.

The "Green Revolution", is a process which leads to higher production through improvement of productivity. In January 1968, several months before the term Green Revolution came into use, I emphasized in a lecture at the Indian Science Congress, the need for improving productivity in perpetuity without associated ecological and/or social harm, in the following words: "Exploitive agriculture offers great dangers if carried out with only an immediate profit or production motive. The emerging exploitive farming community in India should become aware of this. Intensive cultivation of land without conservation of soil fertility and soil structure would lead, ultimately, to the springing up of deserts. Irrigation without arrangements for drainage would result in soils getting alkaline or saline. Indiscriminate use of pesticides, fungicides and herbicides could cause adverse changes in biological balance as well as lead to an increase in the incidence of cancer and other diseases, through the toxic residues present in the grains or other edible parts. Unscientific tapping of underground water will lead to the rapid exhaustion of this wonderful capital resource left to us through ages of natural farming. The rapid replacement of numerous locally adapted varieties with one or two high-yielding strains in large contiguous areas would result in the spread of serious diseases capable of wiping out entire crops, as happened prior to the Irish potato famine of 1854 and the Bengal rice famine in 1942. Therefore the initiation of exploitive agriculture without a proper understanding of the various consequences of every one of the changes introduced into traditional agriculture, and without first building up a proper scientific and training base to sustain it, may only lead us, in the long run, into an era of agricultural disaster rather than one of agricultural prosperity." Later I coined the term "evergreen revolution" to highlight the pathway of increasing production and productivity in a manner that the short- and long-term goals of food production are not mutually antagonistic.

Edward O. Wilson in his recent book, *The Future of Life* (2002), has referred to my concept of evergreen revolution in the following words: "The problem before us is how to feed billions of new mouths over the next several decades and save the rest of life at the same time, without being trapped in a Faustian bargain that threatens freedom and security. No one knows the exact solution to this dilemma. The benefit must come from an evergreen revolution. The aim of this new thrust is to lift food production well above the level obtained by the Green Revolution of the 1960s, using technology and regulatory policy more advanced and even safer than those now in existence."

How do we achieve this evergreen revolution? I would like to indicate some of the essential but integrated steps for achieving the goal of achieving a balance between human numbers and human capacity to produce food of adequate quantity, quality and variety.

Fatigue of the Green Revolution. The growing damage to the ecological foundations essential for sustainable food security, such as land, water, biodiversity, forests and the atmosphere is leading to a stagnation in yield in

Green Revolution areas. Compounding such problems is the threat of climate change leading to potential adverse changes in temperature, precipitation, sea level and ultra-violet β radiation.

An analysis of food insecurity indicators in rural India carried out by MSSRF with support from the World Food Programme (WFP), indicates that the Punjab-Haryana region, which today serves as India's food basket, may become very food insecure in another twenty years. The indicators used in measuring sustainability of food security are: land degradation and salinization, extent of forest cover, ground water depletion, and the nature of crop rotation. In all these parameters, Punjab and Haryana occupy a low position. The rice-wheat rotation has led to the displacement of grain and fodder legumes capable of enriching soil fertility. What is now happening is best described as a trend towards land and water mining, and not sustainable farming.

Forewarned is forearmed. What can we do to launch global agriculture on the pathway of an evergreen revolution, where advances in crop and farm animal productivity are not accompanied by either ecological or social harm?

Converting Know-how into Do-how

The following suggestions are aimed at converting the vast know-how now available into field level do-how:

Integrated attention to the components of food security. Food Security has three major dimensions: Availability of food – a function of production; Access to food – a function of purchasing power/access to sustainable livelihoods; Absorption of food in the body – determined by access to safe drinking water and non-food factors like environmental hygiene, primary health care, and primary education.

The capacity to support even the existing human and animal populations has been exceeded in many parts of the developing world. Hence, the future of food security depends upon population stabilization, the conservation and care of arable land, through attention to soil health and replenishment of fertility, and the conservation and careful management of all water sources, so that more crop can be produced per every drop of water.

Ownership and sustainable use. Much of the degraded and desertified land belongs either to resource poor families, or constitutes over-used and over-grazed common property. Ownership pattern of land and water determines the feasibility of introducing integrated and sustainable land and water management systems. Even where land is individually owned, locally acceptable systems of social management may have to be introduced through legislation, education and social mobilization. Women's access to land is also important. Water, particularly groundwater, should be a social resource and not private property. Creating an economic stake in conservation is vital for ensuring the sustainable use of natural resources.

Environmental refugees. Degradation and diversion of arable land and the depletion and pollution of water resources result in the loss of rural livelihoods.

This triggers the unplanned migration of the rural poor to towns and cities, resulting in the proliferation of urban slums. The rise in the numbers of such environmental refugees threatens peace and security. Norman Myers has chronicled the seriousness of the situation. Hence, there should be a monitoring mechanism for avoiding the loss of rural livelihoods. Development programs should strengthen the linkages between ecological and livelihood security.

Grain mountains and hungry millions: The growing paradox. There are now unique opportunities for launching a Food for Sustainable Development Initiative, in the form of a "grain for green" movement. Such a program could accord priority to the: restoration of hydrological and biodiversity "hot spots", particularly in mountain ecosystems; coastal agro-aqua farms (planting of salkornia, mangroves, casuarina, palms, etc. along with coastal agriculture and aquaculture); water harvesting, watershed development, wasteland reclamation, and anti-desertification measures; recycling of solid and liquid wastes and composting; agro-forestry and other sustainable land use systems in the fields of resource poor farmers.

A Global Food for Sustainable Development and Hunger Elimination Initiative could be launched by the International Alliance against Hunger proposed by FAO. About twenty-five million tons of grains would provide nearly one hundred million person-years of work designed to eliminate poverty-induced endemic hunger and at the same time restore and enhance environmental stocks.

The Food for Sustainable Development Initiative could be managed at the local level by Community Food Banks (CFB) operated by women's self-help groups. Such CFBs can be designed in a manner that they can address concurrently issues relating to chronic, hidden and transient hunger. The merit of CFBs will be low transaction cost and transparency. They can also help to widen the food security basket, thereby saving what could become "lost crops". Wherever animal husbandry including poultry farming is important to provide additional income and nutrition to families living in poverty, CFBs can also operate Feed and Fodder Banks.

Conferring the right to food and thereby an opportunity for a productive and healthy life on those who go to bed undernourished now, is the fundamental duty of the State as well as of the well-to-do sections of the population. Thanks to both the spread of democratic systems of governance at the grassroot level and technological advances, we now have a unique opportunity to foster a Community Centered and Controlled Nutrition Security System. Such decentralized community management will help to improve delivery of entitlements, reduce transaction and transport costs, eliminate corruption and cater to the twin needs of introducing a lifecycle approach to nutrition security, and meeting the challenge of seasonal fluctuations in nutritional status. If such CFBs are operated by women, this will help to bridge the gender divide in the area of nutrition.

Adopting a Whole Lifecycle Approach to Nutrition Security

a) Pregnant Mothers: Overcoming maternal and fetal under- and malnutrition is an urgent task, since nearly thirty percent of the children born in countries in South Asia are characterized by low birth weight (LBW), with the consequent risk of impaired brain development. LBW is a proxy indicator of the low status of women in the society, particularly of their health and nutrition status during their entire lifecycle.

b) Nursing Mothers: Appropriate schemes will be necessary to provide support to enable mothers to breastfeed their babies for at least six months, as recommended by WHO. Policies at work places, including the provision of appropriate support services should be conducive to achieving this goal.

c) Infants (0-2 years): Special efforts will have to be made to reach this age group through their mothers, since they are the most unreached at present. Eighty percent of brain development is completed before the age of two. The first four months in a child's life are particularly critical, since the child is totally dependent on its mother for food and survival.

d) Preschool Children (2-6 years): A well-designed integrated child development service will help to cater to the nutritional and health care needs of this age group.

e) Youth (6 to 20 years): A nutrition based Noon Meal program in all schools (public and private and rural and urban) will help to improve the nutritional status of this group. However, a significant percentage of children belonging to this age group are not able to go to school due to economic reasons. Such school "push-outs" or child laborers need special attention.

f) Adults (20 to 60 years): The Nutrition Safety Net to cater to this category should consist of both an Entitlements program like Food Stamps and Public Distribution System (PDS), as well as a Food for Sustainable Development program (also called "Food for Work" program). The Food for Sustainable Development program can accord priority to the use of foodgrains as wages for the purpose of establishing water harvesting structures (Water Banks) and for the rehabilitation of degraded lands and ecosystems. In designing a nutrition compact for this age group, persons working in the organized and unorganized sectors will have to be dealt with separately. Also, the intervention programs will have to be different for men and women taking into account the multiple burdens on a woman's daily life.

g) Old and infirm persons: This group will have to be provided with appropriate nutritional support, as part of the ethical obligations of society towards the handicapped.

The above whole lifecycle approach to Nutrition Security will help to ensure that the nutritional needs of everyone in the community and of every stage in an individual's life, are satisfied. Such an integrated approach is being adopted under the "Malnutrition-free Tamil Nadu" program, as shown in the following Table:

Pregnant Mothers	Special support to prevent the incidence of low birth weight children
Nursing Mothers	Appropriate support for 6 months
Infants (0 to 2 years)	Program designed to reach the infant through the mother
Child (2 to 6 years)	Integrated Child Development Service
School Children (6 to 18 years)	Nutritious Noon Meal in Schools
Adults (18 to 60 years)	Food for Sustainable Development
Old and infirm persons (over 60 years)	Special Programs

Over 7 million tons of foodgrains (wheat and rice) have been allotted for the above programs during 2001 and 2002 in India.

Developing and spreading a holistic action plan to achieve sustainable nutrition security at the level of each individual. The major components of such an integrated action plan are the following:

a) Identification: Request the local families themselves to identify those who are nutritionally insecure. Trained Community Volunteers of the kind mobilized in Thailand will be useful for this purpose. Invariably, the ultra-poor who do not have any productive asset fall under this category.

b) Education and information empowerment: Empower those who are not aware of their entitlements about the nutritional safety nets available to them and also undertake nutrition education. An entitlements database can be developed for each area and household entitlement cards can be issued, indicating how to access nutritional, health care, and educational programs. The educational programs should also lay stress on culinary habits in relation to the conservation of essential nutrients in cooked food.

c) Overcome protein-calorie under-nutrition: The various steps indicated under the whole lifecycle approach will have to be adopted. The problems of child labor and of persons working in the unorganized sector will need specific attention.

d) Eliminate hidden hunger caused by the deficiency of micronutrients in the diet: Introduce an integrated approach including the consumption of vegetables and fruits, millets, grain legumes and leafy vegetables and the provision of fortified foods like iron and iodine fortified salt and oral dose of Vitamin A. The basic approach should be a food-based one, with emphasis on home and community nutrition gardens, wherever this is socially and economically feasible.

e) Drinking water, hygiene and primary health care: Attend to the provision of safe drinking water and to the improvement of environmental hygiene. Also, improve the primary health care and primary education systems. Non-food factors like health care and education are equally important for nutrition security.

f) Sustainable livelihoods: Improve economic access to food through market-linked micro-enterprises supported by micro-credit. Also, create an economic stake in the conservation of natural and common property resources. Ensure that agreements under the World Trade Organization (WTO) provide a level playing field for products coming from decentralized production (production by masses or farmers' farming) as compared to those emerging from mass production or factory farming often supported by massive inputs of subsidy, capital and technology. Promote job-led economic growth and not jobless growth.

g) Pay special attention to pregnant and nursing mothers and pre-school children: Measure progress through monitoring MMR, IMR, incidence of LBW children and male-female sex ratio. Iron-folate supplements during prenatal care should be accompanied by steps to overcome protein-energy deprivation. A carefully designed maternity and childcare code will help to bring down speedily MMR, IMR, LBW and stunting. Sex ratio is a good index of the mind-set of a society in relation to the girl child.

International alliance against hunger: Establishing 100,000 Community Food Banks by 2005 (60th anniversary of FAO). Community Food Banks (CFB) can be started at the village level, with initial food supplies coming as a grant from governments and donor agencies like the World Food Programme. Later, such CFBs can be sustained through local purchases and from continued government and international support for Food for Sustainable Development and Food for Nutrition programs. The CFB can be the entry point to not only bridging the nutritional divide, but also for fostering social and gender equity, ecology and employment. They can also be equipped to cater to emergencies like cyclones, floods, drought and earthquakes.

The CFBs can be organized with the following four major streams of responsibilities: a) Entitlements: The benefits of all government and bilateral and multilateral projects intended for overcoming under- and malnutrition can be delivered in a coordinated and interactive manner (as for example those intended for overcoming the deficiencies of macro- and micronutrients); b) Ecology: Food for Sustainable Development with particular reference to the establishment of Water Banks, land care, control of desertification and promotion of afforestation. Thus, grains can be used to strengthen local level; c) Ethics: This group of activities will relate to nutritional support to old and infirm persons, pregnant and nursing mothers and infants and pre-school children; d) Emergencies: This activity will relate to the immediate relief operations following major natural catastrophies like drought, floods, cyclone and earthquake, as well as to meet the challenge of seasonal slides in livelihood opportunities due to natural causes as well as human conflicts and mini-wars.

Each of the above four streams of activities can be managed by four separate self-help groups of local women and men. This will help to generate a self-help revolution in combating hunger. The overall guidance and oversight may be provided by a multistakeholder Community Food Bank Council. One CFB for 1,000 persons may need annually about two hundred and fifty tons of foodgrains. The International Alliance against Hunger can sponsor immediately 100,000 Community Food Banks in the hunger, biodiversity and hydrologic "hot spots"of the world. The foodgrains needed for this purpose could be mobilized through the World Food Programme.

S.No.	Sector	Institutional Mechanism
1.	Dairy	Co-operatives
2.	Poultry	Egg and Broiler Coordination Councils
3.	Integrated on-farm and off-farm employment	Biovillages
4.	Power of scale to small producers	Small Farmers' Agri-business Consortium
5.	Technological upgrading of production and post-harvest sectors	Agri-Clinics Agri-business Centers
6.	Group action for micro-enterprises supported by micro-credit	Self-help Groups
7.	Timely and affordable credit	Kisan (farmer) Credit Cards, Integrated Informal and Formal Banking System
8.	Ensuring minimum support price	Food Corporation of India and State Corporations; buy-back arrangements with the private sector

Institutional structures for extending the extrapolation domain of successful experiences and efforts. Without appropriate institutional structures, isolated success stories will remain just talking points. Based on local cultural traditions and socio-economic conditions, institutional structures should be fostered which can take new technologies to the unreached and which can give the power of scale to small producers at the production, post-harvest and marketing phases of farming. The following are some of the structures which have given promising results in India. They are based on a "win-win" situation for all, as otherwise the efforts will not be sustained over time.

Without socially compatible and socially owned institutional structures, the extrapolation domain of successful experiences and development efforts will remain limited. Community involvement will ensure how transaction costs and a high percentage of success and help to convert *unique* examples into more *universal* ones.

National Alliance Against Hunger. FAO has proposed an International Alliance Against Hunger. To make such an alliance effective, every country should form a National Alliance Against Hunger, with the alliance partners reaching the remotest village and hamlet based on Mahatma Gandhi's principle: "To the hungry, God is bread; this God should be present in every house and hut of the country." The strategy of a hunger-free nation could consist of the following three inter-related groups of activities:

a) Ensuring a whole lifecycle approach to nutrition security at the level of each individual by providing the needed horizontal linkages among ongoing vertically structured programs (often operated by different national, bilateral and UN agencies).

b) Organization of a National Board for the Management of Change in the Farm Sector with multistakeholder participation (government, NGOs, academia, women's and consumer groups, business and industry, media representatives and bilateral and UN partners). Such a Board can help to foster consensus on issues relating to: i) Technology (with particular reference to biotechnology and GM foods); ii) Ecology (desertification, water, biodiversity and climate); iii) Trade (World Trade Agreement in Agriculture including sanitary and phytosanitary measures and food safety standards).

c) Promoting the establishment of Community Food Banks at the local level, to serve as the focal point for according concurrent attention to ending poverty induced endemic hunger, micronutrient deficiency induced hidden hunger and transient hunger caused] by human conflicts and natural calamities.

The National Alliance Against Hunger can provide policy oversight and undertake resource mobilization (financial, technical and managerial) for achieving the goal of "food for all". While the above approach would help to alleviate hunger today, we can avoid hunger tomorrow only by sustaining advances in agricultural production through an agricultural production front.

Conclusions

To sum up, agriculture in most developing countries is not just a food producing enterprise, but is the backbone of sustainable ecological and livelihood security systems. In addition, it is the foundation for national sovereignty. Agricultural progress represents the best safety net against hunger and deprivation because of the greatest good it confers on the largest number of people. Therefore, national and international public policies should recognize that food is a powerful tool for socially meaningful development. At the national level, prime farmland should be conserved in the same manner as National Parks and Protected Areas. Agri-reserves are as important for safeguarding food security as Biosphere Reserves are for conserving biodiversity. Unfortunately, the importance of conserving prime farmland for food production is yet to receive the attention it needs and deserves. All good farmland should be declared as agri-reserves and regulations put in place that will make it difficult to convert prime farmland for non-farm uses, without convincing reasons. Agri-reserves and biosphere-reserves can become mutually supportive – agri-reserves being the guardians of food security and biosphere reserves of ecological security. Both food and ecological security are essential for the survival of life on earth and hence the need for land and water for either of them should not be treated in an antagonistic manner.

Community Food Banks can help to make food a powerful instrument of compassion and love, leading to the spread of a caring and sharing culture. The spread of the CFB movement in the hunger hot spots of the world will be the speediest and low cost method of achieving the goal of food for all and for ever. The uncommon opportunity provided by a foodgrain surplus world, should not be missed.

The World Summit for Sustainable Development scheduled to be held at Johannesburg in August-September this year provides a unique opportunity for strengthening existing partnerships and promoting new ones. It is in the long-term enlightened self-interest of the rich that they should match the words they will speak and the sentiments they will express at Johannesburg with action as expressed by a firm commitment of financial and technical resources for fostering sustainable human security of which sustainable food security is the foundation.

I would like to end with a poem by W.H. Auden, in the hope that we will see during this decade the emergence of a global corps of hunger-fighters, in whose dictionary the word "impossible" does not occur with reference to ensuring that every child, woman, and man has an opportunity for a healthy and productive life. ‖

Hunger allows no choice
To the citizen or the police.
We must love one another or die ...
Defenseless under the night
Our world in stupor lies;
Yet, dotted everywhere,
Ironic points of light
Flash out wherever the Just
Exchange their messages;
Of Eros and of dust,
Beleaguered by the same
Negation and despair,
Show an affirming flame.

63. MSS at the 2000 World Food Prize ceremonies with the 2000 World Food Laureate Dr. Evangelina Villegas of Mexico, Des Moines, Iowa, October 18, 2000

References

de Soysa, Indra and Nils Petter Gleditsch. 1999. *To Cultivate Peace Agriculture in a World of Conflict.* Oslo, Norway: International Peace Research Institute (PRIO).

Swaminathan. M.S. 1994. *Uncommon Opportunities: An Agenda for Peace and Equitable Development.* London: Zed Books.

Wilson, E.O. 2002. *The Future of Life.* New York: Knopf.

37

Acceptance Speech as President of the Pugwash Conferences on Science and World Affairs and Excerpts from "Science and Achieving the Goal of a Hunger-Free World"

Pugwash Conferences on Science and World Affairs,
San Diego, California

The Pugwash Council is pleased to announce that Prof. M.S. Swaminathan, UNESCO-Cousteau Professor in Ecotechnology and Chairman of M.S. Swaminathan Research Foundation in Chennai, India, will become the President of the Pugwash Conferences on Science and World Affairs, to take effect at the conclusion of the 52nd Pugwash Conference being held at the University of California, San Diego in La Jolla, California, from August 10-14, 2002.

Long recognized for his singular contributions to sustainable agriculture and food security, Prof. Swaminathan helped launch the "Green Revolution" which dramatically increased crop yields and converted India from a net food importer to a food exporter. During his entire career, Prof. Swaminathan has worked tirelessly to ensure greater equity and eradicate poverty, hunger and deprivation around the world.

Outgoing Pugwash President, Sir Michael Atiyah, warmly welcomed the selection of Prof. Swaminathan, noting that the new President would help expand the Pugwash agenda in the years ahead to encompass issues relating to the health and welfare of the world's peoples in addition to the traditional Pugwash goals of resolving conflict and eliminating nuclear and other weapons of mass destruction.

Prof. Swaminathan received his Ph.D. from the School of Agriculture, University of Cambridge, UK, in 1952, and is the recipient of numerous honorary degrees and international awards, including most recently the Franklin D. Roosevelt Four Freedoms Award (2000) and the Planet and Humanity Medal of the International Geographical Union (2000).

The Pugwash Council and the entire Pugwash community of more than 3,000 scientists and policy specialists in more than sixty countries warmly congratulate Prof. Swaminathan and look forward to working with him to advance the goals of Pugwash in the years ahead.

M.S. Swaminathan's response to his election as
President of the Pugwash Conference, at the 52nd Pugwash
Conference, University of California, San Diego August 12, 2002

I feel very humble standing before you to accept the Presidency of this unique organization, considering the eminence of those who have occupied this position before. I am happy that Sir Joseph Rotblat, one of the founders of this organization is with us. We do miss Sir Michael Atiyah, who along with Prof. George Rathjens, has done so much during the past five years to strengthen the contemporary relevance of Pugwash. We are indebted to both of them for their monumental contributions to the organization.

I am particularly happy that we have at this meeting large numbers of International Student/Young Pugwash members. Their participation helps to bridge the generation gap in perceptions and priorities.

I realize that I am the first from a Third World country to be elected President. This imposes on me the obligation to bring to your attention the threats to human security arising from poverty, hunger and unemployment. Clean drinking water is becoming a chemical curiosity in several parts of the developing world. Humankind cannot have a better common future without a better common present. Unsustainable lifestyles on the part of a few and unacceptable poverty on the part of large numbers should not continue to co-exist, if we are to achieve the new Paradise referred to in the Russell-Einstein Manifesto.

My initial acquaintance with the mission of Pugwash was in 1958, when at the second UN Conference on the Peaceful Uses of Atomic Energy held in Geneva, Prof. H. J. Muller explained to me the purpose of Pugwash. Since then, I have followed with great admiration the statements issued after Pugwash Conferences, and rejoiced along with numerous other like-minded persons the recognition given to Sir Joseph Rotblat and to Pugwash in 1995 through the Nobel Peace Prize.

During my tenure as President, I shall do my best to draw public and political attention to the catastrophic consequences of the use of nuclear weapons as well as to the moral and ethical responsibility of scientists for their inventions. Anthropogenic threats to human security are increasing day by day. In addition to the nuclear peril, we should fight the biological and chemical perils now looming large on the horizon. For human induced maladies, only human beings can find remedies. I am confident that together with our young Pugwash members, we can continue to keep Pugwash an affirming flame, and a powerful moral force in the midst of the growing violence in the human heart we are witnessing today.

While there are many forms of threats to a sustainable human future, we should not relax our efforts in getting a Comprehensive Test Ban Treaty on all kinds of nuclear weapons approved by all nations by the year 2005, which marks the 50th anniversary of the Russell-Einstein Manifesto, and the 60th anniversary of the dropping of atomic bombs on Hiroshima and Nagasaki. I consider the launching of an Agenda 2005 program designed to terminate nuclear, biological and chemical perils forever, will meet with widespread support from concerned citizens worldwide who are all bewildered at the direction human civilization is taking following the September 11, 2001 events. We should focus our moral and intellectual resources to fostering harmony within humankind and between humankind and nature.

In all such efforts, we should keep in mind what Mother Theresa once said, "My work may be a drop in the ocean; but the ocean will be less with that missing drop." Let us continue to strive to be an organization which helps to make a difference to human destiny – a destiny where the uncommon opportunities opened up by modern science and technology help to confer on every child, woman and man on our planet the four freedoms which President Roosevelt advocated sixty years ago: freedom from fear, freedom from want, freedom of expression, and freedom of worship.

Excerpts from the Dorothy Hodgkin Lecture
"Science and Achieving the Goal of a Hunger-Free World"
August 14. 2002

Introduction

It is a privilege to deliver this lecture in memory of the great scientist and humanist, Dorothy Hodgkin. According to Max Perutz, she was not only an outstanding chemist but also a saintly, gentle and tolerant lover of people, and a devoted protagonist of peace. Her humanism and sense of social justice were evident from her commitment to world peace, conflict resolution through conversation and consensus, and alleviation of poverty. She was a special friend of India. She had held the Raman Professorship of the Indian Academy of Sciences, which elected her an honorary fellow.

I have chosen the theme "Science and Achieving the Goal of a Hunger-Free world" since to quote George Hebert, lasting peace will come only if we eliminate hunger.

Sweet Peace, where dost thou dwell?
I humbly crave.
Let me once know.
I sought thee in a secret cave.
And ask'd if Peace was there.
A hollow wind did seem to answer. No:
Go seek elsewhere.
At length I met a rev'rend good old man:
...
Whom when for Peace
I did demand, he thus began:
"Take of this grain, which in my garden grows.
And grows, for you;
Make bread of it: and that repose
And peace, which ev'rywhere
With so much earnestness you do pursue,
Is only there.

The same idea can be found in the following words of the Roman Philosopher Seneca who was convinced that where hunger rules, peace cannot prevail: "A hungry person listens neither to religion nor reason, nor is bent by prayers."

A Roman farmer, Varro, explained the principle behind sustainable agriculture by pointing out: "Agriculture is a science which teaches us what crops should be planted in each kind of soil. And what operations are to be carried out, in order that the land may produce the highest yields in perpetuity."

In spite of the recognition that food and drinking water constitute the first among the hierarchical needs of human beings, currently over a billion children, women and men go to bed partially hungry each night and access to safe drinking water is increasingly becoming a luxury in most developing countries. At the World Food Summit convened by FAO in Rome in 1996, it was resolved that by 2015, the number of hungry should be reduced to half the present number. However, a review of the progress made since 1966 in achieving this target, carried out at a FAO Conference held in Rome in June, 2002, revealed that the current reduction is about six million per year, in contrast to the target of twenty to twenty-two million per year. In sixty-seven developing countries, the number of undernourished increased by seventy-seven million since 1996. According to FAO, the total number of hungry in the world now stands at eight hundred and fifteen million, of whom seven hundred and seventy seven million are in the developing countries; twenty-seven million are in the countries in transition from central rule to democratic governance, and eleven million in the industrialized world. [...][1]

"Science for Basic Human Needs" Movement

Since the onset of the Industrial Revolution in Europe, technology has been a major source of economic inequity among nations and among communities within nations. If technology has been a cause of economic and social inequity in the past, today we have an opportunity for making technology an ally in the movement of social, gender and economic equity. Modern information technology provides this opportunity. Knowledge and skill empowerment can now be achieved at a fast pace. However, the technological and skill empowerment of the poor cannot be achieved through programs designed on the basis of a patronizing and a top-down approach. The information provided should be demand- and need-driven and women belonging to the socially and economically underprivileged sections of the society should preferably manage the knowledge centers. Our aim should be the initiation and spread of a Knowledge Revolution for ending economic and gender inequity.

The accomplishment of the tasks requires considerable technical, managerial and financial resources. Scientists of the International Peace Research Institute, Oslo, have studied the causes of armed conflicts during the last 10 years. They found that violent conflicts in most cases could be traced to economic rather than ideological differences. They have hence suggested that investing in agriculture which helps to promote food and livelihood security in many nations is an effective strategy for preventing future wars, eradicating poverty, preventing

[1] **Editor's note**: Several of the papers presented in the sections "Hunger and Ethics", "Science, Technology, and Economic and Social Inequity", "Science and Food and Water Security" and "Emerging Scientific Revolutions and Ecology of Hope" are covered elsewhere in this volume.

environmental destruction and reducing violence. Unfortunately even now, for too high a proportion of national GDP is being spent on arms and military equipment as compared to programs designed for poverty eradication and meeting the basic needs of the underprivileged sections of humankind.

An International Commission on Peace and Food, which I chaired, in its report submitted in 1994, chronicled the uncommon opportunities now available for a better common future for humankind, if funds can be diverted from defense to development designed to end hunger and poverty. Unfortunately, the so-called "peace dividend" still remains only in the realm of possibility.

We now have a huge stockpile of scientific discoveries and technological innovations. This stockpile is more than adequate to help all nations to provide every adult human being an opportunity for a healthy and productive life and every newborn child a happy future. It is therefore a sad commentary on our political, social and spiritual value systems that the number of children, women and men living in poverty today exceeds the entire human population of the year 1900.

Is there then no hope for the future? Mahatma Gandhi showed a path, which we should now re-discover. That path was paved with the principles of *antyodaya* (i.e., attention to the poorest person), *sarvodaya* (a pattern of development where there are no winners and losers, but where there is a win-win situation for all), trusteeship mode of handling one's intellect and surplus wealth, and non-violence.

In other words, the formidable power of science and technology can benefit

Figure 1. Paradigm Shift

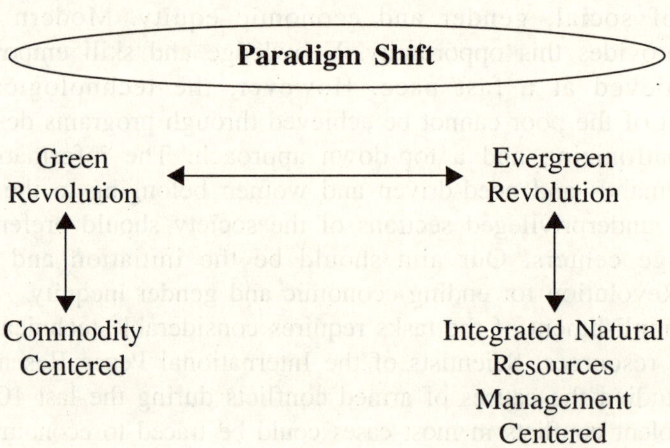

humankind only if we know how to temper it with humanism. Let us hope that this decade will bring forth a new crop of leaders of science who will be able to usher in an era of humanistic science. It is the duty of scientific establishments and science academies to nurture and foster the growth of young men and women research leaders capable of initiating and managing change in goals and strategies in the coming century.

Towards a Hunger-Free World

Hunger is the extreme manifestation of poverty.[2] We now have the capacity to achieve universal freedom from hunger, if we can bring about the requisite fusion among political will and action, professional skill and people power. This requires concurrent attention to achieving an evergreen revolution on the farms leading to a continuous rise in productivity without associated ecological or social harm, and to the equitable distribution of food. Open-ended Food for Eco-development programs in the hunger hot spots of the world will help to achieve the goals of both hunger elimination and ecosystem conservation and enhancement.

The "Green Revolution", a term coined by Dr William Gaud of the United States in October 1968, is a process which leads to higher production through improvement of productivity. In January 1968 several months before the term Green Revolution came into use, I emphasized in a lecture at the Indian Science Congress, the need for improving productivity in perpetuity without associated ecological and/or social harm, in the following words: "Exploitive agriculture offers great dangers if carried out with only an immediate profit or production motive. The emerging exploitive farming community in India should become aware of this. Intensive cultivation of land without conservation of soil fertility and soil structure would lead, ultimately, to the springing up of deserts. Irrigation without arrangements for drainage would result in soils getting alkaline or saline. Indiscriminate use of pesticides, fungicides and herbicides could cause adverse changes in biological balance as well as lead to an increase in the incidence of cancer and other diseases, through the toxic residues present in the grains or other edible parts. Unscientific tapping of underground water will lead to the rapid exhaustion of this wonderful capital resource left to us through ages of natural farming. The rapid replacement of numerous locally adapted varieties with one or two high-yielding strains in large contiguous areas would result in the spread of serious diseases capable of wiping out entire crops, as happened prior to the Irish potato famine of 1854 and the Bengal rice famine in 1942. Therefore the initiation of exploitive agriculture without a proper understanding of the various consequences of every one of the changes introduced into traditional agriculture, and without first building up a proper scientific and training base to sustain it may only lead us, in the long run, into an era of agricultural disaster rather than one of agricultural prosperity". (Swaminathan, 1968, 1993). Later, I coined the term "evergreen revolution" to highlight the pathway of increasing production and productivity in a manner that the short and long-term goals of food production are not mutually antagonistic (Swaminathan, 1996). While the Green Revolution was based on commodity centered research there is need for research on integrated natural resources management, for achieving an evergreen revolution (Figure 1). Edward O. Wilson in his recent book, *The Future of Life* (2002), has referred to my concept of evergreen revolution in the following words: "The problem before us is how to feed billions of new mouths over the next several decades and save the rest of

---

[2] **Editor's note**: This paragraph was used in a previous essay by Dr. Swaminathan but it bears repetition here.

life at the same time, without being trapped in a Faustian bargain that threatens freedom and security. No one knows the exact solution to this dilemma. The benefit must come from an evergreen revolution. The aim of this new thrust is to lift food production well above the level obtained by the green revolution of the 1960s, using technology and regulatory policy more advanced and even safer than those now in existence." ▮

References

Swaminathan. M.S. 1968. The Age of Algeny, Genetic Destruction of Yield Barriers and Agricultural Transformation. Presidential Address, Agricultural Science Section. *Fifty-fifth Indian Science Congress Proceedings*, Varanasi, India: Indian Science Congress. (January)

-----. (Ed.) 1993. *Wheat Revolution: A Dialogue*. Madras: Macmillan India

-----. 1996. *Sustainable Agriculture: Towards an Evergreen Revolution.* Delhi: Konark Publishers.

Wilson, E.O. 2002. *The Future of Life*. New York: Knopf.

64. MSS speaking at seminar on Gender,
MSSRF, Chennai, 2000

38

Toward a Hunger-Free World: The Final Milestone

I. *Introduction*

The award of the Nobel Peace Prize to Dr. Norman Borlaug in 1970 highlighted the interrelationships between hunger and peace. We witness today a growing violence in the human heart due to a variety of reasons. An important reason however is the growing sense of alienation and deprivation prevailing among nearly three billion members of the human family, who, according to the World Bank, live on an income of two US dollars or less per day. It is becoming clear that where hunger persists, which is the extreme manifestation of poverty, peace cannot prevail.

There are both hunger hot spots and food security bright spots in the world. The human right to food is now widely accepted and slowly the political and legal will to implement this right is growing. Brazil, for example, has developed a Zero Hunger Program containing the following components: ensuring availability; improving accessibility; employment generation; minimum wage; agrarian reform; supporting small scale farmers; minimum income programs; nutrition programs; public health and food safety programs; assistance to vulnerable social groups, reducing discrimination; increasing total social spending.

Norman Borlaug's work during the last sixty years has shown that given appropriate synergies between science and public policy, progress in improving food production can be fast. He has underlined the need to address the problems of hungry and thirsty soils in order to get the best of genetic strains of major food crops which have the potential for high yields. Pedro Sanchez (2002) has demonstrated that in many parts of Africa characterized by chronic hunger, soil fertility replenishment through appropriate agro-forestry techniques holds the key to raising the ceiling to yield. After the soil's fertility is replenished, high yielding crop varieties, integrated pest management, conservation tillage, high-value trees, vegetable crops and dairy cattle can all be introduced in the farming system.

Today, feasible and affordable approaches are available to address the following three major kinds of hunger: *Endemic hunger* or chronic protein-energy under-nutrition; *Hidden hunger*, caused by deficiencies of micronutrients in the diet; *Transient hunger*, caused by drought or other natural calamities as well as ethnic conflicts. It is also realized that to overcome these forms of hunger, there is need for an integrated strategy which pays concurrent attention to the following: i) Food availability, which is a function of home production or food imports or both.

Presented as Norman Borlaug Lecture, University of Iowa, October 14, 2003.

65. Norman E. Borlaug, CIMMYT, Mexico

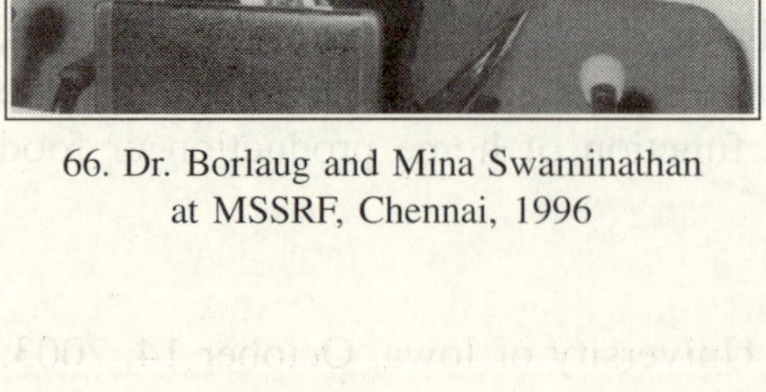

66. Dr. Borlaug and Mina Swaminathan at MSSRF, Chennai, 1996

67. Sartaj Aziz, Norman Borlaug, Anwar Dil, MSS at IFPRI Conference, Washington DC, 2002

ii) Food access, which is a function of purchasing power or sustainable livelihood opportunities. iii) Food absorption, which is a function of access to clean drinking water, environmental hygiene, primary health care and primary education.

In India, detailed Food Insecurity Atlases have been prepared for rural and urban areas by MSSRF and the World Food Programme (WFP), which serve as valuable tools for public policy and action.

At the global level, the following eight UN Millennium Development Goals illumine the path to a world free of hunger and unacceptable deprivation: Eradicate extreme poverty and hunger; Achieve universal primary education; Promote gender equality and empower women; Reduce child mortality; Improve maternal health; Combat HIV/AIDS, malaria and other diseases; Ensure environmental sustainability; Develop a global partnership for development.

There are specific targets for being achieved by 2015. I would like to indicate why we are now in the final milestone with reference to ensuring food for all and for ever.

II. *Bridging the Nutritional Divide*

UNDP's Human Development Report 2001 has introduced a Technology Achievement Index (TAI). TAI is an aggregation of four groups of indicators relating to the creation of technology, diffusion of recent innovations, diffusion of old innovations, and human skills. Creation of Technology has been measured by the number of patents per capita and receipts of royalty and license fees from abroad per capita. The emphasis is thus on the Intellectual Property Rights (IPR) of nations, evidenced by the power of proprietary science. The other indicators relate to digital, extension and educational divides. This Report titled "Making New Technologies Work for Human Development", has however not drawn attention to the fact that bridging the expanding nutritional divide is fundamental to bridging the other divides, particularly that relating to IPR.

The Commission on the Nutrition Challenges of the 21st century in its report titled "Ending Malnutrition by 2020: An agenda for change in the millennium", has pointed out that some thirty million infants are born each year in developing countries with intra-uterine growth retardation, representing about 24% of all new births in these countries (Philip et al, 2000). Low biith weight (LBW) children are characterized by mental impairment. Worldwide there are more than 150 million under-weight pre-school children and more than 200 million stunted children. At current rates of progress in fighting these maladies, about one billion children will be growing up by 2020 with impaired mental development. What will be the impact of such a denial to the child of opportunities for the full expression of its innate genetic potential for mental and physical development on the intellectual property of a nation? Denying the child an opportunity for mental and physical development even at the foetal stage is the cruellest form of inequity. In contrast, overweight is the major health problem among children in most industrialized countries and some developing ones (Table 1). Thus, bridging the nutritional divide is the first requisite for a more equitable and humane world.

Table 1. Share of Children who are Underweight and Adults who are Overweight, Selected Countries, Mid-1990s

Country	Share Underweight (percent)	Country	Share Overweight (percent)
Bangladesh	56	United States	55
India	53	Russian Federation	54
Ethiopia	48	United Kingdom	51
Vietnam	40	Germany	50
Nigeria	39	Colombia	41
Indonesia	34	Brazil	36

Source: Gardner and Halwell, 2000

III. *Growing Disparities*

The nutritional divide is increasing between the rich and the poor within and among nations. The situation is particularly alarming in developing countries. The nutritional paradox of South Asia lies in the co-existence of grain mountains and hungry millions. This is largely due to inadequate purchasing power arising from lack of sustainable livelihood opportunities. The Famine of income is becoming the most important cause of a famine of food at the household level. Pregnant and nursing mothers and children belonging to families living below the poverty line (the World Bank poverty line is an income of one US dollar per capita per day or below) are the worst sufferers. For example, severe anaemia during pregnancy is associated with very high relative risk of maternal death. Maternal mortality rates are as low as 3 to 4 per 100,000 births in industrialized countries, while in many developing countries they are at least 100- to 200-fold higher. Protein-energy malnutrition (PEM) affects nearly 30% of children under five years of age in countries in Sub-Saharan Africa. A comparison of the nutritional status of populations in three Asian countries – China, India and Sri Lanka – provides some interesting insights into the impact of public policy on the nutritional well being of the population.

IV. *Nutrition Profile Among a Few Nations in Asia: Role of Non-Nutritional Factors*

Four parameters - underweight, stunting, wasting and low birth weight - reflect the nutritional status of children below five years of age. The comparative profile of Sri Lanka, China and India is given in Table 2. The data show the importance of non-nutritional factors like education and health care in the nutritional well being of an individual.

Table 2. Comparative Nutrition Profiles of Children

Parameter	Sri Lanka	China	India
	In percentage		
Undernourished	25	11	21
Underweight	38	10	53
Stunting	24	17	52
Low birth weight	25	6	30

Source: *HDI Report 2001* and Planning Department of Sri Lanka

a) Body Mass Index: It gives the nutritional status of adults. Adults having BMI less than 18.5 are considered to be chronically energy deficient. Body Mass Index over 25 indicates overweight. Obese persons will have BMI over 30. The situation in Sri Lanka is given in Table 3.

Table 3. Body Mass Index (BMI) in Sri Lanka

BMI	Men	Women
< 18.5	36.2	33.4
> 25	9.0	12.8
> 30	0.8	1.5

b) Iron Deficiency Anaemia: In Sri Lanka, 58% of the children in the age group of 5 to 10 years suffer from iron deficiency anaemia affecting cognitive capacity and academic performance. In the case of adults, 45% suffer from iron deficiency anaemia. In the case of pregnant mothers, the proportion suffering from anaemia is less, namely, 39%.

c) Mortality rates: Thanks to advances in preventive and curative medicine, mortality has been declining between 1970-75 to 1999-2000 in China, India and Sri Lanka (Table 4). IMR and MMR are still high, although there is considerable variability among states within the country. The State of Kerala, in India for example, has figures similar to those of Sri Lanka.

Table 4. Mortality Rates

Category	Years	Sri Lanka	China	India
Life Expectancy in years	1970-75	65.1	63.2	50.3
	1995-2000	71.9	69.8	62.3
Infant mortality	1970-75	65.0	85.0	127
	1996-2000	17.0	33.0	70.0
Under-5 mortality	1970-75	100	120.0	202.0
	1995-2000	19.0	41.0	98.0
Maternal mortality*	1995-2000	60.0	55.0	408.0

Source: *Human Development Report 2001* * Per 100,000 live births

d) Female Literacy and Child Health: Education of women and a rapid increase in the rate of female literacy have been achieved in Sri Lanka as a result of the introduction of free education from 1945 onwards. It enabled girls to have equal access to education as boys. The situation is similar to that observed in the Indian State of Kerala.

Both men and women have achieved high literacy rates with 83% for women and 90% for men. They also have very low drop-out rates, viz., 4% for girls and 6% for boys. There is a significant impact of mothers' education on the nutritional status of children (Table 5).

Table 5. Mother's Education and Child Malnutrition in Sri Lanka

Category	Stunting	Wasting	Underweight
All categories	23.8	15.5	37.7
No education	46.0	16.7	57.9
Primary	33.6	18.7	37.8
Secondary	22.6	16.8	39.1
More than secondary	13.0	11.3	24.6

Source: Planning and Development Department, Sri Lanka

e) People Power Revolution in Nutrition: Ultimately the success of various nutrition-related programmes depends upon the effectiveness of the delivery systems. Hence, Sri Lanka is attempting a community-based nutrition intervention program. This is called participatory nutrition improvement project (PNP). This program was started in 1993 with the help of UNICEF. The guiding principle was to mobilize the energies of the community and their commitment to their own and their families' nutritional well-being. PNP is a people-focused project, enhancing the ability of mothers and fathers, through group formation and strengthening, to identify or explore their nutritional problems, identify their nutritional needs and maximize their potential in meeting those needs. Countries like Cuba, China and India have also rich and varied experience in the development of effective delivery systems. In Sub-Saharan Africa, Ghana has made rapid progress in overcoming PEM through community-based nutrition (Gardner and Halwell, 2000). Mobilizing "people power" in the cause of nutritional security is the most effective and sustainable strategy. The example of Thailand illustrates this fact.

f) Thailand's Nutrition Security Compact: During the past ten years, Thailand has achieved remarkable progress in reducing maternal mortality as well as the incidence of LBW children. The strategy consisted of the following components: Eliminate severe, moderate and mild protein-energy malnutrition (PEM); Monitor growth among all pre-school children and provide food supplements where needed; Mainstream nutrition in health, education and agricultural policies; Retrain and retool existing staff and mobilize community volunteers - choose one community volunteer for every ten households and build their capacity; Encourage breast feeding and organize school lunch programs; Promote home gardening, consumption of fruits and vegetables, aquaculture and food safety standards; Introduce an integrated food safety net with emphasis on household food and nutrition security. The positive impact of the above Nutrition Security Compact is evident from the decline of maternal mortality from 230 per 100,000 live births in 1992 to 17 in 1996 (Philip et al, 2000). Thailand's initiative in organizing a Community Volunteer Corps for Household Nutrition Security is worthy of emulation by other nations.

V. *Challenges Ahead*

Among the nutritional challenges facing the countries in transition, the following need priority attention.

a) Low birth weight: For the reasons already mentioned, governments and civil society organisations in developing countries should accord high priority to overcoming maternal and foetal undernutrition and malnutrition. Future intellectual attainments of nations will depend very much on success in this area.

b) Undernutrition and stunting among children: Because of its linkages to mental impairment, stunting should be addressed through an integrated package of health care, education and nutritional measures. Early undernutrition accentuates adult chronic diseases including diabetes, heart disease, hypertension and cancer.

c) Undernourished adults: Judged by a body mass index of less than 17 kg/m^2, over 240 million adults in developing countries are severely undernourished. The nutritional safety net for this category could include programs like food for eco-development i.e., food for work and food for nutrition.

d) Vitamin A and iodine deficiencies: Sub-clinical vitamin A deficiency still affects nearly 200 million pre-school children in developing countries. Sustained efforts are also needed to eliminate iodine deficiency disorders.

e) Pandemic anaemia: Maternal anaemia is pandemic and is associated with high MMR. Anaemia during infancy, compounded by maternal under-nutrition, leads to poor brain development.

f) Lack of access to clean drinking water: This is a major nutritional problem since contaminated water is a major cause of intestinal infections and diarrhoea in children. Access to clean drinking water is becoming a luxury in many developing countries.

g) Access to sustainable livelihoods: Ultimately, it is the lack of purchasing power that is responsible for poor access to balanced diet. In India; the poverty line is defined in nutritional terms. The estimation of poverty is based on the consumption expenditure level below which a household of 5.5 persons, on an average, cannot meet the recommended intake of 2,400 kcal for adults in rural areas and 2,100 kcal in urban areas. In the case of poor households, over 70% of the daily income goes to food. Even by this austere yardstick, over 250 million persons in India live below the poverty line. In the area of income poverty, South Asia is the "hot spot" (Table 6).

Table 6. Regional Comparison of Income Poverty in Developing Countries

	1987	1990	1993	1996	1998
East Asia and Pacific	417.5	452.4	431.9	265.1	278.3
South Asia	474.4	495.1	505.1	531.7	522.0
Asia and Pacific	891.9	947.5	937.0	796.8	800.3
Europe and Central Asia	1.1	7.1	18.3	23.8	24.0
Latin America and Caribbean	63.7	73.8	70.8	76.0	78.2
Middle East and North Africa	9.3	5.7	5.0	5.0	5.5
Sub-Saharan Africa	217.2	242.3	273.3	289.0	290.9
Asia & Pacific as % of world total	75.4	74.2	71.8	66.9	66.8
Total	1183.2	1276.4	1304.3	1190.6	1198.9

Source: IFAD (2001)

VI. *Meeting the Challenges*

1. *Food-based approach to nutrition security.* Such an approach will involve the following steps.

a) Food availability: This is a function of both home production and imports. In many developing nations, the gap between potential and present yields is high in most farming systems, even with the technologies available on the shelf. High priority should hence go to bridging the productivity gap through a mutually reinforcing blend of technologies, services and public policies. Also, mainstreaming the nutritional dimension in the design of cropping and farming systems is essential. There is no time to relax on the food production front. The present global surplus of foodgrains is the result of inadequate consumption on the part of the poor, and should not be mistaken as a sign of over-production. Developing nations should aim to achieve revolutions in the following five areas to sustain and expand the gains already achieved. i) *Productivity revolution*: The scope is great since average yields are still low in most cropping and farming systems. However, the production techniques should be environmentally sustainable, so that high yields can be obtained in perpetuity. ii) *Quality revolution:* This can be achieved through greater attention to post-harvest technologies and bio-processing, as well as to sanitary and phytosanitary measures and *Codex Alimentarius* standards. iii) *Income and employment revolution:* This will call for an integrated attention to on-farm and non-farm livelihoods and to farming systems intensification, diversification and value addition. Post - harvest processing offers scope for generating additional livelihoods through micro-enterprises supported by micro-credit. iv) *Small farm management revolution:* Institutional structures which will confer upon farm families with small holdings the advantages of scale at both the production and post-harvest phases of agriculture are urgently needed. For example, thanks to the cooperative method of organization of milk processing and marketing, India now occupies the first position in the world in milk production. Strategic partnerships with the private sector will help farmers' organizations to have access to assured and remunerative marketing opportunities. In relation to factors of production, water is likely to become the key constraint during this century. Hence every effort should be made to enhance productivity and income per every drop of water. v) *Enlarging the food basket:* During the last century, there has been a rapid decline in the number of crops contributing to global food security. In the past, local communities depended upon a wide range of crops for their food and health security. It is important that we revive the old dietary traditions. Particular attention needs to be paid to leafy vegetables which are rich in micronutrients.

b) Food access*:* Inadequate livelihood opportunities in rural areas result in household nutrition insecurity. India today has over thirty million metric tons of wheat and rice in government godowns; yet poverty induced hunger affects over 200 million persons. Macro-economic policies, at the national and global level, should be conducive to fostering job-led economic growth based on micro-enterprises supported by micro-credit. Where poverty is pervasive, suitable measures to provide the needed entitlement to food should be introduced.

c) Food absorption*:* Lack of access to clean drinking water, as well as poor environmental hygiene and health infrastructure, lead to a poor assimilation of the food consumed. Nutrition security cannot be achieved without environmental hygiene, primary health care and clean drinking water security. Culinary habits also need careful evaluation. Some methods of cooking may lead to the loss of vital nutrients.

d) Transient hunger*:* Ferro-Luzzi et al (1994) have carried out a detailed study of seasonal cycling in body weights related to changes in weather. Any strategy for nutrition security should provide for steps to meet such transient hunger. The Indian State of Maharashtra introduced nearly twenty-five years ago an Employment Guarantee Scheme to assist the poor to earn their daily bread during seasons when opportunities for wage employment are low. Similarly there is need for mainstreaming considerations of gender, age and occupation in the national nutrition strategy.

2. *Fortification and synergy among dietary components.* Our understanding of low cost and high synergy nutritional systems is growing. Fortification of flour with folic acid and genetic enrichment of staple grains with beta-carotene and iron are now receiving attention. Knowledge relating to the metabolic interrelationships among micronutrients is also growing, as for example among vitamin A-protein-zinc-iron-folic acid-vitamin C. However, in the absence of dietary interventions, iron-folate supplementation often fails to bring about a complete correction of anaemia. Hence, the attack on undernutrition induced hunger and micronutrients-deficiency-caused hidden hunger should be an integrated one. Such an integrated strategy should accord concurrent attention to food availability, access and absorption. In addition, there should be provision in the strategy for overcoming seasonal or transient undernutrition caused by loss of opportunities for livelihood during seasons of drought, floods or other natural calamities.

3. *Genetic enrichment of nutritional quality.* While the problems relating to the food and environmental safety aspects of genetically modified foods are yet to be fully resolved, there is little doubt that an integrated approach to Mendelian and molecular breeding is likely to make a food-based approach to nutrition even more effective in the future. The quantity and quality of proteins, carbohydrate, fats, vitamins and minerals can all be improved now. The scope for the genetic enhancement of nutritional quality will be evident from the following examples.

a) Quality Protein Maize (QPM): Scientists have long had an interest in improving maize protein quality. Quality Protein Maize (QPM) refers to enhanced levels of the two essential amino acids, lysine and tryptophan, in the endosperm protein. Using Mendelian breeding methodologies supported by rapid chemical analysis of a large number of samples, scientists led by S. Vasal and Evangelina Villegas at the International Maize and Wheat Research Centre (CIMMYT) in Mexico were able to slowly accumulate modifier genes to convert the original soft opaque-2-endosperm into vitreous hard endosperm type (Vasal et al, 1984). This conversion took nearly three decades. These remarkable new varieties look and taste like normal maize but the nutritive value of their protein is nearly equivalent to cow's milk. They also produce yields as much as 10% higher than the best local hybrid maize varieties and are more tolerant to biotic and abiotic stresses. QPM, which is a product of Mendelian breeding promises improved nutritional value and cost savings for a wide array of products ranging from infant food to corn chips and feed for non-ruminant livestock. The impact of this breakthrough is likely to be felt throughout the food industry and has great promise in the developing world both for human and animal nutrition.

b) Beta-carotene Rich Rice: A promising development in the field of genetic engineering is the success in breeding a nutritionally enriched rice variety now popularly referred to as "golden rice". This genetically modified rice contains genes that produce high levels of beta-carotene and related compounds, which are converted in the human body into the crucially needed vitamin A. Vitamin A Deficiency (VAD) causes more than a million childhood deaths each year and is the single most important cause of blindness among children in developing countries. Rice plants do produce cartenoid compounds (that our body converts into vit-A) but only in the green parts of the plant and not in the part of the grain normally eaten. Dr. Ingo Potrykus and Dr. Peter Beyer of Germany at the Swiss Federal Institute of Technology inserted genes from a daffodil (*Narcissus pseudonarcissus*) and a bacterium (*Erwinia uvedovora*) into rice plants to produce the modified grain, which has sufficient ß-carotene to meet total vit-A requirements in a typical Asian diet (Ye et al, 2000). If golden rice, currently still in the laboratory stage becomes a success in the field, it will help to strengthen the food-based approach to nutrition security.

c) Iron enrichment: Iron-deficiency anaemia is the most widespread nutrient deficiency in the world, affecting an estimated two billion people worldwide. Between 40 and 50 % of children under the age of five in the developing countries are iron deficient and iron deficiency accounts upto 20% of all maternal deaths. It also impairs immunity and reduces the physical and mental capacities of people of all ages. In short, iron deficiency is a major public health problem worldwide with enormous social and economic costs. Rice fortified with iron was created through the introduction of proteins from the kidney beans *Phaseolus vulgaris* by the same researchers of the Swiss Federal Institute of Technology (Lucca et al, 2000). It is reported that the iron content increased two-fold in the modified crop, currently under testing stage. Japanese scientists have also succeeded in enriching the rice grain with iron. The International Rice Research Institute (IRRI) has developed rice breeding lines high in iron and zinc using traditional plant breeding techniques. This rice is currently being tested by novitiates at a convent in the Philippines to see how well the nutrients are absorbed in the human body.

d) Designer Potato: Advances in plant tissue culture techniques and gene transfer technology have opened up possibilities for modifying the amino acid contents of plants. Potato, which is the most important non-cereal food crop, ranks fourth in terms of total global food production, besides being used as animal feed and as raw material for the manufacture of starch, alcohol and other food products. This crop was genetically modified using a seed albumin gene *Ama*1 from *Amaranthus hypochondriacus* by researchers at the Jawaharlal Nehru University (JNU), New Delhi (Chakraborty et al, 2000). The Ama1 protein is non-allergenic in nature and is rich in all essential amino acids. Its composition corresponds well with the WHO standards for optimal human nutrition (Raina and Datta, 1992). The JNU team was able to use a seed albumin gene with a well-balanced amino acid composition as a donor protein to developing transgenic potato. The genetic enrichment of protein quantity and quality in potato can make a significant contribution to child and adult nutrition, since mashed potato can be fed to young children.

The above are a few examples of the work in progress in improving through conventional and molecular breeding techniques protein quantity and quality in important food crops. Consumer confidence based on an appreciation of the scientific evidence and the regulatory checks and balances will ultimately decide whether or not genetically modified foods (GMOs) will make a significant contribution to feeding the eight billion people who are likely to inhabit our planet by 2020. Marker-aided selection and transgenic approaches are two powerful tools to accelerate plant breeding to produce crop varieties with improved nutritional traits and qualities. An intelligent integration of Mendelian and molecular breeding techniques will help to enhance the nutritive value of staples. By integrating pre-breeding in laboratories with participatory breeding in farmers' fields, it will be possible to breed location specific varieties and maintain genetic diversity in crop fields.

VII. *Building a Sustainable Community Nutrition Security System*

Conferring the right to food and thereby an opportunity for a productive and healthy life on those who go to bed undernourished now is the fundamental duty of the State as well as of the well-to-do sections of the population. Thanks to both the spread of democratic systems of governance at the grassroot level and technological advances, we now have an uncommon opportunity to foster a Community Centerd and Controlled Nutrition Security System. Such decentralized community management will help to improve delivery of entitlements, reduce transaction and transport costs, eliminate corruption and cater to the twin needs of introducing a lifecycle approach to nutrition security and meeting the challenge of seasonal fluctuations in nutritional status. The basic guidelines for such a system are the following.

1. *Adopt a whole lifecycle approach to nutrition security.*

a) Pregnant mothers: Overcoming maternal and foetal under- and malnutrition is an urgent task, since nearly 30% of the children born in countries in South Asia are characterized by low birth weight (LBW), with the consequent risk of impaired brain development. Ramalingaswami et al (1997) have pointed out that half of the world's malnourished children are in India, Pakistan and Bangladesh. LBW is a proxy indicator of the low status of women in the society, particularly of their health and nutrition status during their entire lifecycle (Narayanan, 2001).

b) Nursing mothers: Appropriate schemes will be necessary to provide support to enable mothers to breast feed their babies for at least six months, as recommended by WHO. Policies at work places, including the provision of appropriate support services, should be conducive to achieving this goal.

c) Infants (0-2 years): Special efforts will have to be made to reach this age group through their mothers, since they are the most unreached at present. Eighty percent of brain development is completed before the age of two. The first four months in a child's life are particularly critical, since the child is totally dependant on its mother for food and survival.

d) Preschool children (2-6 years): A well designed integrated child development service will help to cater to the nutritional and health care needs of this age group (Measham and Chatterjee, 1999).

e) Youth (6 to 20 years): A nutrition-based Noon Meal program in all schools (public and private and rural and urban) will help to improve the nutritional status of this group. However, a significant percentage of children belonging to this age group are not able to go to school due to economic reasons. Such school push-outs or child laborers need special attention.

f) Adults (20 to 60 years): The nutrition safety net to cater to this category should consist of both an entitlements program like Food Stamps and the Public Distribution System (PDS) as well as a Food for Eco-development program (also called "Food for Work" program). The Food for Eco-Development program can promote the use of foodgrains as wages for the purpose of establishing water harvesting structures (Water Banks) and for the rehabilitation of degraded lands and ecosystems. Thus, many downstream benefits and livelihood opportunities will be created. In designing a nutrition compact for this age group, persons working in the organized and unorganized sectors will have to be dealt with separately. Also, the intervention programs will have to be different for men and women taking into account the multiple burden on a woman's daily life.

g) Old and infirm persons: This group will have to be provided with appropriate nutritional support, as part of the ethical obligations of society towards the handicapped.

The above whole lifecycle approach to nutrition security will help to ensure that the nutritional needs of everyone in the community and of every stage in an individual's life are satisfied.

2. *Adopt a holistic action plan to achieve sustainable nutrition security at the level of each individual.* The major components of such an integrated action plan are the following:

- Identification: Identify those who are nutritionally insecure through the local community. Trained Community Volunteers of the kind mobilized in Thailand will be useful for this purpose.
- Education and Information empowerment: Empower those who are not aware of their entitlements about the nutritional safety nets available to them and also undertake nutrition education. An entitlements database can be developed for each area and household entitlement cards can be issued, indicating how to access nutritional, health care and educational programs. The educational programs should also lay stress on culinary habits in relation to the conservation of essential nutrients in cooked food.
- Overcome protein-calorie undernutrition: The various steps indicated under the whole lifecycle approach will have to be adopted. The problems of child labor and of persons working in the unorganized sector will need specific attention.
- Eliminate hidden hunger caused by the deficiency of micronutrients in the diet: Introduce an integrated approach including the consumption of vegetables and fruits, millets, grain legumes and leafy vegetables and the provision of fortified foods like iron and iodine fortified salt and oral dose of Vitamin A. The basic approach should be a food-based one, with emphasis on home and community nutrition gardens, wherever this is socially and economically feasible (Gopalan, 2001).

• Drinking water, hygiene and primary health care: Attend to the provision of safe drinking water and to the improvement of environmental hygiene. Also, improve the primary health care system.

• Sustainable livelihoods: Improve economic access to food through market-linked micro-enterprises supported by micro-credit. Also, create an economic stake in the conservation of natural and common property resources. Ensure that agreements under the World Trade Organisation (WTO) provide a level playing field for products coming from decentralized small scale production (production by masses or farmers' farming) as compared to those emerging from mass production technologies or factory farming. Promote job-led economic growth and not jobless growth.

• Pay special attention to pregnant and nursing mothers and preschool children: Measure progress through monitoring MMR, IMR, incidence of LBW children and male-female sex ratio. Iron-folate supplements during prenatal care should be accompanied by steps to overcome protein-energy deprivation. Mina Swaminathan (1998) has proposed a maternity and child care code, which if adopted, will help to bring down speedily MMR, IMR, LBW and stunting. Sex ratio is a good index of the mind-set of a society in relation to the girl child.

3. *Community Food Bank as an instrument of Sustainable Food and Nutrition Security.* Community Food Banks (CFB) can be started at the village level, with initial food supplies coming as a grant from governments and donor agencies like the World Food Programme. Later, such CFBs can be sustained through local purchases and from continued government and international support for Food for Eco-development and Food for Nutrition programs. The CFB can be the entry point to not only bridging the nutritional divide, but also for fostering social and gender equity, ecology and employment. They can also be equipped to cater to emergencies like cyclones, floods, drought and earthquakes.

The CFBs can be organized with the following four major streams of responsibilities.

a) Entitlements: The benefits of all government and bilateral and multilateral projects intended for overcoming under- and mal-nutrition can be delivered in a coordinated and interactive manner (as for example those intended for overcoming the deficiencies of macro- and micro-nutrients).

b) Ecology: Food for Eco-development with particular reference to the establishment of Water Banks, land care, control of desertification and afforestation. Thus, grains can be used to strengthen local level water security.

c) Ethics: This group of activities will relate to nutritional support to old and infirm persons, pregnant and nursing mothers and infants and pre-school children.

d) Emergencies: This activity will relate to the immediate relief operations following major natural catastrophes like drought, floods, cyclone and earthquake, as well as to meet the challenge of seasonal slides in livelihood opportunities.

Each of the above four streams of activities can be managed by four separate self-help groups of local women and men. This will help to generate a self-help revolution in combating hunger. The overall guidance and oversight may be provided by a multi-stakeholder Community Food Bank Council.

The World Food Programme has pioneered many meaningful programs relating to the use of food for eco-development. The United States operates the following programs: i) *Food for Peace* (PL-480) – started in 1949. ii) *Food for Progress* – started in 1985. iii) *Food for Education* – started in 2000. Other than the United States, India is in a position to launch such programs. India has reached a stage in its agricultural evolution when farm production will increase only if consumption is improved.

VIII. *Green Revolution and Ecotechnology in the Third Millennium*

On the eve of the UN Conference on Environment and Development held at Rio de Janeiro in June 1992, the Union of Concerned Scientists published an open letter titled, "World Scientists' Warning to Humanity", which stated that "human beings and the natural world are on a collision course." The letter stated further, "If not checked, many of our current practices put at serious risk the future that we wish for human society and the plant and animal kingdoms, and may so alter the living world that it will be unable to sustain life in the manner that we know". This warning was signed by over 1,600 scientists from leading scientific academies in seventy countries. The list included 104 Nobel Laureates.

Colborn, Dumanaski and Myers (1996) in their book *Our Stolen Future* and James Morgan (1999) in his book *The Last Generation* also provide a picture of the grim future that awaits the generations yet to be born, if we lose further time in restoring harmony between humankind and nature.

It is now widely realized that the genes, species, ecosystems and traditional knowledge and wisdom that are being lost at an increasingly accelerated pace limit our options for adapting to local and global change, including potential changes in climate and sea level. The Hadley Centre of the UK Meteorological Office has recently predicted that even if Governments cut greenhouse gas emissions, sea levels may rise by at least two meters over the next few hundred years. If the global community can limit emissions upto 550 ppm, which is twice the pre-industrial levels and 50% above today's, about two billion persons can be saved from water shortages, low crop yields and increased coastal flooding, especially in India and Africa (See *New Scientist*, October 30, 1999)

The Global Biodiversity Assessment published in 1995 by the United Nations Environment Programme (Cambridge University Press) estimates that about thirteen to fourteen million species may exist on our planet. Of this, less than two million species have so far been scientifically described. Invertebrates and micro-organisms are yet to be studied in detail. In particular, our knowledge of soil micro-organisms is still poor. Also, biosystematics as a scientific discipline is tending to attract very few scholars among the younger generation.

Another important paradigm shift witnessed in recent decades in the area of management of natural resources is a change in the concept of "common heritage". In the past, the atmosphere, oceans and biodiversity used to be referred to as the common heritage of humankind. However, recent global conventions have led to an alteration in this concept in legal terms. Biodiversity is now the sovereign property of the nation in whose political frontiers it occurs. Further, the Trade Related Intellectual Property Rights (TRIPS) provisions of the World Trade Agreement have made it mandatory to cover products of genetic improvement with either patents and sui generis methods of intellectual property rights protection. Under the UN Convention on the Law of the Sea, nations with coastal areas have access to a 200 mile Exclusive Economic Zone (EEZ). The Climate Convention and the Kyoto Protocol provide for both common and differentiated responsibilities to countries. Thus, the global commons can be managed in a sustainable and equitable manner only through committed individual and collective action among nations. In the Asia-Pacific Region, Australia can play a catalytic role in fostering cooperation in both avoiding and mitigating the adverse impact of climate change.

A Chinese proverb warns, "if you do not change direction, you will end up where you are headed". Since we do not want to reach where we are presently headed, what change of course should we bring about in the field of agriculture?

IX. *Ecstasy and Agony*

As we said goodbye to the 20th century we could look back with pride and satisfaction on the revolution which the farm men and women of the Asia /Pacific region have brought about in contemporary agricultural history. While we can and should rejoice about the past achievements of our farmers, scientists, extension workers and policy makers, there is no room for complacency. We will face several new problems, such as the following. First, increasing population leads to increased demand for food and reduced per capita availability of arable land and irrigation water. Second, improved purchasing power and increased urbanization lead to higher per capita foodgrain requirements due to an increased consumption of animal products. Third, marine fish production is tending to become stagnant and coastal aquaculture is facing environmental problems. Fourth, there is increasing damage to the ecological foundations of agriculture, such as land, water, forests, biodiversity and the atmosphere and there are distinct possibilities for adverse changes in climate and sea level. Water famines are likely to become more serious in several parts of the world than food famines. Finally, while dramatic new technological developments are taking place, particularly in the field of biotechnology, their environmental, health and social implications are yet to be fully understood.

Since land and water are shrinking resources for agriculture, there is no option except to produce more food and other agricultural commodities from less per capita arable land and irrigation water. In other words, the need for more food has to be met through higher yields per units of land, water, energy and time. It would therefore be useful to examine how science can be mobilised for raising further the ceiling to biological productivity without associated ecological harm. It will be appropriate to refer to the emerging scientific progress on the farms as an "ever-green revolution", to emphasize that the productivity advance is sustainable overtime since it is rooted in the principles of ecology, economics, social and gender equity, and employment generation.

The Green Revolution has so far helped to keep the rate of growth in food production above population growth rate. The Green Revolution was however the result of public good research, supported by public funds. The technologies of the emerging gene revolution in contrast, are spearheaded by proprietary science and can come under monopolistic control. How then can we harness the power of frontier science to promote an evergreen revolution in our farms?

The 20th century began with the rediscovery of Mendel's laws of inheritance. It ended with moving specific genes across sexual barriers with the help of molecular mapping and recombinant DNA technology. The impact of science and technology in every field of crop and animal husbandry, inland and marine fisheries and forestry has been profound. Let me illustrate this, taking the improvement of wheat production in India as an example.

Wheat cultivation started in the Indian subcontinent over 4,000 years ago. Wheat kernels have been found in the Mohenjodaro excavations dated 2,000 BC. From that period up to August 1947, when the colonial rule ended, Indian farm men and women developed the capacity to produce seven million metric tons of wheat per year. Between 1964 and 1968, when semi-dwarf strains containing the Norin 10 genes for dwarfing were introduced in irrigated areas, wheat production rose from ten to seventeen million metric tons per year. In other words, 4,000 years of progress was repeated in four years (Swaminathan, 1993). During 2002-03, wheat production in India reached eighty million metric tons, i.e., a ten-fold increase in about fifty years.

Similar progress has been made in improving the production and productivity of rice, maize, soybean, potato and several other crops as well as in farm animals in many developing countries around the world. New technologies supported by appropriate services and public policies as well as international scientific cooperation have helped to prove doomsday predictions wrong and have led to the agricultural revolution (the Green Revolution) becoming one of the most significant of the scientific and socially meaningful revolutions of the 20th century. A world without hunger is now within our reach. A hunger-free world will be possible if every nation pays concurrent attention to improving food availability through ecologically sustainable methods of production, to enhancing economic access to food by promoting a job-led economic growth strategy, and to ensuring the biological absorption of food in the body through the availability of safe drinking water and environmental hygiene. Steps should also be taken to enlarge the base of the food security basket by revitalizing the earlier tradition of cultivating a wide range of food crops.

Emerging farming technologies will be based on precision farming methods leading to plant scale rather than field scale husbandry. Farming will be knowledge intensive, using information from remote sensing, Geographical Information System (GIS), Global Positioning Systems (GPS), and information and computer technologies. Farmers in industrialized countries are already using satellite imagery and GPS for early detection of diseases and pests, and to target the application of pesticides, fertilizer and water to those parts of their fields that need them urgently. Among other recent tools, the GIS methodology is an effective one for solving complex planning, management and priority setting problems. Similarly, remote sensing technology can be mobilized in programs designed to ensure drinking water security.

Biotechnology will play an increasingly important role in strengthening food, water and health security systems. Recent widespread public concern relating to genetically modified (GM) food stresses the need for more effective and transparent mechanisms for assessing the benefits and risks associated with transgenic plants and animals. An internationally agreed Biosafety Protocol on the lines recommended in Article 19 of the Convention on Biological Diversity is an urgent necessity. Biotechnology companies should agree to the labelling of GM foods in the market. All food safety and environmental concerns should be addressed with the seriousness they deserve. Broad-based National Commissions on Genetic Modification for Sustainable Food and Health Security could be set up, consisting of independent professionals, environmentalists, representatives of civil society, farmers' and women's organizations, mass media and the concerned government regulatory authorities. This will help to assure both farmers and consumers that the precautionary principle has been applied, while approving the release of GM crops. The recent public opinion survey conducted in the UK has revealed the need for assuring the public that regulatory mechanisms are in place to ensure that only safe and responsible use of genetic modification techniques is promoted (See *Nature*, September 25, 2003).

Biodiversity-rich but biotechnology-poor countries are adversely affected by the prevailing non-adherence to the ethical and equity principles in benefit sharing contained in Articles 8 and 15 of CBD. The primary conservers, largely tribal and rural women and men, live in poverty, while those who use their knowledge and material for producing commercial products become prosperous (Swaminathan, 1999). The invaluable contributions of tribal and rural families to genetic resources conservation and enhancement have been recognized in the Convention on Biological Diversity. Yet, the political will to implement the equitable benefit sharing provisions of CBD is lacking. We need urgent steps to recognize and reward the contributions of indigenous communities to providing material of great importance to global food and health security. The following three validated findings will be adequate to stress the significance of traditional knowledge and conservation efforts to help mitigate handicaps caused by ageing in human beings.

Country	Plant	Property
India	*Trichopus Zeylanicus*	Helps to remove fatigue
India	*Bacopa monnieri*	Helps to improve memory
Tropical Africa	*Prunus africana*	Treatment for benign prostatic hyperplasis.

Article 27(b) of the TRIPS component (trade-related intellectual property rights) of the World Trade Agreement is now under review. All nations should agree to incorporate in this clause the ethics and equity principles enshrined in articles 8(j) and 15 of CBD. The World Intellectual Property Rights Organization (WIPO) has also stressed the need to recognize the intellectual property rights of the holders of traditional knowledge.

X. *Towards Hunger-Free India: Countdown from 2007*

During the last three years, scientists of the M.S. Swaminathan Research Foundation at Chennai and of the UN World Food Programme, New Delhi, have been working together on the causes of food insecurity in rural and urban India using multiple indicators relating to food availability, access and absorption in the body. The Food Insecurity Atlases of Rural and Urban India provide valuable guidelines for public action to end chronic under- and mal-nutrition in the different States of the country.

Releasing the *Food Insecurity Atlas of Rural India* on April 24, 2001, the Prime Minister of India, Mr Atal Bihari Vajpayee, called for concerted action to achieve substantial freedom from hunger by August 15, 2007, which marks the 60th anniversary of India's Independence, thereby helping to fulfil the hope of Mahatma Gandhi that no child, woman or man should go to bed hungry in Independent India. To quote Mr Vajpayee,"The sacred mission of a 'Hunger-Free India' needs the cooperative efforts of the Central and State Governments, local self-government bodies, non-governmental organisations, international agencies and above all, our citizens. We can indeed banish hunger from our country in a short time. Let us resolve today to make this mission substantially successful by 2007, which will mark the sixtieth anniversary of our independence."

1. *The challenge.* The quantitative and qualitative dimensions of the challenge of achieving a hunger-free India are daunting. The incidence of poverty, endemic hunger, communicable diseases, infant and maternal mortality rates, low birth weight children and stunting and illiteracy is high. There are however many examples where progress in the elimination of poverty-induced hunger has been rapid because of a symphony approach in dealing with the multi-dimensional problem of hunger and malnutrition. Successful experiences in the elimination of hunger and poverty have shown that synergy between political will and action and strategic partnerships can help local communities to achieve seemingly impossible tasks. Such "messages and methods of hope" should therefore be documented and spread widely, since they not only inspire confidence that the goal of a hunger-free India can be achieved, but will also help to build the self-confidence of all engaged in the mission of overcoming under- and mal-nutrition.

2. *Basic approach: Food with Human Dignity.* Food with Human Dignity should be the basic approach. The poor should not be subjected to a patronage approach and referred to as "beneficiaries" but should be treated as partners in achieving the aim of ensuring that every child, woman and man in the country has an opportunity for a productive and healthy life. The right to adequate food and clean drinking water should be regarded as a basic human right.

3. *Guiding principles for converting goals into accomplishments.* Hunger is the extreme manifestation of poverty, since the poor spend a high proportion of their earning on food. The elimination of hunger is thus the first requisite for eradicating poverty. Without adequate nutrition, the energy needed for higher work output will not exist in malnourished individuals. The following steps will help to accelerate the pace of progress in eradicating poverty-induced hunger.

a) Decentralization: The desired goal can be achieved speedily and surely only if a decentralized approach to implementation is adopted. "Think, plan and act locally and support at the state and national levels" should be the motto. Elected local bodies, together with the concerned Departments of Government (health, education, women and child welfare, rural and tribal development, etc.) should prepare micro level action plans. They should form a local level "Alliance for a healthy and productive life for all". Elected members of local bodies, particularly the one million elected women members can be empowered to spearhead the freedom from hunger movement, since they are more aware of the problems of nutrition and drinking water. Decentralization will enhance accountability, reduce transaction costs and remove corruption in delivery systems.

b) Lifecycle Approach*:* For ensuring nutrition security at the level of each individual, a lifecycle approach is necessary so that the nutrition needs of an individual can be met from birth to death. Special programs for adolescent girls, pregnant women, nursing mothers, infants (0-2 years) and old and infirm persons should continue. What is needed is the horizontal integration of numerous vertically structured programs. Such a functional integration will help to create a symphony at the level of each village / town / city to ensure that all links in the food availability – access – absorption chain function at a high level of efficiency and effectiveness. Management tools and not additional monetary support will be needed to bring about at the field level such convergence and synergy among ongoing programs.

c) Information, Education and Communication*:* There is need for launching a Nutritional Literacy movement to spread awareness of the adverse consequences of malnutrition-induced intellectual and physical dwarfism among children. The Nutritional Literacy movement should include issues relating to food safety, *Codex Alimentarius* standards, sanitary and phytosanitary measures, etc. Mass media, particularly those in the public sector like Doordarshan and All India Radio can play a very important role in making the Hunger-Free India movement a success. Community radio stations, giving location-specific information, should be encouraged to assist other mass media in spreading messages of hope. It will be useful to set up Media Resource Centres for Freedom from Hunger. Such centers can provide credible and timely information to the print, audio, video and new (i.e., Internet) media.

d) Household Entitlement Card*:* It will be useful to provide every family with an Entitlement Card, giving information on the various government projects which they can access. The information may be disaggregated by gender and age and precise addresses of contact persons and offices may be given. Such information will enable everyone to make the best use of their entitlements. A single step of this kind will help enormously to ensure the effective utilization of all the schemes of central and state governments and bilateral and multilateral donors.

e) Asset Building and Community Development: The poor are poor because they have no assets like land, livestock or fish ponds. They often are illiterate and lack proper dwelling. They survive on wage employment which, particularly in the case of women, does not reach the level of even the prescribed minimum wage. A massive effort is needed to help them to shift from unskilled to skilled work through training in market-driven skills. The on-going micro-finance led self-help revolution will be the speediest way to help them to rise above the poverty level. This will call for establishing effective forward and backward linkages, particularly with technology sources and markets. Insurance and Venture Capital support should also be available to micro-enterprises. We have now an opportunity to leapfrog in achieving our goal of enabling every one to earn his/her daily bread.

f) Capacity Building: Since a decentralised approach involving the empowerment of over three million women and men members of local bodies holds the key to the success of this national movement for food and clean drinking water for all, it is essential that a national consortium of Agricultural, Rural and Women's Universities as well as government and non-governmental training and research institutions is formed for undertaking capacity building in areas such as management, communication and organizational skills with reference to the implementation of the hunger-free area program. The capacity building programs can be organised on a Trainers' Training model, in order to achieve a multiplier effect.

4. *Initiation of a National Food for Social Capital Program.* The social capital of a country is the product of interaction between the human capital and the cultural, political, economic, nutritional and natural environments. Human and social capitals constitute the most precious wealth of a nation. A society committed to building its social capital will try to promote programs which represent a "win-win" situation for all, thereby avoiding winners and losers and the consequent social conflict and disruption. It is now clear that farmers will produce more if there are opportunities for assured and remunerative marketing. Therefore, the initiation of a National Food Guarantee Scheme will help to ensure that all who are hungry today due to lack of livelihood opportunities or other constraints are able to have food for a productive life. Such a National Food Guarantee Scheme can serve as an umbrella for all ongoing projects like those mentioned earlier. In addition, it can provide foodgrains for initiating a wide variety of social support initiatives like Food for Health (TB, HIV/AIDS, Malaria etc.), Food for those employed in ICDS, Nutritious Noon Meal and other similar projects. In other words, food can become a powerful currency for achieving the goal of a hunger-free India. Using food as a currency has twin advantages, namely, there could be greater off-take of foodgrains from farmers, thereby providing them with an incentive to produce more, and secondly, for meeting the immediate needs of the poor, destitutes, migrant labor and all who are under-nourished today.

5. *Consultative Group for Freedom from Hunger.* Both at the national and state levels multi-stakeholder Consultative Groups for "Agenda 2007: Hunger-Free India" could be organized comprising representatives of the concerned Government of India Ministries and Departments, professional experts, National Commission for Women, civil society organizations, business and industry, mass media and bilateral and multilateral donors, with the Union Planning Commission serving as the nodal agency for such a Consultative Group. Such a CGFFH would help to foster strategic partnerships as well as synergy among political leaders, professionals and peoples' organizations.

XI. *No Time to Relax on the Food Production Front*

While the alleviation of hunger by improving access to income and balanced diets and safe drinking water should receive high priority, there is no time to relax on the food production front. We need to bring about productivity, quality (including food safety), profitability and sustainability revolutions in farming based on a Farming Systems Approach. There is an urgent need for enhancement of investment in agriculture and rural infrastructure development. There is also need for conferring on small producers the power of scale through cooperatives, self-help groups and other socially viable methods of group endeavor both at the production and post-harvest phases of farming. Our spectacular progress in the dairy sector is largely through such management innovations. There is need for a movement for trade and quality literacy including an understanding of sanitary and phytosanitary measures and *Codex Alimentarius* standards. There is also need for launching a Water Literacy Movement. Above all, there is need for a paradigm shift from jobless to job-led growth in order to ensure that every poor person is enabled to earn his or her daily bread.

Unless the agricultural competitiveness of developing countries improves, earnings from the farm sector in international markets will not grow significantly. In a predominantly rural and agricultural country like India, agricultural progress (i.e., crop and animal husbandry, fisheries, forestry and agro-forestry and agro-processing) is the most effective social safety net against hunger and poverty. Hence, the ongoing fatigue of the Green Revolution in wheat, rice and other major crops should be converted into an evergreen revolution designed to promote productivity improvement in perpetuity without associated ecological harm. A focus on improving factor productivity in agriculture should be designed to cover all the major agro-ecosystems. Agricultural and rural development, if given adequate and appropriate attention, will help predominantly agricultural countries to take to the path of job-led economic growth.

XII. *Shaping Africa's Agricultural Future*

During the twenty years, Dr Norman Borlaug has devoted his enormous energy and expertise to helping African nations to become self-reliant in food, his goal has been to stimulate an African Green Revolution. The African continent, being the center of origin of Homo sapiens, is rich in ecological, ethnic, agricultural, cultural and climatic diversity. Amidst this diversity, there is also unity in terms of uncommon opportunities for enhancing through mutually reinforcing packages of technologies, services and public policies, the productivity, profitability, sustainability and employment intensity of major farming systems. Available scientific data indicate that the untapped production reservoir, even with the technologies now on the shelf, is high in most African countries and farming systems. Therefore, progress in achieving a productivity revolution need not wait until new technologies become available. Integrated steps in the areas of soil health and fertility enhancement, water conservation and management, the conservation and sustainable and equitable use of agro-biodiversity, and greater emphasis on post-harvest technology and agro-processing will help most African nations to leapfrog in agricultural progress and agrarian prosperity. Unity of goals but diversity of approaches based on local socio-cultural, socio-economic and agro-ecological conditions will be needed to achieve the desired goals.

Technology, training, techno-infrastructure and trade are some of the major components of an agricultural renaissance. Opportunity for assured and remunerative marketing has now become the most important component of agricultural progress. Another urgent need is effort to attract and retain youth in farming. The technological and skill empowerment of women in agriculture is vital for agricultural progress. There is rich traditional wisdom in relation to crop and animal husbandry, fisheries and forestry in African nations. It is essential to conserve such dying wisdom and blend them with frontier science and technology.

Symbiotic partnerships are vital for speedy progress and for the efficient and economic use of available resources. The partnerships could take the following forms: *Lab to Lab* (consortium of appropriate scientific institutions); *Lab to Land* (scientist – farm family interaction); *Land to Lab* (deriving benefit from traditional knowledge and ecological prudence); *Land to Land* (lateral sharing of knowledge and experience among farming families). Such linkages will help to accelerate progress in the development, testing and dissemination of appropriate technologies.

According to the International Food Policy Research Institute, a one percent growth in agricultural productivity in Africa reduces poverty by 0.6 percent. A one percent increase in yield can help six million more people to raise their income above one US dollar per day. Also, the higher the percentage of people who depend upon agriculture alone for their livelihoods, the greater is the incidence of hunger and poverty. Improvements in agricultural productivity and non-farm employment opportunities are therefore the most effective safety nets against hunger and poverty.

Access to technology and markets will shape Africa's agricultural future. Enhancing technological capability and bridging the growing technological, gender and genetic divides are urgent tasks. Technology development and delivery have to receive equal attention. Farmer Participatory Knowledge Systems (FPKS) will be needed for this purpose. Such systems can help to convert generic information (e.g., those relating to weather forecasts and market conditions) into location-specific action points.

Research, education and extension should have organic linkages. Advanced centers on the one hand, and techniracy (i.e., learning by doing) on the other, are needed. Personnel policies which can help to attract and retain youth in farming are important. Both on-farm and non-farm employment need attention.

Strengthening the ecological foundations essential for sustainable agriculture, i.e., improving productivity in perpetuity without associated ecological harm or an evergreen revolution is the need of the hour. Ecosystems which need special attention are:Land: Anti-desertification measures; Seawater farming for coastal area prosperity; Biodiversity: Agro-biodiversity sanctuaries; Mountain ecosystems; Crop-livestock integrated production systems.

Technologies that can help to improve productivity, profitability, stability and sustainability of farming systems need to be developed locally. Marker assisted selection, genetic engineering, functional genomics, proteomics, information and communication technology, space technology and nano-technology will also play an increasingly important role. Precision farming techniques will have to be adapted to local conditions. Animal husbandry, inland and marine fisheries offer scope for improvement. Seawater farming can help to enhance coastal area prosperity. Social prestige and recognition are vital to build the morale of farm scientists and to foster scientific excellence and social relevance in their efforts. African nations could select an eminent scientist from their respective countries to head a Scientific Advisory Committee to the Head of State or Government.

Africa is diverse politically, socially, ecologically and economically. Hence a "one-size fits all" prescription will be disastrous, as has become clear from many top-down programs. We need to reverse the science strategy paradigm and begin planning in consultation with farm women and men. Participatory research for developing location-specific technologies and participatory knowledge management systems are vital for bridging the gap between potential and actual yields with the technologies already available in the numerous international, regional and national research institutions in various parts of Africa.

Technologies should aim at three time dimensions – i) *Immediate*: Technologies already developed and available for immediate dissemination. ii) *Medium-term*: Technologies in the pipeline which need testing, incubation and adaptation. iii) *Long-term*: Strategic research aimed to develop new technologies through the use of the new genetics and other areas of frontier science and technology. Anticipatory research is also needed to meet potential changes in climate, particularly temperature and precipitation.

Science and Technology are not magic wands with which hunger and poverty can be removed. However, they are the catalysts of change and progress. Technology has been the prime mover of economic progress in the industrialized world and hence technological empowerment of farm women and men is essential for moving agriculture forward. As stressed earlier, the tools of science and technology have to be tailored to solving the problems of i) *Endemic hunger*, caused by poverty and lack of access to balanced diets; ii) *Hidden hunger*, caused by the deficiency of micronutrients in the diet; iii) *Transient hunger*, caused by natural calamities or ethnic conflicts

Peace and security are essential for progress. However, in the African context, progress in improving farm productivity has been poor even in countries which have had peace and political stability. Sustainable food security for all and for ever has to be the vision for Africa. This has to be achieved through mobilizing science and technology supported by appropriate public policies for achieving sustainable intensification and diversification of farming systems and value addition to primary products. Synergy between technology and public policy is vital to convert potential into production.

XIII. *Collapse of Cancun Trade Talks:*
Implications for Predominantly Agricultural Countries

The collapse of the Cancun session of WTO has been attributed to failure in reaching agreement between industrialized and developing nations in the fields of domestic support, export subsidies and market access in relation to farm commodities. The failure of the Cancun negotiations is both a blessing and a curse. The sad part of the collapse relates to the continuation of the unequal trade bargain entered into by developing countries at Marrakesh in 1994. Until further agreements are reached, the unfavorable trade bargain will continue to damage the interests of the farmers of the Third World. On the other hand, the Cancun fiasco provides a breathing space for Third World countries to take steps to enhance their agricultural competitiveness. A mismatch between production and post-harvest technologies persists and the infrastructure for handling perishable commodities continues to be poor. There has been hardly any systematic effort to spread genetic, quality and trade literacy among small farm families. This is true in many countries in Asia and Africa.

The breathing spell provided by the delay in arriving at an agreement designed to make trade in agriculture both free and fair should be utilized for building the infrastructure for a highly efficient and quality conscious farming ethos.

Road to free and fair trade. First, all boxes (grey, green and amber) may be abolished and the do's and don'ts with reference to trade distortion and unfair trade practices may be spelt out in clear and unambiguous terms. This will help all countries to decide on the pattern of support to be extended to farmers, which will be non-trade distorting. Second, as an alternative, a fourth box relating to Sustainable Livelihoods (Livelihood Security Box) may be introduced, which will empower developing nations facing the challenge of providing livelihoods to the rural population to place restrictions on imports, where there is convincing evidence that such imports will erode job/livelihood opportunities in their countries. Since over 66% of the population of many developing countries including India depend upon agriculture (crop and animal husbandry, fisheries, forestry and agro-forestry and agro-processing) for their livelihoods, trade which leads to the destruction of rural jobs/livelihoods will further enhance poverty and hunger. The result will be social disintegration because of a further increase in the rich-poor divide. Globally, the continuation of the present situation where a few million farm families in industrialized countries, supported by heavy inputs of technology, capital and subsidy, compete with over a billion small farmers, having little access to technology, credit and adequate post-harvest infrastructure, will not help to make free trade an instrument of poverty eradication. The percentage of population dependent on agriculture for their livelihoods should be the major criterion for eligibility for using the provisions of the proposed Livelihood Security Box. The minimum could be 50% of the population.

XIV. *Ending Hunger: Reaching the Final Milestone*

The following steps will help to achieve the human quest for a hunger-free world.

- Define food security for operational purposes as "physical, economic, social and ecological access to balanced diets and clean drinking water to every individual" and make access to food and drinking water fundamental human rights in the national constitution.
- Adopt a whole lifecycle approach ranging from pregnant women to old and infirm persons, by bringing about convergence and synergy among numerous on-going vertically structured food safety net programs
- Take steps to bridge the prevailing gap between the potential and actual productivity of major farming systems through mutually reinforcing packages of technology, services and public policies based on an inter-disciplinary constraints analysis.
- Promote farmer-participatory research and knowledge systems and reach the unreached in information and skill empowerment.
- Strengthen livelihoods through an integrated approach to on-farm and off-farm employment and bring about a paradigm shift from unskilled to skilled work through greater attention to post-harvest technology.
- Foster community-centerd Gene, Seed, Water and Food Banks. Give priority to the establishment of community Food and Feed Banks in the hunger hot spots and to enlarging the food basket through the inclusion of under-utilized crops.
- Pay concurrent attention to nutritional as well as non-nutritional factors like sanitation, health care and education.
- Facilitate trade in farm commodities on fair and equitable terms, i.e., provide a level playing field to developing countries.
- Recognize the critical role of women in both food production and hunger elimination and take priority steps to empower them socially, legally, economically and technologically.
- Strengthen the World Food Programme by enabling it to effectively address chronic and hidden hunger, in addition to transient hunger or emergencies.
- Establish a Global Virtual Academy for Food Security to spread information on hunger bright spots like the Zero Hunger Program of Brazil, Agenda 2007 – Hunger-Free India movement, and similar efforts around the world, so as to enlarge the sphere of hope and political commitment.

In the short run, UN World Food Programme can be assisted to organize an "International Bank for Nutrition for All" for the purpose of launching an integrated attack on hunger in all the hunger hot spots of the world. Such a Bank will need an initial allocation of 100 million metric tons of foodgrains. This is not beyond the capacity of grain-surplus nations.

The persistence of chronic hunger, is related to lack of sustainable livelihood opportunities to the poor. Global trade, economic, political and philanthropic institutions should therefore keep "jobs/livelihoods for all" as the bottom line of their national and international policies. The on-going micro-credit supported micro-enterprise revolution in countries like Bangladesh, India, China and several other developing countries should be allowed to succeed. It is only by overcoming the famine of employment opportunities at the national level that the famine of food at the household level can be overcome.

Above all, the critical role of women in all aspects of the food production-processing-marketing chain needs explicit recognition. In South Asia and Sub-Saharan Africa, the two major hunger hot spots of the world, women play even a greater role in food production and in fetching drinking water, than many other areas in the developing world. In Sub-Saharan Africa, for example, women own only 1% of the land and receive 7% of the farm extension services and less than 10% of the credit given to small scale farmers. Yet, they produce 80% of the food and do the vast majority of the work to process, transport, store and market them. No hunger elimination strategy will succeed if adequate attention is not paid to the social, legal, economic and technological empowerment of women. The highest priority should go to reducing the number of hours of work and enhancing the economic value of each hour of work in the case of rural women, through knowledge and skill empowerment. Hunger elimination is the first step towards a world characterized by gender equity and justice.

XV. *Conclusions*

The following are some of the major changes needed in technology and public policy if the goals of poverty alleviation and hunger elimination are to be achieved.

a) Availability of food: Shift from Green Revolution to evergreen revolution technologies based on the principles of ecology and integrated natural resources management.

b) Access: Encourage the diversion of assetless families from farm to non-farm enterprises, i.e., from the primary to the secondary and tertiary sectors of the economy; this will call for a shift from unskilled to skilled work and from individual to group endeavors and through institutional devices like cooperatives and self-help groups. In population rich but land- and water-hungry countries, "production by masses and not mass production" technologies will be the most relevant from the social and gender perspectives.

c) Absorption: This will call for urgent attention to the provision of clean drinking water, education, health care and sanitation facilities.

If technologies and public policies can be directed towards addressing the above issues in an integrated manner, today's "hunger hot spots" will soon become "food security bright spots".

Ninety-two percent of the world's hungry are not victims of extreme events like ethnic conflicts, drought and other natural and human made calamities. Ninety-five percent of the hungry live in the tropics, and a high proportion of them are women and children. Over 500 million out of the 800 million who are undernourished today are in Asia, with a high concentration of the hungry in South Asia. In general, the larger the percentage of people who depend on agriculture alone for their livelihoods, the greater is the incidence of hunger and poverty. Nearly twenty-five years ago, China introduced an integrated strategy for the rapid growth of the farm (household responsibility system) and non-farm (rural township enterprises) sectors of employment. This has paid rich dividends, since the incidence of hunger and poverty has come down speedily (See *An Overview of the Development-Oriented Poverty Reduction Program for Rural China,* China Financial & Economic Publishing House, Beijing, 2003).

Sustainable intensification and diversification of farming systems (i.e., ever-green revolution) and value addition to primary products through agro-processing and improved post-harvest technology, constitute the best safety nets against hunger and poverty in all countries where over 60% of the people live in rural areas.

Asset reform policies designed to help the poor should place emphasis on the knowledge, skill and technological empowerment of the poor in addition to providing land to the landless. The M.S. Swaminathan Research Foundation has established for this purpose a Virtual Academy for Food Security and Rural Prosperity. The Fellows of this Academy are rural women and men and the knowledge empowerment system involves the integrated use of the Internet, cable TV, community radio and the vernacular press.

There are feedback linkages among nutrition, farm production and market development. Progress can be fast if intentions are backed up by initiatives and implementation structures. Above all, there is need for a mind-set change in relation to agriculture. Agriculture should not be regarded just a food producing machine, but as the soul of rural livelihood and ecological security systems. In the immediate future, agriculture-led rural development is an effective and affordable pathway for the sustainable end of hunger and poverty. This will call for strategies which are sensitive to the environmental, economic, gender, social, technological and educational dimensions of the agricultural and rural development challenges. Technology, training, techno-infrastructure and home and external trade are the major triggers of progress.

We urgently need bold political initiatives like the Zero Hunger Program of Brazil and the Employment Guarantee Scheme and the Universal School Noon Meal Programmes of the States of Maharashtra and Tamil Nadu in India. What is needed is not just political will as expressed in speeches, but political action as reflected in resource allocation and public policies.

The UN Millennium Development Goals represent the final milestone for a hunger-free world. Not reaching these goals will only result in greater violence and unrest in many parts of the world. It is in the enlightened self-interest of rich nations and the rich in all nations that they join together to achieve speedily what is readily achievable in the fight against hunger and poverty. The following demonstration program can be the first step in walking the talk.

There are untapped technological opportunities for enhancing the productivity and profitability of major farming systems in Africa on an environmentally sustainable basis. "Seeing and harvesting are believing" to resource poor farming families. I therefore suggest the immediate initiation of a Science and Technology Demonstration Program, which can help to introduce appropriate Science and Technology inventions and interventions for unleashing latent productivity gains leading to an enhancement of household food and income security.

The farming systems that could contribute most to increased agricultural productivity and improved food security are: Maize-based farming systems, cereal-root crop mixed farming systems, irrigated crop-livestock –fish integrated production systems, and farming systems involving agro-forestry techniques.

Such a S & T demonstration program should be based on integrated attention to the following components of the production-processing-marketing-consumption chain (experience has shown that if all links in this chain are not attended to concurrently, the anticipated productivity gains will not be forthcoming).

- An integrated nutrient and soil fertility enhancement program
- Integrated pest management program
- Water harvesting and efficient and economic use through micro-irrigation and fertigation systems of delivery of water and nutrients
- Biotechnological applications like improved genetic strains, biofertilizers, and bio-pesticides and, where appropriate, the cultivation of tissue-culture propagated banana and fruit trees
- Use of improved farm implements and appropriate mechanization for improving labor productivity, reducing drudgery and ensuring timely farm operations
- Introduction of appropriate post-harvest processing, storage and marketing techniques, with priority attention to perishable farm commodities like fruits, vegetables, fish, milk, eggs, meat and their animal products
- Promotion of off-farm employment through the introduction of technologies for adding economic value to primary products and through agri-business enterprises based on micro-credit
- An information and communication program to provide location-specific information relating to meteorological, management and marketing factors and to promote genetic, quality and trade literacy among rural farm families
- Establishment of farmers' field schools for integrated pest, disease and weed management, integrated nutrient supply, scientific animal husbandry, aquaculture and the other aspects of production and post-harvest technologies
- Promotion of institutional structures like cooperatives and self-help groups which can confer the power of scale to small producers at the production and post-harvest phases of farm operations and make marketing producer-oriented
- Promotion of contract cultivation by private industry in order to ensure market and income security

I suggest that Interdisciplinary Spearhead Teams may be constituted to prepare business plans for the proposed demonstrations in the "best bet" farming systems. Nothing succeeds like success and hence the sites for the initial demonstrations should have the substrate scientific, political, social and ecological conditions essential for achieving the goals of this program. Such technical teams may be guided by local Farmers' Advisory Groups, consisting of both men and women farmers. The local Farmers' Advisory Council may also undertake monitoring and evaluation of the program. Local farm families should have a sense of participation and ownership of this program at all phases, ranging from program design to implementation.

In the demonstration areas, every effort should be made to bring about convergence and synergy among the various programs designed to achieve the following UN Millennium Development Goals:

- Eradicate extreme poverty and hunger through a paradigm shift from unskilled to skilled work and through farming systems intensification, diversification and value-addition
- Achieve universal primary education
- Promote gender equality through technological and skill empowerment of women
- Improve maternal health and nutrition, so as to avoid the birth of babies characterized by low birth weight (LBW children suffer handicaps in brain development)
- Combat HIV/AIDS, malaria and other diseases
- Ensure conservation and enhancement of basic life support systems, i.e., land, water, forests, biodiversity and the atmosphere

To mobilize the necessary technological, financial, managerial and institutional resources essential for the successful implementation of this demonstration program designed to harness the best in frontier science and traditional wisdom for enhancing the productivity, profitability and sustainability of major farming systems, it is necessary to organize local level consortia or coalitions of government, non-government, community, research and educational and mass media and donor institutions. Such a demonstration-cum-technology verification and incubation program will help to stimulate productivity advances in perpetuity (even without marine fish). ∎

References

Chakraborty S., N. Chakraborty, A. Datta A. 2000. Increased nutritive value of transgenic potato by expressing a non-allergenic seed albumin from *Amaranthus hypochondriacus*. *Proceedings of National Academy of Sciences, USA* 97: 3724-29.

Colborn, T., D. Dumanaski and J.P. Myers. 1996. *Our Stolen Future.* New York: Dutton.

FAO. 2002. *World Agriculture: Towards 2015/2030 Summary Report*. Rome: FAO.

Ferro-Luzzi, A., F. Branca and G. Pastore. 1994. Body Mass Index defines the risk of seasonal energy stress in the Third World. *European Journal of Clinical Nutrition* 48 (Supplement 3): 165-78.

Gardner, Gary and Brian Halwell. 2000. Overfed and Underfed: The Global Epidemic of Malnutrition. World Watch Paper 150. Washington, DC: World Watch Institute.

Global Science Panel on Population & Environment. 2001. *Population in Sustainable Development – Analyses, Goals, Actions, Realities*. Vienna: Remaprint.

Gopalan, C. 2001. *Combating Vitamin A Deficiency and Micronutrient Malnutrition Through Dietary Improvement*. Chennai: MSSRF. (Mimeographed)

IFAD. 2001. *Assessment of Rural Poverty in Asia and the Pacific Region*. Rome: IFAD.

Larkins, Brian and Edwin T Mertz. (Eds.) 1994. *Quality Protein Maize 1964-1994. Proceedings of The International Symposium on Quality Protein Maize,* Sete Lagoas, MG, Brazil. (December 1-3, 1994)

Lucca, P., R. Hurrell, I. Potrykus. 2000. Development of iron-rich rice and improvement of its absorption in humans by genetic engineering. *Journal of Plant Nutrition* 23: 11-12.

Measham, A. R. and Meera Chatterjee. 1997. *Wasting Away: The Crisis of Malnutrition in India*. Directions on Development Series. Washington, DC: The World Bank.

Morgan, James. 1999. *The Last Generation*. Florida, USA: Krieger Publishing Co.

Narayanan, Rama. 2001. How secure are our mothers and children? Paper presented at a workshop on "Human Health Security", Chennai, India.

Philip, James et al. 2000. Ending malnutrition by 2020: An agenda for change in the millennium. *Food and Nutrition Bulletin*, United Nations University, Tokyo. Vol 21 (13), Supplement, 88 pp.

Raina, A. and A. Datta. 1992. Molecular cloning of a gene encoding a seed-specific protein with nutritionally balanced amino acid composition from *Amaranthus*. *Proceedings of National Academy of Sciences, USA* 89: 11774-778.

Ramalingaswami, V., U Johnson and J Rohde. 1997. Malnutrition: A South Asian Enigma. In *Malnutrition in South Asia*, ed. by Stuart Gillespie. Rosa Publication No. 5. Nepal: UNICEF.

Sanchez, Pedro A. 2002. Soil fertility and hunger in Africa. *Science*, Vol.295. (March 15, 2002)

Swaminathan, M.S. (Ed.) 1993. *Wheat Revolution: A Dialogue*. Madras: Macmillan India

——. 1996. *Sustainable Agriculture: Towards an Evergreen Revolution*. Delhi: Konark Publishers.

——. 1999. *A Century of Hope: Towards an Era of Harmony with Nature and Freedom from Hunger.* Chennai: EastWest Books (Madras).

Swaminathan, Mina. 1998. Maternity and Child Care Code. Forum for Creche and Child Care Services (FORCES), New Delhi.

UNDP. 2001. *Human Development Report (2001) – Making New Technologies Work for Human Development*. London: Oxford University Press.

Vasal, S.K., E. Villegas, C.Y. Tang, J. Werder, M. Read. 1984. Combined use of two genetic systems in the development and improvement of quality protein maize. *Kulturpflanze* 32: 171-85.

Ye, X., S. Al-Babili, A. Kloti, J. Zhang, P. Lucca, P. Beyer, I. Potrykus. 2000. Engineering provitamin A (ß-carotene) biosynthetic pathway into (carotenoid-free) rice endosperm. *Science* 287: 303-05.

AART II.
LIFE AND WORK OF M.S. SWAMINATHAN: CONTEMPORARY PERSPECTIVES

68. MSS, 1999

69. The 70th Birthday Celebration, 1995

70. Farewell to M.S. Swaminathan, Los Banos, 1987

Section 1.
Excerpts from Selected Interviews

1.1 "How to Eradicate Hunger." Interview by Parshuram Ray, *Humanscape*, September 2000, pp. 5-11.

In an exclusive interview with Parshuram Ray at the Swaminathan Foundation in Chennai, Dr. Swaminathan spoke about various challenges and opportunities in the fields of agriculture and food security. He also provided a blueprint for a hunger-free India. Parshuram Ray is an environmental activist, researcher and writer based in New Delhi.

Parshuram Ray: *What, in your opinion, are the major challenges and problems confronting India as regards agriculture and food security?*

M.S. Swaminathan: The first major challenge is to retain prime farmland for agriculture. Good farmland is being used for non-farming purposes. If this happens on a continuing basis, we will be in serious difficulty. So, the conservation of land and water is the Number One issue. Second, how to retain young people on the farms. Whether educated or uneducated, the youth are increasingly not taking to farming. It is all left to their parents or grandparents. Students from agricultural universities do not go back to the farms, they do something else. They take up jobs in the city. Now, this has serious consequences because our country is predominantly young. The youth must be motivated and they can be motivated only if farming is more economically rewarding and intellectually satisfying. This is where modern ecological agriculture offers great opportunities. We must do our best to attract and retain our youth in farming. Third, the question of the cost, risk and return structure of farming, the economics of farming – import-export policies, trade policies, the impact of the World Trade Agreement. If there is a lot of dumping of cheap food in India from outside, then I think our livelihood will be destroyed.

Therefore, the three challenges are: We have to produce more food from less land, but we must conserve the land – prime farmland – for agriculture. We must produce more jobs, livelihood, employment opportunities and, above all, we must increase the income of the farm-family, because out of over one hundred million farm-families in our country, seventy-five million are very small farmers, marginal farmers. The smaller the farm, the greater the need for marketable supplies from the farm. So we need more intensification, diversification and value-addition in farming.

Question: *What are the available opportunities in this field?*

M.S. Swaminathan: We have a very large, untapped yield-reservoir, even at currently available levels of technology. First of all, we have great opportunities for increasing our production immediately by taking advantage of the untapped yield-reservoir available even without big technologies on the shelf. We don't need big technologies for, say, a crop like rice. Our average yield is less than 2.5 metric tons per hectare. China's average yield is five metric tons, Korea seven metric tons, Japan also seven metric tons. We can almost double our yield in rice without additional technology. What are needed are appropriate services (inputs, supplies, services like seed and water and so on), the timely availability of inputs and assured and remunerative marketing opportunities. If there is an assured and remunerative marketing opportunity, our farmers will do the rest because they are good economists.

Another major opportunity we have today is seeing whether we can take advantage of our capability to produce almost all forms of fruits, vegetables, and flowers in our country. We have a diverse climate, diverse soil conditions, therefore we can grow temperate, sub-tropical, sub-temperate and tropical fruits and vegetables. This is a great opportunity. We now produce over one hundred and twenty-five million metric tons of vegetables and fruits. But our post-harvest technology is poor. There's a mismatch between production and post-harvest technology. Our processing, cold storage, refrigerator vans, and packaging – these are still very poor. Where we have made advances is in post-harvest technology, for example in Maharashtra, in the case of grapes, strawberries and alphonso mangoes. The opportunities for us to market good quality fruits and vegetables in India are great. It is a sad commentary today that we have to import apples from Australia. I find, in the Chennai market, that all the apples are from Australia even though we produce good quality apples in Himachal Pradesh and J&K. It's sad that our government policies are such that we are not able to take full advantage of what we have. For example, in 1981, in a committee chaired by me, we recommended a National Horticulture Development Board along the lines of the National Dairy Development Board. Look at how professionally run the NDDB is. It is thanks to Dr. Kurien, Dr. Amrita Patel and all their colleagues that our country has increased our milk production from twenty-five million metric tons to seventy-five million metric tons. We have now become the largest milk-producing country in the world. All this is possible. We would have had the same success in horticulture if bureaucrats had not run the National Horticulture Development Board. I think it is the tragedy of our agriculture that our agricultural policies are being made, and agricultural departments run, by people who do not know agriculture, who do not know the problems of the farmers. This is self-inflicted damage. You can't blame any outside country for this. We, our own political system and our own mindset – our colonial mindset – are responsible.

The third great opportunity in our country is becoming a leading grain exporter. As I said earlier, from the present eighty-five million metric tons of rice, we can go up to a minimum of one hundred and fifty million metric tons. We can't absorb it; therefore, we have to reduce it. We import pulses from Australia and Myanmar. It's again tragic that with the so-called pulse technology mission we should hold the person who's responsible accountable for this. There's no accountability in our country, otherwise we should be able to produce more pulses. It's tragic to have to import pulses from outside. We are the home of the pulses. India is the center of origin of the pulse crop. Both for home production and for external consumption or export we have enormous opportunity in this country. But there are three important things, which are to be done. First, we must professionalize our agricultural departments; professionals must run them. Second, we must end the mismatch between production technology and post-harvest technology. Post-harvest processing is what adds value to the crop. Therefore, production, processing and marketing should be looked at as integrated links in one chain. Third, we must develop long-term import-export policies, which will not damage the livelihood security of our producers. Ours is a country of small producers. Agriculture is the largest productive enterprise, with over one hundred million families involved in it. Our milk revolution would have been completely destroyed had we allowed indiscriminate import of milk and milk powder and butter from North America and Europe. Because there it is produced through highly mechanized large-scale farming. Our small producers cannot compete. In any case, they have enough money to subsidize – very heavy subsidies are being given by the European Union on agriculture. Therefore, we will have to have a proper policy. The so-called National Agriculture Policy statement has not yet come. As and when it does, I hope it will address the real issues.

Question: *Do you see any positive trends that could give us hope in the fields of agriculture and food security?*

M.S. Swaminathan: The most important positive trend in our country is the enormous enthusiasm and willingness of farm men and farm women to take to new technologies. They have no hesitation in taking to any new technology provided it will improve their livelihood and well-being. As I said earlier, the cost-risk and return-structure of farming influence their decision-making. The most hopeful trend is the emergence of new technologies. We have new tools in the form of information technology, including Internet in local languages. We have seen, in our own Center, the enormous impact of communication technology on reaching the unreached, as we say. We have a great deal of opportunity here because we have great competence in this country in space technology, space satellites, communication and television. We have great capacity in the fields of information technology and biotechnology. We also have lots of traditional wisdom, for example in the case of medicinal plants, which are the hope of health security systems not only of India but of the world. We have a lot of traditional knowledge and material but we ourselves are lagging behind. We go on complaining that other countries are patenting neem or turmeric or haldi or karela and so on. There's no point in complaining. We must put our own house in order. We must immediately enact three important Acts. First, the National Biodiversity Act, which provides a mechanism for conservation and sustainable use and equitable sharing of benefits. Second, we must also pass the Plant Varieties Protection Act soon. The farmers' rights must be completely protected. Third, we must have a Geographic Appellation Act by which nobody can patent basmati or anything of this kind. Today we are defenseless. In a world of increasing global competition we are fighting our battles without any defense. As I said, we can't blame others. One day somebody asked me: "What about the terminator?" I said, "We can terminate the terminator provided we terminate our lethargy first, otherwise we will not be able to terminate anything at all."

Question: *Can you tell us about some positive cases, innovative experiments and good initiatives, which could become the seeds for your "pro-nature, pro-poor, pro-woman development" paradigm in India?*

M.S. Swaminathan: There are a large number of positive cases in our country. For example, in the case of water management. Anil Agarwal and his group have brought out dying wisdom. Even in desert areas of Rajasthan, such as Jaisalmer, Barmer, there still is drinking water, whereas in Cherrapunji, which has the world's highest rainfall, there is no drinking water in the summer months. Like Ralegaon Siddhi, there are many water-harvesting, watershed-management projects. Our own bio-villages and information villages have been quoted in the Human Development Report of the UNDP. Also by the US National Academy of Science as the finest example of taking technology to the poorest of the poor, to women on a pro-women, pro-poor, pro-nature basis. We have a number of examples spread over the country, work done mainly through non-governmental organizations or voluntary civil society organizations with the motivation. Many of them also have the technological capability. So we have positive cases in every area. Go to a place near Balarshah. Sardar Ajit Singh, a wonderful man, was a forest contractor who came from Pakistan with nothing, in 1947. If you go to his orchard, you will see one of the world's best orchards. It is like an oasis in a desert. It is in the Chanderpur district of Maharashtra.

I can give you hundreds of examples in animal husbandry, in horticulture, in fisheries, in agriculture, in our country. But these are all isolated examples. A country progresses only when the average is high. Some of our pandits produce ten to twelve metric tons of rice, but our average is low. If a country has one Nobel laureate, like C.V. Raman, it does not improve your science as a whole. It is really the average that makes the country and our average is poor. Going on talking about Anna Hazare or the *Chipko Andolan* will not help us. A mass movement has to begin. The unique must become the universal. That has to be our aim and that is what our Center is dedicated to.

Question: *Do you think that hunger can be banished from India? Can you give us a blueprint or plan of action for the eradication of hunger from India?*

M.S. Swaminathan: We have been working on how to eradicate hunger totally from this country and we feel that we are not at the end of our work. We have reached the final milestone, as you may call it. We can immediately take seven steps to realize that end.

First, use the gram sabhas. The finance minister called this year the Year of the Gram Sabha. Recently, the prime minister said that gram panchayats alone could save us. Let these not remain mere words. In my view, every gram sabha should identify the ultra poor, those who are going hungry, in their village. I find that if you leave it to them they will not come up with fake data because they have to do it in the publicity way. Though the Planning Commission says thirty-five percent of people are below the poverty line, we find that only ten to fifteen percent of people are really going to bed hungry. Largely because they have no money. Also, a large number of them are dalits, they have no land, no livestock, no assets. So first identify the ultra poor.

Second, information empowerment of the poor. We find that the governments of our country invariably have at least one hundred twenty to one hundred forty schemes supposed to reach the poor. The poor are not aware of this. So, what we have done in our Information Village is to set up a database and dalit women operate it. Anyone can come here and find out about the schemes available to them. In a democracy, information empowerment is required which is why Aruna Roy and others in Rajasthan have started the Right to Information campaign. If all the available schemes were to be really implemented there would be no hunger. I can name one hundred forty-five schemes available in a small place like Pondicherry, but nothing reaches the poor at all.

Third, protein-calorie undernutrition. The government has three schemes targeted at the PDS. They have recently started one called Annapurna for older people. I have written to the prime minister to implement a new scheme for pregnant mothers because every third child born in our country is underweight. This affects brain development, among other things. It should cost us only two to three million metric tons of food grain. All pregnant women who are undernourished should be treated through this scheme.

Fourth, micro-nutrients such as vitamin A, iodine and so on, now iodized salt, are available. The silent hunger, hidden hunger caused by a deficiency of micro-nutrients, must immediately be eliminated.

Fifth, sanitation – what we call biological absorption. A child may eat a lot but it is not able to absorb its food. We may make food available in the market, and provide the ability to buy it, but a child must have the capacity to absorb it too.

Sixth, livelihood opportunities – micro-credit, micro enterprises.

And lastly, special attention must be afforded to women and small children, particularly pregnant women, nursing mothers and young, pre-school children.

To attend to all these seven points, the additional cost is not very great, as per our calculations. Only, there must be an efficient implementation of ongoing programs, plus a few critical gaps. The BJP government and its allies have stated in their agenda that they will work for a hunger-free India within the next five years. We must work toward this.

Question: *What is your vision for the 21st century?*

M.S. Swaminathan: In my view, the 21st century is going to be an ecology century of hope, caused largely by the communication revolution in India, the print media, mass media, electronic media and panchayati raj institutions. We have a unique opportunity to manage all our natural resources sustainably, to ensure a hunger-free India and to make good Gandhiji's vision that every man, woman and child in this country has an opportunity to lead a productive and healthy life. That is much more important than putting men in space, or on the moon. Let the Americans do that. This is not our priority. We must get our priorities right.

1.2 "M.S. Swaminathan: A Better Harvest." Interview by Tushaar Shah, *Management Review* (Bangalore, India), December 2000.

Tushaar Shah: *Over twenty-five years ago, in your Sardar Patel Memorial Lectures, delivered over the All India Radio, you spoke on "Our Agriculture Future". Today, at the dawn of the new millennium, has your prognosis changed?*

M.S. Swaminathan: In 1972, the prices of fossil fuel suddenly shot up and with it there was enormous escalation in fertilizer prices. People were therefore very pessimistic about the future of agriculture in the country, though the Green Revolution had been hailed only a few years before. So in my talk on "Our Agriculture Future" in 1973 I said that what we needed was an integrated approach: a farming systems approach looking at our land and water resources. I also spoke about the concept of Sarvodaya Agriculture, which is a high synergy society of farmers, managing and caring for natural resources like land and water, through group cooperation and group farming.

In the same year, I also delivered the Coromandel lecture on "Agriculture in Spaceship Earth", based on Buckminster Fuller's concept of a limited or finite carrying capacity. I felt that what we needed in our country was an economic ecology or "Do Ecology" as opposed to a "Don't Ecology". I warned environmentalists not to be carried away by what was happening in Europe or North America, because for them the struggle was to preserve the very high standard of living they had already achieved. They could afford to keep saying, don't, don't, don't. For us the struggle is to get some standard of living for a majority of the people. So, we need a more positive Do-Ecology, or economic ecology. For example, in our JRD Tata Ecotechnology Center, we are trying to promote aquaculture. You could go to the Supreme Court and get all such efforts stopped. But sixteen million children are born every year. A hundred million new jobs are needed. Where are the livelihood opportunities? Modern industry is not labor absorbing, it is only labor displacing.

So coming back to your question, my prognosis remains unchanged. The basic maladies of Indian agriculture are the same twenty-seven years later. Similarly, the remedies I advocated then were ecologically sustainable and I stand by them. The emphasis is on both social and environmental sustainability, in terms of technology development and dissemination – what we now call Ecotechnology.

Question: *In recent years, you have been deeply involved with the problems of poverty, livelihoods and employment. Indeed, the earliest initiative of the MSSRF was to create one hundred million jobs through an environmentally sustainable program of rural development. How far has the MSSRF been able to influence large policy framework in this direction? What are its current programs and priorities?*

M.S. Swaminathan: Agriculture is the mainstay of India's livelihood security system, apart from providing food and nutrition. It's a far different picture from that of the developed countries. The US, for instance, has nine hundred thousand farming families, while we have a hundred and five million farming families. In the US, one hundred thousand people produce seventy million metric tons of milk, while in India, eighty million women produce seventy five million metric tons of milk. This bears out Gandhiji's concept of technology of production by the masses. Indian agriculture has to provide not only more food for the urban population, but more income and more jobs or livelihoods for the rural people. That position has not changed in nearly three decades. In fact, the numbers have increased. Although the contribution of agriculture to the GDP has gone down to twenty-nine percent or so, the onus of employment is still on the agriculture sector. According to the latest [Five Year] Plan document nearly sixty-five to seventy-five percent of the people have to depend on agriculture for their livelihood. That itself shows the growing rural-urban disparity. So we must look upon agriculture not merely as a method of increasing food production for the urban population, but more importantly for the livelihood security of the poor, particularly of the landless poor and of small and marginal farmers. Secondly, there is a very strong gender dimension to agriculture planning, what is now generally known as the feminization of poverty. Women play a vital role in agriculture, doing most of the underpaid or unpaid jobs, jobs involving drudgery. And any natural resource degradation affects women more, because fetching water, fodder and fuel is traditionally women's work. These problems require the adoption of a farming systems approach rather than a commodity centered approach to agriculture planning. In other words, we need to concentrate on crop-livestock-fish integration or agro-forestry integration, and the reduction of drudgery. The third point on my agenda was the need for environmental sustainability in the production process. There, I advocate two different approaches. One is the integrated nutrient supply, organic manures, biofertilizers, green manures and limited amount of manure fertilizers. The second is the social synergy organization for the conservation of water, land and air. An individual farmer cannot do it. While cooperative societies may be the best form of organization for the dairy sector, conservation of natural resources requires more cooperation than a cooperative society.

Question: *With the advent of globalization, new institutional arrangements are being created for governing agricultural trade. In India, for instance, there is much concern about Intellectual Property Rights, TRIP and new WTO regulations. There is also concern about the impact of multinational corporations on Indian agriculture. How can India make the most of these changes?*

M.S. Swaminathan: Right now, as a result of the WTO and indiscriminate import of agricultural commodities, our farmers are struggling to defend their home market, forget capturing new markets. Markets abroad have become difficult, have introduced stringent sanitary, phytosanitary measures, and so on, while even in the year 2000, our rice farmers are drying their grain on the road. The road and the roof are their only dry areas. Then the rats get a large part of the crop. There is very little infrastructure. How can they compete with the Iowa farmer? I once calculated that an investment in post-harvest technology is one hundred thousand dollars in an Iowa farm; in an Indian farm it is one dollar. Where is the level playing field here?

Our problems are due to poor quality and poorly organized marketing. We have always prided ourselves on being the largest producer of fruits and vegetables, but the fact remains that except for Maharashtra, which has made some investments in grapes and Alfonso mangoes, our marketing is very poor. Our average fruit vendor makes Rs. 30-40 per day and has to bear the costs of spoilage and rejects. How can s/he survive? Another frequent occurrence is that at the time of harvest the prices crash, and these are all perishable commodities. A primary onion producer may get as little as Rs. 16 or 17 per quintal. An average small farmer may bring about five hundred quintals to the market; and out of that more than fifty percent is his cost of production. What is the take home pay for them?

When I was Secretary, Agriculture and Rural Development, I introduced the rural godown scheme for both perishable and non-perishable commodities all over the country. The aim was to avoid distress sales and panic purchase, particularly in drought prone areas, give visibility to storage, promote land use planning and reduce spoilage of perishable commodities. Some State governments were interested, and are now reviving the scheme.

Another handicap is that agriculture in industrialized countries is a technology-, capital-, and subsidy-driven enterprise. It is very low labor absorbing and high technology and subsidy absorbing. The fact is that in the United States, Japan and Europe, ninety-seven to ninety-eight percent of the people, who are employed in other sectors, subsidize the two to three percent who are agriculturists. In our case, twenty-five to thirty percent are subsidizing seventy percent. The US last year gave $26 billion of support to farmers, the highest in fifteen years. They can continue giving high subsidies, because they have written the text of the WTO and they know the rules of the game. Because the developed countries do the drafting in almost all global negotiations, they play a proactive role, while developing countries play a reactive role. For example, if you look at the aggregate measure of support in the WTO you will find things like research exempted. Billions of dollars are going into technology development, which are not included in any part of the aggregate measure of support to the farmers. In this way they find ways of giving greater support to their farmers. In Switzerland, if you are a farmer, seventy-five percent of your money comes from the government. These incentives are offered partly because agriculture is a foreign exchange earner. In the United States it is the number two foreign exchange earner, next to the armament industry. And while only two to three percent are involved in agriculture, another thirty to thirty-five percent of the American population is involved in the secondary sector (processing, business, industrial activity, production of fertilizers, pesticides, and so on) and the tertiary sector (packaging, forwarding, and shipping), which is particularly important in terms of marketing. Therefore, the WTO is not a level playing field at all, especially in agriculture.

However, in spite of these handicaps, globalization should give us an opportunity provided the industrialized countries are honest about their subsidies, because we do have some advantages that we have not exploited. For instance, our exports can find a market in countries which have very limited arable land. Even if China stabilizes its population at 1.4-1.5 billion, they will have to import because land availability is very low and their yields are already high, five tons of rice per hectare. On the other hand, we hardly get two tons, so our untapped yield reservoir is high. Even without genetic modification, we can double our total production. But we must improve the rural infrastructure, in terms of milling and so on. And it is vital that we pay more attention to dryland farming.

The increasing import of pulses and oil seeds into this country is an index of the neglect of dryland farming. The most important cash crops of dryland areas are pulses, oil seeds and millets. Now when you import a hundred thousand tons of pulses from Australia, which has highly mechanized large farms, our farmers are not going to get a proper price for groundnuts and pulses. This is a political issue, the markets are flooded with oil seeds and pulses, especially at festival time, to earn the blessings of the people.

Question: *But are you suggesting that we can be competitive in irrigated foodgrain products even if the West does not withdraw its subsidies?*

M.S. Swaminathan: Even if the West does not withdraw its subsidies we can still be competitive because their wages are very high. To some extent they compensate by using no labor at all in the farm. One couple and their son can manage an Iowa farm of a thousand acres because of their highly mechanized farming systems. But the relative cost of everything is high there. So if we can improve our water management efficiency in particular (because water management efficiency enhances nutrient use efficiency), and use appropriate varieties, we will get a very high return, reduce the cost of production, and increase the value. But in order to seize these opportunities, we will need a complete restructuring of the extension department in India.

I agree with Dr. Kurien when he says the extension must be under the control of the farmers. The development approach of the present extension is a patronage approach. How can you call the farmers "the beneficiaries"? You and I are the beneficiaries. We should be the "target group" not the farmer. There is an urgent need for a shift from patronage to genuine partnership in dealing with resource-poor farming families.

Question: *Coming back to globalization what has been your thinking on Intellectual Property Rights especially for Indian agriculture?*

M.S. Swaminathan: Indian agriculture has got both a bright side and a flip side. On the bright side we have one of the most dynamic and extensive plant breeding programs in the world, particularly in the public sector. Even in the international centers most of the top breeders are Indians: Gurdev Khush is a world leader in rice breeding, Sanjay Rajaram from Banaras is number one in wheat breeding. But we must protect our own work. Our outdated patent laws have now been revised. And we must bring in a Bio-diversity Bill. We are the only developing country that is rich in bio-resources as well as biotechnology. We are a mega bio-diversity area and, as with Information Technology, we have tremendous capacity in terms of brilliant young people.

Question: *I believe it is not good enough for our scientists to get a patent here?*

M.S. Swaminathan: That is one thing in the revised TRIP, if a patent in one country is accepted as a patent in other countries, then you will have the universal patent. The United States is in favor of this because for them it is an advantage if their patents are honored in India or China. In the Okinawa statement, G8 countries have recommended harmonization of intellectual property rights.

Question: *Registering patents seems to be a habit in the developed world, but not with our scientists.*

M.S. Swaminathan: They are just not used to it. I myself never gave any thought to patent rights fifteen years ago. We were working in the area known as the public good research, and the idea of patenting our intellectual property never even crossed one's mind. It was a different atmosphere altogether. But all that will change now. People like Dr. Mashelkar, head of the Council of Scientific and Industrial Research, have been trying to promote the idea that patent literacy is a must for every scientist in this country.

Question: *What are the opportunities and threats arising from the recent breakthroughs in genetic engineering? Are there opportunities for increasing the productivity of dry land farming through this route? What are the implications for our agriculture future?*

M.S. Swaminathan: There have been a number of technological changes in the last twenty-seven years, for example, many varieties of hybrid rice are being produced. Genetic modification has become the technology of the future. So much so that the private sector, which was never much involved in research or production of food crops like wheat and rice, is now finding these very attractive. Now the wheat and rice revolutions in Punjab have altered the concept of cash and food crops. Now, these food crops, and not sugarcane or tobacco, are the major cash crops.

Genetic engineering has been coming over a period of time from 1953 when Crick and Watson identified the double helix structure of DNA. From that time onwards people have been wanting to map the human genome. They wanted to manipulate and bring in genes from totally unrelated species, what is known as parasexual or crossing the sexual barriers. When I was an active breeder, the maximum I could do was to bring genes from breeds which would cross with each other. I could do tissue culture, but essentially with breeds which were sexually compatible. Now we can move genes across sexual barriers. For example, in our own Center here, some of my young colleagues, Ajay Parida and his group, have identified the genes for seawater tolerance in mangroves, which as you know, grow in estuaries, and have transferred it to mustard, tobacco and now rice. When I got my Ph.D. in genetics in 1952 from Cambridge, I wouldn't have dreamt that it would be possible to transfer genes from a tree to an annual crop.

India has a great opportunity in genetic modification. What we need is a risk-benefit analysis methodology by which one can identify what the risks are. The only risk I see is there could be some allergens. After all, I am taking a gene from mangrove and putting it into rice or mustard. It is possible our stomachs are not used to that protein. So you could have an allergic reaction. Another problem may be that the markers used are antibiotics. If the human system gets a lot of these, then it may not respond to antibiotics when you need them. So there are some genuine human health concerns. Environment concerns, such as the possibility of some new weed proliferating, are more remote possibilities.

To my mind, there are two serious questions concerning genetic engineering. One is the food safety concern, which has to be studied in greater detail; and the other is the ethical question. The question of ethics more closely involves the human side, like cloning. There is a whole series of issues on ethics, and that is why UNESCO has a universal declaration on the human genome and human rights. So I would say at the moment we should develop our capacity to the full. Because, in a way, what has happened over the millennia is some kind of genetic engineering – whether it is animals, birds, or pollination of plants across great distances from other plants – this is why there is so much variability in nature, so much mutation and recombination. What is new is that you can do it much more precisely. In breeding I have to make a cross and grow thousands of seedlings, hoping that I will get the combination that I need. Today I need not hope for the combination. I can make that combination at the molecular level. For example, if I want the genes for brown plant hopper resistance or stem borer resistance, now I don't take the whole thing, I take only that piece of DNA and transfer it. So there is a great deal of precision in what you are doing now.

What is important is to discuss it with the public, and establish a more transparent, credible method of risk and benefit analysis. For instance, in England they had the *bovine spongiform encephalopathy* (BSE) or mad cow disease, and the public found that the scientists had not been honest with them. The mad cow disease was swept under the carpet and then there was another problem with chicken in Belgium. The American public is also suspicious of corporates because the tobacco companies suppressed vital data for a long time, until they were punished. So they feel that multinationals will be self-serving, that they won't put all their cards on the table. So you must have a mechanism by which the public is assured that the risk-benefit analysis procedure is above board, that it is done in an objective way with the public good in mind and not their own personal agenda. Fortunately, the scientists are also going back to the drawing board, re-examining their methods and trying to foresee and plug all possible dangers or risks. We must have a national commission on genetic modification for food security. And we must ensure that this commission has very broad multi-stakeholder participation: representatives of non-governmental organizations, women's organizations, consumer forums, media representatives, the experts concerned and the government regulation authorities. So you will have openness, transparency and credibility. And, of course, there must be labeling for some time. The Americans were very much against labeling because they thought consumers would be put off by the genetically modified (GM) label. But the consumers have every right to know what they are eating. So we should label GM food until consumers have the confidence that it is like any other food. This may take another five years, until then the companies also must be patient, they must try to educate the public. So we must have a multi-pronged strategy: labeling of GM food in the market, transparent mechanism of risk-benefit analysis, and orienting our research agenda in such a way that the genuine concerns of food allergy and food safety are addressed carefully. If you do that, then India can convert its rich bio-resources into economic wealth through biotechnology.

One of the projects funded by the Rockefeller Foundation in Switzerland and Germany has created Golden Rice, rich in vitamin A or beta-carotene. They transferred the genes for high vitamin A from a species totally unrelated to rice. So a young child who cannot eat much rice, will get his requirement of vitamin A from just a few spoons of rice, and problems arising from lack of vitamin A like night blindness, can be immediately controlled. Similarly, Professor Ashish Dutta and his colleagues at the Jawaharlal Nehru University at Delhi have transferred genes from amaranths or *saag*, which are very rich in certain kinds of amino acids and proteins, to potatoes, creating highly nutritive potatoes. So all this work is going on, which can open up completely new vistas in terms of quantity, quality, and resistance to both biotic and abiotic stresses.

Question: *You've accomplished so much, in your years at the Indian Council of Agricultural Research (ICAR), with the Government of India, at the International Rice Research Institute (IRRI) and at the MSSRF; as a scientist and as an institution builder. What were your priorities when you set up MSSRF?*

M.S. Swaminathan: Early in 1988, I left IRRI so that I could set up an institution of my own before I was too old. Although I felt that this country did not need any more institutions and universities for agriculture, I noticed a gap. I had become deeply interested in coastal agriculture, and I found that although different departments (forestry, fisheries, tourism, and so on) were doing something, there was no real organized effort to look in an integrated way at natural resources management along the coastal areas. So I wanted to develop a Coastal Systems Research (CSR) methodology. And the government of Tamil Nadu was very willing to give us this land, in the Taramani institutional area.

Then the question was what kind of research priorities: one was CSR, which is an adaptation of the farming systems research methodology, except you look at the land surface and the sea surface in an integrated way. Coastal aquaculture, capture fisheries, culture fisheries, coastal-forestry, agro-forestry and coastal agriculture – all of these needed to be studied intensively.

I was also deeply interested in the issue of climate change, which I felt was inevitable. Because, whatever one may say, the industrial countries are not going to shift to sustainable lifestyles. Their fossil fuel consumption will continue to be high. We could endlessly argue about who is the greatest polluter, but the sufferers will always be the poor. On the one hand, the poor nations, and on the other, the poor in all nations. Arguments about who is responsible for global warming do not help the poor. I thought we should do anticipatory research. So the theme of my lecture at the 1989 Global Climate Conference in Geneva was mitigational anticipatory measures to meet the challenges of climate change and sea level rise. The Japanese government felt that it was such an important area of research that they provided half a million dollars to fund the project. So in 1990 we started this work on coastal biodiversity and biotechnology, conservation of mangrove forests round the coast, creating an awareness of the importance of mangroves, transferring the salinity tolerance genes from the mangrove to annual plants and so on. Later on we extended this work to the Gulf of Mannar area, adjoining Sri Lanka, which was once upon a time very rich in biodiversity though the dugongs and the seagrass beds are all disappearing now. So efforts are underway to save the Gulf of Mannar.

Another area in which I was always interested in is the blending of traditional technologies and frontier science. I also wanted to do more work on how to introduce ethics and equity in benefit sharing, in the use of genetic resources. I was then Independent Chairman of the FAO governing council. So at the same time we set up an international commission on plant genetic resources and developed the concepts of farmers' rights, rights of conservers, the contributions of traditional conservation, traditional wisdom, traditional knowledge and so on.

The fourth area we chose was called reaching the unreached, that is largely how to reach children and women. This program, run by my wife Mina, works in the areas of gender discrimination, gender justice, female foeticide, nutrition of children, mobilizing theater, traditional tools, and so on. We also helped the UN Commission on Nutrition to develop their report on child malnutrition. One of our aims, when we established this Center ten years ago, was to get the best in modern science to those who have been bypassed by science and technology. So we took up four areas in technology: Biotechnology, Information Technology, Space Technology (particularly GIS Mapping, Natural Resource Survey Mapping and so on), and Renewable Energy (particularly biomass, biogas and solar).

A fifth area that emerged was Management Technology; it was defined as putting all the pieces together in a proper way so that there was synergy among them, and also blending traditional wisdom and traditional knowledge with modern science.

We have also set up knowledge centers to disseminate information. One such center is Veerampatinam, a coastal fishing village in Pondicherry. All the fishermen go out to sea in catamarans, so the monsoon is a very worrying time for them. Now there is data from the US Navy that gives you the size of the wave in each area, each beach – 1,000 meters from the coast, 2,000 meters from the coast and so on. So they download it in the morning from the Internet, and broadcast it on loudspeakers all over the village. Everybody is excited because they never thought that it would be possible to know accurately what the condition of the sea would be – not the Indian Ocean, but their own stretch of sea – before they got into their catamarans. So modern information technology is very powerful, provided the users control it. And this information is driven by demand, not by supply.

Our program in Informatics (modern information technology) has been an adventure, a new form of social synergy between scientists and men and women who are economically and socially underprivileged, many of whom live in rural areas. At the same time it has provided opportunities for our young men and women to express their creativity.

In recent years, the UNDP has adopted our slogan, which is that sustainable human development must be pro-environment, pro-poor, pro-women and pro-employment. So we have had a wide impact on the world's thinking.

Question: *That is probably why* Time *magazine acclaimed you as one of the twenty most influential Asians of the 20th century.*

M.S. Swaminathan: I had the privilege of traveling widely in Asia when I was in IRRI, to develop Rice Institutes in a number of countries: in Burma, Vietnam, Cambodia, North Korea and of course, the National Rice Institute of China, where I received an award from the Chinese government for significant contributions to the country. What I have been impressed with is, wherever you go, the rural people are extraordinarily receptive to ideas. The picture of the Asian farmer as fatalistic and impervious to new ideas is a myth created by the Western world. Take our computer work in the villages of Pondicherry: women who are barely literate take to technology like fish to water. And therefore it is important for all developing countries to adopt the American slogan of Land of Opportunity.

1.3 "Global Agriculture - Facing the Challenges." *New Agriculturist* (UK), April 2001.

New Agriculturist: *What do you see as the challenges facing global agriculture between now and the year 2020?*

M.S. Swaminathan: The first point is that we have to produce more, because today, in spite of the large food mountains there are also millions of hungry. The paradox of hunger in the midst of plenty, of food mountains and hungry mouths co-existing, is a very great shame, a national shame. Secondly, it is also vital to ensure that our production is environmentally sustainable so that the ecological foundations like land and water and the bio-diversity of flora and fauna are not destroyed. For example, the Punjab is our breadbasket in India. Today the food security is very high but by 2020, they will be very food insecure because of groundwater depletion. There is pollution and soil salinization taking place and more and more land is becoming unusable. There is no proper crop rotation: it is rice, wheat, rice, wheat, and no legumes are planted in between to replenish soil fertility and soil health.

These challenges can be met only by taking the best in traditional wisdom and technologies, and combining that with cutting-edge science. In order to bridge the divide between the genetic haves and have-nots, we must start at the village level, with the very poor. We need to consider environmentally safe biotechnologies that are available, whether bio-fertilizers, bio-pesticides, or biological control agents like ladybird beetles. We need to look at how they are produced, how they are marketed, and whether they can make village level agriculture more sustainable. We also need to look at non-farm employment, because a large number of people who are going to bed hungry have no land, no livestock, no assets, and often they are illiterate. But how can you give the asset-less some assets? We must look at the options for rural non-farm employment, offering people new skills and new opportunities.

For a next step we have identified what I call "bio-valleys", biologically rich areas where, with modern technologies, we can convert the naturally occurring bio-resources into wealth and jobs that are meaningful to the people. These "bio-valleys" are even more important than "Silicon Valley" because bio-resources are the ones which can provide the largest number of jobs, whereas the Silicon Valley type economy does not produce many jobs: it results in jobless growth, not job-led growth! Agriculture offers job-led growth.

Then, we have the idea of "bio-partnership". This is very important in terms of ethics and equity. At the moment bio-piracies are a growing problem; people come and take away the knowledge of indigenous people. For example, there are many plants, which indigenous people have found to have medicinal properties. These are the intellectual property of the people who discovered them. If someone else comes and collects those plants, they are taking that intellectual property. In the area of bio-diversity, we can take the example of rice. The International Rice Research Institute tells us that there are over one hundred thousand rice varieties. Of these, communities preserve some ninety thousand; they are not in the wild, they are in the community. Now, fortunately, the Global Convention on Biological Diversity for the first time has introduced ethics and equity in knowledge sharing and community conservation.

I also emphasize the need for a gender dimension in agricultural research, for the simple reason that a large number of agricultural activities, particularly those involving much drudgery and very low pay, are done by women. But if given practical help, the rural poor, particularly women, take to new technologies like fish to water. To give you an example from digital technology, in fishing village I have visited, the women were very keen to know about the weather because they are artisan fishermen. We therefore established a computer-aided and Internet connected rural knowledge center in the villages. Now, they are downloading information on sea conditions through the Internet, from the Naval Oceanographic Office of the USA; this is up-to-date data, not on the global conditions but on what the condition of the sea in their own area will be. The world is becoming information rich but action poor. How to bridge the gap between "know-how" and "do-how" is the whole purpose of my work with the rural poor.

Finally, by 2020 the "divides", what we call the demographic divide, the digital divide, the genetic divide, and the nutritional divide will widen, unless we take strong steps between the industrialized countries and the developing countries. In developing countries, agriculture is not merely an engine for producing food for the urban population, it's the very backbone of the livelihood security system of the rural masses. In India, for example, nearly six hundred million out of one billion depend for their livelihood on agriculture: crops, livestock, fish, trees, agroforestry systems, agro-processing and so on. Therefore, agriculture has a larger meaning in terms both of human well being, and also peace and security within the country. And we see this everywhere, for example in all the ethnic conflicts in Africa between those who have land and those who do not have land. So, I think agriculture has a critical role in both national and global security. It could be the engine of internal conflicts if agriculture does not go well.

And agriculture is also the engine for the triggering of industrial growth. In India, for example, many of our industries are based upon agricultural raw material, large industries like textiles. Cotton is very important, providing jobs for sixty million people, who entirely depend on the cotton crop. Similarly, with the jute and the sugar cane plantation crops. In addition, rural purchasing power is important for urban industrial growth. Starting from the Industrial Revolution, technology has been the prime mover of European prosperity, in contrast to those who were not part of that revolution. Today, I'm saying that this technology needs to be an ally in the equity movement.

1.4 "The WTO Pact On Agriculture Will Kill Our Farmers." Interview by Shobha Warrier, January 25, 2001.

Eminent agriculture scientist Dr. M.S. Swaminathan - who was conferred the prestigious Indira Gandhi Prize for Peace, Disarmament and Development for 1999 for his outstanding contribution in the domain of plant genetics and ensuring food security to hundreds of millions of people in the developing world – is an agitated man.

His usual air of optimism is conspicuous by its absence. The WTO Agreement on Agriculture is the sole cause of his uneasiness. At the recently concluded Science Congress, Dr. M.S. Swaminathan demanded a White Paper on the Agreement and also offered a solution to solve the problem that stares India and many other developing countries in the face.

He spoke to Shobha Warrier about the 'dreaded' WTO Agreement on Agriculture. He also briefly touched upon the impending General Budget 2001.

Shobha Warrier: *Once our markets are flooded with cheaper foodgrains when the WTO Agreement on Agriculture comes into force, what do you think will be the plight of our farmers?*

M.S. Swaminathan: We have not prepared ourselves for the consequences of the WTO Agreement on Agriculture. The previous UNCTAD (United Nations Conference on Trade and Development) did not include agriculture. Cheaper goods will hit our markets from April 1, but the government has taken no concrete steps to ready our farmers for the tough times ahead. It is only the Uruguay Round of Negotiations, after the United States and other industrialized countries insisted, that included agriculture within its ambit. Now, an international organization will directly control and influence the domestic policies of various countries. Normally, UN organizations leave domestic policies untouched. But this is an exception. The World Trade Agreement on Agriculture deals with domestic support, export subsidies, market access, trade-related intellectual property rights and so on. We have voluntarily agreed to place the livelihood of millions of Indians under international coordination. In India, agriculture is not just a food-producing industry. It is largely a livelihood-generating industry. This, however, is not the case in industrialized nations where hardly two to three percent – maybe a maximum of four percent – of the population is involved in agriculture. Agriculture is the backbone of our livelihood and food security system. It has been six years since we signed the WTO Agreement, but we have not really prepared ourselves to take advantage of any potential benefits.

Question: *Are there any potential benefits for a developing country like India in this Agreement?*

M.S. Swaminathan: There would be some if industrially developed countries remove their subsidies completely. Because of the high wages there, our goods will be always cost-competitive. But they have not reduced subsidies. In fact, they have increased them. Last year, they gave $300 billion as subsidies under various clauses: the Green Box, the Blue Box and so on. They have neither reduced export subsidies nor increased market access. On the other hand, they have put additional restrictions under the pretext of sanitary and phytosanitary conditions. There have been attempts to add other environmental and other social concerns like child labor and so on. We have not seen the benefit of the World Trade Agreement. On the negative side, our government in its wisdom has put many items under the Open General License. So imports have been terrible in terms of oil seeds and other products. The WTO pact can actually kill our farmers. From April 1, New Zealand will be able to flood our market with milk at about Rs. 8 per liter. Our poultry industry, which provides income to a large number of "small people", will be killed. In the United States, they don't like chicken legs. But in India they are preferred. So, they will be hygienically packed in the US and dumped in Indian markets. That then will put our broiler industry to the sword. Already the oilseed industry is in distress because our oilseed import is going up enormously. Now, because of farmers' distress and political pressure, trade in our country is becoming like a relief operation.

Question: *You have been warning the government about repercussions of the WTA for a while now. Weren't your warnings taken seriously?*

M.S. Swaminathan: A professional can only point out. Whatever I have pointed out in the last few years, is coming true. There is no use in saying, "I said so". I don't like to do that. The government must be having their own reasons for not increasing tariff or allowing indiscriminate imports. They were importing wheat when fifty million metric tons of our own wheat were perishing in the godowns. We just don't understand the rationale of the government.

Question: *How will your request to the government to publish a White Paper on the World Trade Agreement on Agriculture serve any purpose?*

M.S. Swaminathan: At least we will know the facts. There has to be a reason behind their moves.

Question: *Why did the Commerce Ministry not increase import tariff?*

M.S. Swaminathan: The Japanese put two hundred percent tariff on rice import to prevent American rice from coming to Japan. But in our country, they give some excuse or the other, and the common man remains confused. Even I cannot understand the rationale often. But having served in the government, I know there may be a rationale. Therefore, it is important in a democratic society to know the facts. More so because agriculture is not just another profession in India. Modern industrialists have an overriding influence in the government policies. But let me ask, how many jobs does the organized sector provide? The truth is that they don't provide jobs; they only sell their products. The onus of employment is in the farming sector - horticulture, animal husbandry, agriculture, fisheries, forestry, etc. That is why I call this sector the backbone and livelihood security system of India.

Question: *In your address at the Science Congress, you wanted a Livelihood Box to be included in the World Trade Agreement for ten to fifteen years. What exactly do you mean by Livelihood Box?*

M.S. Swaminathan: A Livelihood Box will give us an option to impose quantitative restrictions on imports, that is, wherever livelihoods will be destroyed by imports.

Question: *Will the Livelihood Box be applicable only to Indian farmers?*

M.S. Swaminathan: No, not at all. Any developing country or poor country can have the Livelihood Box. If you are going to dump cheap *cassava* in an African country where they can produce their own, what will the local farmers do? I am not talking about India alone, because you can't have a box only for India. The Green Box and the Blue Box were developed by the industrialized countries cleverly to protect their subsidies. For example, the Blue Box is a good device as it makes adjustment between production and demand. Under this, the United States gives lot of subsidies for not growing crops so that production can be kept at a level that the market can absorb. We don't have a facility to advise our farmers on land use. Our farmers do not have the competence to decide on their own. Look at the tobacco industry in Andhra. Tobacco is a dying industry, but there are millions of tobacco farmers. Are we doing something pro-actively? Have we prepared them for an alternative? What we need is pro-active advice on land use planning. Industrialized countries are doing it. That is how the Blue Box functions.

Livelihood Box is a method by which the revised World Trade Agreement agrees to insert a box. If a dispassionate analysis shows that imports will kill the local livelihood of the poor and aggravate poverty, the box has to be used. But it cannot be used lightly. There has to be some rationale. Before the WTO Agreement, why were these countries surviving? Why didn't we have so much foreign debt? Only in the last ten to fifteen years, the debt has risen. That was because we were conservative earlier. We did not import anything – including automobiles. Call it *swadeshi*, but there was a considerable amount of self-reliance in this country. Today, how can a small industry survive unless the government gives high subsidies, like they give for khadi, cottage industries, etc.? But how long can they do it? Today, surplus sub-standard paddy is bought at a high price by the government. I don't call it trade. I call it trade relief. When trade becomes a relief operation, is there a scope for this country to come out of the poverty trap?

Question: *Since the WTO Agreement is detrimental to our farmers, can we pull out of the Agreement?*

M.S. Swaminathan: It is not advisable to pull out because a multilateral agreement based on a UN pattern is always better than a bilateral agreement. Otherwise, you have to enter into a bilateral agreement with several countries. But we must have a National Trade Policy. The first thing in this policy should be Home Trade through which we should have regional trade grids. We must divide our country according to ecological conditions. For example, today Assam can supply rice to the entire north-east India. Andhra Pradesh can take care of the rice needs of south India.

Then, you must have a policy for regional trade. In our own SAARC region or ASEAN, we can have agreements whereby we buy something that they produce efficiently and vice versa.

Even if we import palm oil from Malaysia, we can sell our onions, chillies or apples. We must have long-term trade agreements with our neighbors. We must also develop a long-term global trade policy with others.

We must convert this calamity into an opportunity. A nation that does not learn to do this will always remain backward. So, the trade calamity of 2001 should be converted into an opportunity by developing a long-term trade policy and then ask for a Livelihood Box for ten to fifteen years. Meanwhile, we must invest massively in post-harvest technology, sanitary measures, rural roads, cold chains, etc. Thus, we can be globally competitive.

Question: *How effective, according to you, are the oft-repeated plans of increasing the Public Distribution Scheme and provision of insurance cover to farmers?*

M.S. Swaminathan: These are issues that the White Paper should bring out. Insurance also has been talked about for a long time. It is not easy in a country where there are one hundred and five million farming families and over six hundred million engaged in farming. It is a question of billions of rupees. Because the problem is complex, it will be swept under the carpet. You may refuse to face the problem, but the problem will face you. That is what we are facing now. For the last six years, we refused to face the consequences of the World Trade Agreement.

Question: *Although seventy percent of our population is involved in agriculture, this is one area neglected by every government. Is it because agriculture is not an organized sector?*

M.S. Swaminathan: Yes. They do not have a strong voice like the CII or the FICCI. That is why I have been saying that there must be a national federation of agricultural organizations. If you read the budget speech of all the finance ministers, you see maximum sympathy going to agriculture. They raise slogans like *"Jai Kisaan"*. Maximum paragraphs are there for them, but the action is minimum.

Question: *Do you expect anything challenging in the coming budget?*

M.S. Swaminathan: Yaswant Sinha says that he is consulting widely, and wisely. He knows the problems. He says he puts a lot of weight on agriculture in his budgets, but, unfortunately, nothing is followed up. He announced 1999 as the Year of the Gram Sabha. He also talked about democratic decentralization. But no State government has done anything. In the last budget, he talked about Land Use Board, Commission, etc. Again, nothing happened. In my view, the first part of his budget speech should be about the actions taken. He should first talk about the promises made in the previous budget and then inform to what extent did we succeed. Only then, he should move on to new promises.

Question: *Is it because we have been producing surplus food after the Green Revolution that various governments are ignoring this sector?*

M.S. Swaminathan: The truth is, we are not producing surplus food. If all the three hundred million children, women and men who will go to bed partially hungry tonight, eat well, there will be no surplus. Our surplus is only an expression of poverty. Every third child born in this country is of low birth weight because of maternal under-nutrition.

Question: *But isn't there an illusion that we produce surplus foodgrain?*

M.S. Swaminathan: That is because all newspapers depend on advertisements. The government gives sops and not solutions. There is a difference between sops and solutions.

Question: *Doesn't the government have a solution?*

M.S. Swaminathan: Non-professionals advise them. The secretary in the fisheries department will become a vigilance commissioner in his next posting. We have this self-inflicted injury of not having enough professionals to take decisions. WTO is not responsible for our plight. We ourselves are responsible.

1.5 "The Know-how of GIS Should be Converted to Do-how." ***Geospatial Today*** **(Hyderabad, India), 1(5):22-23, January-February 2003.**

Geospatial Today: *How can spatial technologies like Remote Sensing and GIS be used in a country like India for sustainable development?*

M.S. Swaminathan: Both Remote Sensing and GIS are extremely important in order to gather accurate, reliable information. And, also dynamic information. Remote Sensing gives you much more dynamic information like what is the situation today and also what was it like in the past. You can compare the data. GIS helps you to integrate the whole series of information. For example, we have used GIS in our Hunger Free Zone projects GIS methodologies to interpose land data, number of characteristics of the place, household structure, number of schools and so on. You can put multiple criteria and get an integrated picture. In fact, our Rural Atlas is also mainly based upon GIS methodology. We have put nineteen different parameters together for the maps. So it has become a powerful tool for decision-making and planning and we are fortunate to have the capability of Remote Sensing in this country. With MetSat we should be able to get a lot of information on weather conditions. We can prepare different computer simulated models for different weather problems. Drought will come one year or the other. The question is how do you manage it. I think now the space technologies and applications provide greatest strengths to our country for launching a sustainable agriculture movement.

Question: *You have been talking about drought code and weather code. How can that be made possible using these technologies?*

M.S. Swaminathan: Drought code is nothing but long-term time sequence of weather data. We have over one hundred and thirty to one hundred and forty years data on climate available to us. Using that one could see the kinds of problems and try to extrapolate. By analyzing the data we can understand whether the monsoon is getting delayed or whether it will withdraw. You can work out three or four major possibilities of rainfall patterns and then arrive at scenario A or scenario B. This will help the farmer to think of alternative crops.

Then building Seed Banks and Fodder Banks. In our Orissa project at Kalahandi, Koraput, areas which are always prone to drought, we have established different Banks, teaming with the local tribal families. One is a Gene Bank which maintains their traditional varieties. Then there is a Water Bank. Finally, a Food Bank. Whenever there is shortage, these respective banks take care of the needs of the farmers.

Rajasthan, which has a large cattle belt, faces a severe drought situation almost every alternative year and faces severe shortage of fodder. In the Punjab, ten million metric tons of straw are burnt as they do not use it for fodder. We need to provide linkages. In areas where there is no feed available, Fodder Banks should be set up. Especially in all the drought hot spots, it is important to develop these banks. That's where GIS data comes into the picture. Once you have the GIS map of an area with critical parameters for livelihoods and agricultural crops, then planning for adverse weather and good weather is possible. How do you maximize the benefits of good weather for strengthening the ecological foundation for sustainable agriculture: That is, planting trees, agro-forestry, reclaiming wasteland, etc? All these can be done only when there is some moisture in the soil.

The drought management code is how to minimize the impact of drought for human beings, crops and animals. So the weather code and the drought management code are doable strategies, not theoretical strategies. In many laboratories you get very good remote sensing maps. But they are seldom used in the field. So the "know how" of GIS should be converted into "do how" at the field level.

information on weather conditions? We can prepare different computer simulated models for different weather problems. Drought will come one year or the other. The question is how do you manage it. I think now the space technologies and applications provide greatest strengths to our country for launching a sustainable agriculture movement.

Question. You have been talking about drought code and weather code. How can that be made possible using these technologies?

M.S. Swaminathan: Drought code is nothing but long-term time sequence of weather data. We have over one hundred and thirty to one hundred and forty years data on climate available to us. Using that one could see the kinds of problems and try to extrapolate. By analyzing the data we can understand whether the monsoon is getting delayed or whether it will withdraw. You can work out three or four major possibilities of rainfall patterns and then arrive at scenario A or scenario B. This will help the farmer to think of alternative crops.

Then building Seed Banks and Fodder Banks. In our Orissa project at Kalahandi, Koraput, areas which are always prone to drought, we have established different Banks, teaming with the local tribal families. One is a Gene Bank which maintains their traditional varieties. Then there is a Water Bank. Finally, a Food Bank. Whenever there is shortage, these respective banks take care of the needs of the farmers.

Rajasthan, which has a large cattle belt, faces a severe drought situation almost every alternative year and faces severe shortage of fodder. In the Punjab, ten million metric tons of straw are burnt as they do not use it for fodder. We need to provide linkages. In areas where there is no feed available, Fodder Banks should be set up. Especially in all the drought hot spots, it is important to develop these banks. That's where GIS data comes into the picture. Once you have the GIS map of an area with critical parameters for livelihoods and agricultural crops, then planning for adverse weather and good weather is possible. How do you maximize the benefits of good weather for strengthening the ecological foundation for sustainable agriculture: That is, planting trees, agro-forestry, reclaiming wasteland, etc.? All these can be done only when there is some moisture in the soil.

The drought management code is how to minimize the impact of drought for human beings, crops and animals. So the weather code and the drought management code are doable strategies, not theoretical strategies. In many laboratories you get very good remote sensing maps. But they are seldom used in the fields. So the "know how" of GIS should be converted into "do how" at the field level.

Section 2.
Excerpts from Archives

2.1 Dr. Norman E. Borlaug, International Maize and Wheat Improvement Center (CIMMYT), Mexico, November 10, 1970.

The Green Revolution has been a team effort and much of the credit for its spectacular development must go to Indian officials, organizations, scientists and farmers. However, to you, Dr. Swaminathan, a great deal of the credit must go for first recognizing the potential value of the Mexican dwarfs. Had this not occurred it is quite possible that there would not have been a Green Revolution in Asia. …

We must not become complacent because of the wonderful recent publicity that has come to the Green Revolution. Instead, we must all increase our efforts to expand and stabilize wheat production. We must at the same time very aggressively push forward the production of rice, maize, sorghum and millet. Unless there is a more dynamic production effort in the latter four crops the Green Revolution will stall and die. Wheat cannot alone continue to carry the brunt of the attack on – we must get the production of other crops moving more dynamically as well.

We want all of you to know that it has been a privilege for both the entire CIMMYT staff and for me to have had an opportunity to participate in a modest way in your phenomenal increase in wheat production. We hope to continue to collaborate with your brilliant program.

2.2 Felicitations on the award of the First World Food Prize, Washington DC, 1987.

It is with great pleasure that I congratulate you on being chosen as the First General Foods World Food Prize Laureate.

This award recognizes what many in the global food and agricultural community have known for a long time – that your efforts have made a dramatic and lasting impact on improving world food supply. I also applaud your outstanding work in India, where you earned the well-deserved title of "Father of the Green Revolution," and your continued efforts to pioneer the field of agricultural research at the International Research Institute in the Philippines. You can be proud of these accomplishments as well as the dignity and self-reliance you have helped to bring to the people you have served.

Again congratulations, and best wishes for continued success in your work.

– Ronald Reagan
The White House, Washington DC, September 11, 1987

I was so pleased to hear the news of your award. There could not have been a more appropriate choice. The central focus of application of plant genetics technology has moved a long way since our association at the University of Wisconsin Genetics Department almost forty years ago. I can think of no one who has combined the insights of the cutting edges of biological science, with attention to the most urgent of human needs, with the competence, devotion and energy that you have given.

The world is and will be a better place on account of your contributions to its welfare.

– Dr. Joshua Lederberg
1958 Nobel Laureate in Medicine and Physiology
September 25, 1987

Of all the problems which the people of our world must tackle, none is more fundamental than insuring adequate food supplies. Food is basic to good health, child development and learning, productivity, and the ability to live a life free from want and physical deprivation.

To make a difference in the world's capacity to feed itself is indeed an awesome accomplishment, and so I congratulate you most sincerely and warmly on the honor given you as recipient of the 1987 World Food Prize. Your accomplishments as a scientist and a scientific leader should give you great personal satisfaction. More importantly, however, they have had an impact on the lives of millions of people in Asia and other parts of the developing world.

I am especially glad that this recognition comes when you are still actively at work on the new knowledge and applications of knowledge that will be needed to sustain future food supplies in the face of increasing world population and dwindling resources of land and water. May your efforts continue to be as productive and vital as they have to the present time.

– Dr. Frank Press, President
National Academy of Sciences, Washington DC
September 29, 1987

Rosalyn and I want to add our congratulations to those you are receiving from your friends and colleagues throughout the world. The General Foods World Food Prize is an outstanding way to recognize your achievements in improving the world food supply.

You can take pride in knowing that many generations of people will benefit from your contributions. Your accomplishments will be of great advantage to scientists and agronomists of the future who will continue your efforts to meet the needs of our expanding population.

– Jimmy Carter
Former President of USA
October 6, 1987

Dr. Swaminathan is a living legend. His contributions to agricultural science have made an indelible mark on food production in India and elsewhere in the developing world. By any standards, he will go into the annals of history as a world scientist of rare distinction.

– Javier Perez de Cueller
Secretary General, United Nations
October 1987

2.3 Felicitations on his 70th birthday, National Academy of Agricultural Sciences, New Delhi, August 1995.

On the happy occasion of the seventieth birthday of Dr. M.S. Swaminathan who is a living legend in our times, I offer him my greetings and good wishes for a long, healthy and happy life of continued service to the nation and to humankind.

Pleasant, soft spoken, dignified but unassuming, Dr. Swaminathan has a demeanor which conceals his brilliant intellect, unparalleled achievements in science and research, and countless awards, prizes, fellowships and honors showered on him by national and international institutions. The Green Revolution which transformed our chronic food-deficit country with three hundred sixty million people into a food self-sufficient one with nine hundred million has earned for him an indelible place in our national history. The Secretary General of the United Nations, Dr. de Cueller wrote on the occasion of Dr. Swaminathan receiving the first World Food Prize: "By any standards, he will go into the annals of history as a world scientist of rare distinction."

May Dr. Swaminathan march vigorously forward towards higher goals in science and in greater positions in national service.

– R. Venkataraman, Former President of India

I am glad to know that the National Academy of Agricultural Sciences is felicitating Dr. M.S. Swaminathan on his seventieth birthday.

Acclaimed internationally for his pioneering contributions to agricultural science and practical application of his scientific genius for modernization of agriculture and enhancement of food production, Dr. Swaminathan stands out with rare distinction in the scientific community. Our fight against hunger through the success of the Green Revolution owed a lot to his effort. In his pursuit of scientific research he has consistently emphasized on sustainable agriculture and an environment friendly approach. His scientific insights have opened new frontiers of success for our nation.

I have great pleasure in congratulating Dr. Swaminathan on his seventieth birthday. I also send my good wishes for his health and happiness and many more years of service to the nation.

– K.R. Narayanan, Vice-President of India

Dr. Swaminathan is the doyen amongst the agricultural scientists who laid the foundations of India's Green Revolution and its present food security. He is undoubtedly one of the foremost agricultural scientists of this century.

– P.V. Narasimha Rao, Prime Minister of India

I wish to extend my very best wishes to Professor Swaminathan and my heartfelt felicitations. I wish him a long and happy life.

– Dr. Abus Salam, President
International Centre for Theoretical Physics, Trieste, Italy

I regret that I am unable to participate in the National Academy of Agricultural Sciences' celebration commemorating your 70th birthday. I have a previous commitment in Japan that same date, which makes it impossible for me to participate.

I do, however, take this opportunity to congratulate you for the many decades that you have dedicated to agricultural sciences – research and education – which have contributed greatly to increasing food production to alleviate poverty and human misery. I take this opportunity also to wish you many more years of continued fruitful service to humankind, in the broader context.

– Dr. Norman E. Borlaug, Nobel Peace Laureate
CIMMYT, Mexico

On the auspicious occasion of your seventieth birthday, I have great pleasure in sending you warm felicitations and saluting you for all your achievements in a life dedicated to freeing humankind from hunger and malnutrition.

As a scientist and researcher, as a pioneer of the Green Revolution, as an advisor and policy maker, and above all as a leader, I recall with admiration the great efforts undertaken and the strides made by you in the relentless struggle to find sustainable ways to achieve man's basic right to food. You have given leadership to the most prestigious centers of research and agricultural sciences at both national and international levels and have been instrumental in shaping the policies of many organizations devoted to food and agricultural development. In this context, while I have had a long association with you in various capacities, I recall in particular the important role you had once played as the Independent Chairman of the Council of FAO. In my present capacity as the Director-General of FAO, I am also pleased to mention that you have always found time to respond to me whenever I have sought your advice.

On this occasion, I wish you, Dr. Swaminathan, a long and healthy life and convey my best wishes for your continued work in the furtherance of agricultural sciences to the benefit of humanity.

– Dr. Jacques Diouf, Director General of FAO

It gives me great pleasure on the occasion of your seventieth birthday to send you my warmest greetings and deepest sentiments.

I have known you for so many years now that it is difficult to recall when I first met you. Throughout the years you have left an indelible record in my mind of the deep, sustained and strong commitment you have for the global environment and its preservation. You have played a pivotal role worldwide in the conservation of biological diversity and can be called the "guru" of agricultural scientists and the "father" of the worldwide network of agricultural research centers. Your record of achievements would be hard to match, let alone replicate.

M.S., you have made an outstanding contribution to the world and while preserving our environment you have made people its beneficiary. The people of the world will always be indebted to you, not only for your innovative approaches, your research and the many books and articles you have written, but also for the leadership role you have taken in all areas for the preservation of our planet.

– Dr. James Gustave Speth, Administrator
United Nations Development Programme (UNDP)

Dr. Swaminathan's lifelong work to increase the Earth's food supply has been of great significance to the developing world, to which he has brought nothing less than hope for a better life and hope of a brighter future.

– *Dr. Elizabeth Dowdeswell, Executive Director*
United Nations Environment Programme (UNEP)

The President and Council of the Royal Society extend their congratulations on your 70th birthday and their good wishes for many further years of life and of scientific achievement. The Society counts itself fortunate to have, as one of its Fellows, someone who has contributed so much for the benefit of humankind.

– *Sir Michael Atiyah, President*
The Royal Society, London

You have made great contributions to the world agricultural and scientific development.

You are a very good friend in cooperation between China and India, and between IRRI and CNRRI.

You have made great contribution to the friendship between the Chinese and Indian people.

– *Prof. Wang Lianzheng, Vice-Chairman*
Chinese Association of Science and Technology, Beijing

I along with the fellowship of the National Academy of Agricultural Sciences offer our felicitations on the 70th birthday of the architect of modern Indian agriculture. The achievements on the agricultural scenario of the country which have been a pace setter for the developing countries are synonymous with Dr. M.S. Swaminathan. He has rightly been addressed as the Father of the Green Revolution in India.

Thanks to his brilliance and dedication, Dr. Swaminathan rose to occupying the highest positions in the field of agricultural sciences both at the national and international levels. He has made noteworthy contributions to agriculture first as a scientist and later as a science leader of rare distinction. …

A plant geneticist by training, he made original research contributions in the areas of plant cytogenetics, use of mutations in a number of crops like potato, wheat, rice, etc. His work on the use of "Norin" dwarfing genes to overcome the barriers to high yield in wheat is widely quoted. He inspired agricultural scientists, not only in professional pursuit of their choice but also in several other disciplines of agriculture with visionary concepts and ideas such as "crop cafeterias", mid-season corrections in crop scheduling, risk distribution agronomy and alternative cropping strategies for different weather conditions, conservation of genetic resources, National Demonstrations and Lab to Land program, etc. His latest love is the conservation of natural resources as basic life support systems for sustainable development.

Dr. Swaminathan is a teacher par excellence. He has been inspiring the students by his thought-provoking lectures. In spite of the most pre-occupying positions held by him during the last four decades he never permitted interruption in his teaching schedules. … He is an honorary Professor in some of the most prestigious universities in China, United States and India. …

His complete dedication to science, particularly agriculture is glaringly reflected from the fact of his donating his life's earnings for the establishment of M.S. Swaminathan Research Foundation in Madras of which he is the Chairman and continues to contribute to the development of science in India. He is the Director, Centre for Research on Sustainable Agricultural and Rural Development, a model institution in the private sector.

We, the agricultural scientists pray that the Almighty bestow on him His choicest blessings and wish that he may continue to guide the destiny of Indian Agriculture for many more decades.

– *Dr. N.S. Randhawa, Vice President*
National Academy of Agricultural Sciences, New Delhi

On behalf of the Indian Council of Agricultural Research and all the agricultural scientists of the country, I offer our heartiest felicitations to you Dr. Swaminathan, for all that you have done for this country. On your 70th birthday it is a matter of great pleasure for each one of us to wish you all the happiness, prosperity and sound health so that we continue receiving your guidance and patronage for the growth of agriculture in this country. ...

In Dr. Swaminathan we find a human being par excellence with brilliant head and most kind heart.

We would always be indebted to you for what you have done for this country. You are the architect of the Green Revolution. Your contributions are known internationally in terms of what you have done to bring us from begging bowl state to the state of self-sufficiency.

As a teacher, Dr. Swaminathan, I was your student, we have seen that you excel all. ... As a planner we have seen in you a man with vision and dynamism.

We are all aware of what you have done for the development and upliftment of the scientific community in agriculture. The Agricultural Research Service which is so popularly known today and which has helped this nation to absorb the latest technologies to not only achieve the Green Revolution but also white, yellow and now we talk of blue revolution. All these could be possible because of the missionary effort and initiative which you took in initiating a number of programs. And with Lab-to-Land program you have linked scientists with farmers. You are an institution builder. ...

– *Dr. R.S. Paroda, Director General, ICAR;*
Secretary, Department of Agricultural Research and Education
Government of India, New Delhi

We are celebrating the 70th birthday of one of the most nationally and internationally renowned scientists of this country, a great human being, a symbol of intellect, compassion, wisdom, generosity, dedication and affection for all humankind.

Dr. M.S. Swaminathan, Sir, in you today we see, in a true sense, the *Yugpurush* of Indian Science. On behalf of the Fellowship of the National Academy of Sciences and on my personal behalf I convey to you, Sir, our most sincere and profound regards and best wishes for your creative and rewarding future ahead.

The National Academy of Sciences, Allahabad, the oldest of the three Academies has a very special regard for Dr. Swaminathan. He took over as President of the National Academy in 1989. The Academy organized during his tenure a number of scientific seminars and symposia on very important subjects of relevance to the country involving a large number of young scientists and technologists in particular. The subjects he covered during this period included public policy implication of biotechnology for Asian agriculture, restoration of the natural and cultural sites along the river Ganges (because the Academy is located at Allahabad), India's geosphere and biosphere program, and sustainable management of natural resources. He injected a new enthusiasm and scientific strength in the activities of the Academy. With his vision and deep understanding of the problems in the areas of agriculture, environment and natural resources, he guided the deliberations of these seminars, the scientific activities of the Academy to arrive at extremely important recommendations. This has always been a unique feature and characteristic of Dr. Swaminathan. He never stops or leaves any work without bringing it to a logical conclusion. He not only pursues the work himself but also motivates others working with him to do so.

I have had the rare privilege of receiving his guidance, support and affection for a very long time during my career as a scientist particularly in managing and promoting science in this country. I want to mention here two very brief incidents. As a Senior Scientific Officer of the Department of Science and Technology, for the first time I met Dr. Swaminathan when he was Director General of ICAR, and imagine I had gone to discuss a very small project - All India Coordinated Project for development of blue green algae to be used in rice as a biofertilizer. He encouraged me, guided me, to successfully launch the project. Today, after twenty years it is one of the most successful programs of the Department of Biotechnology. The second occasion, I can never forget. Again only as a Principal Scientific Officer, I was dealing with the work of the Science Advisory Committee to the Cabinet of which Dr. Swaminathan happened to be the Chairman. After the first meeting, the formation of the Committee was announced on March 12th, 1981. The first meeting of the Science Advisory Committee he organized on April 30th; then the meeting with the Prime Minister; meeting of the Cabinet Committee of Science and Technology on May 1st, just within a span of one month. But the part which I have never forgotten is that when the former Prime Minister, late Shrimati Indira Gandhi came to address the scientists, after the meeting he took me to her and said, "Here is the young lady who has done all the work for the Science Advisory Committee and prepared the agenda notes." It was such a rewarding occasion – meeting the Prime Minister and being introduced by Dr. Swaminathan. This is the way he has encouraged young people in this country to do science.

– *Dr. Manju Sharma, President*
The National Academy of Sciences, India

2.4 Excerpts from the Resolution by The Board of Trustees, The Ford Foundation, New York, September 25, 1997.

Monkombu S. Swaminathan joined the Board of the Foundation in 1989. An internationally renowned geneticist and agricultural expert, educator, administrator and policy maker, he brought to our deliberations a unique blend of insight, wisdom and perspective. ...

The impact of the life and work of this modest citizen of the world is reflected in the many tributes, honorary degrees and prestigious awards he has received.

M.S. has devoted his life to many of the subjects that are at the core of the Foundation's program work: sustainable development; education; issues of equity and the role of women; and international collaboration and cooperation. While unassuming, M.S. has never been reticent in speaking with conviction at our meetings, and he has always done so in a generous and supportive fashion. He has pointed out critical connections between different aspects of our work and identified opportunities for cross-regional learning. He has described both local realities at the village level and their global implications. He has emphasized the importance of biodiversity in the plant and animal kingdoms and has brought to our attention significant parallels in the realms of human behavior and international relations. As Trustee and as Chair of the Urban and Rural Poverty Committee and later of the Asset Building and Community Development Committee, he helped shape our programs in poverty and development.

In M.S., intellectual power is combined with an arresting humility. His unfailing kindness and thoughtfulness are appreciated by the Foundation staff at all levels. He has always been cognizant of grantees' perspectives and sensitive to their needs. His judgment and thoughtfulness have made M.S. an unusually effective Trustee and have led to key Committee appointments as well as important assignments relating to the governance of the Foundation. Over the years, he comfortably assumed senior trustee status and became a goodwill ambassador and eloquent spokesman for the Ford Foundation. This was especially evident during the visit of the Board to South Africa in 1996.

As we lose M.S. to the rules of retirement, we have no doubt that he will remain active – indeed overextended – as he continues the important work of his center in Madras and pursues other missions around the world. We thank M.S. for all he has given us. We also salute Mina, who has long been effectively engaged in issues relating to children, families and education, and has graced our program trips with her zest and intelligence. The two Swaminathans are a formidable force for good, and we extend to them our gratitude and warmest best wishes.

2.5 Prof. Norman Myers, Oxford University, February 16, 1999.

Dr. M.S. Swaminathan has had a long and distinguished career in developing-world agriculture. ... But his impact is more far reaching than crop plant genetics. Virtually alone among agricultural leaders, he has shown that the challenge of feeding developing countries is closely tied in with a number of other crucial factors, notably poverty, women's roles and the environment. These further factors broaden agriculture to include concepts of food security, basic human needs and social equity. It is in this larger sense that Dr. Swaminathan is far more than an agricultural pioneer. Indeed, he ranks as one of the world's finest exponents of human development in its proper broad sense – a purpose that always embraces food supplies but extends way beyond agriculture.

For these cogent reasons, Dr. Swaminathan has done much to promote the cause of peace over a period of fully five decades. Although it has not always been recognized as such, "food security" is a key component of security concerns. There is abundant evidence from all three main developing regions and especially from Asia, that hunger breeds economic discord and social strife, leading to political upheaval and often to outright violence.

Dr. Swaminathan is widely recognized as the architect of the original Green Revolution in India. While working at the Indian Agricultural Research Institute during 1954-72, while Director-General of the Indian Council of Agricultural Research during 1972-79, while Principal Secretary of the Ministry of Agriculture 1979-80, and while responsible for Agriculture under the National Planning Commission during 1980-82, he led the way in introducing high-yielding rice and wheat. He did more than any other individual to stave off major food shortages and famine in India at a time when its population was soaring from four hundred and twenty million to seven hundred and ten million people within just twenty-five years. In fact, India moved from being the country with the largest food deficit to becoming self-sufficient in food.

Swaminathan's work also started the agricultural advances that subsequently spread right across southern and south-east Asia. In 1965, rice production in the eighteen Asian countries that grew and consumed most of the world's rice was two hundred and thirty-three million metric tons; by 1986 that figure had soared to four hundred and thirty-two million metric tons, a rise of eighty-four percent in just over two decades. He further promoted these advances during his period as Director General of the International Rice Research Institute in the Philippines. It is not going too far to say that as Swaminathan's strategies for agriculture and associated sectors become still more widespread throughout Asia and other developing regions, we may eventually find that he has saved more people from hunger than any other individual.

Food Systems

While pioneering the Green Revolution across tropical Asia, Swaminathan was already seeking to expand it into an Evergreen Revolution. This meant emphasis on the environmental underpinnings of agriculture: soils, water, natural nutrients, biodiversity and climate. In fact, as far back as 1968 Swaminathan was urging the introduction of environmentally sustainable strategies to go alongside the technologies of the Green Revolution. He highlighted the need for three vital contributions: nutrient supplies of organic as well as inorganic kinds; integrated pest management with less dependence on synthetic pesticides; and science-based management of land and water stocks. In other respects too, he urged that chemical input should be complemented and even substituted by biological inputs. This approach would, he hoped, reduce the risk that intensified agriculture could prove unsustainable because of over-loading of croplands, negligent use of irrigation water pollutant impacts from widespread applications of synthetic fertilizers and pesticides, and a tendency to foster soil erosion. To cite just one breakthrough, he developed a program with the Government of Indonesia to tackle crop pests through the strategy of Integrated Pest Management, which not only raised farmers' harvests but also reduced their costs, and enabled the Government to phase out fifty percent synthetic pesticides with their toxic pollutants.

Experience of the past three decades shows how much his prescience was timely, even though it was little heeded at the time. Whereas the Green Revolution was primarily an agriculture based on resources (land, water, chemicals), his alternative approach was based on science and farmers' knowledge. The inherent value of his approach has been amply demonstrated through his precision farming methodologies in many parts of India and other countries of developing Asia. For instance, he has helped Vietnam to become self-sufficient in food, even to become a food exporter; and in Cambodia he has laid the foundations for the country's agricultural re-construction. He has accomplished similar advances in China, Philippines, Indonesia, Myanmar, Bangladesh and Sri Lanka, and further afield in Egypt, Tanzania and Madagascar.

There was another critical factor at stake, this one extending beyond farms to farmers. The success of the Evergreen Revolution, in Swaminathan's view, would lie with small-scale farmers and their traditional practices. For instance: a wheat crop of six tons per hectare requires 150 kgs. of nitrogenous fertilizer, half of which can be readily supplied by green-manure fertilizer – as is demonstrated by small-scale farmers with their indigenous knowledge. The Evergreen Revolution also depends on other grassroots inputs such as capital assets, meager though they often are among rural communities. In turn, this places a premium on relief of poverty; and in turn again, it requires greater attention to women as the main practitioners of agriculture, women also being the majority of the poor. Some sixty to eighty percent of staple food production in developing countries is supplied primarily by women, yet they receive scant credit for their contributions through crop cultivation and marketing networks.

Farmer Systems

Whereas many agriculture experts know how to produce more food, faced with the challenge of producing twice as much food within the next three decades, agriculture must do better than feed more hungry people, it must feed hungry people more. In order to relieve poverty, then, Swaminathan has coined the term "eco-techniracy" to indicate that the poor and the illiterate can master the latest technical skills – provided they have a chance to contribute what they already know and are enabled to "learn by doing". In order to launch a national eco-techniracy movement, he has established a Science Center for Agricultural Technology, followed by a similar grassroots movement in forestry based on a Forest Science Center to promote "people's forestry". This movement has been articulated through three hundred and fifty Farmer Science Centers (Krishi Vigyan Kendras), comprising thirty million farmer families, the bulk of them women. He has further pursued his convictions on this front while serving as Chair of the U.N. Advisory Committee on Science and Technology for Development. In recognition of his contributions, UNESCO has named its Chair in Eco-technology after him.

The above demonstrates how Swaminathan has devised a "whole systems" approach to agriculture – an approach that goes way beyond producing more rice and wheat and serves agriculture embracing crops, livestock, fisheries and forests. In particular he has developed the concept of Food Security, with two key components: a) Every individual shall have the physical, economic, social and environmental access to a balanced diet that includes both macro- and micro-nutrients, and also entails safe household water and sanitation, primary health care, environmental hygiene, and education in basic skills; b) Food shall originate from environmentally benign technologies that conserve and enhance the natural-resource base of agriculture.

Thus Swaminathan has expanded agriculture's purview from food production to food, security – a world of difference.

Further Specific Initiatives

Swaminathan has mounted a series of demonstration projects to pass his farmer-systems message right down to cropfield level. A prime vehicle is the Biovillages Movement in southern India. This strategy seeks to support the economically and socially disadvantaged sections of society, supplying them with means to escape the poverty trap. He believes the poor are poor because they lack assets such as land, livestock and market skills, their only resource being their labor. Accordingly, his approach to poverty has centered on expanding their capacities through building expertise in areas such as agroforestry, carpentry and biogas. These supplementary activities should be an integral part of agriculture systems writ large. The Biovillages have gained so much acclaim that the concept has been adopted by China, where Swaminathan serves as Principal Advisor.

In similar style, Swaminathan has long been concerned with biodiversity, particularly with regard to its role in the lives of farming communities and the rural poor. These people are more concerned with gross natural product than gross national product. They possess abundant knowledge about native plants, especially wild relatives of food crops, and they are critically placed to ensure their survival – yet they hold next to no rights in these biodiversity resources. As part of his pro-nature campaign, and in parallel with his pro-poor and pro-women campaigns, Swaminathan has developed strategies to reward those rural communities that help to conserve plant genetic resources through their "intellectual property" contributions. It was while he was President of the International Union for Conservation of Nature and Natural Resources during 1984-90 that he became one of the early promoters of a global Convention on Biodiversity, finally achieved at the Rio Earth Summit in 1992.

The prime habitats for biodiversity occur in tropical forests, which links up with Swaminathan's work on "people's forestry". He has put his ideas into practice through the Iwokrama International Rain Forest Center in Guyana, this being the largest effort in the world to engage in forest exploitation that is demonstrably sustainable in the eyes of local people and the wider community alike.

Human Development

Swaminathan believes that conventional development pathways are unsustainable because of three basic deficiencies: they damage life-support systems of the environment; they foster inadequate forms of economic growth, notably "jobless growth"; and they aggravate both economic and gender inequalities. Too often, they help those who can already help themselves, while marginalizing those who are helpless through no fault of their own. They promote "economic, social and technological apartheid", which in turn serves as a major source of endemic and hidden hunger.

Equally important, the deficiencies can even serve as a source of social disintegration and civil strife. This linkage became a theme of the International Commission on Peace and Food chaired by Swaminathan; the Commission's report, entitled "Uncommon Opportunities: An Agenda for Peace and Equitable Development," stated: "Conflicts will become serious if the growing marginalization of the poor is not halted." This assertion was documented by numerous examples from throughout the developing world, in accord with Swaminathan's experiences in countries as diverse as India, China, Philippines, Bangladesh, Egypt, Ethiopia, Nigeria, Sudan, Angola, Mexico and Brazil. The reader can find concrete illustrations in his International Commission's Report.

All this highlights another vital but often-overlooked dimension of agriculture in particular and development in general: that failures can lead to violence, while people-focused development can promote peace. It is a mark of Swaminathan's influence that the principles have been adopted by the Government of Tamil Nadu for eliminating poverty-induced hunger through a Hunger-Free Program across the state of sixty-two million people, this being the first effort in India to eliminate both hunger and sources of hunger. Within a wider context, the conceptual approach has been taken up by the United Nations Development program as a primary ingredient of its campaign for human development.

With particular respect to another poverty problem, unemployment, Swaminathan urges the need for economic growth to be "job led", by contrast with what is becoming a prominent feature of economic growth in many parts of the world, jobless growth (even as economies grow, their work forces shrink). As early as 1980 and through the sixth Five-Year Development Plan of India (for which he was a major drafter), he formulated a policy for Production by Masses rather than Mass Production. He has articulated his view through his Biovillages and integrated farming systems with their employment for those lacking assets such as land and livestock. As he has long proclaimed, "Poverty will persist if the human resource is under-valued and land and material resources are over-valued."

The Humanitarian Cause

It is characteristic of Swaminathan's approach that he opposes a "do good" and patronizing attitude to the poor. Example: at the request of the Government of India in 1982, he converted the Wardha District, where Mahatma Gandhi spent most of his later life, into a "Gandhi District", defining it as a place "where no one is below the poverty line – not because of dole-outs and charity, but because everyone is enabled to earn their daily bread."

When the Indian Government introduced two innovative themes into its 1980-85 Five Year Plan, it asked Swaminathan to write both chapters in question, "Environment and Development" and "Women and Development". For his work on women and gender equity, the International Association of Women in Development in 1984 gave him its first award for supporting women's empowerment – a factor that Swaminathan views as paramount in development processes. In particular, he has engaged in a crusade notably through the U.N. Nutrition Commission, against malnutrition in pregnant women. This malnutrition results in infants with low birth weight and poor brain development, deficiencies afflicting every third child in India and other countries of southern Asia. Swaminathan has termed this "the cruelest form of inequity" since the child is deprived even at birth of the chance to lead a full life. Again, he has been a super-pioneer in this regard: as early as 1967 he started to warn against the "intellectual dwarfism" that has now overtaken tens of millions of children in Southern Asia.

To further put his pro-women convictions into practice, Swaminathan has established the Asian Network for Women in Rice Farming, which seeks to address gender-specific issues in the production and processing of rice. Notwithstanding their vital role in rice farming, women have traditionally been assigned unskilled work of extreme drudgery. The Network has been instrumental in expanding women's skills, opening up credit systems for them, and generally adding economic value to their work. Swaminathan's breakthrough on this broad front has received tribute from the CGIAR which plans to establish a worldwide network for the technical empowerment of women in agriculture, based upon Swaminathan's experiences in Asia.

As a measure of his work in fields related to the relief of hunger, Swaminathan has served as Chair of five of India's National Committees: on Alleviation of Poverty, Environment, Population, Eco-Development of Mountain Zones, and Eradication of Leprosy and Preventable Blindness. As regards population, he has broached the issue within three contexts: social development, environmental carrying capacity, and individual choice. All these three are in marked contrast to the centralized, target driven, and contraceptive-based practices of modules to enable local communities to develop their own socio-demographic charters. In recognition of his approach to population, Swaminathan was invited to be a keynote speaker at the International Conference on Population and Development in Cairo in 1994. As for the Committee on Leprosy, his report marked the beginning of a rapid decline of the disease in India.

A final item. It is entirely like Swaminathan that he was one of the first to publicly criticize India's nuclear tests in mid-1998. This was a bold and courageous initiative insofar as the country's leaders and public were solidly behind the tests.

Swaminathan's contributions have been as productive as they have been pervasive. All of them have been directed at the plight of the hungry and the poor. His five decades of work on several fronts - mostly concerned with agriculture in one respect or another, but reaching far beyond agriculture – has been geared in essence to what one might term "a new development agenda for human well-being and peace – peace in the proper full sense of the word, viz., peace for individuals, peace for communities, and peace for nations. It is in this perspective that his achievements are to be evaluated. During my visits to India I have often heard that he is viewed by many as "a modern Gandhi". Indeed, it is hard to think of another individual who has done as much to enhance the lifestyles of so many – surely billions – of disadvantaged people.

2.6 Sartaj Aziz, Minister of Foreign Affairs, Islamabad, September 13, 1999.

I was delighted to see your name in the list of "100 Most Influential Asians of the 20th Century" in *Time* Magazine of 23-30 August 1999.

To be bracketed with political giants like Mao Zedong, Ho Chi Minh, Sukarno, Mahatma Gandhi and M.A. Jinnah is a great achievement in itself, but for me the more gratifying aspect is the recognition of the role of scientists in changing the course of history.

Having watched and admired your outstanding contribution to the Green Revolution in Asia at first hand, I can say without hesitation that this recognition is well deserved.

2.7 Wakako Hironaka, Former State Minister and Director General of Environment Agency, Tokyo, Japan, January 25, 2000.

Not only is Dr. Swaminathan a world class scholar in the field of agriculture and pioneer in the Green Revolution, he is a humanist trying to improve the lives of the people by applying scientific knowledge to rural India. I personally visited one of the "Biovillages" initiated by the Swaminathan Foundation in the outskirts of Madras and saw happy faces of women, men and children working in the field with sure prospect for the better future. Their shining eyes, enthusiastic talk and smiles are the proof of the success of his humanitarian endeavor.

2.8 Dr. Ismail Serageldin, Chairman, CGIAR, Vice-President, The World Bank, Washington DC, January 27, 2000.

I have known Dr. Swaminathan in my earlier capacity as the World Bank's Vice President for Environmentally and Socially Sustainable Development and as Chairman of the Consultative Group on International Agricultural Research (CGIAR). Dr. Swaminathan is a great humanitarian, an international scientist, and a bold and imaginative visionary. The results of his efforts have had a multiplier effect, helping small farmers to produce more food, increase incomes, and protect the environment. This has resulted in the elimination of major sources of discord and conflict. Dr. Swaminathan's contribution to the development of sustainable agriculture, is unassailable, and is universally acknowledged ... Norman Borlaug himself has said that the "Green Revolution" owes as much to Dr. Swaminathan as to himself. Since then, Dr. Swaminathan has redefined and redirected the scope of agricultural research towards the achievement of an evergreen revolution characterized by a harmonious balance between productivity and natural resources management. He has broadened the concept of agricultural sustainability to include not only considerations of ecology and economics but also of ethics, social and gender equity, and employment generation.

Dr. Swaminathan has also been involved in training a whole new generation of young Indian scientists through his innovative "Biovillage" program that seeks to bring modem technology to rural villages in an environment-friendly and socially-sustainable fashion. Dr. Swaminathan's efforts and achievements represent an important and effective initiative by an outstanding scientist to mobilize the potential of cutting edge agricultural science to meet the needs and aspirations of the poor; and to do so in a manner that helps to maintain peace and security!

2.9 Her Majesty Queen Noor of Jordan, Amman, April 2000.

While I had known of Dr. Swaminathan's work for a number of years previously, it was only when I became a member of the International Commission on Peace and Food, which he chaired, that I had the opportunity to witness firsthand his leadership and dedication to improving global well being. Dr. Swaminathan guided the ICPF through five years of international research and debate which culminated in the publication of the report "Uncommon Opportunities: An Agenda for Peace and Equitable Development", which I had the honor to present to the UN Secretary General in 1994. The goal of the report was to help in accelerating the pace of progress in the evolution of a new human ethic conducive to human beings living in harmony with nature and each other. While hundreds of individuals assisted in the compilation of the report, it was Dr. Swaminathan's drive, vision, his own personal experiences in the field in different countries, and his unwavering natural leadership skills which led to its successful completion. Dr. Swaminathan has devoted his life to identifying ecologically sustainable advances in biological productivity, to insuring food security for all, and to improving the lives of the world's poor.`

2.10 R. Martin Lees, Rector, University for Peace, San Jose, Costa Rica, January 4, 2002.

... I have known Dr. Swaminathan since 1979 in several capacities: first in my capacity as Assistant Secretary General and Executive Director of the United Nations Financing System for Science and Technology for Development, then as coordinator of the China Council for International Cooperation on Environment and Development of which Dr. Swaminathan is a founding Member, and later as a member of the Swaminathan Commission on Food, Employment and Peace.

Dr. Swaminathan is not only a world renowned scientist but he has been remarkably effective in transmuting the results of scientific research into policy decisions at the highest levels of government and of the international system. He has also managed in parallel to maintain a determined and consistent focus on the fundamental problems of the poor, the unemployed, the malnourished and the disadvantaged.

I will briefly outline here only four aspects of his remarkable contribution to world peace.

First, as a world-renowned agricultural scientist. Dr. Swaminathan is widely recognized as the father of the "Green Revolution" which proved to be a turning point in improving the condition of hundreds of millions of human beings, not only in India but throughout the world. Abiding hunger and poverty are the breeding ground of conflict: this transformation of agriculture has proved to be a critical contribution to world peace and progress.

The scale of his contribution can be appreciated as follows. He is now continuing his work to transfer the genes which enable some plants to thrive in a salt water environment to plants which are essential as food, such as rice and cereals. Progress has been good and the prospects look promising. Success would open up vast areas to food production which are currently unusable making a further improvement in world food production on a substantial scale.

Second, Dr. Swaminathan has devoted his life to the reduction of poverty through many channels and efforts, in particular, he himself convened and organized a top-level Commission on Peace, Employment and Food. This private Commission undertook to mobilize world opinion to address widening poverty and exclusion through concerted policies aimed at the development of agriculture while stimulating employment in the countryside and thus the purchasing power of the poor. This unique approach was adopted as the basis of policy by the Government of India and may prove to be a seminal concept for national policy as the population of India increases in coming decades.

Third, while acting effectively through scientific research and at the highest levels of policy, Dr. Swaminathan has achieved extraordinary practical results in addressing poverty at the human, grassroots level. In promoting the elimination of poverty, he is also a strong advocate of responsible policies to preserve and rehabilitate the environment and to improve the condition of women. He has established a Research Center and Foundation - which has now achieved a worldwide reputation – to focus many practical applications of science and technology to eliminating poverty in the villages of India through employment, education and empowerment. Such approaches, for example through the establishment and support of a network of "biovillages" in southern India, have application throughout India and, of course, in other countries.

In confronting directly the real and urgent problems of poverty, gender and environmental degradation afflicting hundreds of millions of people, Dr. Swaminathan has demonstrated remarkable leadership in mobilizing efforts, both national and international, to resolve many profound social injustices, such as the condition of landless laborers, access to education, the empowerment of women, the prevention of child labor and access to health care. His prestige is such that his leadership on these issues, both in promoting action at the grassroots level and as an advocate at the high policy level is both respected and effective.

Fourth, not content with his remarkable achievements to date, Dr. Swaminathan is now committing his energies to bringing the benefits of modern information and communications technology to improve the living conditions, education and prospects of the poor in villages throughout India. Pilot projects are already in operation as a prelude to the wide dissemination of the approach. Like the Green Revolution, this information revolution offers the hope and prospect of rural employment and progress on a substantial scale to meet the needs and aspirations of a growing population. ⊞

71. First World Food Prize 1987: James L. Ferguson, Chairman, Executive Committee of General Food Corporation; A.L. Clausi, Chairman, Council of Advisors; Norman Borlaug; MSS.

72. Receiving the Albert Einstein World Science Award, World Cultural Council, London, 1986

Section 3. Excerpts from Press Coverage

3.1 "Tribute to a Hero of the Green Revolution" by Normita Thongtham, *Bangkok Post*, April 28, 1985.

Last Friday, April 19, at its 43rd graduation ceremony, the Asian Institute of Technology paid its own tribute to Dr. Swaminathan when it awarded him an honorary degree, making him a Doctor of Technology. He was also the guest speaker at the ceremony. …

Despite all the international awards he has received, the IRRI job is the first and only assignment the Indian researcher has taken outside his country. "My whole research career was in India, from 1947 to 1982," he said. "India went through a very interesting phase from a begging bowl to a bread basket, and I was happy to be associated with that very important transformation."

But he won't take all the credit for India's Green Revolution. "That transformation in the 1960s and the 1970s was made possible by the commitment of the political leadership," he said, "especially the late Prime Minister Indira Gandhi, who gave tremendous support to science and technology, particularly agriculture science."

Any project needs three important ingredients for success, said Dr. Swaminathan. One is political will, "for it is the political leadership, or the government who provide the overall guidance and overall support, and decide on priorities and allocations."

Next is professional skill – a vast army of professionals who are conversant with modern science and technology, and who are committed to the cause of using science and technology for national development.

"But even though you have political will and professional skill, still nothing will happen without people's cooperation and participation, in this case farmers' participation," Dr. Swaminathan said.

He joined IRRI in 1983 because, he said, "I am a scientist and at IRRI I will have the facilities to continue my research work in plant genetics. Secondly, rice is an important crop for the whole region, not only India. I feel that in the future it is the rice economy of South and South-east Asia which will determine largely the future of agricultural economy, because rice can grow under a wide range of conditions."

Rice improvement holds the key to ensuring food security in South and South-east Asia, Dr. Swaminathan believes. Hence IRRI, which has been responsible for the development of short-straw rice varieties that have doubled and even tripled rice production, is forever doing research to improve existing strains. While Dr. Swaminathan was in Bangkok, news ten days ago from the Philippines said IRRI had developed a new strain of rice that can yield up to twenty-four percent more grain than common varieties. The new strain, called IR-64, yields up to 7,000 kilograms per hectare more than the IRRI-developed IR-36 and IR-42 commonly grown throughout Asia, the news report said.

Productivity

"What is called the Green Revolution technology has largely been based upon improving the yield or productivity per hectare of irrigated rice," Dr. Swaminathan said. "That kind of area provides more than seventy-five percent of the world's rice today but it constantly faces threats like pests, diseases and soil problems," and IRRI has to be always vigilant to defend, maintain and improve the gains already made.

IRRI has also made experiments to extend the benefits of new technology to less favorable environments, like the deep-water areas of southern Thailand and the drought-prone areas of northeastern Thailand. It has developed floating rices for the deep-water areas, and the rain-fed rices for areas prone to drought.

Dr. Swaminathan believes that for the rest of the century, the famine of jobs and job opportunity will be far more serious than the famine of food in South and South-east Asia. "With the populations still going up, a lot of rural people are unemployed or under-employed. Many women have no jobs, yet unless the woman's income is also added, the family income will be very low.

"Agriculture will not go wrong if farmers are helped, and farmers include the whole farming family. Women play a key role both as decision makers as well as farm workers and farm managers but developing countries' extension service in the past did not fully realize the importance of also taking the latest skills and knowledge to women."

To help rice-farming families, IRRI has initiated a project with support from the Asian Development Bank. Called Prosperity Through Rice, it has three components: showing the farmers how to produce rice at minimum cost; improving their income through mixing farming and crop, livestock and fish production; and producing bio-gas from plant waste materials.

Dr. Swaminathan praised the AIT (Asian Institute of Technology) for its excellent remote sensing center, as well as natural resource and water resource engineering that are the bases of agriculture. "But the greatest contribution from institutions like AIT is the production of manpower – the human resource development. Only trained people have a multiplier effect so you can never measure what the impact of this kind of contribution is". IRRI itself plays an important role in education and training; so far it has trained 3,500 students from eighty rice-growing countries including Thailand.

Although the scientist and his colleagues at IRRI are forever on the lookout for pests and diseases, Dr. Swaminathan does not advocate the use of pesticides. "Our approach is based on integrated pest management. IRRI is famous for its pest-resistant varieties. We make crosses to find rices which have a very high degree of resistance to most pests and diseases, or what you call genetic resistance, so that the farmers do not have to buy any pesticides."

IRRI has 75,000 varieties of rice in its collection, called the Rice Gene Bank, from where the rice scientists draw their materials for cross breeding. Many of the genes are of wild rice strains which occur in forest areas and other places which are continually being destroyed and therefore in danger of vanishing forever.

Other important forms of pest control are through biological control and cultural practices, said Dr. Swaminathan. "In nature, most pests have their own enemies so if you spray indiscriminately, you kill the natural enemies. The important thing is not to spray insecticides unnecessarily but to promote the growth of the pests' natural enemies.

"And plant rice in rotation with other crops, particularly a leguminous crop which can fix nitrogen from the air, to break the pest multiplication cycle."

But farmers do not need fertilizer to produce a good yield, said the rice expert. One ton of rice requires about twenty kilograms of nitrogen, he said, so if you want to produce five tons of rice, your crop has to be given one hundred kilos of nitrogen.

"Usually the soils of Thailand and most of South-east Asia have enough nutrition to support one ton of rice. Then we may be able to add other nutrients to produce two tons through organic matter. But if you want to yield five or six or seven tons, then you have to use some inorganic fertilizer."

Dr. Swaminathan does not altogether rule out the need to import insecticides and fertilizers, but in general the policy must be integrated pest management and integrated nutrient supply, he said. And we should not neglect our own local resources such as manure, natural nitrogen from the air, crop rotations and growing of leguminous crops, in addition to the pest management, which he mentioned earlier. All these should form part of any government's integrated strategy in production planning, he stressed. …

Applied Ecology

"Agriculture is what you call applied ecology," explained Dr. Swaminathan. "Ecology means remaining in harmony with nature, and agriculture means using natural resources in a sound and scientific way. Therefore, anybody who's interested in agriculture, not only for today but for the future, has to think of the conservation of nature and natural resources. My main job is with IRRI, but I know the importance of conservation and it is my duty to share that knowledge."

3.2 "AWID Honors Swaminathan." *News from CGIAR*, Vol. 5, No. 2, June 1985.

Dr. M.S. Swaminathan, Director General of IRRI, received the Association for Women in Development Award for "outstanding contributions to the integration of women in development" at the Association's April meeting in Washington. In accepting their award for Dr. Swaminathan, Dr. C. Jean Weidemann of the Midwest Research Institute quoted from Swaminathan's address to the Bellagio meeting, part of which is excerpted below.

"It is essential that IARCs avoid a laissez faire approach (to women's issues) and move positively and aggressively in the direction of assisting women dependent upon agriculture for their well-being. This is particularly important in the context of the increasing emphasis placed by IARCs on attending to the problems of ecologically handicapped farming areas and economically disadvantaged farm families. It is precisely in such situations that the value of women's labor and income to household happiness and survival is immense.

"How can IARCs help in generating greater opportunities for flexible and productive employment when most of them are not concerned with post-harvest technology and the off-farm employment sector?

"An effective way of responding to this challenge is to capitalize upon IARCs single most important asset, namely access to diverse scientific institutions and political systems. By the very nature of their functioning – through networks, cooperative programs, monitoring tours, symposia and conferences, and training activities – IARCs exert an influence on national research systems, which far exceeds their budget or scientific capability and infrastructure. They have equal access to knowledge and material pool in developed countries. They can hence lead a positive movement of helping women, particularly (those) belonging to small farmer and landless agricultural labor families, through the organization of workshops which can help to compile a portfolio of research and training tasks for each major farming system and getting interested laboratories and scientists, both in developed and developing countries, to adopt specific tasks. In addition, they can set an example in involving women scientists to a greater extent in all aspects of technology development and dissemination.

"While action on the above lines is feasible and should be taken, it is important to recognize that science is not a magic wand with which gender inequalities in workload and economic returns can be made to vanish. This should be emphasized clearly as otherwise false hopes will be aroused about the capacity of science and technology to remove deep-seated social maladies.

"In the ultimate analysis, it is only the concern, commitment and concerted action of national agricultural research systems and policy makers that can lead to meaningful results in imparting a users' perspective in research priorities and strategies. The major role of IARCs should be to trigger a self-propelling and self-replicating pattern of involvement of NARS in R&D efforts designed to give equal attention to the needs of men and women farmers and agricultural labor. Prospects of external funding should not be the main motivating factor for the participation of NARS in networks or studies in this field. There are numerous examples to show that involvement without conviction and commitment leads to the collapse of bilateral or multilateral donor-aided programs when the external input is withdrawn. Enduring benefits will result only when a proper blend of political will, professional skill, and people's participation is achieved within each country."

3.3 "Swaminathan: A Rare Breed of Man" by Normita Thongtham, *Bangkok Post*, July 9, 1987.

The world population is approaching five billion but among them it will be difficult to find another Monkombu Sambasivan Swaminathan. As director general of the Philippine-based International Rice Research Institute he is a banker of a different kind: he collects rice genes, of which Thailand has contributed 4,000 varieties. But that is not the reason why he is a rare breed of man.

The Indian cytogeneticist was recently named the recipient of the General Foods World Food Prize, a newly created international award to recognize, encourage and reward outstanding individual achievement in improving and increasing the world food supply. He will receive the coveted prize on October 6 at the Smithsonian Institution in Washington DC, it was simultaneously announced in Manila and New York recently by General Foods, Inc., sponsor of the award.

Instead of enumerating his own achievements when he was named recipient of the General Foods World Food Prize, he paid tribute to the small farmers of the world, saying, "I would have liked to see the prize go to a farmer, for it is the farmer who toils in the sun and the rain so that the rest of us can exist."

He seemed to have forgotten, that without him, the world's farmers would still be struggling with less productive and disease-prone wheat and rice varieties.

When he came to Thailand in 1985 to receive the Doctor of Technology honorary award from AIT, Dr. Swaminathan told this writer of his self-appointed mission to help fellow human beings, "because it brings the greatest satisfaction, mentally and spiritually. It would be short-sighted to think that we can be happy while millions of people are impoverished and unhappy." The sincerity in those soft-spoken words was most convincing, but his achievements speak even more eloquently.

3.4 "Indian Scientist to Receive General Foods World Food Prize."*Food Technology*, September 1987.

Dr. Monkombu Sambasivan Swaminathan, a sixty-one-year-old internationally renowned agricultural scientist from India, will receive the first annual General Foods World Food Prize on October 6, 1987 at ceremonies to be held in Washington, DC, in connection with a day-long international colloquium on world food issues, cosponsored by the Smithsonian Institution and General Foods.

The $200,000 prize was established to recognize contributions to the improvement of the quality, quantity, and availability of the world food supply. James L. Ferguson, chairman of the executive committee of the General Foods board of directors, said the prize is intended to "serve as a catalyst to stimulate interest, develop knowledge, and encourage international cooperation in the pursuit of solutions" to world food problems.

Dr. Norman Borlaug, a Nobel Peace Prize laureate and chairman of the General Foods' selection committee, describes Swaminathan as "a craftsman of the highest order and architect of the Green Revolution in India".

Swaminathan is best known for introducing semi-dwarf wheat in India to avoid the problems encountered when native varieties toppled over following fertilizer-stimulated growth.

In announcing Swaminathan's selection, Borlaug said, "He has performed with consistent, energetic deftness the roles of innovator, statesman, scientist, and administrator – focusing always on the goal of improving the quantity, quality, and availability of food to national, regional, and worldwide communities." His lifelong efforts have been directed toward and have personified a better life for the farmer, enough food, and good health for all.

Swaminathan was selected from among seventy-four individuals, representative of twenty-two countries on five continents, and eighteen disciplines in food and agriculture, which were nominated by professional societies or other qualified nominating bodies for the award. [...]

3.5 "Hunger Fighter Never Ceases: Inspired by Gandhi, Indian Works to Halt Famine" by Risuko Nakamura, Japan Times, Tokyo, 1992.

Nearing the end of his hectic schedule, Dr. Monkombu S. Swaminathan (67) was very tired on his last night in Japan. And although he had an early morning flight back to India the next day, he willingly accepted the abrupt request for an interview, confessing, "I like media. They can spread the message and influence the public to make the world better. Very powerful." ...

At a seminar titled, "Ending World Hunger – The Opportunity for Global Partnership" held October 5, Dr. Swaminathan stressed that the poor are the first to suffer when the environment isn't properly developed. Pointing out the inseparability of environmental and hunger issues, he said the growing damage to the ecological foundations of agriculture further widens economic marginalization. According to Dr. Swaminathan, the polarization of the rich and the poor is the main cause of existing hunger.

Another important item on Dr. Swaminathan's agenda in Japan was taking part in the October 2 press conference announcing ISME's cooperation with the production of Masashi Yamamoto's film "Kumagusu".

More than a year ago, director Yamamoto ("What's Up Connection," "Robinson's Garden") began to work on a film about the life of 19th century naturalist and biologist Kumagusu Minakata. Kumagusu's reputation as an internationally minded scholar with an encyclopedic knowledge of many academic subjects has recently inspired "Kumagusu gensho," a phenomenal interest in him.

Unfortunately, the shooting of "Kumagusu" was halted mid-way, due to a lack of funds. However, with the blessings of ISME and Dr. Swaminathan, it was announced that production work would resume and the collaboration's goal would be two-fold: to raise not only funds for both ISME and the "Kumagusu" production, but also the public's awareness of environmental issues.

Director Yamamoto jokingly said at the recent press conference that the common thing between mangroves and the "Kumagusu" production is that both are on the verge of extinction. In fact, he's still unsure when, or even if, the film will be completed.

Dr. Swaminathan sagely commented, "We shouldn't be overly ambitious. We just do our duties even though they only make a small change." And for someone whose ultimate goal is to create a world without hunger, Dr. Swaminathan knows about ambition and patience.

He also knows the virtues of global cooperation. At the seminar, he urged Japan to demonstrate at the 1995 UN Summit (its 50th anniversary) that hunger, "the oldest and most persistent foe of humankind," can be solved by both traditional and modern technologies based on principles of ecology, economics, and equality.

He spends most of his time on village work and agricultural study. With what little time he has left, he plays with his grandchildren. While showing the pictures of his grandchildren, he said, "That's my trouble – I don't have enough time!" But in his loving, grandfatherly expression, one finds perhaps the true motive of his devotion. He's staying busy to make a better world, not only for us but also for the next generation.

3.6 "Pioneer of the Green Revolution Fights Poverty with the Internet." *UnescoPress On-line*, June 29, 1999.

M.S. Swaminathan, who helped harness emerging biotechnology techniques in the 1960s to bring about a Green Revolution to alleviate poverty in rural India, is applying a similar strategy again, this time using the Internet. In an address at the opening of the World Conference on Science (WCS) in Budapest on June 26, he said that "new technologies supported by appropriate services and public policies have helped prove the doomsday predictions wrong". And while the agricultural ("Green") revolution "is becoming one of the most significant of the scientific and socially meaningful revolutions of this century", Swaminathan holds out a similar hope for the Internet. His goal today is much the same as thirty years ago – to bring the latest scientific advances into the poorest communities to help overcome the poverty and malnutrition that divide humankind.

For Swaminathan, "time is running out in many developing countries. There is enormous poverty and social disintegration. According to the Oslo Peace Research Institute, of the over ninety ultra-national conflicts in the last eight years, most are related to economic reasons of poverty, unemployment and despair." And, he says, the potential for scientific knowledge is not being used to change this picture. "The new social contract between science and society, the main theme of this conference, is not happening in most of the developing world."

One of the most serious problems, he says, is the twenty-five to fifty percent of children born with low birth weight in developing countries. Where it does not kill, this undernutrition hampers "the full expression of innate genetic potential for physical and mental development," further limiting possibilities for change through the acquisition of knowledge, even if that knowledge were to be delivered.

In the 1960s, the Green Revolution took high-yield hybrid strains of wheat produced by new gene transfer technology and planted them in the poorest farms with the lowest yield. The experiment paid off. While wheat production had reached seven million metric tons at the end of colonial rule in 1947, production based on the new strains rose from ten to seventeen million metric tons between 1964 and 1968. "Four thousand years of progress was repeated in four years," says Swaminathan. And from 1998 to 1999, wheat production in India exceeded seventy million metric tons, a ten-fold increase in fifty years.

The research behind this Green Revolution earned Swaminathan over $500,000 in science prizes, which he decided to plough back into alleviating poverty and malnutrition among the poor, again using the latest technology. Founded about eight years ago, his M.S. Swaminathan Research Foundation (MSSRF) is now setting up experimental Village Knowledge Centers in twenty isolated rural villages, using the Internet. About half the population in many of these villages has a total family income of less than $25 per month. The idea of the project is to bring knowledge on demand to meet local needs.

Project volunteers first poll villagers to find out what knowledge they want. High on the list are women's health information, advice on growing local crops and protecting them from diseases, the daily market prices for these crops, local weather forecasts and clear information about the welfare programs the Indian government provides to aid poor families. This information, together with targeted dietary information, based on an analysis of local nutrition deficiencies, is then entered into databases in a dedicated World Wide Web site. Most important, the information is presented in local languages. Telephones are rare in these rural areas, so access to the Web site is community-based. And where a village is not connected to the electricity grid, renewable energies can be used to power the project computer.

To take part in the project, each village has to provide a room to house the computer as well as the salaries of trained women officers. The choice of women is also a deliberate strategy to redress the "feminization of poverty" through technological empowerment as recommended at the 1995 Fourth World Conference on Women, held in Beijing (China). Women, says Swaminathan, "are the poorest of the poor, being both socially and economically handicapped". The Village Knowledge Centers, he says, have already proved a tremendous source of empowerment. The project has opened up a new world and has increased self-esteem of these women.

Now, says Swaminathan, China and countries in some other developing regions such as West Africa, are interested in setting up similar schemes. But his project has also had an unexpected impact at the leading edge of science and Internet use. Drawing on the experiences of the Village Knowledge Centers, Bruce Alberts, President of the US National Academy of Science (NAS), told WCS delegates that he sees the Internet as a means for scientists to step outside the lab and to interact with ordinary citizens While the public needs a better understanding of science, he says, scientists also need a better understanding of the public.

3.7 "M.S. Swaminathan" by Anthony Spaeth, *Time Magazine (Asia),* August 23-30, 1999.

If you travel by land from any Asian metropolis, it doesn't take long to hit a timeless landscape carpeted with fields of rice, wheat, millet or maize. Whether you're at the terraced rice paddies of Banaue in northern Luzon or the wheat bowl of central China, farmers tend fields as their ancestors did, harvesting grain for their families and countrymen.

But beneath the soil of those seemingly unaltered tableaux is a high-tech invention that changed not only Asia but the world. The seeds planted today by farmers from Punjab to Pusan are nothing like those used by their ancestors. If they were, the entire continent would either be starving or enslaved to the outside world for food or financing.

That turn of history, one of the truly astonishing transformations of the century, is now known as the Green Revolution. It relied heavily on the work of a diminutive Indian geneticist named Monkombu Sambasivan Swaminathan. As godfather of the Green Revolution, Swaminathan, 74, is modest about his own achievements but forthright about his work's impact on his native land and the planet Earth. "Our history," he says, "changed from that time."

Swaminathan, together with colleagues in India and around the world, managed in a few short years to demolish the dire Malthusian world-view that was so prevalent, and pertinent, four decades ago. Asia's populations were growing uncontrollably. None of the largest countries was self-sufficient in food. China lost as many as thirty million people to famine from 1958 to 1962 during and after the Great Leap Forward, and postwar India lived a "ship-to-mouth" existence, subsisting on food-grains imported from the US. Too many mouths, ever more pregnancies, the same farmers growing the same crops – something had to give.

Instead of tragedy, though, a miracle was born in the mid-60s at Swaminathan's laboratory in New Delhi – and, a few years later, at the International Rice Research Institute in the Philippines, which he later headed. Swaminathan brought into India seeds developed in Mexico by US agricultural guru Norman Borlaug and, after cross-breeding them with local species, created a wheat plant that yielded much more grain than traditional types. Scientists at IRRI accomplished the same miracle for rice. Imminent tragedy turned to a new era of hope for Asia, paving the way for the Asian economic miracle of the 1980s and 1990s.

As with all great revolutions, though, the seed was but the starting point. Swaminathan combined all the great components of a revolutionary: vision, dedication, energy and follow-through. ...

Swaminathan's poor, over-populated homeland was importing vast amounts of grain. "Importing food was like importing unemployment," he recalls. "Seventy percent of our people were employed in agriculture. We were supporting farmers in other countries." By 1966, Swaminathan was director of the Indian Agricultural Research Institute in New Delhi, spending his time in fields with farmers trying to help improve their productivity. Fertilizers were a dead end: when the wheat plant's pod grew more seeds, its stalk collapsed under the weight. With help from the Rockefeller Foundation, Swaminathan found a crossbred wheat seed, part-Japanese and part-Mexican, that was both fruitful and staunch. (He would later marry this plant to an Indian variety to produce the golden-colored grain favored by Indians.)

That was the breakthrough in the Green Revolution, but there was a lot more work to be done. Indian farmers, immersed in traditional ways, had to be convinced to grow the new wheat. In 1966, Swaminathan set up 2,000 model farms in villages outside New Delhi to show farmers what his seed could do. Then came the hardest part. He needed the government to help – specifically, to import 18,000 tons of the Mexican seed at a time of fiscal hardship. Swaminathan lobbied then Prime Minister Lal Bahadur Shastri. "He probably thought nothing could be worse," Swaminathan recalls. "Famine was imminent. There was a willingness to take risks." The first harvest with the new seeds was three times greater than the previous years.

But the revolution was still incomplete. Only the Punjab State had the right irrigation for the new technologies, the state-run food collection and distribution networks were notoriously inefficient, and new fertilizers and pesticides were needed, along with credit lines for small farmers. Political leadership was vital to solve that tangle of problems, and Swaminathan found it in Shastri's successor. "Indira Gandhi was a strong nationalist," he recalls. "She wanted an independent foreign policy, and food was a political weapon." Gandhi bluntly asked him how India could be free of imports and gave Swaminathan a free hand to organize a new agricultural program. Today, India grows some seventy million tons of wheat a year, compared to twelve million tons in the early 1960s.

Swaminathan now believes farmers must adopt more eco-friendly methods, and he's using his influence to spread the message. And, although populations continue to mushroom, he maintains that still greater harvests are possible. All that's needed, he says, is "inspiration, perspiration and luck". The greatest stroke of luck for hundreds of millions of Asians has been Swaminathan's revolution.

3.8 "*Time* (Asia) Recognizes Swaminathan Among Most Influential Asians." *Crop Science-Soil Science-Agronomy News*, American Society of Agronomy-Crop Science Society of America-Soil Science Society of America, October 1999.

M.S. Swaminathan has been ranked by *Time Magazine* (Asia Edition; August 23, 1999) as among the twenty most influential Asians of the century. Recognized as the father of the Green Revolution in India, Dr. Swaminathan, who received Honorary Membership in both ASA and CSSA in 1996, used his skills in genetic engineering and his powers of persuasion to make famine an unfamiliar word in Asia. ...

Dr. Swaminathan has often been noted for his understanding of the entire food system. His service in government is testament to this understanding: in several political leadership positions, he established programs of ecological rehabilitation, rural development, and technology transfer. His programs effectively helped farmers reap their fair share of credit and income while conserving national resources.

"Ultimately, it is the political will of the country to have policies in place that will stimulate production by small farmers. Without it, all research, technology ... any external advice will go in vain," said Dr. Swaminathan. ...

In a discussion with C.S. Prakash from the Center for Plant Biotechnology Research at Tuskegee University in Tuskegee, Alabama, Dr. Swaminathan said: "The next Green Revolution in India can come only from biotechnology, and it is important that India catches up with this technology. I recognized this back in 1981 when I recommended to Indira Gandhi that she open a new Department of Biotechnology in the Indian government."

The impact of Dr. Swaminathan's work has reached far beyond the borders of his homeland. His worldwide reputation has made him an easy choice to chair many prestigious international conferences, including the United Nation's World Food Congress in Rome in 1974. His foresight and inspiration had led to the establishment of renowned organizations to promote international scientific collaboration. …

Dr. Swaminathan's enthusiasm for passing on knowledge has earned him a reputation as a lucid educator and a profound humanitarian. In a career dedicated to alleviating human suffering, Dr. Swaminathan has completed the work of many lifetimes and still continues, via his leadership of the M.S. Swaminathan Research Foundation, established with his World Food Prize money.

3.9 "The Volvo Environment Prize." AB Volvo, Press Information, Göteborg, Sweden, October 26, 1999.

Access to know-how is a key factor for sustainable economic, ecological and social development. But how can this know-how benefit everyone? How do we build bridges between those with access to information, and those who need it most – the poor, the illiterate, those whose prime concern in life is to earn enough to survive each day?

For the underprivileged in the villages near Pondicherry in southern India, advances in modern information technology have opened the door to a new world filled with possibilities. Behind the project – the Knowledge System for Sustainable Food Security – is one of the world's leading agricultural scientists, Dr. M.S. Swaminathan. Back in the 1960s, he championed India's transformation from a nation with a chronic food shortage into a country that became self-sufficient in its food requirements. In 1989, he founded the M.S. Swaminathan Research Foundation in Chennai in southern India, where through a number of internationally renowned projects, he demonstrated just how the conditions of poor people can be improved in a sustainable way. Now at the age of 74, he is still immensely active and enthusiastic over the potential for equal development opportunities offered by the new technology. Dr. M.S. Swaminathan was awarded the Volvo Environment Prize in October 1999 for his invaluable innovations and unwavering dedication.

"My feeling is that the world has enormous opportunities today because of modem science and technology," says Dr. Swaminathan. "Take information technology, modern biotechnology, space technology. Or take renewable energy technology, whether it is solar, biomass, biogas or wind. And finally, take management technology, putting all the pieces together in the form of eco-technology, which means ecological, economical and socially equitable technology."

However, in the fast-growing world of global know-how, access to information is far from equal for everyone.

"Knowledge should be available for people's well-being. As Einstein said, if the product of a brain does not help the deprived, it will be more of a curse than a blessing. That's why modern technological empowerment of the poor has become so important. There we have a great opportunity through modern information technology. And the people that are to be reached in particular are women and children."

Dr. Swaminathan feels that our ability to create long-term support potential for all the peoples of the world is the one vital parameter determining whether or not we can ensure sustainable development. We cannot blame poor people if they cut down the forests near their homes, if there is no alternative fuel for them to use.

"Apart from the earth-centric concept of sustainability – the earth and its life supporting-systems must be preserved – at the same time the human inhabitants, the humanistic form of sustainability is going to determine the future of our planet.

"Knowledge System for Sustainable Food Security is a very interesting example of how to bridge the gap between those who have information and those who need it most. In each of the twenty villages included in the project, there is a so-called Village Knowledge Center with a computer linked to the Internet and various databases providing information in the local language. The village inhabitants can receive information about such things as women's health care, advice on ecological ways of growing local crops and protecting them from diseases, daily market prices for these crops, local weather forecasts and so on. The system also provides them with information about programs provided by the Indian government to aid poor people, that is to say know-how that can assist people in their daily lives. These Knowledge Centers are managed by a few of the village women who have been trained for this purpose. If the information centers had been managed by men, many women would have hesitated to turn to them. By allowing women to take responsibility for the new technology, we're ensuring that everyone has the same access to information. And at the same time, we're helping to boost women's self-confidence and promote their place in society. We also know that measures that favor women also bring about improvements for the family as a whole. Women have an immense need for information on various health-related issues. Women often do not want to confide in the local doctor, who is usually a man, particularity when it comes to reproductive health problems," says Dr. Swaminathan.

This meant that many women simply did not receive any help at all. However, because the information centers are run by women, women feel more comfortable about asking questions and taking advice on issues relating to their health.

"In the area covered by these villages, there are more than one hundred and forty different state-run development projects aimed at the underprivileged. The local population has been largely unaware of many of these projects, not least owing to the high illiteracy rate. Now however, people can turn to the information centers in their villages to find out which projects are running and which are relevant to their specific needs."

Something as commonplace as having access to local weather forecasts may be of crucial significance. "For the people in the fishing-villages along the coast, for instance, the large- scale weather forecasts shown on TV have no relevance at all," says Dr. Swaminathan. "What they need to know before venturing out in their small vulnerable boats is what the weather is going to be like in their particular stretch of water. Being forewarned of an approaching storm may mean the difference between bringing home a catch or not returning at all. Through the information centers, they can receive an up-to-date local weather forecast every morning. So we find that these knowledge centers are having an impact on their total life. Not only in economic life, not only in terms of benefiting from their entitlements, but also in the information that affects their quality of life. We want to make this virtual college into a new social contract between those who have information and those who need it badly – especially the rural illiterate and the unprivileged, and particularly women."

The Knowledge System for Sustainable Food Security can be a source of inspiration of immense global significance. This is the view of experts such as Dr. Bruce Alberts, President of the US National Academy of Sciences, who visited the villages and the M.S. Swaminathan Research Center in January this year.

"Drawing on this concept, I envision a global electronic network that connects scientists to people at all levels - farmers' organizations and village women for example. The network will allow them to easily access the scientific and technical knowledge that they need to solve local problems and enhance the quality of their lives, as well as to communicate their own insight and needs back to scientists."...

Dr. M.S. Swaminathan is awarded the 1999 Volvo Environment Prize for his unique contributions in the wider environmental field. The 1.5 million kroner prize will be awarded at a ceremony in New York on October 26.

The motivation of the prize committee was as follows: "The Volvo Environment Prize is awarded to Dr. M.S. Swaminathan because of his achievements as a plant breeder and administrator which have led to dramatic increases in crop yields, his international leadership in agriculture and resource conservation, his deep concern for the poor and disadvantaged, and his continuing research and leadership to ensure that they get the opportunities they need to develop in ways that enhance the natural environment on which they depend."

3.10 "The Modern Gandhi's Message of Hope for Agriculture and Environment" by Heather Slater, Crawford Fund Seminar, Belconnen, Australia, November 24, 1999.

World renowned specialists in agriculture, the environment and development addressed an important international seminar attended by over two hundred senior representatives from government, industry, NGOs, science, academia and the general public in Parliament House in Canberra today (November 24). Organized by the Crawford Fund and entitled "The Food and Environment Tightrope", the seminar was opened by the Minister for Foreign Affairs, The Hon. Alexander Downer.

Chairman of the Crawford Fund, The Hon. Tim Fischer, welcomed keynote speaker Professor M.S. Swaminathan as an acknowledged world leader in the field of sustainable food security, an international authority on biotechnology in food, a humanitarian and, according to *Time* magazine, one of the twenty most influential Asians of the century. "The Crawford Fund is honored that Professor Swaminathan is able to accept our invitation to come to Canberra. In addition to delivering his address to you today, he has a series of high-level meetings scheduled with Federal MPs, senior civil servants, and fellow scientists, I am sure that they will gain much from their discussions with him," he said.

Having set out the seemingly impossible challenge for agriculture to feed the world's burgeoning population with less arable land and irrigation water without destroying the environment in his keynote address, "Walking the Tightrope", Professor Swaminathan, winner of the World Food Prize and numerous international environment awards, delivered a message of hope. "A world without hunger is now within our reach if every nation pays attention to improving food availability through ecologically sustainable methods of production, to enhancing economic access to food by providing a job-led economic growth strategy, and to ensuring the biological absorption of food in the body through the availability of safe drinking water and environmental hygiene," he said.

Professor Swaminathan provided examples of the profound impact of science and technology (S&T) in every field of crop and animal husbandry, inland and marine fisheries and forestry. "Science can be mobilized for raising further the ceiling to biological productivity without the associated ecological harm that has occurred in the past. This productivity advance can be sustainable over time if it is rooted in the principles of ecology, economics, social and gender equity and employment generation," said Professor Swaminathan.

"Future agricultural production technologies should be based on the foundation of integrated natural resources management and this is where Australian experience and expertise will be of particular value," he said. Professor Swaminathan discussed the three major revolutions in S&T that are the basis for his optimism for the future of agriculture and the environment: a) The gene revolution which is developing new processes and products for agriculture, industry, the environment and human and animal health; b) The ecotechnology revolution which promotes the blending of the best in traditional knowledge and technology with frontier technologies such as biotechnology, space and information technologies, renewable energy and new materials; c) The information and communication revolution which allows a very rapid growth in the assimilation and dissemination of relevant and timely information.

The Professor provided examples of how a range of organizations such as his own research institute, the centers related to the Consultative Group for International Agricultural Research and Australia's own Australian Center for International Agricultural Research are bringing these revolutions to the poor.

Professor Swaminathan congratulated Australia's development assistance program in its role in promoting the agricultural renaissance of his homeland India: "The attitude of the development agency should be one of sustained partnership and not of patronage. The policy of the Government of Australia to strengthen concurrently both national agricultural research centers and international agricultural research centers is the most meaningful pathway of development assistance."

Professor Swaminathan called on governments to make the safeguarding and strengthening of the livelihood security of the poor as a major goal of liberalized trade. "The current trend of increasing the rich-poor divide will have to be stopped if social conflicts are not to increase," he said. "Australia has provided innovative leadership in developing international institutional structures and development assistance procedures and should intensify its efforts to foster relevant partnerships in the Asia-Pacific Region for launching and sustaining an evergreen revolution on farms," he continued.

In a discussion about the international scene and the Climate Convention, the professor called on Australia to play a catalytic role in fostering cooperation in both avoiding and mitigating the adverse impact of climate change in the Asia-Pacific region.

Referring to the various international conferences and conventions held in the last decade Professor Swaminathan highlighted that most of the talk is over. "We are really walking a tightrope in terms of achieving sustainable solutions to the problems of population, poverty and environmental degradation. It is now up to nations to act so that the uncommon opportunities now opened up by science and technology and democratic systems of governance for creating a food secure world are not missed," he concluded.

3.11 "M.S. Swaminathan: Transforming India from Begging Bowl to Breadbasket" by Ajoy Sen, *Asian Entrepreneur*, January 2001.

On a winter afternoon in 1967, Prime Minister Indira Gandhi was impressed by what she saw outside a village near Delhi – a huge landscape carpeted with wheat, maize and millet crops. Earlier, little or nothing grew in those fields.

A large crowd of farmers at Kanjhawala village started singing a welcome song. Indira Gandhi – India's most charismatic leader after Mahatma Gandhi and Nehru – was amused because for a change, she realized the villagers were not singing in her honor. Instead, they were singing in honor of the short, bashful man sitting next to her on the makeshift platform. When the singing had finished, the village pradhan (chieftain) approached Gandhi and asked her to put a medal around the neck of India's top agricultural scientist. This scientist has become the godfather of millions of Indian farmers by transforming the country into a breadbasket with plenty to eat and sell. …

He has had the satisfaction of seeing his work bear fruit. Asked for his definition of success he notes that, "You have a vision and if you are able to develop the ingredients for the success of that vision, I will consider it a success. If something goes wrong, I am not discouraged."

His present aim is "to show, through bio-villages and other projects, the methodology by which agriculture can be an instrument, not only for more food but also for more jobs and income for small farmers. The bottom line is jobs for Indians."

This has been a key source of motivation. "The advice from my mother was most important," Dr. Swaminathan recalls, "She used to say work hard and never get discouraged. Work without expectations of reward."

However, the rewards and the awards have been many, but despite that, the appreciation of the Khanjhawala farmers on that winter afternoon in 1947, remains a highlight. "The love the farmers displayed to me on that occasion was the most touching event in my life," says Dr. Swaminathan.

3.12 "Genetically Modified Foods Fill Developing World Silos" by Andrew S. Holbrook, *Harvard Crimson,* March 12, 2001.

M.S. Swaminathan, architect of the Indian "Green Revolution" in the late 1960s that doubled crop yields in just four years, painted a confident picture on Friday at Harvard University of sustained rises in agricultural output in the developing world, fueled by genetically modified foods and other new technologies. The Green Revolution introduced superior plant breeds and increased rice and wheat farmers' productivity. It has been widely credited with having prevented famine in Swaminathan's home country.

In a lecture at the Science Center sponsored by the University Committee on Environment, Swaminathan outlined what he called an "evergreen revolution" that would produce continued productivity gains but prevent farmers from exhausting the soil and water supply. "There is no complacency. There is no time to relax," he said. "One must defend the gains that have already been made. ...We must make new gains in farming systems."

Environmentalists and others have objected to a key element of Swaminathan's plan – genetically modified foods – saying the ecological impact is uncertain. Already, Britain, Germany, Switzerland, Brazil and other countries have curbed farmers' use of these modified crops. Though he said no such crops have been approved for planting in India, Swaminathan predicted genetically engineered crops would gain approval there within the next five or six years. He said they would cause crop yields to climb consistently over the next fifty years. Genetically modified foods could also have nutritional benefits, he said. For example, he said that "golden rice" which is enriched in beta-carotene, could help to combat vitamin-A deficiency, which kills a million children worldwide every year. But Swaminathan cautioned that genetically modified products are not cure-alls that will eradicate health problems, such as children's blindness. "There's a lot of hype in the media," he said. "The over statements have hurt the cause. It is not possible to achieve all that."

He called for more research on other crops, as well as poultry and fish, which he said are essential to ensuring the poor eat well-balanced diets. Swaminathan also said scientists will have to work with social scientists to resolve ethical concerns over genetically engineered organisms.

As he talked of "reaching the unreached," Swaminathan said major social divides, especially gender gaps and unequal access to technology, threaten developing countries. He described "micro-planning" projects he and his team of researchers have carried out to begin closing those gaps, such as low-cost greenhouses and wireless Internet connections.

After his lecture, Swaminathan faced questions challenging his reliance on individual farmers in an era of large, dominant multi-national agricultural firms. Though he acknowledged, "you can't beat them," Swaminathan said he believed his small-scale projects would set a self-replicating example for the entire country.

"I was very surprised by how optimistic he was," said McKay Professor of Environmental Engineering Peter P. Rogers, who attended the lecture. According to Rogers, who has studied natural resources in India and elsewhere, India's water table is falling and some studies have suggested demand for water will far outstrip supply within twenty years. Since water is key to agriculture that could mean a gloomier forecast for continued rises in farming output, though Rogers said he did not necessarily dispute Swaminathan's outlook. "He's always been upbeat," Rogers said. "He has reason to be optimistic. He's done it one time before."

3.13 "The Face of Food Security" by Gnana Bharathy, *Indiatogether*, May 5, 2001.

When Monkombu Sambasivan Swaminathan was born on the 7th of August, 1925 in Tamil Nadu, not many would have imagined the great changes that would sweep India in his lifetime. The struggle for independence lay unfinished, its promise still two decades from fulfillment. After the dawn of freedom, the country would be recast in the image of its own people, and would struggle to find that first measure of her independence, self-sufficiency in food production. As the population rose dramatically, even this much would become too much to ask of the new nation, and in the accomplishments of the geneticist from Tamil Nadu, the quest for food security would be fulfilled.

Food security is a notion that many of us, drawn from middle and upper class India, have only noted in passing as a matter of some concern; its potential ravages rarely threatened us directly. Nevertheless, there is much for us to celebrate in Dr. Swaminathan's life, providing as it does for us a glimpse of how the accomplished and the able can dedicate their lives to both fulfilling professional endeavor and gratifying social service. More than an exciting scientist, Dr. Swaminathan has led a life of celebrated humanitarianism along the way, receiving repeated and remarkable honors for his work.

A major reason why this has been possible is that Dr. Swaminathan has, over the years, consciously and consistently broadened the purposes to which he has applied his learning as an agriculturist. Despite his conventional education as a geneticist at Coimbatore Agricultural College, the Indian Agricultural Research Institute, and Cambridge University, he has gone far beyond the traditional understanding of agriculture – as a food creation process. In doing so, he has embraced ideas such as food security, social equity, gender equality, and the satisfaction of basic human needs. The value of these ideas cannot be overemphasized; they go a long way in reducing economic disparities and minimizing social conflict.

By linking these ideas to his scientific expertise in induced gene mutations and the creation of high-yield hybrid varieties, Dr. Swaminathan fostered and produced the Green Revolution, turning India from her status as the nation with the world's largest food deficit to a state of self-sufficiency in less than twenty-five years. The expansion of his ideas, and their application to other nations in Asia, doubled the output and consumption of traditional staples, especially rice, in southern and south-east Asia in a span of twenty years.

3.14 "Swaminathan to Head Pugwash Movement." *The Hindu* (Delhi), March 18, 2002.

Food Scientist, M.S. Swaminathan, has been elected to lead the Pugwash Conferences on Science and World Affairs for the next five years. This is the first time an Indian has been chosen to preside over the movement launched nearly five decades ago by Albert Einstein and Bertrand Russell.

At the 51st annual session of the movement here [Agra], it was announced that Prof. Swaminathan would take charge of the assignment later this year. The Pugwash movement won the Nobel Peace Prize in 1995 for its contributions to promotion of nuclear disarmament. The outgoing president, Michael Atiyah, welcomed the selection. "Prof. Swaminathan would help expand the Pugwash agenda in the years ahead to encompass issues related to health and welfare of the world's people in addition to the traditional Pugwash goals of resolving conflict and eliminating nuclear and other weapons of mass destruction," he said.

Section 4.
Selections from Citations of Honorary Degrees and Awards

4.1 "Award of the Silver Jubilee Commemoration Medal" by The Indian National Science Academy, 1973.

After getting a Ph.D. degree in genetics from the University of Cambridge in 1952, Dr. Swaminathan joined the staff of the Indian Agricultural Research Institute, New Delhi where he along with a band of devoted research scholars made significant contributions, during the fifties, in the fields of genetics and cytogenetics with particular reference to wheat. This school of plant cytogenetics soon became one of the leading ones in the country and Dr. Swaminathan's work during this period was recognized by the award of the Shanti Swarup Bhatnagar Award for Biological Sciences in 1961. Later, he and his co-workers concentrated on breaking the ceilings to high yields in crop plants and they used all the techniques genetics offered for achieving their goals as speedily as possible. Thus, introduction from Mexico of dwarf wheats, conventional hybridization and selection, back crossing, exploitation of hybrid vigor, induced mutagenesis and induced polyploidy, were all employed with great effectiveness in different crops and different situations. This work led to the initiation of the High Yielding Varieties program in 1966, a program designed to introduce a synergistic package of practices in farmers' fields.

Dr. Swaminathan's contribution to the science of genetics was recognized by his nomination as a Vice-President of the International Congress of Genetics in 1963 and by the award of the Mendel Centenary Award by the Czechoslovak Academy of Sciences in 1965, he being the only Asian geneticist chosen for this award. His contributions to applied botany were recognized by the Indian Botanical Society through the award of the Birbal Sahni Medal in 1966.

Dr. Swaminathan's passion is the scientific upgrading of Indian agriculture. He has fostered cooperative endeavor among a wide group of scientists belonging to different disciplines and institutions. He has stood for an integrated approach to the study of problems and has regarded the goal rather than the discipline as sacred. He initiated the concept of National Demonstrations, through which scientists demonstrate themselves the validity of their experimental findings in farmers' fields. It was these demonstrations which led to a widespread clamor for seeds of the dwarf wheats in 1966. He and his colleagues organized an entire village in Delhi as a "seed village" in 1965.

Dr. Swaminathan is a scientist who takes great pleasure in teaching and in training students. He believes in the dictum that a teacher is known by his students. Judged by this yardstick, his students are doing outstanding work in different countries.

His practical contributions to agriculture were recognized by the Government of India through the award of Padma Bhushan in 1972. His work in seed improvement led to the Swedish Seed Association electing him as its first honorary Fellow from India. The Sardar Patel University at Vallabh Vidyanagar awarded him an honorary D.Sc. degree in 1970. He was chosen by the University Grants Commission to deliver the first Zakir Husain Memorial Lecture in 1970.

The Ramon Magsaysay Foundation of the Philippines awarded the Magsaysay Award for Community Leadership in 1971 for his contributions as "scientist, teacher of farmers and students and administrator towards generating a new confidence in India's agricultural capabilities." The significance of this award will be clear from the fact that the first award for Community leadership was given to Acharya Vinobha Bhave.

Dr. Swaminathan has regarded helping the country to convert scientific findings into practical accomplishments as the foremost duty of every Indian scientist. Early in 1972, delivering the Rajendra Prasad Memorial Lecture of the Indian Society of Agricultural Statistics on "Can we face widespread drought again without food imports?" he warned that unless serious anticipatory steps are taken, complacency will lead the nation into serious difficulties, a prediction which came true in that very year.

Dr. Swaminathan served as Secretary of the Academy from 1969 to 1972 and gave much of his time to the operation of the Basic Sciences Project. The Academy is happy to recognize his many-faceted services to Indian Science through the Silver Jubilee Commemoration Award.

4.2 "The Borlaug Award" by Coromandel Fertilizers, India, September 4, 1979.

Monkombu Sambasivan Swaminathan
Servant of Agriculture

In profound appreciation of his catalytic role in providing deep insights and inspiring fellow scientists to set goals, share experience in the process of social change and transformation to a society which treats of Man at the center.

For evolving a strategy for agriculture rooted in science, but tempered by concern for ecology and human values.

For the amplitude of his perceptions which has encouraged community effort directed to a synthesis in the movement of agriculture.

4.3 Honorary Doctor of Science degree, University of Wisconsin, 1983.

Citation: For over a quarter of a century, M.S. Swaminathan has been a major force in shaping the Indian government's policies in agriculture and environmental affairs. Equipped with post-doctoral work at the University of Wisconsin, he has dedicated his career to preventing famine and ameliorating hunger in his homeland and throughout the world. As a geneticist, he worked for the development of strains of wheat, rice and coarse grains that would flourish in the ecological setting of India. ...

As a result of his having put science at the service of his people, India, which seemed doomed to mass starvation by this period of its history, now feeds itself by its own efforts. Dr. Swaminathan has been at the center of one of the notable human achievements of our generation.

Conferral: Monkombu Sambasivan Swaminathan: For your brilliant work in science, which has helped to sustain a great nation, the University of Wisconsin confers on you its honorary degree, Doctor of Science.

4.4 Doctor of Science (Honoris Causa), University of the Philippines, April 8, 1984.

In recognition of his outstanding contributions to India and the Third World as scientist, educator, agricultural planner and administrator, community leader, international public servant and leading exponent of modern peasant agriculture;

For his sterling achievements in the field of science, particularly in basic and applied plant genetics that resulted, among others, in the breeding of sturdier, more productive and better quality food crops;

For bringing science to bear on one of the most acute problems facing India in this century – the threat of mass starvation and death in the mid-1960s because of a severe drought – by causing the introduction to India of the Mexican dwarf Sonora 64 and Lerma Rojo 64 wheat varieties, improving on them, and convincing India's tradition-bound farmers to adopt them, thereby making possible the Green Revolution on the Indian subcontinent and making India self-sufficient in wheat for the first time in its modern history;

For his enviable record of leadership in the field of Science, having served as president of no less than nine Indian professional societies, and as a fellow in as many as nine academies of Science all over the world and for his leadership in international bodies, including service as Chairman, UN Advisory Committee on Science and Technology for Development, as President, International Federation of Agricultural Systems for Development (IFARD), and as Director General of the International Rice Research Institute;

For his work as a gifted agricultural educator, who has imparted the virtues of selfless service and self-reliance to the scores of doctoral and masteral students who have studied under his wing, as well as to the thousands of farmers who have benefited from his teaching;

And for giving others the splendid example of his life as scientist and humanist – a life of integrity and humility, of self-abnegation and total commitment to the welfare of others, especially the poor farmers of Asia, bringing to his work a missionary zeal which is infectious and a sense of dedication which is inspiring, the Board of Regents, by unanimous vote of its members and upon recommendation of the University President and the Committee on Honorary Degrees, today confers upon you the degree of Doctor of Science, Honoris Causa. ...

4.5 Honorary Doctor of Technology, Asian Institute of Technology, Thailand, April 19, 1985.

Mr. President, I am honored to present Dr. Monkombu Sambasivan Swaminathan, Director-General of the International Rice Research Institute, for the conferment of the honorary degree of Doctor of Technology. The Board of Trustees has approved the award in recognition of Dr. Swaminathan's outstanding contributions to the academic and technological development of this region of the world. Indeed, Mr. President, his achievements have been so numerous that I cannot possibly mention all of them in the short space of time allowed to me in this address and I beg his forgiveness for these necessary omissions.

The Asian Institute of Technology exists to find solutions to the technological needs of Asia. There is probably no more important need than the requirement to attain and, in some cases, maintain food self-sufficiency for this region. If present population growth rates continue there will be twice as many Asians alive in twenty-five years time as there are today. The problems of feeding this vast population will be enormous, particularly since most of the land, which is good arable land, has, with few exceptions, already been put to the plow. The only possible solution to this problem lies in the intensification of agricultural production. Either more crops per year must be grown or the yield of existing crops must be increased. It is in this second area that Dr. Swaminathan has made such outstanding contributions. He has, in fact, devoted the whole of his professional career to improving the yield of the staple food grains of this region, and he has been tremendously successful in this endeavor. ...

4.6 "Award for Outstanding Contributions to the Integration of Women in Development" presented by Celia Jean Weidemann on behalf of the Association for Women in Development, Washington DC, April 27, 1985.

This is the first year that the Association for Women in Development is presenting awards for outstanding contributions to the Integration of Women in Development. It is a great privilege for me to announce the award being given to Dr. M.S. Swaminathan, Director General of the International Rice Research Institute (IRRI) in the Philippines. Dr. Swaminathan deeply regrets that unforeseen meetings with IRRI's donors made it impossible for him to be here with us today.

As many of you know, IRRI is one of the networks of thirteen International Agricultural Research Centers (IARCs) devoted to the study of various crops and livestock. One of the Centers, CIMMYT in Mexico, was in large part responsible for the Green Revolution. The Centers receive multi-lateral, bi-lateral, national, and private funding, often from foundations such as Ford and Rockefeller.

The Centers shape in large measure the agricultural research agenda for the world and have a profound influence on national level research in agriculture. The Centers generate basic research directed primarily at increasing crop yields, which is then adapted by national research centers. It is rare indeed for a Center Director to concern himself with what might be viewed as a very user-oriented research perspective.

I first met Dr. Swaminathan in 1982 at the annual meeting of the Directors General of the International Agricultural Research Centers. I was giving a report with Barbara Knudson on some work which Jane Knowles, Susanna Lastarria, Susan Poats, and I were doing to engage the Centers' interest in Women in Development issues. Dr. Swaminathan, it turns out, had already taken the leadership and was planning an international conference for the following year on the topic of Women in Rice Farming Systems. In 1983, I attended that conference of approximately one hundred women from rice producing countries around the world. It was my experience of Dr. Swaminathan that prompted me to nominate him for this award. I heard his opening address and expected this would be the last we would see of him at the week-long conference. Instead, with his office down the hall and the demands of administering an international center tugging at him, he chose to spend most of every day in the conference room with us – listening, contributing, and learning what the participants had to say about women and rice farming.

In the evening, he sat with a working group on his porch and searched with us for ways to overcome constraints faced by women in using the rice technology developed by IRRI. It was one of my few experiences of a man in a position of global influence being truly committed to WID at a deeper level than political expediency, and of publicly and continuously working on that issue. Dr. Swaminathan has labored tirelessly on this topic, with particular attention to the poor and landless women rice farmers in Asia. With the support of the Rockefeller Foundation, under the leadership of Dr. Joyce Moock, he was instrumental in organizing a seminar in March 1985, on Women and Agricultural Technology to focus the attention of all of the Center Directors on WID issues in agricultural research. Earlier this month, he worked with the Ford Foundation and International Development Research Center of Canada to sponsor at IRRI a Project Design Workshop on Women in Rice Farming Systems. IRRI is the only international agricultural research center with two positions devoted to WID issues. Dr. Swaminathan is currently working with the national level agricultural research networks throughout Asia to focus their attention on the impact of technological change on women in the rice sector. ...

It is with deep affection and great pride that we recognize Dr. M.S. Swaminathan as a global leader, who is working with us to integrate WID into the world's agricultural agenda.

4.7 The Tyler Prize, Los Angeles, 1991.

M.S. Swaminathan's mission in life has been to increase biological productivity on an ecologically sustainable basis. One of the world's most eminent agricultural scientists, Dr. Swaminathan has single-mindedly devoted himself to research in genetics and breeding to discover genetically superior strains of wheat, rice and coarse grains to enhance the productivity and stability of food crops in India and other parts of the Third World. While searching for methods to achieve higher production, he has been sensitive to the need for populous, land-hungry countries like India to protect forest land from being destroyed and to adopt "land saving agriculture".

4.8 "Award for International Cooperation on Environment and Development" by China Council for International Cooperation on Environment and Development (CCICED), Beijing, September 23, 1996.

China Council for International Cooperation on Environment and Development hereby bestows this, its highest award, upon M.S. Swaminathan for his outstanding contributions to the lofty cause of environmental protection and development, and for his signal accomplishments in the field of international cooperation.

4.9 "Planet and Humanity Medal" by The International Geographical Union, Vienna, 2000.

The International Geographical Union presents the medal "Planet and Humanity" to Professor M.S. Swaminathan in recognition of his unique success in outstanding scientific research and its application, leading to Asia's Green Revolution. His endeavors to combat hunger and food shortages by promoting new seed varieties and applying these with ecologically sound principles and sustainable agriculture are all part of his profound humanitarian ethos, which reminds scientists and politicians worldwide of their responsibilities for stewardship of nature and humanity on our common Plant Earth.

4.10 Honorary Doctor of Science, University of Massachusetts, Boston, June 2, 2001.

Today the University of Massachusetts, Boston, is conferring the honorary degree of Doctor of Science on Prof. Monkombu Sambasivan Swaminathan.

In his letter to Prof. Swaminathan, Dr. David J. MacKenzie, Chancellor of the University, says that Swaminathan was chosen for this honor "in recognition of your nonpareil achievements as a scientist, global citizen and leader in the application of science to major world problems. Your list of honors is as extensive and deep as that of any other living person and there is perhaps nothing I might say in praise that has not been said before many times over." In particular, the university wanted "to recognize your efforts for the underprivileged and disenfranchised."

Here are a few more excerpts from the Chancellor's letter.

"By recognizing your leadership and India's, we express our hope that India's role on the world stage increase and that India's particular genius for embedding values of the spirit in daily life become more prominent on that stage.

"We share your values in all that you have accomplished for emerging societies and emerging peoples within those societies. By honoring your work we seek to further those values, here, in India and everywhere.

"We recognize also the global nature of your achievement. Benefits of your work extend worldwide and have been recognized in countries around the globe. We honor the magnificent inclusiveness of your concerns, by nation, socio-economic group, gender, inter-generational, and including both human and natural environments." ▯

Section 5.
Selected Prefaces, Forewords, Introductions and Reviews of M.S. Swaminathan's Books

5.1 *Biotechnology in Agriculture: A Dialogue*, ed. by M.S. Swaminathan. Madras, India: Macmillan India, 1991.

Technology, training and trade have been prime movers of economic prosperity in industrialized countries. Progress in technology, particularly in the broad area of biotechnology covering fields such as tissue culture, hybrid seeds and farm animals, genetic engineering, biopesticides, biofertilizers, vaccines, sero-diagnostics, aquaculture and bioprocessing has been spectacular during the eighties. Past experience with new agricultural technologies in India reveals that it is important to subject new tools and techniques to a proactive impact analysis from the points of view of ecology, economics, equity and employment, if they are to lead to sustainable improvements in the quality of human life.

In the sixties, a major challenge was achieving food security at the national level. Today the challenge is achieving food security at the household level. Achieving this goal will be possible only by generating more opportunities for skilled jobs on the one hand, and by linking the livelihood security of the farm family with the ecological security of the farm, on the other. Biotechnology has opened up opportunities for working towards these goals. A pro-poor bias will however be necessary to ensure that the economically and socially disadvantaged sections of the rural population derive benefit from such opportunities.

This book is a chronicle of an interdisciplinary dialogue held at Madras from January 22 to 26, 1991 on methods of taking the benefits of biotechnology research to the rural poor. I am grateful to Ms. Mini Krishnan, Editor, Macmillan India for her patience, perseverance and thoroughness in editing the verbatim records while helping me to prepare a publication which, I hope, will be of particular benefit to scientists and scholars striving to impart a pro-poor thrust in technology development and dissemination.

Foreword by M.S. Swaminathan
November 1991

5.2 *From Stockholm to Rio de Janeiro, The Road to Sustainable Agriculture*, by M.S. Swaminathan. Madras: M.S. Swaminathan Research Foundation, 1991.

In January 1968, when there was evidence that India was on the threshold of making impressive progress in improving wheat production as a result of the popularity of the semi-dwarf wheats introduced from Mexico, I made the following observations in my Presidential Address to the Agricultural Sciences Section of the Indian Science Congress held at Varanasi (Swaminathan, 1968): "Intensive cultivation of land without conservation of soil fertility and soil structure would lead ultimately to the springing up of deserts. Irrigation without arrangements for drainage would result in soils getting alkaline or saline.

Indiscriminate use of pesticides, fungicides and herbicides could cause adverse changes in biological balance as well as lead to an increase in the incidence of cancer and other diseases, through the toxic residues being present in the grains or other edible parts. Unscientific tapping of underground water would lead to the rapid exhaustion of this wonderful capital resource left to us through ages of natural farming. The rapid replacement of numerous locally adapted varieties with one or two high yielding strains in large contiguous areas would result in the spread of serious diseases capable of wiping out entire crops, as happened prior to the Irish potato famine of 1845 and the Bengal rice famine of 1942. This process has already begun during 1967 with ergot of bajra and may happen with reference to rusts of wheat and virus diseases of rice during 1968. Therefore, the initiation of exploitative agriculture without a proper understanding of the various consequences of every one of the changes introduced into traditional agriculture and without first building up a proper scientific and training base to sustain it, may only lead us into an era of agricultural disaster in the long run, rather than to an era of agricultural prosperity."

Thus, it was clear from the beginning that the indiscriminate use of chemicals and non-adoption of ecologically sound crop-management practices would do harm in the long term in sustaining high yields. How then can we foster an ecologically sound Green Revolution? Fortunately, timely scientific efforts initiated by the Indian Council of Agricultural Research and Agricultural Universities made it possible to combine advances in production and productivity in wheat and rice with yield stability. The task now is the addition of the dimensions of ecological sustainability and equity in the Green Revolution pattern of agricultural advance.

The first UN Conference on the Human Environment held at Stockholm, Sweden, in 1972 aroused global consciousness on the need for subjecting all developmental activities to well defined ecological ground rules. It also brought home the point that so long as widespread poverty persists there will be no hope for achieving harmony between humankind and nature. Twenty years later, in June 1992, a UN Conference on Environment and Development will be held at Rio in Brazil to adopt an Earth Charter, an agenda for action in the 21st century and probably, global conventions on climate, biological diversity and forests.

The World Commission on Environment and Development chaired by Mrs. Gro Harlem Brundtland of Norway pleaded for making all development ecologically sustainable. Sustainable development may mean different things to different countries. To the rich nations, sustainable development may imply preserving for posterity the already high standards of life achieved in their countries. To the poor nations, it may mean a pathway of economic development which can help to meet the minimum needs of the people without harm to the environmental capital stocks. The World Bank used a per capita annual consumption of US $370 to fix the poverty line and found that 1.115 million were below the poverty line in the world in 1985. Of these, 630 million or 18 percent of the total population of the developing world suffered from extreme poverty, with a per capita consumption capacity of $275 per year. In the United States the poverty line is determined by an annual income of less than $14,000 for a family of four. Thus, there is much variability in the assessment of what constitutes poverty.

Whatever may be the per capita income levels used to estimate poverty, there are certain common trends which characterize poverty. Most of the poor in developing countries live in rural areas and depend for their livelihood directly or indirectly on crop and animal husbandry, forestry and agro-forestry, fisheries and agro-industries. They have few productive assets. A large proportion of their income goes towards purchase of food, since food occupies the first position among the hierarchical needs of human beings. Nutrition security involving physical and economic access to balanced diet and safe drinking water is their urgent need. Women in particular suffer from greater deprivation and discrimination in the labor market. Wherever this happens, household nutrition security is the first victim. The poor are also the first victims of environmental breakdown.

There is widespread expectation that the 1992 UN Conference on Environment and Development may help to reverse these trends. Can it help to enlarge the concern for global food security to include efforts to ensure food and nutrition security at the household level? Can it help to link economic entitlements with ecological obligations? Can it help to link in a symbiotic manner the livelihood security of farm families with the ecological security of farm holdings? What kinds of re-orientation are needed in technology and public policy to achieve a better quality of human life within the carrying capacity of supporting ecosystems?

The present publication is a compilation of a few recent papers (1990-91) which deal with issues related to ecologically sustainable advances in biological productivity. It is hoped they will be of use to agricultural scientists and scholars struggling to convert the term "sustainable" now added extensively before the term "development" from a cliché to an achievable concept.

– *Introduction by M.S. Swaminathan*
October 14, 1991

5.3 *Information Technology: A Dialogue*, ed. by M.S. Swaminathan. Madras, India: Macmillan India, 1993.

Technological empowerment is vital for adding value to the time spent in farm work by rural women and men. New technologies are also essential for achieving continuous improvements in the quality of rural life, while living within the carrying capacity of the supporting ecosystems, a process commonly referred to as sustainable development. Taking the benefits of new technologies to the economically and socially disadvantaged sections of the rural population is thus a pre-requisite for promoting a new paradigm of rural development based on concurrent and integrated attention to the imperatives of ecology, economics, employment and equity. For working towards this goal, pro-active interaction among all the stakeholders is necessary before new technologies are introduced. The M.S. Swaminathan Research Foundation has been organizing for this purpose an interdisciplinary dialogue each year under the general title "New Technologies: Reaching the Unreached". The first dialogue in this series, held in

January 1991, related to biotechnology. The proceedings were published by Macmillan India in November 1991. The second dialogue, held in January 1992, was on Information Technology. This book contains a record of that meeting.

Information, in the field of agriculture, to be of benefit has to be tailored to local agro-ecological and socio-economic conditions. It has to be backed up by relevant input supply services and public policies. Synchronization in time and space between knowledge and input delivery systems is essential to impart credibility to the extension message. Mechanisms for converting the know-how being imparted through the information system into field level do-how, such as training and institutional structures for producer-oriented marketing, have to be put in place. All this calls for a systems approach to knowledge and skill dissemination systems.

Fortunately, a wide range of information systems, both traditional (folk media, demonstrations) and recent (computers, remote sensing and telecommunication satellites), are now available. A blend of the two would help to promote a proper match between the message to he conveyed and the medium or media-mix chosen.

The emerging ecological agricultural systems are more knowledge-intensive since they aim to substitute knowledge and biological inputs for chemicals and capital. An effective information system is essential to promote and sustain ecological farming. This book should hence be of widespread interest.

– *Foreword by M.S. Swaminathan*
March 1993

5.4 *Wheat Revolution: A Dialogue,* ed. by M.S. Swaminathan. Madras, India: Macmillan India, 1993.

Wheat has been cultivated for several thousand years in India. Wheat kernels have been found in the Mohenjodaro excavations. The strain cultivated during the Mohenjodaro period was later described as *Triticum aestivum*, sub-species *sphaerococcum*. As the name indicates, this sub-species has spherical grains. In addition, it has a dwarf plant stature. From the days of Mohenjodaro up to the dawn of India's independence in 1947, the country developed the capacity to produce about seven million metric tons of wheat. In 1966, the country imported nearly ten million metric tons of wheat, largely through the PL 480 program of the United States, to fill the gap between supply and demand.

Alarmed at the growing gap between the rates of growth of population and of food production, India's first Prime Minister, Jawaharlal Nehru, said in 1948: "Everything else can wait but not agriculture." Several measures to stimulate food production, including land reform, irrigation, fertilizer production, strengthening of research and organization of a national extension service, were initiated in the fifties. Production of wheat and rice went up but productivity per unit area of land remained practically stagnant. Enhanced production came from an increase in both total cropped area and irrigated area. Wheat production went up to twelve million metric tons in 1964 which, from the point of view of monsoon behavior, was a good agricultural year.

In order to enhance productivity in irrigated areas, the Government of India initiated the Intensive Agriculture District Program (IADP) in 1961. The aim was to introduce good seeds and a package of agronomic practices which can help to optimize the benefit from irrigation water. Unfortunately, the early IADP experience was not encouraging. It was found that the package of practices promoted had one important missing ingredient, namely varieties which can respond well to good irrigation and soil fertility management. It is this missing ingredient that was provided in 1966 through the High Yielding Varieties Program (HYVP) in wheat, rice, maize, sorghum (jowar) and pearl millet (bajra). Wheat production rose to nearly seventeen million metric tons in 1968. To commemorate this quantum jump in wheat production, the then Prime Minister, Indira Gandhi, released a special stamp titled "Wheat Revolution" in July 1968. Soon afterwards, an American agricultural expert, Dr. William Gaud, described the phenomenon as "Green Revolution". I had then described what was happening as "land-saving agriculture", since the pathway of production improvement was higher productivity. After several thousand years, the stagnation in yield was broken in wheat, rice and other major food crops. How did all this happen?

In January 1990, Dr. Norman E. Borlaug, who developed the wheat varieties which constituted the critical ingredients of India's wheat revolution, wrote suggesting that we should hold a meeting of "key researchers and Government policy makers who were involved in the wheat revolution for some reminiscing about the main events and people". The aim was not only to review past events and accomplishments, but also to discuss future trends in productivity, supply and demand. Such a get-together was organized at Madras on March 8 and 9, 1990 with the kind help of Dr. R.S. Paroda, then Deputy Director General (Crops) in ICAR. In addition to Dr. Borlaug, Shri C. Subramaniam, who as Union Food and Agriculture Minister introduced the new agricultural strategy in the mid-nineteen sixties, and Shri B. Sivaraman, who was the then Agriculture Secretary (unfortunately he passed away in 1992), participated. In order to capture the reminiscences of the wheat scientists who could not participate in the Madras dialogue, a special meeting was arranged at the Indian Agricultural Research Institute, New Delhi, on March 10, 1990. This book contains a detailed record of the Madras dialogue together with a summary of the main points made at the New Delhi meeting.

The discussions at Madras and New Delhi were divided into the following three major groups of issues: a) Package of Technology, including new varieties, agronomic practices and post-harvest handling; b) Package of Services, including the timely supply of seeds, fertilizer, water and credit; c) Package of Public Policies, including input and output pricing, assured and remunerative marketing, building up of grain reserves and strengthening of rural techno-infrastructure.

Fortunately, most scientists who were active in wheat research in the sixties participated. The participants missed the smiling and enthusiastic faces of Dr. B.P. Pal and Mr. V.S. Mathur, who had passed away. Still, it was a unique gathering of scientists whose hard and determined work changed the public mood from one of despair to hope with reference to India's agricultural destiny.

The quest for the breeding of crop varieties which can respond to higher levels of plant nutrition started in 1952 when at the instance of the late Dr. K. Ramiah, a program for incorporating genes for fertilizer response from *japonica* rice varieties into *indica* strains was initiated at the Central Rice Research Institute, Cuttack, under the sponsorship of FAO and ICAR. I worked in this project during 1954 under the guidance of the late Dr. N. Parthasarathi. Our major aim was to select from segregating populations of indica × japonica crosses, lines which showed the ability to utilize effectively about 100 kg of N per hectare. With this quantity of nutrient supply, about five metric tons of rice per ha can be produced. This program led to varieties like ADT-27 in Tamil Nadu and Mashuri in Malaysia. Several genetic problems arose, rendering the speedy selection of high-yielding rice varieties from indica × japonica crosses difficult. With the advent, in the early sixties, of the semi-dwarf, non-lodging, relatively photo-insensitive varieties based on the Dee-gee-woo-gen dwarfing gene identified in China, interest in transferring genes for fertilizer response from japonica varieties waned. Semi-dwarf indica rice like Taichung Native 1, IR 8 and Jaya provided the initial material for the High Yielding Varieties Program.

When I joined the Indian Agricultural Research Institute (IARI) in late 1954, I started a research program for developing non-lodging and fertilizer responsive varieties of wheat, on the lines of the work earlier initiated on rice. At that time the research strategy adopted had three components. First, crosses were made between the semi-dwarf, stiff straw *compactum* and *sphaerococcum* sub-species of *T. aestivum* with cultivated bread wheats. Second, attempts were made to induce erectoides mutants in commercial wheat varieties through the use of radiations and chemical mutagens. Third, studies on the potential for increasing straw stiffness through different chemical treatments were initiated. Unfortunately, in all these three approaches, short and stiff straw was always associated with short panicles with fewer grains. The reason why straw stiffness became such an essential prerequisite for favorable response to water and fertilizer was the tendency, among the then cultivated tall wheat varieties, to lodge or fall down when fertilizer was applied. Also, such lodging made it difficult to give irrigation during the grain development phase, when the crop will benefit much from water availability. Thus, with the earlier tall varieties, it was difficult to get economic response to the application of mineral fertilizers and adequate irrigation water. In wheat, yields stagnated at less than one ton per ha. This is why the breeding of non-lodging varieties was accorded such high priority during the fifties, when the country had taken to the path of expanding the area under irrigation and manufacturing mineral fertilizers.

During the late fifties, scientific publications on the work done under the leadership of Dr. Orville Vogel in Washington State of the United States of America, on the transfer of dwarfing genes from the Norin-10 wheat variety to North American winter wheats, started appearing. When requested, Dr. Vogel was kind enough to send me seeds of Gaines, a semi-dwarf winter wheat variety with red grains. He further suggested that I should approach Dr. Norman E. Borlaug in Mexico for seeds of semi-dwarf varieties in a spring wheat background (winter wheat needs long hours of sunlight to flower and set seeds).

In March 1962, a few dwarf spring wheat strains entered by Dr. Borlaug in the International Wheat Rust Nursery and sent by the U.S. Department of Agriculture were grown in the fields of IARI. Their phenotype was most impressive. They had reduced height and long panicles, unlike the earlier hybrids between *aestivum* and *compactum* and *sphaerococcum* and the induced *erectoides* mutants which had short height and small panicles.

In May 1962, I wrote to Dr. B.P. Pal, who was then Director of IARI, requesting him to arrange for the visit of Dr. Borlaug to India and for obtaining a wide range of dwarf wheat material possessing the Norin-10 dwarfing genes. I explained in my letter why we should take up immediately a dynamic dwarf breeding program based on genes which do not exert a pleiotropic effect on the panicle. My letter was forwarded by Dr. Pal to the Ministry of Agriculture, and the Ministry requested the Rockefeller Foundation, in 1962, to arrange for the visit of Dr. Borlaug and for the supply of dwarf wheat material. Dr. Borlaug visited India in March 1963 and sent a wide range of material in October 1963. I presented a detailed report on the trials conducted with the Mexican wheat material at the All India Wheat Research Workers' Conference held at Delhi in August 1964. I also proposed, in 1964, that we should start a National Demonstration Program in farmers' fields, both to verify the results obtained in research plots and to introduce farmers to the new opportunities opened up by semi-dwarf varieties for improving very considerably the productivity of wheat. When small farmers, who with the help of scientists organized the National Demonstration Program, harvested over five metric tons of wheat per hectare, its impact on the minds of other farmers was electric. The clamor for seeds began and the area under high yielding varieties of wheat rose from four ha in 1963-64 to over four million ha in 1971-72. A small Government program became a mass movement. The rest of the history is recorded in this book in the words of those who played a key role in the realm of science and public policy.

I observed, in 1967, a tendency among farmers in north-west India with relatively large holdings to use large quantities of fertilizers and to grow in large and contiguous areas a single genetic strain. In my Presidential Address to the Agricultural Sciences Section of the Indian Science Congress held at Varanasi in January 1968, I dealt with the need for adding the dimension of ecological sustainability in efforts to improve yield.

The remarkable speed with which the high-yielding varieties were identified from the initial Mexican material and later developed within the country was the result of the multi-location testing and inter-disciplinary research organized under the All-India Coordinated Wheat Research Project of the Indian Council of Agricultural Research. The country owes a deep debt of gratitude to the successive coordinators of this most successful exercise in organized cooperation – Drs. A.B. Joshi, S.P. Kohli, M.V. Rao, A. Austin and J.P. Tandon. We salute the late Dr. B.P. Pal, who initiated both organized wheat breeding and coordinated varietal testing programs in the country. We also recall with gratitude the invaluable services rendered by the late Dr. R.G. Anderson, who served as Joint Coordinator of the All India Wheat Research Project between 1964 and 1970. Dr. Ralph W. Cummings, Field Director of the Rockefeller

Foundation in New Delhi between 1958 and 1966, responded to our requests with speed and our thanks go to him. Above all, numerous young research workers and scholars participated with enthusiasm in the different aspects of the wheat program, including the organization of a Seed Village in Jounti village of Khanjawala block of Delhi state.

Twenty-five years ago, the country officially recognized the beginning of the Wheat Revolution. The role of science in triggering such a revolution was recognized by including in the special stamp a portrait of the IARI library. Dr. J.P. Tandon, Project Director Wheat of ICAR, has written a short note for this book summarizing the progress made through the All India Coordinated Wheat Improvement Project since 1968.

I am most grateful to Dr. Norman E. Borlaug and Mr. Christopher Dowswell for suggesting this dialogue and for participating in it actively. My particular thanks go to Shri C. Subramaniam, who came all the way from Bombay to share his reminiscences and insights. …

I would like to end this Preface by quoting what I wrote in the May 11, 1969 issue of the *Illustrated Weekly of India*: "Brimming with enthusiasm, hard-working, skilled and determined, the Punjab farmer has been the backbone of the wheat revolution. Revolutions are usually associated with the young, but in this revolution, age has been no obstacle to participation. Farmers, young and old, educated and uneducated, have easily taken to the new agronomy. It has been heart-warming to see young college graduates, retired officials, ex-army men, illiterate peasants and small farmers queuing to get the new seeds. At least in the Punjab, the divorce between intellect and labor which has been the bane of our agriculture is vanishing."

It is to the Indian farm men and women, who proved the prophets of doom wrong and who helped the country to take to a non-PL 480 pathway of achieving food security, that this book is dedicated.

– *Preface by M.S. Swaminathan*

5.5 *Uncommon Opportunities: An Agenda for Peace and Equitable Development.* Report of the International Commission on Peace and Food, Chaired by M.S. Swaminathan. London: Zed Books, 1994.

The International Commission on Peace and Food (ICPF) began its work at a time of uncommon opportunities for rapid transition on a global scale to democratic ways of life, freedom of the media, and an independent judiciary; food, literacy, health and jobs for all; and sustainable life styles based on harmony between humankind and nature. The end of the Cold War and the consequent shift of expenditure from defense to development, the onset of the Information Age and the organization of information superhighways, the growth of mechanisms for regional cooperation, a renewed faith in the United Nations, and commitment to multilateral negotiations and action – all provided new opportunities for achieving global peace and prosperity. Rapid progress in the application of science and technology in agriculture, industry, and human health strengthened the hope that the concept of "food, education, health and employment for all" can be converted from rhetoric into reality, provided the peace dividend can be deployed to achieve these goals.

ICPF had the advantage of building on the work of earlier independent commissions, particularly those chaired by Willy Brandt, Gro Harlem Brundtland and Julius Nyerere. We began our work with the conviction that, as Willy Brandt said, "the problems created by men can be solved by men". The ancient Indian scripture, *Bhagavad Gita*, expressed the same sentiment by stressing, "Whatever man aspires to, that he can accomplish." The achievement of food self-sufficiency by India – in a period when famine threatened millions of lives – through the power of political will and scientific action in a democratic society illustrates this truth.

In spite of such uncommon opportunities for a better common present, during the last six years the world has been witness to an escalation of ethnic and mini-conflicts, growing intolerance of diversity, the spread of small arms, an increase in violence – starting with violence to oneself through drug addiction – and damage to the life support systems of land, water, flora, fauna and the atmosphere. In addition, the current pattern of economic and technological growth is leading to gross social inequity and rising joblessness. The rich-poor divide is widening both within and among nations. At the same time, the spread of knowledge and information, particularly through the electronic media, has led to an era of rising expectations. The economically and socially underprivileged sections of human society are now aware that their future is not a question of fate, but of political and technological choice. If they are not given access to the information superhighways and other technological highways to progress, a new form of social and economic apartheid will spread, which will be even more serious in its consequences to the future of human civilization than the skin-color-based apartheid which has just ended.

True education is leadership in thought. There are periods in human history when the progress of decades can be accomplished in a few years. It is the view of the Commission that we have arrived at one of these great, creative moments. Can the political leadership of all countries seize this moment and convert it into an opportunity for promoting job-led economic growth, rooted in the principles of ecology and equity? Can we make the gender inequity and the coexistence of extreme poverty on the part of a billion children, women, and men, and extreme affluence and unsustainable lifestyles on the part of another billion, an anachronism of the past? Can we foster love of diversity and pluralism in terms of religion, race, language, color or political belief?

In our view, there are uncommon opportunities now to find the solutions to these basic challenges facing contemporary human society. We have tried to deal with them in a concise manner in this report, which concludes that commitment to the principles of democracy and human diversity, equity in terms of gender, economic condition and sustainable lifestyles, is fundamental to achieving both a better common present and a better future.

Personally, it has been a privilege working with a highly distinguished group of women and men, whose sole concern has been a better life for all. I wish to thank every member of the Commission, as well as the many others who have helped us, for their time and vision. The Member Secretary, Mr. Garry Jacobs, however, merits a special mention for his total dedication to the principles which inspired the setting up of ICPF. Throughout the last five years, he has devoted his time and extraordinary talent to the work of the Commission. He has been the principal catalyst of the Commission's work. In his heavy responsibilities, he has been ably supported by Robert J. MacFarlane and Robert van Harten. I wish to record our sincere gratitude to this unique trio.

Our work has been largely supported by grants from UNDP and UNESCO. UNESCO also provided the services of Dr. Maurice Goldsmith, Director, International Science Policy Foundation, to edit this report. It has been a privilege working with Dr. Goldsmith, an editor of extraordinary wit and wisdom. I am indebted to Mr. Boutros Boutros-Ghali, Secretary General of the United Nations; Dr. Federico Mayor, Director General of UNESCO; and Mr. James Gustave Speth, Administrator of UNDP, for their support and encouragement. A special word of thanks is also due to Mrs. Robin La Brie-Jackson, who has helped throughout in managing the work of the Commission with great competence and courtesy. Finally, I must acknowledge the inspiration and encouragement provided by Mr. T. Natarajan, Secretary, The Mother's Service Society, Pondicherry, India, from the very inception of the idea of the Commission. His emphasis on leadership in thought that leads to action has served as a guiding principle for the work of the Commission. A more complete list of the many individuals and institutions who have contributed generously of their time and resources to our work is contained in the Acknowledgements appended to this volume.

During 1994 and 1995, several major international events will take place. The International Conference on Population and Development at Cairo in September 1994, the World Summit for Social Development at Copenhagen and the Fourth World Conference on Women at Beijing in 1995, and the designation of 1995 as the International Year for Tolerance – all provide unique opportunities for promoting sustainable and equitable development. The past fifty years since the birth of the UN have marked striking progress in every field of human endeavor, except in the areas of ecological and human security. We should now concentrate on these aspects of security, so vital for sustained human happiness and fulfillment. It is our hope that this report will be of some help in accelerating the pace of progress in the evolution of a new human ethic conducive to human beings living in harmony with nature and with each other.

– *Preface by M.S. Swaminathan*
July 6, 1994

5.6 *Ecotechnology and Rural Employment: A Dialogue*, ed. by M.S. Swaminathan. Madras, India: Macmillan India, 1994.

J.B.S. Haldane in his book *Daedalus or Science and the Future* (1923) emphasized that unless our ethical outlook evolves to keep pace with development of science and technology, social chaos and disaster would result. Increasing rich-poor divide, damage to basic life-support systems, jobless economic growth and ethnic conflicts provide evidence to support Haldane's visionary statement.

It is becoming increasingly clear that the social dimensions of a scientific problem should be studied in a proactive manner, if we are to ensure that the anticipated social benefits materialize. The M.S. Swaminathan Research Foundation and its Centre for Research on Sustainable Agricultural and Rural Development have therefore been organizing each year an interdisciplinary dialogue involving the principal stakeholders in a scientific enterprise to consider methods of taking the benefits of modern technologies to those who normally tend to get bypassed by technological innovations. The following dialogues have been held so far: a) 1990: International Dialogue on Plant Genetic Resources. This dialogue helped to articulate the concept of farmers' rights – held jointly with the Keystone Center, USA; b) 1991: Biotechnology. This dialogue resulted in the organization of biovillages in India and China; c) 1992: Information Technology. This interaction helped to standardize methods of taking the benefits of the information age to rural families. The proceedings of the first dialogue were published by the Keystone Center. The proceedings of the dialogues on biotechnology and information technology have been published by Macmillan India.

The fourth interdisciplinary dialogue in this series was on Ecotechnology and Rural Employment. This dialogue was held at Madras from April 11 to 15, 1993. Its major aim was to compare the experience of China and India in fighting the growing famine of jobs. Inadequate work opportunities result in low household income and purchasing power. This in turn becomes the main cause of under-nutrition at the household level. China has succeeded in creating nearly one hundred million skilled jobs in the non-farm sector during the past ten years. How did this happen? What kinds of technologies, services, techno-infrastructure and public policies are needed to achieve a rapid expansion of employment opportunities both in the farm and non-farm sectors? The present book provides valuable insights into methods of accomplishing the goal of work for all.

Most estimates show that the global population may grow from 5.5 billion today to 8.5 billion over the next thirty years. Much of this increase will occur in the developing countries, where most of the world's poor live. The human carrying capacity of land and water has already been exceeded in many countries of South and South East Asia. The expansion and intensification of global agriculture to feed the growing population has resulted in the degradation of nearly 1.2 billion hectares of soil since 1945. Water scarcity is becoming more widespread and groundwater resources are getting depleted. Under such circumstances, the integration of ecological principles in a job-led economic growth strategy is a must for lasting economic and social benefits. The dialogue participants therefore addressed the interlinked issues among the physical, social, economic and ecological parameters relevant to the technological empowerment of the poor. The term ecotechnology symbolizes the addition of sustainability concerns to technology development and dissemination.

The book contains suggestions for promoting an ecologically sustainable job-led economic growth strategy, based on a pro-nature, pro-poor and pro-women orientation to technology and public policy. Such a change in mind-set leads to each individual creating his or her own income. The change starts with a production process or a process of value-addition through self-employment. To enable every woman and man to create her/his own income-generating activities, we would need to create appropriate institutions and adopt right policies. I had summarized some of the policy changes needed in a sub-chapter under the employment chapter in the Sixth Five Year Plan document (1980-85) under the title "A new deal for the self-employed". I hope this book will lead to the formulation of a charter for the self-employed.

– *Foreword by M.S. Swaminathan*

This is the fourth in our series of Dialogues. Our Centre started functioning at Madras in late 1989. First in the series was plant genetic resources, which is now very much in the forefront of international debate in terms of intellectual property rights; the GATT (TRIPS) negotiations involve a consideration of the rights of indigenous people; it is also the International Year of the Indigenous People. We had a very intensive discussion here under the Keystone International Dialogue Series on Plant Genetic Resources that has been published by the Keystone Center. In fact, the Madras Dialogue, as it is called, made a very important contribution to the ongoing international debate on this issue.

Our second Dialogue related to biotechnology. The series looks at taking the benefits of modern technology to the unreached and is called "Reaching the Unreached". Rich farmers, the corporate sector and large companies have enough funds and resources to benefit from any technology. That is why the disparity between the rich and the poor is growing, so dramatically illustrated by the UNDP in its cover on last year's Report on Human Development. The gap between the rich and the poor is increasing. Our concern is, how do we take the benefits of modem science and technology to the unreached, i.e., families without assets like land, livestock, fish ponds, forest trees, etc., and to small and marginal farmers like those in India and China who run very small farms. Our discussion on biotechnology led to a joint China-India project on biovillages. I am particularly happy that the former Lt. Governor of Pondicherry, Dr. Har Swarup Singh, who helped us to start the project, is also here today. The third was on Information Technology and was held last January; we talked about how to take modern information on science and technology to the rural areas, particularly how to integrate meteorological, management and marketing information into a location-specific package. This Dialogue resulted in the concept of information villages.

These Dialogues are designed to undertake not a post-mortem analysis but a pro-active one on what ought to be done to carry the benefits of the new technologies to the rural poor. In the past, the Green Revolution was criticized by social scientists, who said that the benefits have not reached the poor, the socially and economically under-privileged. When I say "socially", we think of rural women in the forefront. So that is our target, bringing about a pro-nature, pro-poor and pro-women orientation in terms of technology development and dissemination.

The pattern of this Dialogue is that discussion leaders will first introduce a topic and this will be followed by a free-ranging discussion.

As regards the outputs of the Dialogue, the first one is a book to be published by Macmillan, and the second is to help public policy-makers, political leaders, developmental administrators and also financial and lending institutions in promoting a sustainable job-led economic growth strategy. This meeting is co-sponsored by NABARD, CAPART, CSIR, UNDP and the Sasakawa Peace Foundation. Many institutions, both national and international, are looking for ideas on how to combine the four major goals which seemingly cannot be combined. One is economics, because if it is not economically viable, it will not be replicable. So economic viability is important. Second is ecology, because most of the technologies in the rural areas are based upon the utilization of natural resources like land and water, flora and fauna, and therefore if it is unsustainable utilization, it will not be possible to keep using them in the long run. The third very important item is equity in gender, social and economic terms. And, finally, all these three – ecology, economy and equity – leading to employment.

Skilled employment is the key not only to quality of life improvement but also to household nutrition security. Nutrition security involves physical and economic access to balanced diets and safe drinking water to all. Economic access will be possible if there is the requisite purchasing power. This is why "jobs for all" becomes a prerequisite for achieving food and health for all.

We are fortunate to have with us Dr. Song Jian, Prof. He Kang and several Chinese colleagues here. China has created over one hundred million new non-farm jobs during the last eight years through the Spark and Rural Township Enterprises. China has succeeded in taking the benefits of modern agribusiness to resource-poor farming families. They have developed institutional structures which can confer the benefit of scale to small-scale enterprises. India has experimented with many approaches in rural job generation. A careful consideration of the experiences gained in China and India will therefore be of immense benefit. This is the primary goal of this Dialogue.

The term "ecotechnology" is defined for the purpose of this Dialogue as "technologies which combine the best in frontier technologies with the best in traditional technologies and in the ecological prudence of rural and tribal communities". I hope this Dialogue will help us to define how we can foster a dynamic ecotechnology movement.

– *Introduction by M.S. Swaminathan*

5.7 *Sustainable Agriculture: Towards Food Security,* by M.S. Swaminathan. Delhi: Konark Publishers, 1996.

A recent study by Dr. K.N.N.S. Nair and me on the food security challenges in the Asia-Pacific Region revealed that public action for ending endemic hunger should keep in view the following: a) Food security should be considered at the level of the individual, rather than of the household, since women and girl children tend to suffer more from poverty-induced under-nutrition; b) Non-food factors like income, environmental hygiene, primary health care and literacy are equally important; c) Poverty is the primary cause of under- and malnutrition. Based on these considerations it will be appropriate if food security is redefined as follows: "Sustainable food security involves strengthening the livelihood security of all members within a household by ensuring both physical and economic access to balanced diet, including the needed micronutrients, safe drinking water, environmental sanitation, basic health care and primary education."

It should be our goal to achieve food and livelihood security for all through integrated packages of technology, techno-infrastructure, services and public policies. There is need for paying particular attention to delivery systems capable of reaching the unreached.

The present book consists of papers/lectures relating to the broad theme of sustainable food security written during the last two decades. No attempt has been made to update the data quoted at different times, since this would have involved extensive re-writing, resulting in a different paper/lecture, rather than the one presented at a particular point of time.

The year 1995 marks the 125th birth anniversary of Mahatma Gandhi, who pointed out just prior to our independence that "to those who are hungry, god is bread". Food and safe drinking water are first among the needs of human beings and hence the first priority of every nation should be achieving sustainable food and livelihood security. Calculations show that we will have to create opportunities for over one hundred million new livelihoods during the next ten years. Most of these have to come from the farm and off-farm sectors in rural areas.

The urgent need for safeguarding and strengthening the ecological foundations essential for sustainable agricultural advance, has been stressed in the first volume. What we need soon is a National Act for Sustainable Food Security with provisions for concurrent and integrated action on the following: a) Protecting the natural resource base essential for continuous advances in terrestrial and aquatic productivity; b) Preventing both the diversion of prime farm land for non-farm use and the unsustainable exploitation of ground water; c) Promoting eco-technologies based on a blend of traditional and frontier technologies, combining their respective economic and ecological strengths; d) Providing the necessary training, techno-infrastructure and trade back-up to enable all farmers, irrespective of the size of their holding or input purchasing power, to benefit from ecotechnologies; e) Undertaking asset reform so that landless labor families have access to productive assets; f) Introducing public policies which will enable all children, women and men to have access to balanced diets, safe drinking water, primary health care, primary education and environmental hygiene; g) Developing and spreading a Hunger-free Area Program all over the country, on the model of the one hundred percent literacy districts, to ensure that the "Food for All" goal becomes a reality.

— *Preface by M.S. Swaminathan*

5.8 *Sustainable Agriculture: Towards an Evergreen Revolution*, by M.S. Swaminathan. Delhi: Konark Publishers, 1996.

The prefix "Sustainable" is now being invariably used before terms relating to various aspects of human engineered development. Until twenty-five years ago, the sustainability of a human enterprise was measured solely in economic terms. Starting with the United Nations Conference on the Human Environment held at Stockholm in 1972, environmental parameters were added to measure sustainability. The integration of ecological principles in economics took a concrete form in the shape of "Agenda-21" adopted at the UN Conference on Environment and Development held at Rio de Janeiro in 1992. Since then considerations of environmental sustainability have been shaping the agenda for both technology development and public policy formulation.

The growing rich-poor divide, jobless economic growth and the increasing feminization of poverty, all documented very well in UNDP's Human Development Reports of 1992, 1993, 1994 and 1995, have highlighted the fact that development which is not equitable will also not be sustainable in the long-term. Such inequitable development will lead to increased violence and promote social disintegration. Thus, we now know that tools for measuring sustainability will have to be enlarged in a manner that concurrent attention can be paid to economic viability, environmental sustainability and social equity. Equity will have to be measured both in gender and in economic terms. Such a paradigm shift in policy formulation in areas of concern both to science and society calls for a change in mindset among all professionals leading to viewing problems in a holistic or systems manner. Such a change in mindset is essential for ending the prevailing mismatch between the capabilities of uni-directional minds and the needs of multi-dimensional problems. A fragmented approach is not only widely prevalent with reference to ecological issues but also in other areas of great social concern. For example, the question of women's empowerment is often considered in isolation, without relating it to men's enlightenment and involvement.

The present book is a compilation of eight lectures/articles from among those I had delivered or written during the last three decades. Earlier, Concept Publishing Company, New Delhi, had published some of my papers/lectures in the form of two books titled *Science and the Conquest of Hunger* (1981) and *Science and Integrated Rural Development* (1982). Students and scholars of Agricultural Universities and Training Institutions have been suggesting that some of the more recent lectures should be similarly compiled. A paper presented at the All India Wheat Research Workers' Conference held in New Delhi in August 1964 on the performance of the semi-dwarf wheat varieties is also included, since this is the first record of extensive experiments with varieties which subsequently triggered the wheat revolution.

The term "Green Revolution" has been both widely used and abused. It means different things to different people – some good and some bad. In scientific terms, the term implies enhanced crop production through higher productivity per units of land and water. Green Revolution technologies help in harnessing the photosynthetic pathway of development, which makes agriculture the largest solar energy harvesting enterprise in the world. There is no option open to us except to produce more farm products from less land and water in the decades ahead. Therefore, we need an evergreen revolution, rooted in the principles of ecology, equity, economics, employment generation and energy use efficiency.

The confusion concerning the social and ecological value of the Green Revolution has arisen because of the failure to distinguish between greed and Green Revolutions. The fast expanding greed revolution unfortunately leads to the unsustainable exploitation of national resources and to the excessive and unscientific use of chemical pesticides and mineral fertilizers. Such a greed revolution is what is responsible for the current social and environmental problems associated with coastal aquaculture. Our future lies in curbing the greed revolution and fostering an evergreen revolution in farms and factories.

– *Preface by M.S. Swaminathan*

5.9 *Agrobiodiversity and Farmers' Rights*, ed. by M.S. Swaminathan. Delhi: Konark Publishers, 1996.

The commitment of the international community to ensure food and nutrition security to every child, women and man on our planet has been reiterated at the following international conferences held since 1989: 1989 Bellagio Declaration: Overcoming Hunger in the1990s; 1989 Convention on the Rights of the Child; 1990 World Summit for Children, New York; 1991 UNICEF/WHO-Conference on Ending Hidden Hunger (Micronutrient Malnutrition); 1992 International Conference on Nutrition (ICN, 1992), Rome; 1992 UN Conference on Environment and Development (UNCED, 1992), Rio de Janeiro; 1993 World Conference on Human Rights (WCHR, 1993), Vienna; 1994 International Conference on Population and Development, Cairo; 1995 UN Conference on Social Development, Copenhagen; 1995 Fourth World Conference on Women, Beijing.

Despite the high level political reaffirmation of the need to achieve speedily the goal of "Food for All", over eight hundred million human beings suffer now from hunger and malnutrition. Hidden hunger, arising from the deficiency of the essential micronutrients, is widespread. Endemic hunger is now more due to the lack of adequate purchasing power at the household level than due to the lack of availability of food in the market. Both poverty as well as non-food factors such as environmental hygiene, sanitation and lack of safe drinking water are becoming major contributors of food insecurity at the level of communities and individuals. A majority of the food insecure live in Asia.

There is also increasing concern about the earth's capacity to produce adequate food for the growing population. The increase in human population, now mostly confined to the developing countries, enhanced purchasing power leading to greater capability to buy food and increasing urbanization will lead to large demand for food as well as more diversified food products in the coming millennium, while, on the one hand, there will be need for intensification and diversification of agriculture, particularly in the developing countries, the ecological foundations essential for sustainable advances in crop and animal productivity are getting eroded. This situation has led to the ringing of alarm bells by experts concerning the earth's capacity to produce adequate food to meet growing needs. Lester Brown has concluded: "Constraints imposed by the earth's natural systems, the environmental degradation of land and water resources and the diminishing backlog of yield-raising agricultural technologies are slowing the growth in world food production, raising questions about the earth's population carrying capacity. The earth's capacity to produce enough food to satisfy expanding demand is now emerging as the overriding environmental issue as the world approaches the 21st century."

The year 1988 will mark the bicentenary of Malthus's essay on population. The onward march of science and technology, supported by appropriate public policies, has so far helped humankind to keep Malthusian forebodings on the population-food supply equation at bay. It is, however, becoming increasingly clear that without concerted national and global action to arrest environmental degradation and to check population growth, the Malthusian specter may still come true in the next century. The Food and Agriculture Organization of the United Nations (FAO) has, therefore, convened a World Food Summit in Rome from November 13 to 17, 1996.

A Science Academies Summit was held at the M.S. Swaminathan Research Foundation, Madras, from July 8 to 11, 1996, to discuss the world food situation and propose a strategy for harnessing the uncommon opportunities that exist now for working towards a food and nutrition secure world. The Madras Declaration adopted by the Summit participants redefines food security as follows.

National policies for food security should ensure: a) that every individual has the physical, economic, social and environmental access to a balanced diet that includes the necessary macro- and micro-nutrients, safe drinking water, sanitation, environmental hygiene, primary health care and education so as to lead a healthy and productive life; b) that food originates from efficient and environmentally benign production technologies that conserve and enhance the natural resource base of crop and animal husbandry, forestry and inland and marine fisheries.

The Madras Declaration contains a ten-point action plan for achieving at the national, regional and global levels the concept of food and nutrition security.

Among the many scientific components of agricultural progress, breeding of new varieties of plants, animals and micro-organisms has been an important one. The "Green Revolution" was triggered in Asia by high-yielding varieties of wheat and rice and hybrids of maize, sorghum and pearl millet. Success in breeding has, in turn, come from the intelligent use of genetic variability. However, the very process of improving crop productivity through genetic and recombination has led to increasing genetic homogeneity in cultivated varieties and hybrids. Genetic homogeneity enhances genetic vulnerability to biotic and abiotic stresses, necessitating the increased use of pesticides. Indiscriminate use of chemical pesticides and mineral fertilizers harms soil, water, flora and fauna. In addition, they lead to new health hazards among the human population including the suppression of natural processes of immunity.

Genetic variability provides the feedstock for plant and animal breeding and genetic engineering enterprises. While, on the one hand, human capacity to produce novel genetic combinations through recombinant DNA technology is increasing, the rate of loss of genes, species and habitats, rich in biodiversity, is also increasing.

The Convention on Biological Diversity (CBD) adopted at the UN Conference on Environment and Development (UNCED) at Rio de Janeiro in June 1992, is a clear index of the concern and commitment on the part of the international community for the conservation and sustainable and equitable utilization of bio-diversity. Biodiversity occurring in plants, animals and micro-organisms of economic value, is now referred to as Agrobiodiversity.

The future of global food security depends on the success of our efforts in the conservation and enhancement of agrobiodiversity. The technical conference convened by the FAO in Leipzig, Germany, in June 1996, to consider issues relating to the conservation of genetic resources has urged all nations to implement a global plan of action for the conservation and sustainable utilization of plant genetic resources for food and agriculture.

Since its inception in 1991, the Global Environment Facility (GEF) has accorded high priority to halting gene erosion and to promoting the conservation of habitats rich in biodiversity. Some outstanding examples of GEF's efforts to preserve for posterity the "genetic paradises" on our planet are given later in this report. Particularly, two noteworthy examples are the Iwokrama International Rain Forest Conservation Program in Guyana and the promotion of on-farm conservation of land races of economic plants in Ethiopia. The loss of every gene and species limits our options for the future in terms of opportunities for genetic recombination. It is unfortunate that such a scientific handicap is occurring at a time when we need more options in genetic recombination, as a result of potential changes in temperature, precipitation, sea level and ultravlolet-β radiation, as well as in consumer preferences and food processing technologies. The emerging "green health" movement is leading to the discovery of new uses for medicinal plants. Genetic engineering offers new opportunities for the development of effective Integrated Pest Management (IPM) and Integrated Nutrient Supply (INS) systems.

The Madras Declaration has emphasized the need to move away from a "favorable area" and "favored crops" mindset to enlarging the contents of the food basket by including the excluded in terms of under-utilized plants and neglected ecosystems. All this calls for redoubled efforts in the deployment of resources for the conservation and sustainable and equitable use of agrobiodiversity.

Convention on Biological Diversity

The objectives of this Convention, to be pursued in accordance with its relevant provisions, are the conservation of biological diversity, the sustainable use of its components and the fair and equitable sharing of the benefits arising out of the utilization of genetic resources through appropriate Material Transfer Agreements.

For the purpose of the Convention, the term "biological diversity" has been defined as the variability among living organisms from all sources including, inter alia, terrestrial, marine and other aquatic ecosystems and the ecological complexes of which they are part. This includes diversity within species, between species and ecosystems.

Agrobiodiversity

Agrobiodiversity refers primarily to genetic variability in cultivated plants and domesticated animals together with their progenitors and closely related wild species growing and evolving under natural conditions. Plants and animals harvested from the wild are also included in this term.

Besides contributing directly to national economies, agrobiodiverstiy also provides employment and livelihood to a large section of the economically active population. On a global scale, it is noteworthy that nearly two and a half billion people rely heavily on wild and traditionally cultivated plant species to meet their daily needs. Tribal women and men also serve as repositories of knowledge systems on their utilization.

While informal crop improvement by farm families is largely based on selection of superior types, scientific plant breeding draws heavily upon planned hybridization among landraces, locally adapted cultivars and wild relatives of crop plants. Such planned pyramiding of genes results in superior varieties combining high productivity with better quality traits, genetic resistance to diseases and pests and greater tolerance to stress environments, including drought, salinity, water logging and extreme temperature regimes.

Centers of Diversity

Agrobiodiversity is not evenly distributed over the globe and tends to be more concentrated in the tropical and sub-tropical regions of the world. Substantial information is now available on the centers of origin of cultivated plants and the areas of their distribution. Explorations and germplasm collection programs, initiated by N.I. Vavilov in the 1920s, have helped to add information on the centers of diversity of domesticated plants.

Biosafety and Bioethics

The value of landraces, primitive cultivars and wild relatives of cultivated plants has been greatly enhanced by the recent advances in biotechnology such as genetic engineering and recombinant DNA technology. Genes can be moved across sexual barriers and this possibility has enhanced the value of the distant relatives of cultivated plants.

Progress in our capacity to create novel genetic combinations (i.e., "playing God") has also raised key issues in the areas of biosafety and bioethics. An important aspect of ethics relates to the incorporation of the principles of equity in sharing benefits with the tribal and rural families who have conserved and often enhanced through selection and information the basic genetic raw material used in genetic engineering research.

In his book, *The Diversity of Life*, E.O. Wilson has warned that Homo sapiens is in imminent danger of precipitating a biological disaster of a greater magnitude than anything we have witnessed so far in our evolutionary history. There is hence no time to relax, if we are to ensure that the Malthusian prophecy of famine and pestilence do not come true in the coming millennium. Legal, educational and participatory measures of program implementation and benefit sharing will all be needed for promoting a peoples' movement for agrobiodiversity conservation.

Conserving Traditional Knowledge and Wisdom

It is now widely acknowledged that tribal and rural women and men have not only conserved biodiversity of great economic, cultural and social value, but have also enhanced them through selection and value addition through knowledge and information. Most of the domesticated plants and animals of today were identified centuries ago. The medicinal plants we now use were identified by our ancestors among the wild flora for specific uses such as food, fiber, oil and medicine. Articles 8 (F), 10 (C and D) and 15 (7) of the CBD all recognize the role of indigenous and local communities in the conservation and improvement of genetic resources. An extremely urgent task is the preservation and revitalization of this priceless heritage. In addition, there is need to introduce a recognition and reward system for indigenous wisdom and knowledge. In this connection, it will be appropriate to recall the verse of T.S. Eliot: "Where is the wisdom that we have lost in knowledge, Where is the knowledge we have lost in information." It is thus appropriate to conserve not only agrobiodiversity, but also the traditions which led to the enrichment of our knowledge on valuable genetic material. Tribal taxonomy differs from modern biosystematics in the sense that in most cases the tribal nomenclature has its roots in the end use of the material. It is thus value-added biosystematics.

While dealing with genetic conservation methods, it is important to distinguish the roles of in situ and ex situ conservation. In situ conservation helps not only to preserve genetic variability, but also to its enrichment through mutation, recombination and selection. Thus, it involves both conservation and continued evolution. In contrast, ex situ conservation is more approximately defined as preservation. Hence, it would be useful to refer to in situ methods as conservation and ex situ methods as preservation. An integrated conservation strategy will involve both in situ and ex situ methods as well as in vivo and in vitro techniques.

An important feature of many plant communities is their resilience and ability to resume successional processes even after release from long-term disturbances like grazing that may have been intense for decades. Under conditions of scarcity of financial and technical resources, a rational, scientifically based decision on priorities in conservation would have to rest on an assessment of the importance of the community, its current health and its likely resistance to further change.

In consonance with GEF's adoption of an ecosystem approach to the management of biodiversity, the present paper deals at length with issues relating to the conservation of agrobiodiversity in the coastal, marine, arid and semi-arid, mountain and forest ecosystems. Finally, the catalytic and strategic role of the GEF within the current and planned initiatives aimed at the protection of globally significant agrobiodiversity is discussed.

This book contains the proceedings of a Technical Consultation held at M.S. Swaminathan Research Foundation from January 15 to 18, 1996 for developing an Implementation Framework for Farmers' Rights. The proceedings will show that the long-term conservation of agrobiodiversity will become feasible only if the concept of Farmers' Rights becomes an operational reality.

– *Preface by M.S. Swaminathan*

The comments of the preceding speakers can be grouped into ten general issues which need to be addressed in our discussions of Farmers' Rights:

1) As we plan to draft legislation, we have to more clearly articulate the definition of Farmers' Rights specific to various situations and different types of farmers. This is essential because Farmers' Rights could apply in at least three different situations. The first of these is related to farmers who are cultivators of crops, and who buy their seeds from seed companies. A second is the GIFTS concept: the contribution of information and indigenous knowledge systems. The third situation purely concerns the genetic material which farmers have conserved and enhanced.

2) The next question which has been raised relates to the relative roles of WTO, UPOV, CBD and FAO and the requirements variety protection – homogeneity, novelty and true breeding. From the biodiversity standpoint, we want diversity. So, how do we strike a balance between the requirements of PVP on the one hand and the future of plant breeding on the other, between homogeneity and heterogeneity. What kind of coordinating mechanisms do we have both at the national and international levels? Mention has been made about different ministries of government, not only in India, but also in industrialized countries, dealing with these issues. So, how many actors are there?

3) Any act that has multiple purposes or multiple goals becomes a lawyers' act, leading to litigation. In such a case, no internal consistency can be maintained. Therefore, should we discuss two acts, a sui-generis PVP act, and a Farmers' Rights protection act to be introduced concurrently in the Indian Parliament? The provisions of these acts can be clearly articulated to minimize the scope for confusion and litigation.

4) How do we maintain a global gene fund? Some methodology of providing support has to be worked out. This is not an issue of charity, but of moral and ethical obligation. If we agree that we must have some fund to recognize, reward, revive in situ conservation habits and compensate for the loss of yield, then how do we create that fund? In the United Kingdom, there is a surcharge of thirteen percent on seeds, which goes to the breeder. Here, we suggested 15% : 10% to the breeder and five percent to the Community Gene Fund. We would have to work closely with the seed industry to develop such a system.

5) If we agree to develop two separate acts in a country like India, to be introduced simultaneously in Parliament, one for Farmers' Rights and the other for PVP, then a number of issues have to be discussed because there have to be international ground rules. These include compulsory licensing, the regulation of seed imports, varietal release and quality control. The mechanism for royalties and the administrative system will also have to be considered.

6) Another issue concerns the patenting of micro-organisms. Of course, they are patentable if they are modified by genetic engineering. However, we need to consider what other policies apply here.

7) What should be the position relating to varieties released prior to the enactment of legislation? How far back do we go? Can such a measure be feasibly implemented? We can, of course, control the hybrids because of the parental material, at least in the private sector. In the public sector, the parental material is everywhere.

8) There are also questions relating to the ex situ collections of germplasms made before the Biodiversity Convention came in. This discussion is a matter which could not be resolved before Rio, and it was ultimately left out of the agreement. The large, value-added collections are with CGIAR.

9) Is Farmers' Rights a universal concept or is it something which is more relevant to the South? Of course, the North will have to contribute funds, but what is the universal concept of Farmers' Rights? Since we are an international group, while we discuss largely Indian affairs, it is important to deal with some of the generic issues too.

10) The final issue pertains to the methods of disseminating information and generating awareness, in terms of in situ conservation of biodiversity, recognition, reward and compensation. Indonesia has compulsory introduction of these issues in all syllabuses. I think it is very important that some mechanisms of this kind are introduced, by which sensitization of people on a large scale is carried out. What are people's obligations, duties, responsibilities and entitlements? What is the role of the mass media?

– *M.S. Swaminathan's Summary of Issues Raised*

5.10 *Gender Dimensions in Biodiversity Management*, edited by M.S. Swaminathan. Delhi: Konark Publishers, 1998.

The Convention on Biological Diversity (CBD), adopted at Rio de Janeiro in June 1992 and to which India is a party, defines biological diversity as follows: "Variability among living organisms from all sources including, inter alia, terrestrial, marine and other aquatic ecosystems and the ecological complexes of which they are part; this includes diversity within species, between species and of ecosystems."

CBD stresses the need for respecting, preserving and maintaining the knowledge, innovations, and practices of indigenous and local communities embodying traditional lifestyles relevant for the conservation and sustainable use of biological diversity. It also calls for the promotion of their wider application with the approval and involvement of the holders of such knowledge, innovations, and practices and for the equitable sharing of the benefits arising from the utilization of such knowledge, innovations, and practices (Article 8J).

The need for gender and social equity in sharing benefits from the commercial use of biodiversity cannot be overemphasized, if we are to succeed in preventing further genetic erosion. The Global Biodiversity Assessment published by the United Nations Environment Programme in 1996 has adopted a working estimate of 13.6 million species for the total number of species occurring on earth. Out of this, only about 1.75 million species are currently known to science. India is classified as a megabiodiversity area from the point of view of species richness and agrobiodiversity. However, two of the major biodiversity rich areas, north-east India and the Western Ghats region, are also classified as "hot spot" areas from the point of view of threats to biodiversity. There is thus need for urgent action to strengthen the involvement of the general population in the conservation of biodiversity.

Biodiversity rich countries are characterized by agro-ecological variability. In addition, there is a strong positive correlation between cultural diversity and agrobiodiversity. Women, in particular, tend to conserve and improve plants of value in strengthening household nutrition and health security. It is, therefore, imperative to give explicit recognition to the role of women in genetic resource conservation and enhancement.

Very little research is in progress on the role of women in the conservation and sustainable management of biodiversity. The M. S. Swaminathan Research Foundation (MSSRF) was established in 1989 in Madras, India, for the purpose of promoting a job-led economic growth strategy based on a pro-nature, pro-poor, and pro-women orientation to technology and public policy. MSSRF has established a Resource Centre on Gender and Development for internalizing gender considerations in its research and educational programs and for assisting others who are engaged in a similar endeavor.

While giving operational content to the concept of sustainable development, we should ask some of the questions Dudley Sears, asked about twenty years ago: "The questions to ask about a country's development are: What has been happening to poverty? What has been happening to unemployment? What has been happening to inequality? If all three of these have become less severe, then beyond doubt this has been a period of development for the country concerned. If one or two of these central problems have been growing worse, especially if all three have, it would be strange to call the result development, even if per capita income doubled." (Sears, D, 1969. "The Meaning of Development." *International Development Review*, Vol.11, No.4).

In dealing with issues relating to biodiversity, which, under CBD, is the sovereign property of a nation, we should also ask similar questions. Is the biodiversity management system conducive to the reduction of poverty, promotion of gender equity, and the generation of livelihood opportunities?

Today, both in situ conservation through a protected area network and ex situ preservation through a variety of mechanisms receive support from public funds. Only in situ on-farm conservation by tribal and rural families, which results in genetic material of immense value in plant breeding and genetic engineering, goes unrecognized. This is an activity vital for public good and hence must be supported from public funds. We should ensure that the proposed Indian legislation relating to biodiversity conservation and plant variety protection is based on principles of gender and social equity.

In January 1997, the FAO Regional Office in Bangkok invited MSSRF to prepare a comprehensive report on the Gender Dimensions in Biodiversity Management in India. The draft report was discussed at a workshop held at MSSRF in June 1997. Several case studies were also presented. Arising out of the Workshop, this book includes salient material from the report, the case studies, and the recommendations made by the participants.

Gender studies in relation to biodiversity management are just beginning to attract attention. We must substitute pious platitudes, concerns, and feelings with serious experimental and analytical work. It is hoped that this book will stimulate such work and also be an addition to the as yet scanty literature on gender and biodiversity.

– *Introduction by M.S. Swaminathan*

5.11 *A Century of Hope: Towards an Era of Harmony with Nature and Freedom from Hunger*, by M.S. Swaminathan. Chennai, India: EastWest Books (Madras), 1999.

As we say goodbye to this century and millennium, there is much to celebrate. There has been spectacular progress in every area of human endeavor – science, technology, agriculture, industry, medicine, and, above all, in democratic systems of governance. During the nineties, there has been explosive development both in biotechnology and information technology. However, the pleasure of advancement is diluted by the pain of a growing awareness that the present pathways of "progress" are not ecologically, economically, and socially sustainable. The various Summits and Conferences held during this decade under the auspices of the United Nations and its specialized agencies have revealed that: the rich-poor divide is widening, leading to the extensive co-existence of unsustainable lifestyles and unacceptable poverty; there is growing damage to the basic life support systems of land, water, forests, biodiversity, oceans, and the atmosphere and, prospects of adverse changes in temperature, rainfall, sea level, and ultraviolet-β radiation are becoming real; there is increasing feminization of poverty and agriculture; gender inequity is increasing; economic growth is taking place at the expense of employment, leading to jobless economic expansion; there is growing intolerance of diversity and pluralism in human communities, resulting in numerous mini-conflicts leading to the proliferation of small arms, horrible terms like "ethnic cleansing" are again dominating political vocabulary; color apartheid is being replaced by technological apartheid; above all, the human population is growing in developing countries at a rate that often exceeds the population supporting capacity of the ecosystem.

While such a negative list can be enlarged further, what are the bright spots in this depressing scenario? Apart from the spread of democratic systems of governance and independent media – both print and electronic – the most hopeful gain is the widespread use of the term "sustainable development" by political and social leaders and by scientists and technologists. It is clear that if we are to achieve a transition to a sustainable civilization in the 21st century, we should create an environment where every child, woman, and man can have an opportunity for a productive and healthy life. This, in turn, calls for physical, economic, and ecological access to balanced diets and safe drinking water. Every third child born in India and several South Asian countries is characterized by low birth weight (LBW). LBW children, resulting from maternal and foetal under-nutrition, suffer setbacks in brain development and several serious health problems in adult life. Such enforced handicaps at birth represent the cruelest form of inequity in the emerging knowledge age.

Thus, food and nutrition security at the level of each individual is vital for a productive life. Food security in turn depends upon the security of livelihoods. Ecological security is the foundation on which both food and livelihood security rest. The conservation and enhancement of natural resources and the eradication of poverty, thus, become essential components of a sustainable food and livelihood security system. The biovillage movement launched by the M.S. Swaminathan Research Foundation, therefore, aims to impart a pro-nature, pro-poor, and pro-women orientation to the strategy of job-led economic growth.

Gender equity is the down-to-earth index of a civilization. Seen in this context, the growing imbalance in sex ratios in India is a matter for serious concern. Female foeticide and infanticide are still prevalent in several parts of India. Educational and economic empowerment and enhancement of the self-esteem of women living in poverty are essential for ending this deplorable situation. This is where the initial experience with the computer-aided knowledge centers for sustainable livelihoods operated by women in villages in the Union Territory of Pondicherry is encouraging. These knowledge centers not only help to empower both women and men with information relevant to their daily life, they also actuate the raising of the social status and self-esteem of women. Modern information technology is a powerful tool for fostering social and gender equity.

In a recent lecture, Dr. Bruce Alberts, President, US National Academy of Sciences made the following remarks: "I want to highlight a wonderful example that points the way forward. The M.S. Swaminathan Foundation has established an experimental network in India that will soon connect more than twenty isolated rural villages to a wireless Internet service. About half of the population in most of these villages has a total family income of less than $25 per month. The project is designed to provide knowledge on demand to meet local needs using the World Wide Web, and it does so through a bottom-up process. The process starts with volunteer teams that help poll the villagers to find out what knowledge they want. Particularly popular thus far are women's health information, advice on growing local crops and protecting them from diseases, the daily market prices for these crops, local weather forecasts, and clear information about the bewildering array of programs that are provided by the

Indian government to aid poor families. To participate, each village must provide a public room for the computer system, as well as the salaries for a set of trained operators. In return, the village receives the needed hardware and maintenance for the communication system, specially designed Web sites in the local language that convey the requested information, and training programs for those villagers who have been selected to run their local knowledge system. ... Drawing on this concept, I envision a global electronic network that connects scientists to people at all levels – farmers' organizations and village women, for example. The network will allow them to easily access the scientific and technical knowledge that they need to solve local problems and enhance the quality of their lives, as well as to communicate their own insights and needs back to scientists."

This leads to a broader question of technology and equity. Starting with the industrial revolution in Europe, the gap in economic wealth between industrialized and developing nations began to increase. Technology became an important factor in enlarging economic and gender inequity. Modern information technology provides an opportunity to halt this trend and to make technology an ally in the movement for economic and gender equity. Frontier technologies like biotechnology, space, communication, information, and renewable energy technologies can all be of great help in fostering sustainable advances in agriculture and industry. Given appropriate public policies, the risks and benefits associated with new technologies like genetic engineering can be carefully assessed. Knowledge and skill empowerment as well as appropriate management and institutional structures are vital for ensuring that technological benefits reach the unreached.

It has been shown that people who until recently were technologically backward – such as the Australian aborigines – routinely master industrial technologies when given opportunities to do so. This is also our experience in the biovillages. Women belonging to the socially and economically underprivileged sections of rural society master modern computer technology within a couple of months. As long back as 1972, in my Princess Leelavathi Memorial Lecture delivered at the University of Mysore, I called for a techniracy movement, based on the pedagogic method of learning by doing. The techniracy concept led to the establishment of Krishi Vigyan Kendras (Farm Science Centers) in different parts of the country. Formal literacy is not an essential pre-requisite for learning new skills in Krishi Vigyan Kendras.

India has the opportunity for demonstrating the pathway to a sustainable and humanistic civilization, rooted in the principles of ecology, gender and social equity, economics, energy conservation, and employment generation. Knowledge is a continuum. We should neither worship nor discard technologies because they are old or new. What is important is to use knowledge and skills in a manner that children can be born for happiness and not just for existence. We need an ecology and technology of hope and not of despair.

History teaches us that all human societies contain inventive people. Only, some environments provide more favorable conditions for stimulating and utilizing inventions than do others. There is an urgent need to recognize and reward the invaluable contributions of tribal and rural women and men to genetic resources conservation and improvement. We need to provide space to everyone in the knowledge society of the 21st century.

The coming century needs: a) an ecology of hope and not of despair; b) environmentalists who can show how to 'do' rather than merely say 'don't'; c) an ever-green revolution on the farms leading to productive and sustainable farming systems and happy farming families; d) a socio-economic environment where children are born for happiness; e) efficient and successful business and industry leading to business leaders and industrialists considering themselves as "trustees" of their surplus wealth and using it for public good; f) the highest priority being accorded to the elimination of both unsustainable lifestyles and unacceptable poverty; g) the emergence of humanistic science where scientific research is based on ethical values and is designed to meet basic human needs.

– *Preface by M.S. Swaminathan*

5.12 *Rio de Janeiro to Johannesburg* by M.S. Swaminathan. Chennai, India: EastWest Books (Madras), 2002.

In 1992,1 wrote a book titled *From Stockholm to Rio de Janeiro*, briefly reviewing the progress made in achieving the quest for harmony between humankind and nature from the time of the UN Conference on the Human Environment held in Stockholm in 1972 up to the UN Conference on Environment and Development held at Rio de Janeiro in 1992.

The UN Summit on Sustainable Development now scheduled to be held at Johannesburg, South Africa, from August 26 to September 4, 2002, was originally to be held from September 2 to 11, but the dates were brought forward. The reason was that the dreadful date, September 11, marked the entry of humankind into an era where the fear of one another has reached unprecedented heights. The Johannesburg Summit is being convened ten years after the Earth Summit held at Rio de Janeiro in 1992 and thirty years after the UN Conference on the Human Environment held at Stockholm in 1972, to take stock of the current status of our common ecological future and to examine how far the concept of sustainable development has been converted from a desirable goal into practical accomplishments.

The Stockholm Conference resulted in the establishment of the UN Environment Program (UNEP) at Nairobi, as an instrument for shaping global governance of environmental issues. Unfortunately, UNEP could not live up to its early expectations due to constraints of financial and political support. It is high time UNEP is developed into a World Environmental Organization, on the lines of the World Trade Organization. Even if the name is not changed, UNEP's mandate and political backing need to be enlarged.

At Rio, several international Conventions like those relating to Climate and Biodiversity as well as an Agenda-21 containing guidelines for sustainable development were adopted. Later, a Convention on Desertification was taken up. The UN Convention on the Law of the Sea also came into force. More recently, an International Treaty on Plant Genetic Resources for Food and Agriculture was ratified in the forum of FAO. Fortunately, the Global Environment Facility

(GEF) set up just prior to Rio has been receiving good financial support and is rendering extremely valuable service. GEF has saved many biodiversity and hydrologic hotspots from total destruction and has been effectively serving as the financial mechanism for the Climate and Biodiversity Conventions.

In spite of all these positive signals, sustainable development is still a far cry, particularly if the social dimension is added to those of ecology and economics. There is a growing violence in the human heart, to some extent due to a feeling of social exclusion and injustice on the part of those who feel that they are losers in the present pattern of development. Mahatma Gandhi asked many decades ago: "How can we be non-violent to nature, if we are going to be violent to each other?" It is obvious that the principles of ethics and equity enshrined in the Convention on Biological Diversity have not been extended to human diversity. The increasing intolerance of variance and pluralism in human societies underlines the need for a Convention on Human Diversity that will help to foster understanding and appreciation of differences in terms of gender, color, religion, languages, ethnicity, or political belief.

We see today increasing conflicts in regions characterized by gross social and economic inequity. Thousands of young men and women in Johannesburg, as in many other cities of the world, have no jobs and no hope for a healthy and productive life. However, they often have guns, leading to a threatening climate of violence. It is likely that at the Johannesburg Summit more money will be spent on protecting the participants than what the wealthy nations are likely to pledge for protecting the planet. Like other UN meetings, Johannesburg has poverty eradication as one of its goals. If speeches and statements would have led to poverty eradication, we will not be witnessing the increasing rich-poor divide today. According to UNDP, two hundred ultra-rich persons in the world currently earn more than the two billion living in poverty.

Damage to biological and human heritage is also continuing unabated. Since Rio, some highly disturbing and anti-conservation and anti-sustainable development terms like "ethnic cleansing", "bio-piracy", and "bio-terrorism" are frequently appearing in the mass media. The threat of nuclear warfare is again looming large on the horizon, although this had receded to the background after the end of the Cold War and the breaking of the Berlin Wall. While until recently the eradication of smallpox was considered a triumph of modern immunization technology, steps have recently been taken in the United States to inoculate several groups of professionals against smallpox as a precaution against a possible terrorist attack. Similarly the "dirty bomb" scare has led to severe restrictions on the handling of radioactive materials in research laboratories.

What then is the role of the Johannesburg Summit? Can it shed some light where only darkness prevails now? What should be the outcome of the Johannesburg Summit?

'Do' Ecology: A Message of Hope for the Johannesburg Summit

If I were to describe my hope in two words, it is "do ecology". In 1973, soon after the Stockholm Conference on the Human Environment, I elaborated this concept in my Coromandel Lecture, 'Agriculture on the Spaceship Earth': "The environmental policies advocated in the richer nations are designed to protect the high standard of living resulting from the unprecedented growth in the exploitation of natural resources during the last century. It is of necessity a policy based on a series of 'don'ts'. The poor nations, in contrast, are faced with the desire and need to produce more food from hungry and thirsty soils, more clothing, and more housing. They hence need a 'do ecology' and not just a 'don't' philosophy."

In operational terms "do ecology" involves the creation of new eco-jobs. Since eco-jobs are knowledge intensive, the knowledge era, initiated by the digital, space and biotechnological revolutions, provides uncommon opportunities for realizing the goal of jobs for all.

The M.S. Swaminathan Research Foundation (MSSRF) organized an interdisciplinary dialogue on the theme "Environment, the New Economy, and New Employment" at Chennai from January 28 to 31, 2002 with support from the Global Environment Facility. Its main purpose was to identify opportunities for fostering environmentally and socially sustainable job-led economic growth. Today we know that the famine of jobs or livelihood opportunities resulting in inadequate family income is the basic cause of the famine of food at the level of individuals. Where hunger rules, peace cannot prevail. Sustainable employment or livelihood opportunities hold the key to peace and progress.

Though our planet is politically divided by frontiers and concepts of sovereignty, our fates are ecologically intertwined. While ecology and communications unite us, economics is unfortunately serving as a divisive force. Trade is becoming free but not fair. Globalization should be designed in a "win-win" mode for all and should not result in creating a large number of losers and some winners. This can be achieved if we make opportunities for sustainable livelihoods/employment opportunities for youth as the yardstick for measuring the beneficial impact of globalization. "Jobs for all" should be the bottom line of trade arrangements, including export and import policies. WTO should subject its agreements to a livelihood impact analysis.

Thanks to the onset of the age of innovation and invention, we have exceptional opportunities for fostering job-led economic growth. For this promise to be realized, political will and action as reflected in priorities in investment in human resource and infrastructure development are a must. There is no single or simple solution to the problem of overcoming the growth of unemployment in the world. The problem is so vast and yet the opportunities are so great that we should foster the fusion of political will, professional skill, and national and global partnerships, to achieve the goal of sustainable livelihood opportunities for all. We can learn from successes, as for example from the integrated approach to on-farm and non-farm employment adopted in China through the Rural Township Enterprises program. The cooperative dairy movement in India, based on small-scale production and on animal nutrition based largely on agricultural residues, is another good example of production by masses, since this provides over fifty million women with secure livelihoods. Global, regional, and national partnerships will be necessary to achieve a paradigm shift from jobless to job-led growth.

Youth constitute the majority of population in most developing countries. If governments acknowledge the opportunity to work and thereby earn one's livelihood as a basic right as well as a basic prerequisite for internal and international peace and human security, they should design strategic interventions which will help to build the self-esteem and competence needed to undertake self-employment. Unity of goals, but multiplicity of approaches, will be needed to achieve success. The complex problem of youth employment and its multiple dimensions will have to be dealt with in a desegregated manner, as for example, first generation learners (i.e., the children of illiterate parents) as well as "school push-outs" (often girls) and handicapped youth will have to receive special attention. Without peace and security, there can be no progress in any field of human endeavor. Therefore, if youth employment efforts go wrong, nothing else will have the chance to go right.

I would like to suggest a framework for generating sustainable employment and livelihood opportunities for all. The basic strategy would include enhanced opportunities for skilled employment in the primary, secondary, and tertiary sectors of the economy; an integrated approach to technology, training, techno-infrastructure, and domestic and export trade; public policies for enlarging the space for remunerative and market-driven self-employment; and, consortiums of public and private sector companies for fostering assured marketing opportunities through buy-back arrangements.

To operationalize this strategy we would need to: a) promote location-specific jobs, based on the natural and human resource endowments of the area; b) carry out livelihood/job impact analysis of all technology-driven development projects; c) form in each compact area a "Jobs for All Consortium" consisting of appropriate representatives of government and non-government organizations, business and industry, service oriented civil society associations, women's groups, financial institutions, bilateral and multilateral donors, and the mass media; d) organize a network of training and capacity-building institutions capable of imparting market-driven skills; e) develop marketing infrastructure needed to link primary producers with domestic and international consumers, and promote an Employment Infrastructure Fund as well as Venture Capital Fund for small-scale decentralized production enterprises; f) encourage, whenever appropriate, a "production by masses" approach, rather than "mass production" and labor-displacing technologies; support such decentralized small-scale production with key centralized services in order to reduce transaction costs; g) develop institutional structures which can confer the power of scale to small producers both in the production and post-production phases of the enterprise; h) facilitate the organization by young entrepreneurs of effective structures like Eco-enterprises Parks, Biotechnology Parks, Food Parks, Horticulture, Poultry, and Dairy Estates, Renewable Energy Parks, etc.; i) assist in creating more non-farm employment opportunities in rural areas and provide opportunities to young farm, commerce, home science, veterinary, fishing, and engineering graduates to organize socially relevant and economically viable services; j) help unemployed youth to acquire the necessary self-confidence and skills to embark upon initiatives which can help to bridge the digital, genetic, gender, and other divides;

k) devise short-term non-degree training programs at appropriate universities and technical institutions to help build the capacity of youth to harness both traditional and frontier technologies. The aim should be to assist the country to progress rapidly in the technological transformation of crop and animal husbandry, fisheries, agro-forestry and forestry, agro-processing, and small rural and urban enterprises.

The report of the Brundtland Commission was titled *Our Common Future*. The period from Rio to Johannesburg has amply demonstrated that there can be no happy common future for humankind without a better common present. However, extending the extrapolation domain of successful experiences and examples will be possible only if we understand the basic principles underlying such efforts. We should emulate the example of the DNA molecule (the basic unit of heredity and life) and design programs capable of replication, recombination (i.e., marriage of successful experiences), and mutation (mid-course correction when needed, or even total recasting of strategies where appropriate). Thus, unique successes can be converted into universal progress.

Food and drinking water occupy the first position in the hierarchical needs of human beings. I have, therefore, chosen sustainable food and water security as the index of our progress towards achieving the humanistic goals of Stockholm, Rio de Janeiro, and Johannesburg. Appropriate in this context is George Herbert's verse:

Sweet Peace, where dost thou dwell?
I humbly crave,
Let me once know.
I sought thee in a secret cave,
And ask'd if Peace was there,
A hollow wind did seem to answer, No:
Go seek elsewhere.
At length I met a rev'rend good old man:
Whom when for Peace
I did demand, he thus began:
"Take of this grain, which in my garden grows,
And grows, for you;
Make bread of it: and that repose
And peace, which ev'rywhere
With so much earnestness you do pursue,
Is only there."

– *Introduction by M.S. Swaminathan*
UNESCO Chair in Ecotechnology, Chennai

73. MSS, 1974

5.13 ***Groves of Beauty and Plenty: An Atlas of Major Flowering Trees in India,*** **by M.S. Swaminathan and S.L. Kochhar. Delhi: Macmillan India, 2003.**

From the very early times human beings have co-existed with nature, and forests have played a crucial role in their survival, providing the basics of life including food to eat (such as roots, succulent herbage, seeds, fruits and nuts and mushrooms at the forest floor), shelter (from wind, rain, snow and burning sunshine), wood (fuel for warmth, tools and weapons), some clothing and a host of other auxiliaries of life. In addition, they have helped to maintain a balance between carbon emissions and absorption because of their role in carbon sequestration. Even today humanity's well being and ultimate survival are inextricably linked with natural resources, especially trees, for varying reasons, such as economic, ecological, and aesthetic, i.e., by providing a setting for recreation, sports, and beautifying the landscape, be these in city roads, parks, gardens, public places and even the compounds of our homes. We have at present in use over 4,500 products that are derived wholly or in part from trees.

More than a thousand different kinds of trees occur naturally in India and many linguistic scholars and nature lovers have been struck by the beauty, richness and almost mystic allure of tree life. We find charming descriptions of most of our beautiful native trees in Kalidasa's *Ritu Samhara*, Valmiki's *Ramayana*, and sage Vyasa's *Mahabharata*. They have also been featured in our religions and ancient paintings and sculptures. In India, as perhaps in many other countries of the world, trees have been in veneration since ancient times. Most Hindu temples have specific trees designated as "temple trees". Some other trees, to mention a few, have been introduced at times from near and far off lands and have become an integral part/or component of our landscape, e.g., the famous Flamboyant or Flame tree (*Delonix regia*) and Traveller's Palm (*Ravenala madagascariensis*) - both from the Malagasy Republic (earlier known as Madagascar); the Frangpani (*Plumeria spp.*), the Rain tree (*Samanea saman*), the Cannonball tree (*Couroupita guianensis*), and the Jacaranda with delicate fern-like foliage and mauve flowers (*Jacaranda mimosaefolia*) - all from the New World tropics; the orange-flowered African Tulip or Squirt tree (*Spathodea campanulata*) and the Sausage tree (*Kigelia pinnata*) - both from tropical Africa; the Bird of Paradise (*Stretilitzia spp.*) from South Africa; the dazzling Golden Shower tree (*Cassia fistula*) from southeast Asia; and the Bottle Brush (*Callistemon spp.*) from Australia.

The number of trees is too large, making it impossible for all species to be included here. This Atlas covers close to one hundred and twenty-five most striking and widespread trees (both native and exotic) from this bewildering profusion. The Atlas is an irresistible and vivid compendium portraying botanical details through more than four hundred color photographs. The text includes, besides the common and botanical names together with their synonyms, information on many aspects such as the names of taxonomic families, chromosome number, etymology (how trees have derived their names), their native place and geographical distribution, mythological notes, religious significance, references in our folklore, a detailed botanical description, phenology (the time of leaf fall, flowering, and fruiting), preferred soil types, propagation and their economic importance. A brief account of closely related species is also given to make the Atlas user-friendly. Local vernacular names in different Indian languages have been provided wherever available. A separate comprehensive index both for English common and botanical names is given at the end together with selected reading material for further reference.

It is hoped that this Atlas would be useful to students and scholars of botany, horticulture, forestry, environmental studies, and indeed to any one who has interest in nature and natural resources, be they amateurs, gardeners, nurserymen, administrators, or scientists. Tourists traveling in the tropics will encounter a world of strange and beautiful trees not only in the hotel complexes, streets, parks and gardens but also in the wild. By using the colored photographs (showing details of foliage, flowers and fruits) and the text together, it would be possible for them to identify with ease the individual species that grow in our neighbourhoods.

We wish to end with two quotations to remind ourselves that we are in this planet as guests of green plants. Lord Buddha once said that the tree is the only living organism capable of unilateral love, since it provides shade even to the axeman who comes to cut it. It will be appropriate to end with the following poem by Joyce Kilmore:

I think that I shall never see
A poem as lovely as a tree
A tree whose hungry mouth is prest
Against the earth's sweet flowing breast
A tree that looks at God all day
And lifts her leafy arms to pray
Poems are made by fools like me
But only God can make a tree!

— *Preface by M.S. Swaminathan and S.L. Kochhar*

74. Title cover of *Groves of Beauty and Plenty* (2003)

75. Title cover of *Wheat Revolution: A Dialogue* (1993)

76. Title cover of *Agrobiodiversity and Farmers's Rights* (1996)

77. Title cover of *A Century of Hope* (1999)

78. Title cover of *From Rio de Janeiro to Johannesburg* (2002)

Section 6.
Chapters from Books on M.S. Swaminathan

6.1 "Editors' Preface" from *Science and Agriculture, M.S. Swaminathan and the Movements for Self-reliance*, ed. by S. Ramanujam, E.A. Siddiq, V.L. Chopra, S.K. Sinha, p. iii. New Delhi: Venus Printers and Publishers, 1980.

This volume attempts to present the major contributions of Dr. M.S. Swaminathan to agricultural research, education and extension over a period of three decades, from October 1947 when he joined the Indian Agricultural Research Institute, New Delhi, as a post-graduate student to April 1980 when he opted to retire from service and joined the Planning Commission. The contributions that Dr. Swaminathan made during this period have been recognized by many learned societies and international bodies as well as by Government and farmers. However, even more important than his own personal researches, have been the team spirit, dedication and commitment to rural welfare that Dr. Swaminathan inculcated into all those who came into contact with him. It is a measure of his excellence as a teacher that many flourishing schools exist, which enthused by his example, will carry on his work even when Dr. Swaminathan, because of the pressure of his other commitments is not able to spare much time to guide them. We hope that this compilation, made at the initiative of several learned societies, will serve to highlight the various qualities of Dr. Swaminathan as a scientist, as a leader and as a person committed to the uplift of his disadvantaged countrymen. ...

6.2 "Foreword" by B.P. Pal, Chairman, National Committee on Environmental Planning & Coordination, Department of Science & Technology, New Delhi (October 16, 1980). From *Science and Agriculture, M.S. Swaminathan and the Movements for Self-reliance*, ed. by S. Ramanujam, E.A. Siddiq, V.L. Chopra, S.K. Sinha, pp. v-vii. New Delhi: Venus Printers and Publishers, 1980.

Agriculture in India on which this country with its huge population very largely depends for its food, raw materials and other basic necessities, covers in its broad sweep so many interdependent areas vital for the well-being of the people, that it may be said to be a world of its own. In this agricultural world, Dr. M.S. Swaminathan stands out as a towering figure closely associated with the tremendous changes which have taken place in the last decade and a half, transforming a traditional agricultural system into one based on modern scientific advance.

From a food-importing country, India has progressed to a comfortable position in which there is sufficient food in the country, and the foundation has been laid for it to become even a food-exporting country. While this is the result of notable teamwork by a large number of people including the farmers of the country who have made a remarkable contribution to this agricultural revolution, the great advance in production has been firmly based on solid scientific foundations – by the creation by scientists of new genotypes of crop plants which smashed the old yield barriers and the introduction, on the basis of modern experimentation, of new integrated agronomic techniques, including plant protection measures, thereby ensuring the optimum utilization of soil and water resources. In the evolution of this great developmental process, starting with basic researches and proceeding through applied research and extension methodology to an integrated strategy to cover the whole country, the name of Dr. Swaminathan is closely and intimately linked.

Throughout his career, whether actively engaged in research and teaching or, later, in administration and planning, he has always paid special attention to the importance of the role of students, of learned bodies and even non-formal groups, in nation-building activities including rural reconstruction. It is very fitting therefore that a large number of his students and a number of learned societies should have brought together in this volume, Dr. Swaminathan's many contributions in the field of basic and applied biological research and in agricultural research and education, in the transfer of agricultural technology and rural development, and also in planning and scientific administration.

It gives me special pleasure to write this Foreword because I have known Dr. Swaminathan when he was a student at the Indian Agricultural Research Institute from 1947 to 1949, and was closely associated with him when later he joined the staff of the same Institute in 1954. I have noted with great admiration his outstanding contributions to the basic sciences of cytology and genetics, his notable successes as a breeder of superior varieties of wheat, rice and other crops, and his development into an agricultural planner and administrator whose interest and vision encompasses the entire gamut of agriculture and the problems associated with it. Above all, I have been impressed by his deep humanity and his ceaseless endeavors to see that the fruits of development reach the people, and especially the poor and the under-privileged. When after a brilliant academic career he had the choice of a number of professions open to him and he chose agriculture, it was great good fortune for this country.

Dr. Swaminathan's significant contributions to research and teaching, development and planning, and indeed to all aspects of agriculture, are well-known and have brought him international recognition and many prestigious awards. As a brilliant speaker who can collate vast amounts of data and present conclusions with remarkable lucidity, he is much sought after, both in national and international seminars and symposia. But in spite of all this, he has remained the same gentle and studious person, freely available to all those who seek his advice or assistance, a friend of the student and equally of the farmer.

The present time is a difficult and critical one. While, on the one hand, the very large human population is increasing at a very fast rate, the demands for food, fuel, fodder and other resources are expanding greatly. Also, development must proceed in such a planned manner that there is the least possible damage to the environment and the life support systems which sustain us. It is on leaders of the stature and the vision of Dr. Swaminathan that the country will depend for guidance. The present volume on Dr. Swaminathan and his contributions in various fields is very welcome. It contains an authentic account of a stirring period in India's agricultural advance and a life sketch of one who contributed so much to it. I am confident that it will serve as a source of inspiration to all those who have interest in science and agriculture.

6.3 "Message" by K.R. Narayanan, President of India, New Delhi (May 6, 2002) from *Scientist and Humanist: M.S. Swaminathan,* by R.D. Iyer, p. xi. Mumbai, India: Bharatiya Vidya Bhavan, 2002.

Renowned agricultural scientist, M.S. Swaminathan, has been working with dedication since 1947 for the agricultural development of our country by harnessing both traditional and frontier technologies. He has been consistently remaining in the forefront of India's quest for food security on a sustainable basis. Acknowledged worldwide for his outstanding scientific work, he has significantly contributed to the practical application of agricultural science for higher yields of foodgrains and for ecological security.

While ensuring accelerated agricultural growth through applications of science and technology, Professor Swaminathan has been relentlessly focusing attention on ecological conservation and on gender and social equity. This integrated approach has remained one of the distinguishing aspects of Professor Swaminathan's original and pathbreaking contributions towards the revolutionary growth of agriculture in our country and indeed the world. He articulated his concerns for ecological security and gender and social equity in 1968 and later continued his mission of giving a gender dimension to our development when he, for the first time in the history of planning in Independent India, authored a separate chapter for the Sixth Plan entitled "Women and Development". Subsequently, he coined the term "Evergreen Revolution" not only to stress the augmentation of production but also to refashion the whole strategy of the Green Revolution to make it pro-environment, pro-women, and pro-poor. His approach in pioneering the evergreen revolution is at the very heart of what is now called sustainable agriculture.

6.4 "Foreword" by R. Venkataraman, Former President of India, New Delhi (April 21, 2002) from *Scientist and Humanist: M.S. Swaminathan,* by R.D. Iyer, pp. xiii-xv. Mumbai, India: Bharatiya Vidya Bhavan, 2002.

Arnold Toynbee, the renowned historian, once said, "It is absurd and unjust to classify a person once and for all as being first class or third class when he is only twenty-two years old. There are slow growers who blossom late in life and conversely there are brilliant starters who fail to fulfill their earlier promise." Dr. Swaminathan belongs to a rare group which is brilliant at school as well as in later life. He is one of the brilliant starters winning several gold medals for excellence in studies and who more than fulfilled his early promise. His career has been one of continuous success in every sphere of activity.

Even in his younger days, Dr. Swaminathan was deeply distressed by the poor yield of lands and the poverty of agriculturists. He had learnt that the yield in India was far below that of the United States, Japan and Thailand. The Bengal famine of 1942-43 and the harrowing tales of the misery of millions, induced him to turn his attention to agricultural education. No wonder that when a few years later he was chosen by the Public Service Commission for the Indian Police Service, he preferred to continue with the subject of his devotion, namely, agriculture. Again in 1954, when a University in the United States offered him a regular teaching-cum-research professorship, Dr. Swaminathan declined and returned to India for carrying on his research for increasing food production in the motherland.

And he did achieve his goal in increasing food production in India spectacularly through the Green Revolution. Addressing the Asian Development Bank at Manila, I once described the pathetic conditions of the agriculturist before the Green Revolution with a story: An honest farmer cultivated five acres of land and ran into debt year after year, until at last the debts swelled more than the value of the land he farmed. He, therefore, abandoned the farm and ran away wearing only a loincloth. As he ran aimlessly, he came across a figure standing stark naked. The farmer ran towards the figure and caught hold of its legs and exclaimed, "Brother, I farmed five acres of land and I have run away with a loincloth; how much more did you cultivate that you have run away even without it?" ...

From that distressing situation where we were unable to feed three hundred and sixty million people in 1950 and went about begging for food aid, India has increased its food production from fifty million metric tons in 1950 to over two hundred million metric tons and has become self-sufficient in food to cover over one billion people. The person who achieved it is the miracle man, Dr. Swaminathan. He experimented with wheat and achieved tremendous success. With the Green Revolution, Dr. Swaminathan had the fullest support of his Minister Shri C. Subramaniam who gave every encouragement and assistance to this project.

Details of the technical process involved in the change have been set out by the author, Dr. R.D. Iyer, in this volume. Likewise, elaborate accounts of several other areas where Dr. Swaminathan has made valuable original contributions have each been separately dealt with.

The last chapter captioned "The Man and His Philosophy of Life" is a fascinating account of the qualities of head and heart of Dr. Swaminathan. His close association with Dr. Swaminathan has enabled the author to bring out Dr. Swaminathan's great qualities as a person, a scientist and a teacher. His father, Dr. Sambasivan, was an eminent doctor who rendered free medical aid to the poor, a great nationalist and an able administrator. As Chairman of the Kumbakonam municipality, he improved the condition of the city. Dr. Sambasivan had also the great reputation of having eradicated the notorious Kumbakonam mosquitoes. Gandhiji stayed in Dr. Sambasivan's house twice during his visit to Kumbakonam. His visit reverberated throughout Tamil Nadu.

As a fellow Tanjorean, I have been fully familiar with all these events. Whatever else Dr. Swaminathan inherited or not from his legendary father, he certainly imbibed qualities such as selflessness, compassion, devotion to duty and the country. Soft-spoken, friendly, and modest without the remotest trace of ego, Dr. Swaminathan has grown into a lovable personality.

This splendid biography compiled by Dr. R.D. Iyer, himself an eminent plant geneticist and the former Head of the Central Plantation Crop Research

Institute, adds light and luster to the multi-splendored achievements of Dr. Swaminathan.

The appendices contain a stunning list of Dr. Swaminathan's writings, speeches and reports. They are bound to be of utmost value as reference material to students and researchers. There is also a list of awards and honors conferred on him. To satisfy human curiosity, I may mention that Dr. Swaminathan has so far received twenty-eight National Awards, twenty-four International Awards, thirty-two Honorary Degrees from India, eleven from abroad, and fifteen positions in International Organizations and Committees. Besides he has chaired thirty-one National Committees. The latest feather to his loaded cap is his election as President of the Pugwash movement devoted to human welfare and non-proliferation of nuclear weapons. In the long history of this movement for over fifty years, Dr. Swaminathan is the first Indian to be chosen for this prestigious position.

The author, Dr. Iyer, deserves compliments for a complete and comprehensive survey of the life and achievements of one of the most eminent scientists that India has produced. I am sure this volume will help younger generations to emulate Dr. Swaminathan as a scientist and as a role model of fine human virtues.

6.5 "Swaminathan, The Institution Builder" from *Scientist and Humanist M.S. Swaminathan*, by R. D. Iyer, pp. 125-50. Mumbai, India: Bharatiya Vidya Bhavan, 2002.

When we build, let us think that we build forever. – John Ruskin

One of Swaminathan's greatest strengths is in institution and capacity building. He is a meticulous strategist and administrator. He is a man of vision too, for without the means of looking ahead, no viable institution can be set up.

National Sphere

At IARI

Ever since he joined the staff of IARI in October 1954, Swaminathan established his identity as a tireless scientist, teacher par excellence, careful planner, and a man of action. Pretty soon, the Swaminathan School of cytogenetics and mutation breeding received worldwide recognition through a spate of outstanding publications appearing in reputed journals. This attracted a constant flow of eminent science leaders from all over the world, who came to see this work. Swaminathan too was invited to important symposiums and seminars in India and abroad, which earned him laurels and appreciation as an outstanding scientist and an articulate speaker known for his clarity of thinking and expression. In 1963, he was invited to serve as the Vice President of the International Congress of Genetics held at The Hague, perhaps the youngest ever to occupy this position. Swaminathan also gave opportunities to his colleagues and students to visit leading institutions in India and abroad, and present papers.

In 1958, the IARI post-graduate school was granted the status of a deemed university to award M.Sc. and Ph.D. degrees. Swaminathan was the moving force behind this. He has been one of the chief architects of post-graduate education in agriculture in our country, since the IARI pattern has been replicated in the chain of agricultural universities that have been subsequently established. As head of the Botany Division, he took active measures to augment research programs, laboratory infrastructure, and farm facilities. He got the sections of Biochemistry, Plant Introduction, Plant Physiology, and Seed Technology elevated to the status of independent divisions, enabling their further growth in infrastructure facilities and staff positions. He also started a separate Division of Agricultural Chemicals to undertake research on botanical pesticides (neem, mahua, tobacco, etc.) and pesticide residues in farm products. In addition, he established the Fundamental Genetics Wing for facilitating advanced research in the fields of biometrical, microbial, and molecular genetics.

Another significant achievement was the setting up of a field-irradiation facility called "gamma garden", on the model of a similar facility at the Brookhaven National Laboratory in the USA, for exposing plant materials to chronic doses of Cobalt-60 gamma rays from a source of the isotope placed inside a central column of the garden, which could be operated safely from outside by remote control. Subsequently, a laboratory irradiation facility called the "gamma cell" was established for exposing seeds, microbes, fruit flies, and culture media to acute doses of Co-60 gamma rays. These facilities were made available to scientists all over India, which led to a spurt in the areas of radiation genetics and induced mutagenesis, and in mutation breeding researches in India. He was also instrumental in getting a "swimming pool" type high energy atomic reactor set up at the World Agricultural Fair held in Delhi, where IARI and other scientists carried out experiments on radiation genetics. He always stressed that radiation genetics is only a tool and that mutation and Mendelian breeding should be taken up together. "Do not worship any research tool, but use those which will help you reach your desired breeding goals speedily and surely," was his constant advice to his students.

One more landmark was the establishment of several project directorates and inter-disciplinary research schools, the most significant ones being the Nuclear Research Laboratory designed in modular fashion similar to the Bhabha Atomic Research Centre in Bombay. This was equipped with modern analytical tools such as the nuclear magnetic resonance (NMR) analyzer for estimating micro-quantities of fatty oils in even a single seed of crops like mustard, in a non-destructive manner, so that the selected mutant seed could be planted to raise the progeny for trials.

Another important step taken by Swaminathan was the setting up of a Quality Research laboratory equipped with an automatic amino-acid analyzer for rapid screening of cereals (such as wheat, rice, and maize) for essential amino acids like lysine. The water technology center, the algal laboratory, the pulses laboratory, and the maize, wheat, and oilseeds directorates all came up during this time. Tissue culture research also received much-needed boost with the setting up of an anther culture laboratory for exploitation of haploids in heterosis breeding in rice and other crops. Swaminathan was responsible for the setting up of a Drosophila research unit, for which he sent one of his Ph.D. students to be trained in Dr. O.S. Reddy's laboratory in Osmania University, Hyderabad. He also gave fillip to research in mammalian cytogenetics, including human genetic work on congenital defects, carried out in close collaboration with blood banks and the medical fraternity in Delhi.

During his directorship of IARI, the library, considered the best in Asia for agricultural sciences, was expanded considerably through a generous grant from the Rockefeller Foundation. Thus, when he left IARI in 1972 to take over as Director General of the Indian Council of Agricultural Research, Swaminathan had elevated the institution to international standards in both research and post-graduate teaching.

At ICAR

In keeping with his expansive mind and wide canvas of research and development activities, Swaminathan expanded the sphere of activities of ICAR through the establishment of a large number of new research institutes and national bureaus for plant, animal, and fish genetic resources, national research centers and project directorates for several crops and for agro-forestry, besides many more all India coordinated projects.

In 1972, soon after his taking over as Director General of ICAR, Swaminathan introduced the concept of "techniracy" (technical literacy) to impart training in the latest technical skills entirely through work experience, in order to bypass the problems created by illiteracy. A chain of Krishi Vigyan Kendras (Farm Science Centers) and later Vann Vigyan Kendras (Forestry Science Centers) were established for this purpose. He strongly believed that "if we can make the jump from the pre-literate to the post-literate world, without passing through the intermediate phase, we can save enormous resources and time".

He took pains to reorganize ICAR governance by effecting major changes in personnel policies, decentralization of authority at all levels, making the scientist and the farmer as focal points of attention, and minimizing bureaucratic controls. One major outcome of his efforts to revamp personnel policies was the introduction of the Agricultural Research Service (ARS), as the equivalent of the IAS cadre, through a rigorous All India competitive examination. The various disciplines of agriculture and related fields constituted sub-cadres under ARS. Selections were done by an autonomous Agricultural Scientists Recruitment Board (ASRB), the equivalent of UPSC, as an exclusive body for recruiting agricultural scientists for the ICAR system. ARS signified a paradigm shift (flexible compensation) from a post-oriented to a scientist-oriented strategy. This not only enhanced the prestige of the agricultural scientist and improved the salary structure considerably; above all, Swaminathan's philosophy of "competing with your own past performance" rather than "competing with each other" changed the entire outlook of research scientists, who could achieve professional progress in their own scales through productive work without worrying about vacancies or competition from colleagues.

Decentralization of authority as envisaged by Swaminathan gave enormous powers to the Directors of Institutes in building up facilities, effecting departmental promotions for scientific, technical, and administrative staff up to a certain level, and other policy matters. It was his vision and hope that they in turn would share some of the powers down the line with their heads of divisions/

sections. He always exhorted Directors to "never deny your staff their legitimate rights and privileges", which alone can help in the smooth running of an institution. Of course, he would also remind the staff that "every right accrues from a duty well done. There can be no claim for rights without discharging responsibilities". Any Director who imbibed this "Swaminathan approach" was sure to succeed in his institution and human capacity-building goals, rather than "empire building" around himself.

Revitalizing the Indian Science Congress

For the first time in the sixty-three year history of the Indian Science Congress, Swaminathan introduced the focal theme concept, which was implemented in the 1976 Science Congress held in Waltair, under his general presidentship. The theme chosen by him and ratified by the General Body at the 1975 Delhi Science Congress was "Science and Integrated Rural Development". Swaminathan meticulously planned this theme for a year, by circulating the structural and organizational changes he visualized, so that everyone came prepared for his/her presentation, resulting in meaningful discussions. Finally, a task force was constituted to implement follow-up action on recommendations, which were to be presented at the next year's Congress. On the last day of the Science Congress at Waltair, a concrete plan of action for bringing about a new social contract between scientists and rural communities was presented to the Deputy Chairman of the Planning Commission. Finance Minister C. Subramaniam provided Rs. 15 crore in the budget for 1976-77 to implement the Science Congress Plan of Action. The focal theme tradition is being followed even today, by which the Indian Science Congress has gained the respectability and serious attention not only of the government, the public, and the media, but also of the world scientific circles.

At the Planning Commission

Swaminathan was invited in 1980 by Prime Minister Indira Gandhi to join as Deputy Chairman of the Planning Commission, and later as Member-in-charge of Agriculture, Health, Education, Science and Technology, and Employment, in which capacity he served till April 1982. Swaminathan carved for himself a permanent place in the planning history of India, by introducing the gender, environmental, and self-employment dimensions of development. Even today, he is deeply involved in the drafting of the Tenth Five Year Plan document as Chairman of the Steering Committee for agriculture and allied sectors.

In addition to national planning, Swaminathan has played a key role in science planning too as Chairman of the Science Advisory Committee to the Cabinet, and in promoting and planning biotechnology research and setting up a National Biotechnology Board, of which he was the Founder-Chairman.

Global Outreach

United Nations Committee

As Vice-Chairman of the Technical Advisory Committee (TAC) to the Consultative Group on International Agricultural Research (CGIAR) of FAO, Rome, Swaminathan played a key role in the establishment of the International

Crop Research Institute for the Semi-Arid Tropics (ICRISAT) at Patancheru, near Hyderabad. In December 2001, as a tribute to his contributions, ICRISAT has dedicated a new building – the M.S. Swaminathan Genome Research Center – to him.

He was the moving spirit behind the starting of the International Board for Plant Genetic Resources (IBPGR) in Rome, now renamed as the International Plant Genetic Resources Institute (IPGRI). Similarly, he also helped in the establishment of the International Council for Research on Agro-Forestry (ICRAF) in Nairobi, Kenya, and served as Chairman of its Board of Trustees from 1977 to 1982.

As Chairman of the United Nations Advisory Committee for Science and Technology, Swaminathan played an important role in setting up the International Center for Genetic Engineering and Biotechnology (ICGEB) in New Delhi.

Swaminathan also played a key role in promoting a Global Institute for Cooperation in Water Management at Valencia in Spain, when he served as Chairman of the International Committee on Water Management from 1996 to 1998.

Besides Swaminathan's accomplishments in India, his work has had great impact in other countries of Asia and Africa, which led *Time* magazine to identify him as one of the twenty most influential Asians of the twentieth century.

Swaminathan helped to build a China National Rice Research Institute at Hangzhow during the period 1983-87. Although China is the largest producer of rice in the world, there was no adequate scientific support to back production efforts. Hence, an intensive human and institutional resources development program was carried out from 1982 to 1987. Swaminathan also helped to develop a national genetic resources center at Beijing, a hybrid rice research institute at Changsha in the Hunan province, and an azolla research institute at Fuchow. In recognition of these contributions, both the Academy of Sciences, China, and the Chinese Agricultural Science Academy elected Swaminathan as Honorary Professor, and the Chinese Government gave him a scroll of honor. He has been a member of the China Council for Sustainable Development since its inception in 1990.

Swaminathan was deputed to Vietnam in 1975 along with a team of experts, soon after the unification, when he helped to develop two research institutes – one for rice in the Mekong Delta (considered the "rice bowl" of Vietnam), and another for buffalo, so that Vietnam can increase its rice as well as milk production. As a result of these efforts, Vietnam today is not only self-sufficient in rice, but has also become the second largest exporter of rice in the world. Swaminathan's former student, Dr. Bui Bong has recently been appointed Vice Minister for Agriculture.

During the civil war and the genocidal Pol Pot regime, research institutes in Kampuchea were practically destroyed. Rice being the staple food of the people, in 1985 Swaminathan started a program for rebuilding the country's rice research infrastructure, with financial support from the Government of Australia. Initially, he arranged to train twenty young Kampuchean scientists in different areas of rice research at the IRRI. Then, at the request of Prime Minister Hun Sen, Swaminathan rebuilt the research station near Pnom Penh with their help to provide service to both national and international sectors. As a result, Kampuchea has been transformed from a food-deficit to a food-exporting country.

Burma (Myanmar), which was a major rice producer prior to World War II, started losing this position in the 1960s. Because the poor post-harvest technology, milling recovery was low. Swaminathan undertook steps to build a strong and dynamic rice research institute at Yezin. He also arranged for the training of army officers in collecting and conserving wild rices and other important plant genetic resources in areas affected by insurgency. As a result of these steps, the Burmese Government provided funds for establishing a "Burma Room" at IRRI's Swaminathan Hall.

Swaminathan initiated two major steps to strengthen Thailand's rice and agricultural research. First, a research station was established near Ayuthaya for developing high yield, deep water, and floating rices. Second, in a project initiated under the inspiration of His Majesty the King (known as the King's Project), Swaminathan arranged for the promotion of alternative cropping systems in the place of poppy and other drug producing plants.

Swaminathan was of great help to the Government of Sri Lanka for promoting sustainable rice farming systems in the Mahavalli Command Area, the largest irrigation project in the country. Due to the dynamic leadership of Gamini Dissenayake, Swaminathan was able to introduce modern rice farming practices, including efficient water management, in this area. Later, he also served as the Chairman, Board of Trustees of the International Irrigation Management Institute (now known as International Water Management Institute) in Colombo. In this capacity, he could offer Sri Lanka considerable help in producing more crops per drop of water. The Sri Lankan Government has established a "Sri Lanka Room" in IRRI's Swaminathan Hall.

Swaminathan has helped Pakistan in strengthening rice research both in the West Punjab (Kala Shah Kaku Research Station) and in Sindh. He undertook a large capacity building program, with emphasis on the breeding of salt tolerance and high quality rices. Later, Swaminathan was invited by the President of Pakistan to deliver the Keynote Address at the inauguration of the National Agricultural Research Centre (NARC) at Islamabad.

Following the political transition in Iran, rice production had suffered an initial setback. Hence, Swaminathan took steps to strengthen rice research capacity building activities, by arranging for a large number of Iranian rice scientists to be trained at IRRI. As a result of this effort, Iran was enabled to improve its rice production considerably.

After taking over as Director General of IRRI in April 1982, Swaminathan made a detailed study of the state of rice research and extension in the Philippines, which comprises several thousand islands. It became clear to him that in order to promote sustainable rice production and to derive maximum benefits from IRRI's research, there was need for location-specific research and technologies. To achieve this objective, Swaminathan developed a proposal for the establishment of a national rice research institute to work in partnership with IRRI. The Government of the Philippines accepted Swaminathan's proposal and thus was born the Philippine Rice Research Institute (PhilRice) in November 1985, coinciding with the Silver Jubilee of IRRI. PhilRice has now become a strong research institution, catering to the needs of the various agro-ecological conditions of the country.

In recognition of Swaminathan's outstanding contributions, President Aquino conferred on him the Golden Heart Award of the Philippines (the highest award that can be conferred on a non-Filipino), for: a) Strengthening the International Rice Research Institute (IRRI) as a source of technological knowledge and expertise in the cause of furthering agro-science in the Philippines; b) Resolving a wide range of problems in basic and applied genetics and agricultural research and development in the Philippines; c) Pursuing the establishment of a national rice research institute in the Philippines; d) Initiating technology transfer workshops for the exchange of ideas and information between the Department of Agriculture and IRRI; e) Expanding IRRI's capacity for upstream research to bring the fruits of recent advances in science and technology to Asian rice farmers.

At IRRI, Swaminathan is still remembered with great affection and admiration. A recent correspondent from the institution aptly puts it "...The very caring nature and humble spirit of a man of stature whom IRRI constituents deeply loved and longed for remains unchallenged even up to these days".

Swaminathan's major contribution in Egypt was the establishment of the Egypt Rice Research and Training Center (ERRTC), at Sakha in the Nile Valley, with financial support from USAID to build a strong national rice research program through the Egypt-IRRI Cooperative Research Program. ERRTC was another step forward in building a strong research base for conducting research on all aspects of rice crop and rice-based cropping/farming systems, imparting training in rice production techniques to network and extension workers, disseminating all relevant research findings to farmers, and organizing scientific seed production programs. Swaminathan also helped organize an international symposium on "Rice Farming Systems – New Directions at ERRTC" early in 1987. The proceedings of this symposium, brought out in 1989 by IRRI, is a very valuable compendium covering all aspects of the subject. Swaminathan selected his former student, Dr. E.A. Siddiq, to serve as IRRI's Rice Breeder at Sakha.

In Madagascar, once more Swaminathan was responsible for setting up a rice research institute and the training of a large number of Madagascar scholars at IRRI in rice research and extension. The Madagascar Rice Research Institute, which was inaugurated in September 1987, has helped to substantially improve rice production in that country.

The significant contribution of Swaminathan to Kenya was the establishment of International Council for Research on Agro-forestry (ICRAF) in Nairobi, to promote sustainable land use systems, with support from the International Development Research Center (IDRC) of Canada. He served as its Founder-Trustee and later as Chairman of the Board of Trustees. Swaminathan has served as the chairman of the Committee of the Whole of the UN Conference on

Desertification held in Nairobi in 1977 and promoted the cause of a combined tree-cum-annual crop farming system (agro-forestry) in order to meet the short- and long-term needs of African farmers. He negotiated with President Jomo Kenyatta for acquiring land for setting up an agro-forestry demonstration farm near Nairobi.

In 1982, at the invitation of President Julius Nyerere, Swaminathan developed a blueprint for the establishment of an agricultural university in Tanzania, on the model of Indian agricultural universities, to generate the human resources and research outputs needed for effecting an agricultural transformation of Tanzania, including Zanzibar.

To launch a new strategy in agriculture, Swaminathan conducted a two-week course in 1981 for all the Directors of Institutes in Nigeria at the Administrative Staff College of Nigeria (ASCON) in Topo-Badagiri. This has led to the revitalization of the national agricultural research system in Nigeria.

In the former USSR and Russia, Swaminathan played a key role in revitalizing the Vavilov Institute at St. Petersburg and in initiating a dynamic rice improvement program. He was elected a Foreign Fellow of the N.I. Vavilov Agricultural Academy as well as a Trustee of the Vavilov Institute.

In October 1989, President Desmond Hoyte of Guyana offered nearly one million acres of undisturbed Amazonian prime rainforest area in central Guyana, for an international project under Commonwealth auspices to demonstrate the concept of sustainable management and conservation of tropical rain forests. In February 1990, the Secretary General of the Commonwealth, Sir Sridath Ramphal, invited Swaminathan to assist in converting this generous offer into an implementable project. Swaminathan visited Guyana with a Commonwealth team, surveyed the area and developed a proposal which subsequently led to the birth of the Iwokrama International Center for Rainforest Conservation and Management. Iwokrama is the world's largest project of this kind for learning and demonstrating how a prime tropical rain forest can both be conserved and sustainably utilized. The one million acres were divided into two major zones – nearly 600,000 acres to be maintained as a wilderness preserve in its pristine nature and 400,000 acres to be set apart for sustainable utilization. Iwokrama is now the flagship of the sustainable rainforest management movement. At the request of President Hoyte and later Dr. Cheddi Jagan, Swaminathan served as the Chairman of the Board of Trustees of this Center from 1992 to 1999. Swaminathan obtained support for this project from the Global Environment Facility, the Commonwealth Secretariat, the UK Government, the European Union, and many other donors.

Swaminathan's classic report, *The Commonwealth – Government of Guyana Programme for Sustainable Tropical Forestry*, was released simultaneously on June 5, 1990 (World Environment Day) by the Commonwealth Secretary General in London and by the President of Guyana in Georgetown. Swaminathan's unique contributions to Iwokrama include: a) Designing the Project in such a manner that conservation, sustainable use, and equitable sharing of benefits, were integrated in a symbiotic manner; b) Getting bipartisan political support for the Project, and the Iwokama International Center for Rain Forest Conservation and Development Act of 1996 enacted by the Parliament of

Guyana, which provided the legal framework for implementation of the operational plan; c) Raising adequate donor support by generating donor confidence; d) Explaining the significance of Iwokrama to the Guyanese public, and dispelling their doubts about the utility of the Project; e) Mobilizing the full support of the Amerindian population by appointing an Amerindian Chief as Head of the Experimental Site (Iwokrama in Amerindian language means "a big spiritually powerful fox").

When he wanted to resign in 1992 as Chairman, Iwokrama Board of Trustees, owing to heavy commitments at MSSRF, President Cheddi Jagan called him and said, "There would have been no Iwokrama without Dr. Swaminathan, nor will there be one if you resign." Thus, Swaminathan had to continue as Chairman till 1999, completing nearly ten years of service to the Iwokrama Project.

International Network on Rice Biotechnology

Swaminathan played a notable role in shaping the Rockefeller Foundation-supported Rice Biotechnology Network, now regarded as an outstanding exercise in partnership between advanced research centers in industrialized countries and institutions in the developing countries. Swaminathan organized at IRRI, early in 1984, an Inter-Center Biotechnology Conference which helped to formulate a research agenda for harnessing the tools of biotechnology for public good. This Network was formally launched in 1985 on the occasion of the First International Rice Genetics Symposium organized by Swaminathan and G.S. Khush.

Swaminathan's genius for organization has found its richest realization in his own M.S. Swaminathan Research Foundation.

The M.S. Swaminathan Research Foundation

During his last visit to Delhi in February 1970, Nobel Laureate C.V. Raman surprised M.S. Swaminathan by asking if he would set up a center for biological research in the land belonging to the Raman Research Institute at Bangalore. When a stunned Swaminathan, in all humility, expressed his inability to take up such a major responsibility, Professor Raman extracted a promise from him that he would start his own research center after retirement, without seeking government money. As though to fulfill his word, Swaminathan, soon after his return from the Philippines, plunged into the challenging task of establishing his own research foundation, at the age of sixty-five. Thus was born the M.S. Swaminathan Research Foundation (MSSRF) as an autonomous non-profit Trust. It was registered in 1988 at New Delhi with the World Food Prize money ($200,000), that Swaminathan received in 1987, as the initial core fund. Swaminathan's mission was to organize research and training to promote a job-led economic growth in the rural countryside, based on a pro-nature, pro-poor, and pro-women orientation to technology development and dissemination.

Swaminathan soon began to scout for suitable sites for locating MSSRF, and the choice finally fell on Madras. Why Madras? Why not Delhi, where he had spent a major part of his professional career spanning nearly four decades? The Government of Karnataka offered five hectares of land in Bangalore for setting up the Foundation. However, Swaminathan chose Madras since he wished to give high priority to coastal systems research. The Tamil Nadu Government came forward with a generous offer of one hectare of prime land in the Taramani Institutional Area, in south Madras. Subsequently, the adjacent plot of another hectare was also allotted. The proximity of the vast coastal wetlands enabled him to launch his research project on the conservation and restoration of mangroves to arrest sea erosion and safeguard the livelihood of the coastal population. This was his priority project.

The starting money came from the prestigious World Food Prize. This was followed, in 1991, by The Tyler Prize for Environmental Achievement ($75,000), the Honda Award for Ecotechnology ($150,000) in 1992, the UNEP Sasakawa Environment Prize ($125,000) in 1994, The Volvo Environment Prize ($175,000) in 1999, and the Indira Gandhi Prize for Peace, Disarmament and Development (Rs. 25 lakh). In addition, MSSRF was directly awarded the prestigious Blue Planet Prize ($500,000) in 1996, for outstanding institutional achievements. Besides, other personal donations by the Swaminathan family and a large number of his students, friends, and well wishers have combined to swell this core fund to a comfortable level. The Ford Foundation has also given a major grant. MSSRF has also been receiving financial grants from both national and international organizations as well as the Government of India. The International Tropical Timber Organization (ITTO) of Japan was the earliest one to support the Mangrove Conservation and International Training Programs. Another major support came from the Swedish International Development Agency (SIDA) for the Biodiveristy project and for developing research facilities. In fact, MSSRF's first building, commissioned in April 1993, was funded largely by SIDA together with Swaminathan's prize monies. The India-Canada Environment Facility and the Swiss Agency for Development and Cooperation are presently funding MSSRF in the areas of mangrove wetlands conservation and integrated resource management. Support has been forthcoming from other charitable organizations like the Sasakawa Peace Foundation of Japan, the MacArthur and Summit Foundations of USA, and the Volkardt Foundation of Switzerland. MSSRF has also developed close linkages with many UN bodies like the FAO, UNDP, UNIDO, UNESCO (Swaminathan currently holds the UNESCO-Cousteau Chair in Ecotechnology at MSSRF), International Fund for Agricultural Development (IFAD), Asian Development Bank (ADB), and other multilateral and bilateral donors.

The Bernard van Leer Foundation and the Hunger Projects of India, Japan, and Sweden have financed programs for children and women and in harnessing traditional media for achieving gender justice and empowerment. The group in Japan known as "Friends of MSSRF" formed by Professor Geeta Mehta of Temple University has made a very unique contribution to the cause of environmental protection and poverty eradication, by their generous support to the organization of rural self-help groups engaged in micro-enterprises supported by micro-credit. For this, they have mobilized donations of a large number of computers which are being utilized in the Village Knowledge Centres that act as ready beckoners, at the click of a button, for village folk to access information on weather, market situations, besides general information. These are mostly operated by village women.

The Department of Biotechnology (DBT), Government of India, has been a major funding agency since the inception of MSSRF, for projects on enhancement of mangrove genetic resources and anticipatory research to meet the challenge of coastal salinity arising from sea-level rise. In recognition of the outstanding progress made by MSSRF in the area of molecular mapping of salt tolerance genes and transformation of *Brassica juncea*, DBT has provided special funds for purchasing an automated DNA Sequencer for augmenting molecular genetic research. Sequences of seventy-six ESTs (database on Expressed Sequence Tags) from the mangrove species of *Avicennia marina* have already been entered in the Worldwide EST Database; this is considered to be the first-ever bulk submission of EST from any laboratory in India. DBT has also funded the establishment of a Women's Biotechnology Park on a twenty-acre plot allotted by the Tamil Nadu Government in Kelambakkam, near Chennai. The Park aims to develop an integrated approach towards technology identification, incubation, dissemination, training and re-training, and development of techo-infrastructure through feasibility studies using the criteria of value addition and market demand.

Swaminathan has clearly spelt out the mandate of MSSRF's research priorities in terms of sustainable development rooted in the principles of ecology, social and gender equity, employment generation, and economic viability, as follows: a) Conservation and enhancement of natural assets like water, land, and biodiversity; b) Promotion of sustainable and equitable agricultural and rural development; c) Development of ecologically sound and economically and socially viable "eco-technologies" by integrating traditional and frontier technologies; d) Generation of opportunities for skilled employment for rural women and youth; e) Reaching the unreached and voicing the voiceless in terms of techniracy and gender equity.

Swaminathan has, from the beginning, visualized MSSRF as a scientific and training institution "without walls" – one which relies on partnerships for its vitality and effectiveness. The unique logo he has chosen for MSSRF signifies the "open-ended, many-sided, and continuous evolution of humankind-nature interaction". This is evident in every aspect of MSSRF's activities, right from the design of the building and its surroundings, to the Program Areas selected for investigation.

- Program Area 100 covers coastal systems research, involving a participatory system of managing, conserving, and standardizing methods of rehabilitating coastal mangrove wetlands.
- Program Area 200 spans a wide area of biodiversity and biotechnology including community-based agrobiodiversity conservation and management, molecular mapping and genetic enhancement, monitoring ecosystem health using microbial diversity, conserving and bioprospecting endangered medicinal and mangrove species, and finally the Women's Biotechnology Park.
- Program Area 300 concerns studies on eco-technology and food security being carried out under the aegis of the J.R.D. Tata Ecotechnology Center and the B.V. Rao Center for Sustainable Food Security.

- Program Area 400 includes Project ACCESS (Action for Child Care and Education Strategies and Services), the Uttara Devi Resource Center for Gender and Development, and Voicing Silence, the women's theater project for women's development.
- Program Area 500 involves education, communication, and capacity building covering knowledge systems for sustainable food security, design and development of databases and providing CD-ROM services, analysis of scientometrics literature, library and information services, and organizing workshops and conferences/training programs. *The Hindu* Media Resources Center for Eco-technology and Sustainable Development falls under this Program Area.
- Program Area 600 deals with special projects concerning preparation of expert reports on policy issues and responses, assessment of resources, best practices, gaps in gender, science and technology, and lastly, opportunity in calamity, for rehabilitating communities ravaged by cyclones, floods, or earthquakes.

The coastal area research program generated considerable interest and enthusiasm among coastal communities in Orissa, since the rehabilitation of degraded mangroves has provided good protection against cyclones. Rapid propagation techniques, including micro-propagation, were standardized and community mangrove nurseries were established by local women and men similar to the social forestry ones, which provide additional income to the rural poor. A *Mangrove Atlas of Tamil Nadu* has been prepared and similar ones are in the pipeline for Andhra Pradesh and Orissa. A project, funded by the National Biosources Board, has been taken up for chronicling the status of bio-resources conservation in selected coastal ecosystems rich in biodiversity. Another part of this project relates to molecular characterization and enhancement of coastal bio-resources.

The India-Canada Environment Facility (ICEF) had financed the largest project on coastal wetland-mangrove ecosystems covering Tamil Nadu, Andhra Pradesh, Orissa, and West Bengal for five years. The expertise gained in this project is now being extended to the Gulf of Mannar area to prevent over-exploitation of coral reefs, mangroves, and seagrass beds, and to the Rann of Kutch, for promoting an integrated silviculture-cum-aquaculture movement to create an economic stake in conserving this biosphere reserve. A prototype model of a green belt and water-harvesting ponds for converting coastal wetlands into eco-gardens has been successfully demonstrated at Koodankulam near the nuclear power plant being set up by the Department of Atomic Energy.

Community-based agrobiodiversity conservation and management programs have been implemented in Kerala (Wayanad), Orissa (Jeypore, Simlipal), and Tamil Nadu (Kolli Hills).

The Community Agrobiodiversity Centre at Kalpetta in Wayanad is now housed in a beautiful building located in the farm area of 13.5 acres donated by the Swaminathan family. The main activities of this Center are strengthening community conservation systems through crop diversification and income generation, participating with grassroots networks for capacity building of various formal and informal institutions in conservation and sustainable use of biodiversity, and improving conservation strategies for life-saving plants and animals. Biosystematic studies are being carried out on both edible and medicinal Dioscoreas, including rare and endangered species, using both morphological and molecular markers. Youth, students, and teachers are being trained in taxonomic identification of rare and threatened species for their ex situ and in situ conservation, with the help of the Forest Department. Conservation and organic cultivation of both traditional and medicinal varieties of rice, now totaling one hundred and ten, including thirteen medicinal rices, have been promoted with tribal farm families, who are encouraged to grow vegetable crops along with paddy cultivation using eco-friendly methods.

In the Jeypore Tract of Orissa, five community medicinal plant gardens have been established with the active involvement of village practioners and healers. Community Seed Banks manned by women were set up to support and revitalize traditional crops and practices. MSSRF has conducted interactive training sessions for forest officials on conservation of forest resources and traditional knowledge systems, along with field visits. Participatory management of the Simlipal Biosphere Reserve in north Orissa has been taken up with support from the Summit Foundation of the USA.

In the participatory plant-breeding program in the Jeypore Tract, the farmer-to-farmer variation was substantial and straw yields of landraces were equally high, with high harvest indices. Analysis of data on the seventeen demonstrations, conducted in the kharif season of 2000, revealed that the formal method has demonstrated a consistent advantage over the traditional. Despite variations in the plot sizes, the landraces had performed well, although there was a vast gap between realized and realizable yields. It would be feasible to organize participatory pure-line breeding with tribal farmers, and formal methods of cultivation on farmer-to-farmer extension deserve priority.

A similar participatory plant-breeding program has been taken up at Kolli Hills with twenty farmers, using both formal and farmers' cultivation practices, with landraces of Samai millets.

Under integrated gene management practices, a comprehensive in situ, in situ on-farm, and ex situ conservation approach is followed, involving farm families and integrating various strategies like the establishment of community gene banks, seed banks, and medicinal plant gardens, and participatory community conservation of traditional crop varieties in the field.

At Chennai, the Community Gene Bank's priority is to get recognition and reward to tribal and rural women and men for their contributions to genetic resources conservation and enhancement. A community herbarium is also being maintained, containing rare, endangered, medicinal, and traditional cultivars.

Under a NATP (National Agricultural Technology Project) sponsored program, on-farm conservation of traditional crop varieties, with participatory approach in establishing village seed banks has been taken up in Wayanad as well as in centers in Tamil Nadu.

Under the B.R. Barwale program, agrobiodiversity conservation activities are in progress in Dharmapuri, Kolli Hills, Kerala, and Jagatjori, where training of an Agrobiodiversity Conservation Corps of volunteers is organized in various aspects of biodiversity conservation. Government High School students in Kolli

Hills have formed the M.S. Swaminathan Nature Club to collect basic information on biodiversity to prepare a checklist of medicinal plants and their uses, indigenous knowledge of villages, and faunal distribution. Peoples' Biodiversity Registers have been maintained at Kolli Hills and Wayanad giving details of plants, birds, insects, and animals.

The J.R.D. Tata Ecotechnology Center has been established with an endowment grant from the Sir Dorabji Tata Trust. The Tata Trusts also funded a building for this Center, which was inaugurated by the President of India in July 1977. The J.R.D. Tata Center aims to evolve methodologies for operationalizing sustainable development of agriculture and rural development through participatory research with grassroots institutions. Attention is focused on the Biovillage concept which was originally proposed by Dr. Jitendra Prakash in 1991 at an international meet on "New Technologies: Reaching the Unreached", organized by Swaminathan. The term was used to denote the integration of recent advances in biological technology with the best in traditional technologies to enhance the livelihood security of rural people. Resource use efficiency thus assumes importance. The successful application of the Biovillage concept in the Pondicherry villages has encouraged MSSRF to replicate the experience in Kannivadi near Dindigul in Tamil Nadu, where a Knowledge Centre has also been established with a website providing information on price and stock of vegetables in the Oddanchathram market. Biovillage activities have been extended to Gayaspur and Mallikapur in Orissa. It is also proposed to convert all the two hundred and fifty villages in Pondicherry into Biovillages by August 2007, to mark the diamond jubilee of India's Independence.

Through a multidisciplinary approach, the Kolli Hills Biovillage is being developed as a model for creating an economic stake in the conservation of biodiversity by linking primary conservers with markets, and enhancing the productivity of traditional landraces of minor millets (Swaminathan prefers to call them "nutritious millets") through improved cultural practices. The installation of two mechanical processing mills to de-husk millets has encouraged the local people to include millets in their daily diet. Value-added products like de-husked millet grains and flour are neatly packed, with labels displaying the nutrient composition and recipes, and sold in super-markets by the Self-Help Groups.

The Kolli Hills farming communities followed a traditional farm practice of crop diversification by growing French beans and coriander after harvesting millets, thus ensuring nutritional security and soil health. Capacity building and training programs were organized in precision farming methods like biofertilizer application to millets. At the Kannivadi Center, women, after being trained at the Science Center in Wardha in techniques to recycle banana wastes into enriched bio-manure and producing paper and partition boards, enhance family incomes in resource-poor rural households. Rural women also breed ornamental fish, and produce quality seed of high-value vegetables like hybrid tomatoes, radish, and capsicum using low-cost greenhouses.

Computer-aided and Internet-connected information systems have helped to bring down the massive transaction costs which rural communities face. The Village Knowledge Centres aim at transferring generic knowledge into location-specific knowledge that can be acted upon. Participatory training is organized for selected rural women and men on computer applications to identify the need-driven information to different economic sectors. The villagers document information in Tamil on soil health, pest management, post-harvest management, market details, hybrid seed production, health aspects, and government development schemes. Precision Farming activities like seed villages, pulse villages, and lysimeter experiments are in progress, which are of great relevance to the rural community.

A project on Sustainable Management of Natural Resources for Food Security and Environmental Quality, funded by the Sir Dorabji Tata Trust, has been launched by MSSRF in collaboration with Ohio State University and in partnership with the Tamil Nadu Agricultural University (TNAU), the Punjab Agricultural University, and the Jawaharlal Nehru Krishi Vishva Vidalaya. A development project has been initiated at Annavasal block in the Pudukkotai district of Tamil Nadu, representing the degraded alfisols soil in vialogam series. A network of various agencies involved in agriculture has been formed with the help of the Agriculture Department and TNAU. Experts from Ohio University have identified certain issues like the binding of iron with phosphorus in this region. Involvement of villagers and scientists in evolving a joint strategy to tackle problems of rain-fed agriculture in this region will be the key to success of this Project.

At Chennai, the J.R.D. Tata Center is involved in bio-pesticide production and in supporting research and training program. Out of four green manure plants selected at Kolli Hills for bio-pesticide studies, intensive work is being done on *Dodonaea viscosa* whose hexane extract was found to stimulate egg-laying but arrest hatchability.

An important activity of the B.V. Rao Centre for Sustainable Food Security has been the compilation of a *Food Insecurity Atlas of Rural India*, with the collaboration of the World Food Programme (WFP). This innovative Atlas was formally released by Prime Minister Vajpayee in April 2001 at New Delhi, followed by a two-day Consultation on "Towards Hunger-Free India" organized jointly by MSSRF, WFP, and the Planning Commission, at the end of which a ten-point agenda for action to make India hunger-free was released. Currently, work is in progress for putting together a *Food Insecurity Atlas of Urban India* and a *Sustainability of Food Security Atlas*.

MSSRF is today a beehive of activity. Every second day there are training programs, workshops, seminars, symposiums, consultations taking place on a vast variety of topics. Indian and foreign visitors are often amazed at the output of work produced at the Foundation. The history and growth of MSSRF during the decade 1990 to 2000 has been described in the publication, *A Social Vision for Science*.

Swaminathan has frequently said that excellent buildings and equipment alone will not make a great research institution, and that poverty will persist so long as human resources are undervalued and buildings and material resources are overvalued. In consonance with this philosophy, MSSRF has developed personnel policies to attract and retain outstanding young men and women professionals. In Swaminathan's words: "Matching the market compensation packages is not a viable or desirable strategy for MSSRF. Following 'the market' will be counter-productive and change the culture of the organization. What MSSRF will strive for is to be a caring organization, committed to the professional growth of its staff and scholars." Anyone visiting MSSRF will see the power and wisdom of this approach.

6.6 "M.S. Swaminathan: The Man and His Philosophy of Life" from *Scientist and Humanist: M.S. Swaminathan,* by R. D. Iyer, pp. 151-55. Mumbai, India: Bharatiya Vidya Bhavan, 2002.

It is difficult to describe M.S. Swaminathan – the scientist, the administrator, the social worker, the visionary, the philosopher – in terms that do not smack of hagiography. If there is a single fault that he can be accused of, it is his inability to say 'no' to anyone who approaches him for help or advice. Being a totally relaxed person, he brings a calming effect on the visitor too, and listens carefully and calmly. One can easily get carried away by his courtesy and compassion. Nevertheless, behind that soft exterior is a man of determination and indomitable spirit, who goes about his tasks with missionary zeal.

It was Swami Chinmayananda, the greatest modern exponent of the *Bhagavad Gita,* who once wrote, "A man firmly established in wisdom is tranquil, and his equipoise is never broken even when he is investing his entire energies in the world outside for the service of mankind." Those prophetic words seem to so aptly fit the personality of Swaminathan whose hallmark is tranquility. In spite of all the years spent away from home, Swaminathan is a staunch swadeshi in his thinking and living. In spite of having occupied the highest chairs in India and abroad, and having been decorated with innumerable awards and accolades from all corners of the globe, Swaminathan has always led a simple, unostentatious life, and continues in the same rhythm even today.

His main inspirations have been Mahatma Gandhi and Bhagavan Sri Ramana Maharshi. From these great masters, Swaminathan has imbibed a spirit of tolerance, compassion for the poor and downtrodden, and an egoless state of mind. Above all, the Gandhian philosophy of "Be the change you want to bring about" is deeply ingrained in his thought and action. To his students, and to the farming community also, he has always stressed the need to "learn by doing". The emphasis was on the practical applications of science for the welfare of the poor. The Green Revolution was the result of making farmers take to the new dwarf wheat varieties and the improved technology. It is his belief that it is the small farmers (who are in the majority) who will bring about the change, the transformation, once they learn how to do things the better way.

It has been said of Ramana Maharshi of Arunachala that each look of his contained a shower of grace, every smile of his a divine assurance that Man is God. In the last analysis, the ego does not exist, as it is only a figment of our imagination. When one has learned to be still, one is. Ramana hardly ever spoke; he conveyed profound truths through silence. Yet, those penetrating eyes moved the whole world.

In his esteem for Gandhiji's and Ramana Maharshi's ideals, Swaminathan was greatly influenced by his uncle, Professor K. Swaminathan, who molded his outlook and view of life from childhood. Professor Swaminathan brought about a fusion of the thoughts of Mahatma Gandhi and Ramana. Writing about his uncle in 1995, Swaminathan said: "If there is one lesson to be learned from the life and work of KS, it is that true happiness comes from giving and not acquiring, from caring and not from exploitation, and from self-transcendence and not self-gratification or self-glorification." According to him, the hundred volumes of *The Collected Works of Mahatma Gandhi*, edited by Professor K. Swaminathan, constitute the greatest epic of our times.

As a teacher, Swaminathan has no parallel, in his sincerity and thoroughness of approach, clarity of thinking and expression, and his amazing coverage of the subjects he taught, giving the latest references. I have had the privilege of attending all his classes in Cytogenetics, Radiation Genetics, and Mutation Breeding at IARI, during the late 1950s to the early 1970s, and there was never a dull moment in his well-attended lectures. In a fine article titled, "Teachers Must Be Passionate" written for the *Home and School* magazine in October 1968, Swaminathan has dwelt at length on the qualities needed to be a good teacher:

"The first concerns the attitude of the teacher in the classroom. By this I mean the degree of association or the way the teacher identifies himself or herself with the task in front of him or her. This is very important, as it can make a great impression on the student. When I think of my own teachers, the only people who come to my memory are those who were passionate in their teaching. It is not what they taught that was important, but the way they did it, the seriousness of purpose, and their identification with the students. As a teacher myself, I have covered much of the same ground for many years; but if I consider a course as a monotonous repetition, then I can neither make the students enthusiastic nor can I feel enthusiastic myself. On the other hand, if every lesson is regarded as an experiment and a new experience, then it becomes exciting for both sides.

"Of course, teaching is also partly a gift, but partly it is an attitude of the mind. The seriousness with which the teacher undertakes his or her responsibilities will determine its effectiveness. An earnest teacher establishes a two-way feedback relationship between the student and the teacher. The student soon becomes an admirer or follower of the teacher. Hence, it becomes all the more important that the teacher should set a worthwhile example. Teachers should neither dissuade students from asking questions, nor try to escape answering them. To be told that 'this will be taught in the next class', does not reflect credit on a teacher. Another thing is, one should not bluff."

Swaminathan is also an excellent research guide who takes great pains to help his students at all stages of their theses programs. Often, when he sat with a student to examine his cytological slides, if he found the microscope dirty, he would take out his own handkerchief and start cleaning it himself with a smile. Invariably, the ashamed and surprised student got the message, and the instrument would be clean the next time!

Has he ever been provoked, angry over anything or anybody, one might wonder? It is difficult to imagine Swaminathan losing his temper. I can recall only one occasion when he raised his voice in a C.V. Raman-like harangue: "Let dogs bark. The caravan of Science will march on." This unexpected outburst was in response to repeated insinuations and aspersions cast on his colleagues, questioning the validity of their findings by a section of the audience in the IARI auditorium in 1971 when he was the Director. This reaction stunned the audience, who dispersed quietly. On an earlier occasion, when a colleague in the Botany Division of IARI used data from Swaminathan's thesis and published it in a foreign journal without even acknowledging the source, he was shocked. Instead of admonishing the offending scientist, Swaminathan just put up a note on the notice board, that theses could be borrowed from the library only with the written permission of the concerned author and the guide. The offender later became a Director of another ICAR Institute! Such was the magnanimity of Swaminathan. In fact, he was often accused of "backing the wrong horse", but he believes that we have to fight our own battles, and "in life, one gets what one deserves and not what one desires, and He knows what one deserves" as Swami Chinmayananda has said. In his Zakir Hussain Memorial Lecture delivered at Delhi University in September 1970, Swaminathan said, "Psychologists believe that within each of us, there dwells a captive spirit struggling to find fulfillment, and each one has the need to succeed by his own efforts. Therefore, we should work with rather than work for others. The latter situation produces an expectation of gratitude on the one hand, and resentment at having to be obliged, on the other."

Swaminathan brings to bear on his work a missionary zeal which is infectious and a sense of dedication which is inspiring. It is hard to keep pace with the lightning speed of his thought process, and the follow-up action. In spite of so much pressure of work, Swaminathan knows the art of relaxation. Even in the middle of important meetings and seminars, he manages to catch fleeting "cat-naps", but his mind is ever alert. He invariably manages to bring life and meaning to every meeting he presides over, through his amazing powers of grasping the most essential points. Soon after each meeting, he will dictate the proceedings to his secretary and meticulously finalize it the same day or the next. Likewise, he finds time to answer every letter he receives, promptly and precisely. He has fine-tuned these Spartan habits from the very beginning since he realizes the value of not only his time but also that of other people. And, what a pleasure it is to read his letters! I have received many of them, written from hotels abroad on their letterheads, which shows he can find time for anyone and everyone, whatever be the demands of work.

I have never heard him condemning anyone for his/her faults, even of his critics who have always been around. Borlaug once wrote to him, "There is no remedy for jealousy. The only remedy is to ignore". Swaminathan believes in Gandhiji's faith in humans: "Do not lose faith in humanity. Humanity is a vast ocean. If a few drops of the ocean are dirty, the ocean does not become dirty." In the current turmoil of global unrest due to widespread terrorism, Gandhiji's message of tolerance rings so true and relevant.

To my mind, Swaminathan combines in himself the genius and courage of a C.V. Raman, the charisma and dynamism of a Jawaharlal Nehru, the deep compassion and non-violence of a Mahatma Gandhi, and the egoless state of a Bhagavan Ramana – a rare recombinant indeed!

6.7 "Foreword" by Maurizio Laccarino, Professor of Microbiology, University of Naples; Former Assistant Director General for Natural Sciences, UNESCO. From The *Man Who Harvests Sunshine: The Modern Gandhi: M.S. Swaminathan*, by András Erdélyi, pp. 5-6. Budapest, Hungary: Tertia Kiado, 2002.

"Recall the face of the poorest and the weakest man whom you have seen, and ask yourself if the steps you are going to contemplate are going to be of any use to him. Will he gain anything by it? Will it restore to him control over his own life and destiny?"

This sentence, attributed to Mahatma Gandhi, was quoted by Professor Monkombu Sambasivan Swaminathan in a keynote speech that he gave at the World Conference on Science (Budapest, June 1999). I believe that this concept has been the guiding principle of Professor Swaminathan throughout his long and very productive scientific life. He studied in the Netherlands, the United Kingdom and the USA, and traveled widely afterwards. Although Indian culture is deeply rooted in him, he can be considered a citizen of the world, like many great minds of human civilization. Like the late Abdus Salam he links the western and third world cultures. He worked on problems of the third world and achieved results that are important for industrialized countries.

Through this book the reader gets first-hand information from one of the actors of the Green Revolution, an important event in the history of science, of science policy and world politics. After the Second World War, the danger of lack of food, with consequent social unrest and widespread wars, was real and it was considered imminent. Swaminathan isolated the semi-dwarf wheat that changed the prospects of India. As a consequence, this country, so much humiliated and looted throughout its history, became self-sufficient in cereal production and able to feed an increasing population. Most armed conflicts can be traced to economic, rather than ideological differences. Thus, investments in agriculture are important to achieve food security and therefore prevent future wars: from this point of view, research in agriculture is a very effective contribution to world peace.

The book consists of a brilliant, continuous interview with Professor Swaminathan. We are taken through his experience as a State Secretary for Agriculture in the Indian government, as a Director General of the International Rice Research Institute in the Philippines, and as the founder of his own research foundation in Madras.

In the Middle Ages, developing countries had their own technological systems of knowledge. Later on, modern science and technology started in the already rich Europe and gave rise to the European industrial revolution, which was instrumental in the acquisition of extra resources and labor, thus leading to an increase in the technological gap, already existing in those times, between rich and poor countries. The colonial empires changed the education system in the colonies and almost ignored science and technology in the teaching curricula, thus widening the differences. Today, what is needed by developing countries is not technology, but the scientific knowledge and institutions which will enable them to re-acquire their own autonomy and self-sufficiency. This has been the philosophy that guided Professor Swaminathan in his professional life and this was the message of the UNESCO World Conference on Science.

The importance of moral values has today much declined as compared to the past. In the minds of many, globalization means trying to get as much as possible in economic terms, without taking care of the needs of the poor and of poor nations. If we do not act soon the gap between rich and poor will increase. Yet, how can the world hope to survive in the face of a massive upheaval in developing countries? To avoid a disaster we need a socially equitable development of the Third World. Science and Technology have been, and will continue to be, major factors for industrial development and economic growth. We therefore need great scientists that not only achieve important results, but also organize and perform research and set an example in their own countries. This book honors a great man, but at the same time indicates the way for the politicians of developing countries.

6.8 "Foreword" by Poonam Ahluwalia, Executive Director of Youth Employment Summit. From *M.S. Swaminathan: One Man's Quest for A Hunger-free World*, by Gita Gopalkrishnan, pp. vii-ix. Washington DC: Youth Employment Summit, 2002.

It was my lucky day when I met Dr. M.S. Swaminathan eight years ago. I remember the day very clearly. I had only read about him up until then, about how he had engineered India's greatest triumph over poverty and hunger. The awe I felt when I was first introduced to him could be compared to an astronomer's excitement at finding a new star. Here I was in front of the man who had changed the face of India. The world is a different place now because of him. I want to share the awe I have of this person with today's youth – who now wake up to and approach the world differently because of him, his commitment, his resolve. My wish is that this biography will serve to bring new hope and inspire change among youth.

Dr. Swaminathan's work has made an impact, not only in my beloved India, but also around the world. His deep understanding of food security and sustainable development has ensured a real change in the lives of millions. In the mid-1960s, he planted the seeds of hope, but it was only the first step, and perhaps the easiest. What came next was a revolution in India, then Asia, and expanded to encompass many other parts of the world. Countries became self-sufficient in food production only after a couple of years of sowing Dr. Swaminathan's seeds and implementing his farming techniques. As a scientist and as a policymaker, he has led the change in improving the sustainability of agricultural practices, and he continues to promote improvements in production and productivity in order to meet the demands of future populations.

Throughout his career, Dr. Swaminathan has in various ways been able to positively impact the lives of youth. This comes from the interconnectedness of his life's mission of freeing the world from hunger with the issue of youth employment. The agricultural sector will continue to employ many youth in developing countries, even with the advent of new technologies, and not in spite of. As Dr. Swaminathan argues, production and productivity need to improve, and this will happen because of advances in communication and information systems, and here is where new jobs for youth lie. The infrastructure to support improvements in production, and particularly jobs that complement farm production, will depend on technological advancements. Given that hunger is not always a problem of availability, but a matter of access (in other words, purchasing it), the need to provide employment opportunities to youth in order to better their condition is critical if they are going to be healthy and productive members of society.

In the relatively short time I have known Dr. Swaminathan I have learned many things. The most important one is his response to all my requests, which has always been "Why not?". With those two words he makes everything possible. He made it possible for me to discover that I too have a role to play in breaking the cycle of hunger and poverty in the world. I have chosen the Youth Employment Summit as a means to that end, and I thank Dr. Swaminathan for that inspiration. I appreciate every minute that he has taken to show me the bigger picture, to counsel me – not only on how to do things better, but how to do better things – and to recognize my accomplishments however big or small, which serves to drive me further.

6.9 "Introducing Swaminathan" from *M.S. Swaminathan, One Man's Quest for a Hunger-free World* by Gita Gopalkrishnan, pp. 2-6. Washington DC: Youth Employment Summit, 2002.

Who is M.S. Swaminathan? He is an agricultural scientist, a plant geneticist. But he is much more than just that. He is an environmentalist, an ecologist, an administrator, and educator, an advisor to governments around the world, a social worker, a philosopher, and a visionary whose dream is to rid the world of hunger and poverty.

I wanted to know more about him and went to seek him out at the M.S. Swaminathan Research Foundation in Chennai, India. It was an impressive white building with a red tiled roof. I walked in through a short corridor and saw on either side greenhouses full of plants. I found myself in a hexagonal-shaped area surrounding a courtyard with a garden. There was something unusual about the garden. It was not the normal landscaped one with a lawn and flowerbeds. As I looked at it again, from the depths of my memory came the terms describing the five ecological zones mentioned in classical Tamil literature: mountains, forests, pastoral lands, coastal plains, and arid deserts. I now realized this central garden had been planted with horticultural species typical of each of these regions. There was not much wood used in the structure of the building, I presume, to save the fast depleting forest wealth in the country. The water brought down by the fury of the monsoon rains was collected and stored and used for the gardens and other purposes. There were solar panels on the roof, the computers ran on solar energy. This building was clearly designed on a sound environmental foundation.

An international seminar was on that day. Delegates were arriving; they represented several nationalities. The whole place was buzzing with activity. An affable gentleman in his seventies was greeting everyone coming in. He looked calm and collected in the middle of all the excited goings-on around him. I was told he was the man I had come to meet – M.S. Swaminathan, the founder of this research organization.

It was getting on to 10 a.m. when the seminar was scheduled to start. But, the chief guest, an important member of the government, hadn't arrived. Time is flexible in India. Punctuality is not one of our virtues. And government ministers have to be given some latitude. However, M.S. Swaminathan seemed to be of a different mold. The delegates were all ushered in and they took their seats. He made a brief announcement: "Our chief guest has been delayed, he is on his way. We will not wait, he will join in when he comes." The proceedings started. …

I spent several weeks getting to know Swaminathan and his philosophy of work and living. He is a couple of years over seventy-five but has the energy and enthusiasm of a twenty-five year old. He is forever on the move. One day, I was told he was away in Manila; the next week he had gone to Budapest. Two days he was in Delhi, back to his base in Chennai for just an evening, and off to New York. He travels to out-of-the-way locations in India – to a small town called Kottakal in the south-western state of Kerala to inaugurate a tissue-culture laboratory of rare medicinal plants and, a thousand miles away, to the rice-growing Jeypore Tract in the eastern state of Orissa to look at some research work his Foundation is involved in there. He never shows any signs of fatigue. I asked him how he manages it. "Well," he said with a smile, "if you are committed, and excited by what you are doing, tiredness never sets in."

And excited and committed he is to his cherished cause. Swaminathan strongly believes that the peace our world is sadly so much in need of today can be achieved only by ridding it of hunger and poverty. It is obvious that hunger arises out of poverty, and the most important cause of poverty is unemployment. It is a sad state of affairs that when we have progressed on so many fronts, when man has conquered the skies and the oceans, there are so many people in the world without jobs. Swaminathan puts this down to lack of imaginative thinking and lack of enterprise on the part of governments. New and creative ways can, and must, be found for the unskilled, the illiterate, the landless, and the excluded – particularly youth and women of all ages – to be able to earn their living.

The activities of the M.S. Swaminathan Research Foundation (MSSRF) are directed towards pro-nature, pro-poor, pro-women, pro-employment programs. Its headquarters in Chennai is staffed by young and enthused scientists, ecologists, economists, specialists in gender studies, sociologists – of both sexes. MSSRF runs a series of projects in several states of India, working closely with rural and tribal peoples.

"Sustainable development" is the catchphrase. It means that growth and progress can be shown to be reliable and dependable and can be maintained at an even and steady pace. Swaminathan insists that sustainable development must be firmly rooted in the principles of ecology, social and gender equity, employment generation, and economic potential. In farming, Swaminathan defines sustainable development as producing high yields in perpetuity, without associated social or ecological harm.

What have been the influences that shaped Swaminathan's life and work? Are there pointers in his upbringing, education, research work, his personality and ethical principles, his philosophy of living, that the youth of today can relate to? In exploring all these aspects, I had opportunities to talk to members of his family and to his colleagues at MSSRF. And in the midst of his very busy schedule, Swaminathan gave me quality time.

6.10 "The Man" from *M.S. Swaminathan, One Man's Quest for a Hunger-free World*, by Gita Gopalkrishnan, pp. 70-81. Washington DC: Youth Employment Summit, 2002.

Let me try and describe M.S. Swaminathan. He is seventy-seven years old, of medium height, bald with a wispy white fringe, with a broad forehead, moves about quickly, has a friendly smile and a firm handshake. He prefers vegetarian food, is a non-smoker, and abstains from drinking alcohol. His lifestyle is very simple, no ostentatious frills. He drives around in a small Indian-made car, the Maruti. When he is in Chennai, he is at work by 8 a.m. His door is always open to anyone in his Foundation who wants to share some information or clear a doubt. He has a very paternal attitude towards his people. He has a phenomenal memory and remembers to ask them about some minor personal problems they may have told him about. The atmosphere is a very democratic one. When it is time for a cup of tea, "the professor" as he is addressed at MSSRF, invites his visitor to accompany him to the cafeteria where he pays at the counter. In India, where government and corporate culture see to it that big bosses are waited upon, this is not common.

He is full of energy - attending meetings, seminars, consultations; delivering addresses at universities, at public forums, at scientific congresses. He networks with so many people all over the world that he has good friends on every continent. Every other week there is an article by him in the print media on topical matters, like the situation in Afghanistan or the senseless sectarian violence in India, always connecting such issues to his hopes for peace and sustainable development in the world. He is in correspondence with a great many people on a great many subjects – he never fails to answer any letter, whether it is official or personal, whether it is from a renowned scientist or from a ten year-old schoolboy. His wife, Mina, says he has long had the habit of carrying letter paper and postal stationery with him and dashing off his missives, waiting at airports, during flights, and from hotel rooms. Not for him the laptop, he writes longhand when he is on the move and dictates to his secretaries when he is back at base. He catches up with his reading and his thinking on long international flights. Jet lag doesn't seem to worry him at all. He arrives home in the wee hours – flights from the west reach Chennai at unearthly times – and by mid-morning is at his desk. It is his ability to catnap, snatch some minutes of sleep whenever and wherever he can, which refreshes him.

He always stays calm and unruffled. Mina talks of her husband being totally free of tension and stress. It is not hard work that saps a man's energy as much as pressures and anxiety do. He has his work cut out and he is fully engrossed in it. Nothing else matters: not criticism, not praise. One cannot have achieved the status Swaminathan has without collecting detractors and defamers on the way. He takes them in his stride. He learnt from Norman Borlaug, he says, to just ignore them. He has never tried to stifle criticism, nor has he tried to take any vindictive action against the perpetrators of slander. His philosophy has always been "to your own self be true, you will then not be false to others." Similarly, praise and adulation do not go to his head. He has evolved a good technique to keep his ego in check. When he hears or reads something greatly complimentary about himself, he says he assumes it is about someone else!

It is his immense faith in, and goodwill towards, the ordinary human being that give Swaminathan the strength of action, the energy and enthusiasm for all his work, Mina asserts. More than anything else, people matter to him. He never forgets names and faces, even those of students of long ago or of a fleeting recent acquaintance. He is at his best when he is interacting with others, whatever age or status, whatever gender, whatever profession or nationality. He is always thinking of ways and means to help people help themselves. He builds up their confidence, makes them believe in their own potential. He never stints the hours he spends talking to a young student or guiding a research worker or listening to someone's problems or success story, even at the expense of time for himself or his family. There are so many men and women today in top positions in India and overseas who owe their successful careers to the help and encouragement he has given them. Swaminathan trusts people implicitly and believes that is the only way to bring out the best in them. And when he has been let down, as has happened, he shrugs it off and goes on. He treats everyone with the same courtesy and never fails to acknowledge the smallest kindness he has received. I was told by his cousin of a family gathering where he was the only one to have gone into the kitchen and thanked the cook for the delicious fare she had provided.

Mina Swaminathan has a picturesque way of describing her husband. She compares him to the river Ganga, the greatest of India's rivers. The river flows down from the Himalayas across the northern plains, oblivious to how people treat her. Some use a tremendous amount of her water, some carry away small potfulls. Others just splash a few drops on their faces while yet others pollute her waters. The Ganga continues to flow, serene and unperturbed.

Perhaps not surprising in a person who has spent a good part of his childhood in Kerala, with its beautiful landscape of sea and beaches and rivers and hills and fields of paddy and groves of coconut palms, Swaminathan has a great love for nature. One of his favorite pieces of verse is this by Joyce Kilmore:

I think I shall never see
A poem as lovely as a tree
A tree whose hungry mouth is prest
Against the earth's sweet flowing breast

A tree that looks at God all day
And lifts her leafy arms to pray
A tree that may in summer bear
A nest of robins in her hair
Upon whose bosom snow has lain
Who infinitely lives with rain
Poems are made by fools like me
But only God can make a tree.

Trees and plants have always been close to his heart. He recollects an old Amerindian proverb he was told about in Guyana: "The sky is held up by trees. If the forest disappears, the sky – the roof of the world – collapses and nature and man perish together." He believes that the attitude towards trees and plants is a down-to-earth measure of a civilization, since centers of civilization and centers of origin of economic plants often overlap. To a person who is interested in the secrets of the past, a tree narrates in which years it experienced the stress of drought and when it was nourished with abundant water. In Swaminathan's hometown of Chennai, there is a great banyan tree. Believed to be over five hundred years old, this tree occupies a space of forty thousand square feet, with over one hundred aerial roots descending vertically from the massive horizontal branches. It is an ecosystem in its own right, harboring a variety of birds, rodents, reptiles, and insects. It is reported that on one occasion nearly three thousand people sat under its canopy.

Says Swaminathan, "At the heart of agriculture, there is a simple truth that in the growing of green plants, we can capture the energy of the sun, our most precious asset, and turn it into food, fuel, fodder, and other products. Hence, wherever man plants trees, grows vegetables in small plots, covers a tiny patch with grass or flowers, or covers the roofs of mud huts or brick buildings with vines, he is playing a role in our agricultural development. When Charaka, the venerable physician of ancient India, was asked by his teacher to bring plants which were quite useless, he returned empty-handed, remarking that there was no such plant. When we all understand this, agriculture in our country will have come of age."

"Agriculture is what you call applied ecology. Ecology means remaining in harmony with nature, and agriculture means using natural resources in a sound and scientific way. Therefore, anybody who is interested in agriculture, not only today but also for the future, has to think of the conservation of nature and natural resources," he once explained. In 1984, Swaminathan was named the president of the International Union for the Conservation of Nature and Natural Resources and vice-president of the World Wildlife Fund. Two others who have held such high positions in these prestigious international organizations are Prince Philip, the Duke of Edinburgh, of England and Prince Bernhard of the Netherlands.

An example of his deep concern for environmental protection is the issue of Silent Valley, the only surviving tropical rain forest in peninsular India. The flora and fauna of this area are quite unique. Twenty-three mammalian species, including three endangered ones – tiger, lion-tailed macaque, and Nilgiri langur –

have been recorded. There are unique herbal and medicinal plants. This whole region has also been found to be a reservoir of useful genes in rice, conferring resistance to some major pests.

In the 1970s, the Silent Valley Hydro-electric Project (SVHP) had been mooted to augment the energy requirements of Kerala. Water from this project was expected to provide irrigation for raising an additional crop in ten thousand hectares. Also, the implementation of the project over a period of six to seven years would boost the economy and employment prospects for the local people in one of the underdeveloped parts of Kerala. SVHP was expected to be implemented over about eight hundred and forty-five hectares, of which eight hundred and thirty would be submerged under the reservoir, with forest land comprising five hundred and thirty hectares.

Environmentalists all over the world were up in arms. The natural evolutionary processes extending over a period of fifty million years which resulted in this unique biological wealth would be lost, if human interference on the scale proposed for implementing SVHP was allowed. Swaminathan was asked by the Indian government to study the problem and give his recommendations. His report was typically very balanced and at the same time constructive in finding plausible solutions, keeping in view all the conflicting forces, vested interests, political and bureaucratic pressures bearing on the issue. In conclusion, he said: "If steps are not taken to satisfy the legitimate socio-economic aspirations of people of this area, mere talk about ecology and environment will be met with cynicism and with the question, 'Who is more important – man or monkey?' If we proceed with the implementation of SVHP without taking advantage of alternative methods of providing energy, employment, and irrigation, will future generations forgive us for destroying a fifty-million year old genetic heritage, at a time when the solar energy option is no longer an illusion? Alternative pathways available for providing power, irrigation, and jobs at no ecological risk will, in my view, help to achieve the desired social goals more speedily and economically. It should not be beyond our political, intellectual, or financial capabilities to find solutions which can enable the present-day human population of this area to experience a better quality of life without destroying a priceless biological endowment." The Silent Valley project has not been carried out.

Perhaps he was remembering this verse from the *Atharva Veda*, one of India's ancient philosophical treatises:

What, O Earth, I dig out of thee
quickly shall that grow again
May I not, O pure one,
pierce thy vital spot
or thy heart.

Swaminathan has great concern for children. He is very fond of quoting Condorcet, the 18th century French philosopher: "People have a duty towards those who are not yet born; that duty is not to merely give them existence but to give them happiness." One of the major handicaps that children in several developing countries face is that of low birth weights, giving rise to physical and mental stunting. This has more far-reaching repercussions than one can imagine. Delivering the Keynote Address at the World Conference on Science, Budapest, in 1999, Swaminathan called on the UN agencies like the World Food Programme (WFP), FAO, UNICEF, WHO, and UNDP to address this issue with urgency. He wanted them to launch, in collaboration with interested national governments, a global program to fight maternal and fetal under-nutrition, thereby helping to minimize the numbers of low birth-weight children. He suggested that this program be called the Global Movement for Children for Happiness.

The educational needs of children, particularly in a country like India where agriculture is the major activity, is another issue Swaminathan has been interested in. He believes that an awareness of its biological surroundings must be created in a child right from its infanthood. In primary school, the emphasis should be on learning about nature through direct observation, and on using nature to develop scientific skills, aptitudes, and habits of thought. He deplores the belief that education is something that must take place within a building, in barren and bare classrooms.

In rural India, mainly due to economic necessity, there are a large number of school dropouts in the older age group, almost as many as those in school. Most of these children are already working. Swaminathan has this innovative idea of carrying educative experience out to their workplaces in their fields. This is not just having open-air classes or teaching arithmetic in terms of bags of manure or hectares sown. They can be taught the latest technologies and developments in agriculture by progressive farmers who can give them work experience for part or most of the day.

Extending this concept to the urban underprivileged, such children can be placed for regular work in industrial establishments for a couple of days in the week. This will give them real work experience and preliminary training for future careers. Similarly, urban children can also be sent to live in villages with farming families for some weeks to learn more about agriculture and farm operations through direct observation and participation.

Swaminathan insists that this kind of work experience should be given academic recognition. A boy or girl who can run a simple project such as a small poultry unit or a kitchen garden, would have learnt responsibility, independence, and initiative as well as arithmetic, science, and economics and the conventional skills of reading and writing.

Women's empowerment is another of his crusades. "It's a man's world" is indeed very true, for women are regarded as the weaker sex all over the world – not just physically. Getting the top job is almost impossible for a woman, whether she is a scientist or a corporate executive or a university professor. As a geneticist, Swaminathan says he "really can't see why there should be such a difference just because X and Y chromosomes differ". Custom, tradition, ignorance, and prejudice have relegated women, especially poor and rural women, to the biological functions of child rearing and home making.

Swaminathan believes in the adage that "if we truly seek to help the poorest, we must serve women". Poor women run the family with virtually nothing by way of resources. If one member of the family has to starve, it appears to be taken for granted that it would be the mother. However, given the opportunity to fight against hunger and poverty, poor women use all the means at their command to do so. They have an intense desire to move up, they are hardworking, they are concerned about their human dignity, they are willing to make sacrifices for the good of their children.

Swaminathan advocates an integrated program of literacy and "techniracy" as the most meaningful way of achieving the empowerment of rural women in agriculture. Many of the back-breaking jobs such as paddy transplanting are done by women, he points out, remembering the women workers at his family's fields at Monkombu. They have to break off in between to breast-feed their babies, cook for the family, and tackle numerous other chores.

In his Foundation, scientists and sociologists have been looking at issues like the impact of new technology and new skills to help underprivileged women earn more income per hour, yet reduce the number of hours they work. If a woman earns less than a dollar digging and hewing for eight hours, she can earn four or five times that by producing hybrid seeds, working for fewer hours too. A team from the J.R.D. Tata Ecotechnology Centre, a part of MSSRF, has started a scheme in a small village in south India. Called the Seed Village Project, poor women are being taught to grow quality hybrid seeds of sunflower, cotton, okra, tomato, and eggplant. Appropriate market linkages have been provided. Traditional knowledge has been blended with modem science and it has been ensured that the project is ecologically viable and socially acceptable. The village community is committed to it. The enthusiasm and sense of achievement of the women must be seen to be believed. Actually, way back in 1964, Swaminathan had organized a similar project in the Jounti village near Delhi. The Jounti Seed Village helped to rapidly produce good quality seeds of high-yielding varieties of wheat, thereby hastening the onset of the wheat revolution in India.

Swaminathan has long recognized the gender component in biodiversity management. It is tribal and rural women who are responsible for sowing, hoeing, crop maintenance and harvesting, food processing and storage. They observe, select, and conserve plant seeds and species for future propagation. Research in one single village in Thailand revealed two hundred and thirty different plant species, many of which had been rescued by local women from a nearby forest before it was cleared. Women are very knowledgeable about indigenous plants, trees, and animals. They are the ones who use forest wealth, like fuel and fodder and fibers. The "medicinal plants" we talk about now have been grown and used for centuries all over the world by women. Swaminathan has been fighting for acknowledging and rewarding the conservation and enhancement of plant genetic resources by such marginalized folk.

Could it be that this sensitive man has so much empathy for children and women, because he lost his father when he was at a very vulnerable age and saw his young mother lead the limited life of a widow in the conservative India of more than six decades ago?

Honoring him with a Doctor of Science degree in June 2001, the University of Massachusetts, Boston, talked of "the magnificent inclusiveness of [your] concerns, by nation, socio-economic group, gender, inter-generational, and including both human and natural environments." ∎

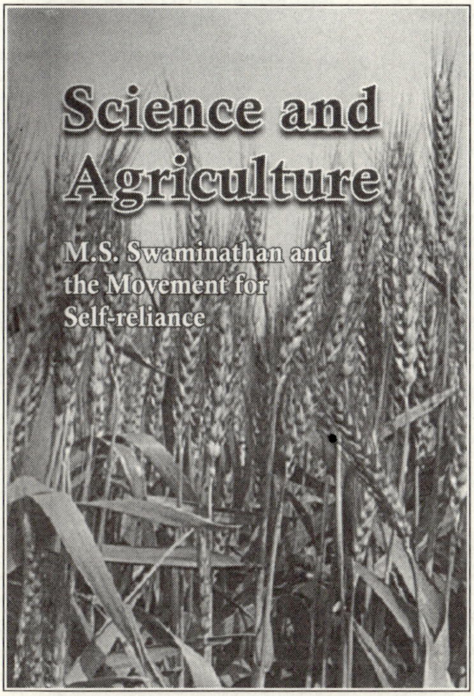

79. Title cover of book by S. Ramanujam et al (2002)

80. Title cover of biography by R. D. Iyer (2002)

81. Title cover of biography by Gita Gopalkrishnan (2002)

Index

Accelerated research, need for, 167-68, 200-02, 263
Afghanistan, 260-61, 439
Africa, 124, 149, 175, 176, 236, 283, 308, 321, 327, 336, 371, 373, 405, 425, 431, 435, 438, 439, 457, 460, 598
New Partnership for Africa's Development (NEPAD), 439
Sub-Saharan, 321, 460, 479,
shaping agricultural future of, 149, 167, 306, 428, 477-80
Agenda-21 of Rio de Janeiro, 311, 315, 317, 372, 567, 579
Agriculture, 133, 307, 323, 413, 438, 440, 448-49
and animal husbandry, 96, 283
and industry, 257, 258, 279, 323, 370, 380
and medicine, 380
challenges, 325-27
coastal, 443, 500
commercial, 258, 480
cultivation practices, 328, 333, 603
diversification, 173, 194, 211, 256, 312, 316, 329, 334, 342, 344-45, 348, 352, 399, 464, 489, 569, 603, 605
diets, 256, 324, 345
farming systems, 480, 483,
labor, 150, 204, 323,
rural employment, 323,
dryland research, 134, 166, 167
dual role, 323-24
exploitive, 441, 455
food security and employment, 320-37
future of, 90, 160, 171, 232, 241, 494, 542,
global, 149, 231, 332, 442, 502-04
growth of, 96, 324, 516, 589
harmony and natural balance, 146
increasing exports, 328-29
integration with marketing and processing, 140, 330
knowledge and skills sharing, 95-98, 182, 204, 293, 317
maximum potential, 329-30
Nobel prize in, 299-301
opportunities, 98. 199, 316, 327-30
political commitment, 190
private investment, increasing, 329
production, increasing, 133, 320-23, 327, 328, 352, 479
progress, 89, 96, 149, 152, 195, 270, 278, 433, 483
resource-based planning, shift from commodity-based to, 107, 189, 198, 214, 296, 329, 419, 495
risks in, 324, 441, 471
Spaceship Earth, on, 130-47, 189
systems of, 150, 333, 350, 438
traditional, 416, 441, 455
trends, 148, 324, 471
urban, 383
Agricultural assets, 139-40, 152-58, 162, 164,
Agricultural endowments, 109-10
Agricultural liabilities, 158-63
Agricultural machinery, 333
Agricultural revolution, 97, 106, 184, 472
Agricultural science & technology, 96-98, 163-64, 185, 249, 270, 276, 279, 290, 291, 294, 384
Agricultural transformation, 89-98, 189, 197, 274, 320, 330
Agricultural universities, 97, 215, 500
Agro-biodiversity, 263, 359, 361-68, 424, 428, 430, 431, 478, 568-74
and farmers' rights, 568-74
Agro-forestry, 236, 242, 249, 255, 270, 348, 443, 457, 477, 480, 484, 495, 500, 509, 555, 583, 593, 595, 598
Ahluwalia, Poonam, 610-11
Alberts, Bruce, 386, 577
Alien species invasion, 398, 400, 401-02,
All-India Coordinated Wheat Improvement Program, 30

Anaemia, Maternal, 462
Animal wealth, 157, 195
Anther culture, 284, 286-87, 593
Aquaculture, 104, 121, 122, 123, 138, 139, 158, 171, 292, 306, 329, 331, 336, 344, 345, 418, 443, 462, 494, 553
  coastal aquaculture, 123, 187, 199, 315, 471, 500
  inland, 103, 330, 334
Aquino, Corazon C., 308, 597
Arid areas, 133, 157
Asia, 138, 149, 153, 167, 196, 236, 239, 274, 275, 277, 291, 293, 297, 303, 309, 321, 327, 340, 375, 392, 405, 424, 428, 460-85, 502, 511, 517, 519, 523, 524, 529, 536, 537, 538, 546, 550, 551, 565, 569, 593, 611
The Asia Magazine, 31
Asian Development Bank, 530, 590, 600
Asset building, 230, 267, 346, 476, 518
Auden, W.H., poem on hunger, 449
Awareness, 93, 130, 131, 149, 162, 175, 182, 216, 226, 263, 267, 293, 310, 316, 341, 407, 438, 475, 502, 534, 574, 576, 617
Bangladesh, 229, 276, 295, 322, 342, 467, 482, 520, 522342
Banks and companies, role in agricultural development, 183-84, 198, 203, 245, 509
Basic human needs, 229, 373, 379, 382, 386, 387, 453-54, 519, 546, 579
  20th century balance sheet, 369-72, 453-54
Basic unity of the living world, 84
Benefit sharing, 424-27, 473, 501, 572
Bengal famine 1942-43, 97, 103, 589
Bhutan, 229, 237, 434
Biodiversity, 210, 237, 240, 245, 251-252, 263, 264, 311, 336-37, 341, 342, 343, 347, 349, 350, 357, 359, 361, 367, 377, 378, 391, 473
  and human security, 391-98
  gender dimensions of management, 356, 359-61, 399-400, 402
  mega-biodiversity areas, 368
  resources, 212
  rich but biotechnology poor countries, 378
Bioinformatics, 262, 263, 381
Biological diversity, Convention on (CBD), 356, 357, 358, 363, 365, 366, 367, 378, 379, 393, 398, 400, 402, 419, 422, 423, 424, 426, 428, 432, 473, 556, 570, 572, 574, 575
Biological heritage, 580
Biopartnerships, 379, 395, 396, 427, 503
Biosafety and Bioethics, 571-572
Biosphere management, 393-95
Biotechnology, 262, 278, 473, 283-90, 284, 204, 378
  and agriculture, 553
  global environmentally sustainable, 331
  medical, 262, 264
  parks, 254
  policy, 262
Body Mass Index, 461, 463
Borlaug, Norman E., 75, 77, 146-48, 291, 293, 299-301, 415, 457, 557, 558-59
Brain drain, 306
Brayne, F.L., 40
Breeding for soil tolerance and pest resistance, 146
Brown, Lester, 313, 569
Brundtland Commission, 255, 304, 309, 583
Brundtland, Gro Harlem, 304, 370, 554
Buddha, 292
Cancun Trade Talks
  collapse of, 480
  free and fair trade, 480-81
Carbon dioxide, 350, 371
Carson, Rachel, 85, 226
Centro Internacional de Mejoramiento de Maiz y Trigo Mexico (See CIMMYT)

Century of Hope – Era of Harmony with Nature and Freedom from Hunger, 576-79
Central America, 124, 157, 275
Cereal production, 283, 322, 409, 609
CGIAR, 307, 358, 365, 366, 368, 381, 425, 594
Chandler, R.F., 298
Children, 91, 97, 127, 182, 184, 185, 188, 194, 203, 205, 211, 232, 236, 238, 254, 275, 311, 316, 404, 413, 444, 446, 453, 459-60, 501, 535, 544, 561, 617
  blindness, 466, 544
  developing countries, 371, 383, 463, 467, 483,
  dwarfism/stunting, 462, 475
  low birth weights, 173, 236, 265, 267, 372, 383, 462, 577
  selling of children, 261
  Tagore's poem on, 238
  undernutrition/malnutrition, 229, 227, 462, 501
  for happiness, 236, 372
Chile, 67
China, 52, 121, 171, 295-98, 313, 324, 460, 461, 440, 489-90, 496-97, 563, 595
  China Council for International Cooperation on Environment and Development, 525-26
  Chinese Academy of Agricultural Sciences, 298
  Rural township enterprises program, 253
CIMMYT, 29, 30, 45, 47, 72, 407, 409, 411, 414, 416, 417, 420, 465, 511, 550
Clean drinking water, 166, 209, 232, 239, 240, 244, 266, 268, 270, 311, 317, 322, 334, 341, 352, 376, 377, 385, 429, 442, 445, 451, 449, 463, 469, 472, 474, 476, 477, 481, 492, 542, 555, 565, 569, 577, 58
Climate, 88, 145, 153, 226, 230, 329, 334, 354, 378, 397, 418, 509,
  change, 229, 258, 296, 349-50, 391, 420, 429, 442, 500,
  data, 509,
  climate induced disasters, 260,
  climate of violence, 252, 580,
  Global Climate Conference, 251, 371, 391, 471, 500-01, 543, 554, 579
  Virtual university for climate management, 259-60
Coastal eco-systems, 500
Comilla Project, 29
Commission on the Nutrition Challenges of the 21st Century, 459
Common heritage, change in concept of, 470
Community development, 151, 267-68, 518
Community Food Banks, 256-57, 261, 269, 443, 446-48, 469-70
Community Food and Water Security System, 428-29
Community Gene Banks, 363, 603
Community Grain Banks (CGBs), 245
Community nurseries, 166, 168, 197
Community nutrition security system, 467
Consultative Group for Freedom from Hunger, 269
Consultative Group on International Agricultural Research (See CGIAR)
Consumption, 371, 383, 407, 418, 435, 436, 440, 445, 462, 463, 464, 468, 470, 471, 490, 500, 546, 554
Convention on Biological Diversity (CBD), 252, 357, 366, 378-79, 424, 473, 570, 574-75, 580
Cooperation, 120, 189, 213, 218, 246, 263, 277, 294, 307, 315, 326, 335, 353, 371, 404, 420, 471, 472, 494, 495, 518, 529, 533, 534, 535, 543, 559, 560, 595,
  village level, 180-81
Cousteau, Jacques, 373
Cultural practices, changes in, 181-82
Crop planning, 100, 159, 167, 171, 179, 195
Crop substitution, 125-26
Debt and debt servicing burdens, 373

Decentralization, 266, 475, 508, 593
Demographic
divide, 503
goals, 318
profile, 211, 227, 231, 311
trends and famines, 312-13, 319, 369
Desert, reclamation of, 97-98, 154, 157, 164, 165-66, 210, 226, 233, 245, 251, 296, 311, 315, 327, 372m 382, 415, 441, 442, 446, 455, 469, 492, 553, 580, 598, 612
Developed countries, 140, 144, 179, 213, 296, 305, 307, 324, 339, 357, 370, 359, 423, 439, 440, 494, 496, 505, 532
Developing countries, 105, 131, 132, 150, 162, 179, 213, 214, 227, 233, 235, 241, 252, 254, 279, 283, 284, 293, 295, 296, 304, 305, 307, 311, 313, 321, 370, 440
urban bias, 306
Development, contemporary pathways to, 326, 328, 374, 575, 581, 588
Dietary components, fortification and synergy among, 465
DNA 81, 102, 192, 209, 255, 258, 262, 264, 288, 380, 381, 570, 583, 601
"Do" Ecology, 252-53, 494, 580-81
Drought, 100, 111, 120, 125, 133, 151, 154, 159, 163, 166, 167-68, 172, 176, 180, 221, 240, 244-45, 258, 263, 269, 281, 290, 295, 296, 299, 311, 327, 397, 428, 439, 446, 457, 469, 496, 509, 530, 548
Dry Farming Scheme, 166
Dryland research, 167-68
Dwarfing gene, 72-78, 120, 409, 413, 415, 515, 558, 559
Ecological seasons, strategies for, 170-71
Ecological security and sustainability, 214, 230, 243, 244, 311, 317, 318, 341, 342, 400, 448, 483, 553, 555, 577, 589,
food and, 244
systems, 257, 438, 440
Ecology
ecological agricultural systems, 347, 383, 489, 556
ecological fire, 296, 305
Ecology of hope, 232, 257-58, 380, 453, 579
Economic access to food, 249-50
Economic divide, 373, 453
Economic ecology, 164-65
Economics & human welfare, 147, 324
Economies, globalization of, 232-33
Economy through recycling, 176-77
Ecotechnology, 258-59, 339-42
and rural development, 562-65
JRD Tata Ecotechnology Center, 494
Revolution, 381-82
Education, 94-95, 182-83, 561
and agricultural advance, 95-96
for rural development, 204-05
higher and professional, 205-06, 434-36
innovations in, 206-09
Egypt, 119, 159, 180, 313, 520, 522, 597
Ehrlich, Paul, 312-13, 318
Embryo culture, 286
Employment, 151, 161, 205, 208, 215, 253, 255, 268, 274, 277, 279, 308, 320-27, 346
guarantee program, 310
off-farm, 385, 481, 484, 532, 581, 584,
on-farm, 329, 385, 481
policy, 193
rural, 127, 139, 216, 241, 242, 247, 248, 249, 253, 257, 316
self-employment, 90, 92, 95, 204, 205, 254, 312, 332, 375, 385, 564, 582, 594
unemployment/under-employment, 89, 101, 196, 203, 205, 210, 253, 256, 335, 581
youth, 254, 582, 610,
Youth Employment Summit, 611, 613
Environment and development, 226,

235, 296, 304, 311, 331, 541, 552, 554, 555
World Commission on, 554
Environmental degradation, 210, 339
poverty-induced, 371
Environmental policy, 146, 227, 232, 253
Environmental refugees, 442-43
Erdélyi, András, 609-10
Ethics and equity in collection and use of plant genetic resources, 356-68
Ethics, biodiversity, economic opportunities, 336-37
Euphenics, 85
Extension services, reorganizatioon of, 243-44, 482
Evergreen Revolution, 31, 52, 53, 148-49, 231, 240, 245-49, 270, 331, 342, 346-49, 384, 441, 475-79, 519-520, 525, 544-45, 567-68, 589
Famines and scarcities, 151, 304, 311, 312-13, 348, 471
FAO, 132, 151, 159, 251, 276, 283, 296, 308, 411, 413, 438, 440, 576
Anti-Hunger Program, 440
Global Plan to Action, 426-27
Farmers, 90, 99, 100, 101, 116, 123, 127, 139, 147, 149, 151-52, 169, 181, 184, 188, 189, 190, 196, 197, 198, 199, 202, 308, 497
contributions to conservation and development of genetic diversity, 422-33
insurance, 507
organizations, 332, 378, 386, 464, 489, 541, 545, 597
rights, 326, 330, 361, 382, 569-74, 356
subsidies to, 440
technical literacy/techniracy, 195, 278, 374, 407, 418, 427, 516, 538, 549, 559
Farming systems, 224, 270, 399, 477, 494, 495
Farming technologies, 376, 472-73
Female literacy and child health, 244, 311, 370, 461
Fertility projection, 369
Fertilizers, 72, 77, 96, 110, 124, 130, 135, 140-41, 144, 147, 152, 164, 168, 176, 181, 197, 200, 221, 229, 273, 313, 326, 328, 347, 413, 415
Fishes, 103, 135, 138-39, 171, 182, 187
increasing production of, 121-23, 133, 135, 157, 418, 471, 593
fish culture, 121, 138, 171, 494, 533, 558
Floods 100, 125, 130, 159, 160, 170, 178, 295, 296, 303
Flood Avoidance Code, 170, 180
Floristic diversity, 156-57
Flowering plants and trees, 87, 584-85
Fodder and Feed Banks, 256, 269
Food, 80, 84, 93, 96, 97, 99, 103, 105, 106, 126, 127, 133, 146, 162, 166, 173, 232, 266, 320, 459
access, 464
absorption, 464
and nutrition, 214
availability, 46
chains, 150-52, 175, 209
consumption, 109, 150
production, 79, 85, 110, 126, 138, 197, 211, 270
requirements, 133, 196-97, 216
security, 214, 229, 239-50, 257, 311-12, 442, 491, 553
technology, 177, 202
Food and Agriculture Organisation (See FAO)
Food for Nutrition Program, 312, 349, 446, 469
Food for Self-Reliance, 310
Food for Sustainable Development Initiative, 437-49
Food for Work Program, 256, 268, 438
Food guarantee scheme, 255, 257, 268, 476
Food insecurity, 215, 239, 459
Food Insecurity Atlas, 239, 265-69, 459
Food processing, 249, 257, 324, 347, 570

Food secure world, 338-55, 543
international agenda, 330-34
micro-level indicators, 334
nutrition security, 469-70
sustainable development, and 239-50
Ford Foundation, 29, 518, 551, 600
Forest Management Policy, 392
Fuller, Buckminster, 53-54
Gandhi, Indira, 179, 226, 234, 302, 407, 504, 517,529, 538, 543-44, 558, 600
Gandhi, M.K., 33, 34, 38, 39, 90, 91, 92, 96, 102, 227, 235, 252, 274, 275, 293, 495
GATT Agreement, 214
Gaud, William S., 300
Gender dimensions of biodiversity management, 402, 431, 574-76, 618
Gender equity, 226, 317, 339, 346, 352, 374-75, 420, 427, 446, 470, 473, 481, 523, 525, 541, 575, 577-78, 601, 613
Gender studies, 576, 613
Gene, 88, 192-93
management, integrated, 390-402, 603
mode of functioning of, 85, 86
Gene Banks, 259, 363, 368, 383, 395, 397, 400, 423, 431, 603
Gene fund, 343, 362, 366-67, 400, 425, 427, 573
Gene revolution, 257-58, 380-81, 419, 472, 542
Gene transmutation, 80
Genetic conservation, 222, 326, 343, 422-24, 572
Genetic diversity, 119, 356-57, 362-63, 407, 420, 466
farmers' contribution to, 422-33
Genetic engineering, 202, 264, 284, 289-90, 294, 297-98, 338, 339, 378, 401, 417, 424, 465, 479, 498-500, 538, 570, 571, 572, 574, 576, 578
Genetic Evaluation and Utilization (GEU) Program, 277
Genetic homogeneity, 399, 420, 433, 569
informing systems, 401
Genetic resources, 214, 264, 292-93, 307, 326, 342, 343, 356-68, 378, 382, 395, 397, 400, 402, 407, 419, 422, 424, 425, 426, 427
enhancement and sustainable use, 385
plant, 214, 222, 251, 356-68, 397, 422, 423, 424, 426, 427, 501, 521, 563, 564, 570, 579, 596, 619
sharing of, 357-58
Genetic Use Restriction Technologies (GURTs), 378
Genetic variability, 83, 115, 117, 156, 222, 226, 277, 295, 350, 397, 402, 406, 409, 569, 572
Genetically modified foods (See GMOs)
Genetics, 81, 86, 88, 117, 223
Genomics, 381
Genotype environment interactions, 114
Germplasm collections, 223
Ghana 462,
Globalization, 495-97
Global ecological balance, 145-46
Global Environmental Facility (GEF), 570, 580
Global partnership for development, 459
GM Crops , National Research Center for, 263-64
GM foods, 378
GMOs, 381, 419, 438, 465, 467, 473, 499-500, 544-45
Gobar gas plants, 176, 194
Good Weather Code, 166, 305
Gopalkrishnan, Gita, 611-19
Grain mountains and hungry millions, 443, 460
Grain reserves, 295, 304, 348, 428, 440, 557
Gram Sabha, 199, 236, 245, 255, 269, 492, 508
Grameen Bank, 342
Greed revolution, 331, 568
Green to Evergreen Revolution, 240, 399
Green Revolution, 348, 411, 441,

535, 569
abuse of term, 567-68
and ecotechnology, 470-71
and the small farmer, 273-74
benefits, 100, 231, 471-72, 544
concerns, 99, 101, 247, 564
fatigue of, 270, 347, 441-42, 477,
future, 470
global environmentally sustainable, 331
history of, 225, 347, 411, 415, 417, 570, 589, 607, 609
impact of, 231, 331, 376, 472
in India, 274, 301, 325, 417, 440, 494, 546, 555
process, 440
small farmer, and, 273, 399
sustainable, 224, 331
team efforts, 511
technologies, 224, 315, 419, 519, 530
Gross National Happiness, 237, 434
Guyana, 521, 570, 598-99, 615
Habitat destruction, 391-92
Haldane, J.B.S. 86-88, 562-63
experiments in India, 86-88
Hanson, Haldore, 29
Harmony and natural balance, 174-75
Herbert, George, poem on Peace, 85
High altitude research, 169-70
High Yielding Varieties (HYV), 415, 547, 557, 558, 559
HIV/AIDS, 370, 439, 459, 476, 485
Holistic Action Plan, 445-46
Household entitlement card, 267, 384, 445, 468, 475
Human Development, 317, 339, 346, 371, 375, 459, 502, 519, 522, 530
Resource, 193, 224
Human Development Reports (See UNDP Human Development Reports)
Human diversity, 252, 253, 434, 561, 580
Human Genome Project, 258, 380
Human environment, 179, 253
Human heritage, 252, 580
Hunger, 457-58, 482-83, 235
critical role of, 482
endemic, 255, 312
ending, 481-82, 100-101, 307, 443, 489-93,
freedom from hunger, 437-49,
olympics, 309
hot spots, 482
Interdisciplinary Spearhead Teams, 485
transient, 464-65
Hunger-Free Area Program (HFAP), 349, 351, 566
Hunger-Free District Program, 312
Hunger-Free India, 261, 265-270, 377, 474-77, 489-93
Agenda 2007, 477
National Food for Social Capital Program, 476
Tenth Five Year Plan (2002-2007), 266
Hunger-Free World, 261, 271-486, 338-55, 350-51, 404, 452-56; 489-93
Hybridization, 84, 96, 111, 113, 407,
New hybrids, 101, 111, 114
Hybrid rice, 216, 218, 276, 298, 498, 595
Hybrid seed production, 329, 330, 605
India/Indian agriculture, 86-88, 97, 101, 110, 126-28, 211, 224, 230, 236, 237, 249, 274, 313, 67, 86-88, 148-91, 220-23, 458, 460, 461, 472, 494, 557
agricultural research, 219, 226
food requirements, 196-97
Indian Agricultural Research Institute (IARI), 29, 41, 42, 44, 45, 46, 47, 55, 68, 71, 73, 82, 84, 99, 101, 121, 144, 184, 220, 413, 415, 519, 537, 545, 547, 557, 558, 587
Indian Council of Agricultural Research (ICAR), 222-25,
Indian Ocean, 122, 123, 135, 153-54, 158, 501
Individual and social goals, 188
Indonesia, 139, 149, 198, 216, 275, 277, 324, 520, 574
Informatics, 501

Bio-, 262, 263, 381
Information empowerment, 344, 384-86, 445, 468, 493
Information technology, 235, 243, 249, 259, 263, 267, 335, 341, 349, 380, 386, 405, 435, 453, 491, 497, 501, 535-56, 563, 564, 576, 577, 578
revolution, 259, 382
Insect resistance, 114
Institutional structures, 248, 249, 254, 375, 381, 447, 464, 484, 543, 556, 565, 578
Integrated Gene Management System, 392-93, 398, 401, 402, 603-04
Intellectual Property Rights (IPR), 210-17, 263, 337, 343, 358-59, 368, 370, 419, 435, 459, 471, 474, 497, 504
Intensive Agricultural District Program, 150
Intercropping, 351
International Agricultural Research Centers (IARCs), 307, 365, 368, 431, 531, 532, 550
International Alliance Against Hunger, 439, 443, 446, 447
International Center for Cooperation in Water Management, 370
International Center for Maize and Wheat Improvement, Mexico (See CIMMYT)
International Commission on Peace and Food (ICPF), 235, 320, 560
International Council of Scientific Unions (ICSU), 309, 375, 386
International Council for Research in Agro-forestry 595, 597-98
International Crops Research Institute, 126, 167
International Food Policy Research Institute (IFPRI), 232, 478
International Food Security System, 197, 198
International Fund for Agricultural Development (IFAD), 463, 600
International Institute of Tropical Agriculture (IITA), 478
International Peace Research Institute, 235, 387, 437, 453
International Rice Research Institute (IRRI), 78, 216, 276, 277, 286
International Treaty on Plant Genetic Resources, 251, 579
Iron deficiency anaemia, 461, 466
Iyer, R.D., 55, 5891-6069
Japan, 149, 505-06
Jawaharlal Nehru University, 500
Kerala, 87, 89, 156, 169, 198, 204, 317, 461
Kesavan, P.C., 55
Khush, Gurdev S., 216
Know-how to Do-how, 442-48
Knowledge, traditional and wisdom, 258, 349, 357, 364, 379, 380, 382, 391, 395,424, 427, 470, 473, 478, 491, 501, 542, 572-73, 603, 618
Knowledge Century, 370
Knowledge System for Sustainable Food and Livelihood Security, 382-84
Korea, 281, 322, 324, 409, 490
Krishi Vigyan Kendras, 31, 186, 187, 205, 207, 521, 578, 593
Lab-to-Land and Land-to-Lab programs, 30
Land use, 110, 126, 139-40, 169, 241, 169, 242, 362, 377, 443, 496, 506, 598
boards, 212, 216, 242, 243, 508
Leadership, 293, 299
agricultural, 222, 415, 514
community, 99, 100, 102, 204, 561, 527, 548, 549, 561
international, 543, 549,
political, 529, 538, 561,
professional, 216, 261, 541
Learning by doing, 186, 236, 405, 428, 479
Learning revolution, 89-98, 341, 382
Lederberg, J., 32, 40, 43, 85, 511-12

Legislative measures, 209, 427
Lifecycle approach, 256, 267
Literacy, 89, 101, 162, 185, 206, 227, 230, 265, 312, 370, 371, 474, 578, 593
    female, 244, 311, 461
    nutritional, 267, 475
    quality, 257, 270, 477
    technical/techniracy, 128, 148, 186, 189, 190, 540, 593, 618
    trade, 480, 484, 498
    water, 477
Madagascar, 597
Madras Declaration, 570-71
Maize, 84, 96, 97, 118, 124, 194, 245, 327, 407, 465, 472
    dwarf varieties, 172
    hybrid, 407
    Quality Protein Maize (QPM), 465
Malaysia, 171, 413, 422, 507, 558
Malnutrition, 96, 98, 101, 109, 146, 173, 233, 265, 267, 274, 321, 352, 370, 371, 383, 384-85, 439, 446, 459, 474, 475, 535, 568, 462
    child, 501
    fetal, 236
    female, 523
    Protein-calorie/energy malnutrition (PEM), 101, 148, 162, 256, 460
Malthus, Thomas Robert, 79, 103, 311, 569
Management technology, 349, 501, 539
Mangrove ecosystems, 602
Manpower planning, 90
Marine fisheries, 122, 135, 157-58
    Demersal fisheries, 137
    Pelagic fisheries, 137
Marketing, poor quality, 496
Maternal mortality, 265, 311, 377, 385, 460, 462, 474
Maternal mal/under-nutrition, 372, 463, 467, 508, 617
Marx, Karl, 96
Mendel, Gregor, 79, 472, 96, 99, 99, 406
Mendelian breeding, 406-21, 465
Mendelian genetics, 406-09, 419
Mendelian to Molecular breeding, transition from, 417-18
Meristem culture, 284-85
Mexico, 72-78, 100, 222, 411, 537, 547, 553
Milk, 97, 106, 109, 490, 505
    increasing production of, 121
Molecular breeding, 262, 264, 381, 399-400, 402, 407, 417, 420, 465, 467
Monoclonal antibodies, 284, 288-89
Mortality rates, 265, 311, 377, 460, 461, 462, 474
M.S. Swaminathan Research Foundation (see MSSRF)
MSSRF, 239, 251, 260, 312, 323, 335, 340, 342-46, 356, 357, 375, 376, 381, 386, 393, 442, 450, 474, 483, 536, 539, 555-56, 563-64, 569, 577-78, 581, 599-606
Muller, H.J., 406-97
Multiple cropping, 97, 110, 124, 145, 153, 161, 168-69, 174, 187, 200, 283-84
Mutation breeding, 82, 83,84, 126, 407, 591, 607
    mutations affecting single character, 82
    macro-mutation, 82-83
    non-lethal mutations, 83
    systemic mutations, 83
Mutagenesis, 84, 406, 407, 547, 592
Myanmar (Burma), 154, 218, 219, 490, 596
Myrdal, Gunnar, 105
National Agricultural Policy, 491
National Alliance Against Hunger, 48, 447
National Biotech Policy, 262-64
National Commission for Women, 269, 477
National Commission on Agriculture, 107, 109, 110, 139, 152, 168, 178, 195, 196, 197, 198, 203
National Committee on Science and Technology, 109, 173, 192
National Dairy Development Board, 215, 248, 490

National Dairy Research Institute, 121, 175
National Demonstration Program, 302, 415, 559
National Development Council, 268, 269, 349
National Food Security, 212, 295
National Food Security Act, 212
National Institute of Nutrition, Hyderabad, 174, 202
National Population and Social Development Commission, 317
Natural resources, management, 391, 495
Nehru, Jawaharlal, 42, 46, 151, 217, 556
New Partnership for Africa's Development, 440
  new strains, difficulties encountered with, 113
New Zealand, 505
Nigeria, 598
NGOs, 438, 447, 541
Nutrition (See also Malnutrition), 97, 103, 174, 183, 196, 202, 204, 205, 240, 245, 256, 257, 266, 269, 274, 277, 280, 287, 312, 325, 334, 349, 351, 353, 372, 385, 443, 462, 536, 555
  animal, 253, 465, 581
  benefits, 544
  child, 186
  gardens, 468
  iron deficiency, 466
  Participatory Nutrition Improvement Project (PNP), 462
  plant, 413, 558
  security, soil, 72,
  vitamin A deficiency, 466
Nutritional Divide, 459-60, 503
Nutritional literacy movement, 267
Nutritional profile, 460-61
Nutritional quality,
  genetic enrichment of, 114-15, 121, 126, 153, 172, 173, 465, 467
Nutritional security, 214, 267, 298, 305, 444-46, 447, 463; 467-70, 476, 569
  high synergy nutritional system, 46
  holistic action plan to 468-69
  life cycle approach to, 467-69
  sustainable, 467-69
Nyerere, Julius, 306, 598
OECD countries, 259, 331, 439, 440
Oilseeds, 120, 133, 174, 248, 497
On-farm conservation, 358, 359, 361, 366, 368, 392, 426, 395, 401, 423, 425, 428, 604
  agro-biodiversity, 361-64
Paddock, Paul and William, 105
Pakistan, 283, 290, 596
Pal, Benjamin Peary, 68. 218-25, 587-89
Pathogen, 97, 124, 126, 173-74, 179
  outwitting the, 173
Peace, 85
  and equitable development, 560-62
  disarmament and development, 234
  George Herbert's poem on, 85
  International Decade for a Culture of Peace and Non-violence, 236
  International Peace Research Institute, Oslo, 234-35
People power, 455, 463
  in nutrition, 462
Pests, 96, 111, 123, 124,
  and diseases, 123-24
  biological control, 175-76
Pesticides, 97, 144-45, 149, 175, 181, 183, 201, 334, 376, 415, 441, 472, 496, 519, 520, 530, 568, 570, 592
  bio-pesticides, 213, 249, 346, 262, 326, 331, 484, 502, 553
Philippines, 275, 291, 292, 293, 308, 466, 529, 597, 599
Physiography,
  diversity in, 153
Plant architecture, 117, 120
  changing, 172-73,
  plant breeding research, 307
Plant genetic resources, ethics and equity in collection and use of, 356-68
Plant health, 27

Plant hybridization, 79
Plant protection, 111, 113
Plant Variety Protection Acts, 357
Pollution, 141, 142, 158
air, 145, 146, 193, 296
control, 283
genetic, 264
India, 160
management, 377
pesticide, 144-45, 341
poor countries, 279
soil, 502
water, 139, 296, 305, 327, 419, 442
Population and social development, 317, 319
Population, environment and food security, 311-19
Population growth, 79, 101, 103, 107, 126, 202-04, 369, 148, 162, 163, 202, 211, 317, 318, 321, 325, 329, 339, 343, 349, 353, 419, 471, 550, 569
and food supply, 133, 296-97, 319-20
Population policy, 229, 316, 317-19, 345
Post-harvest losses, 177-78, 257
Potato, 67-71, 466, 472
Indian varieties, 70-71
Potrykus, Ingo, 466
Poverty, 148, 450, 457, 463, 341, 369-70, 494-95
alleviation, 237, 304, 372, 373
gender dimensions of, 374
Precision farming / breeding, 241, 472
Prosperity, policy for, 199-200
Protein content, 82, 117, 284, 287
Protoplast culture and fusion, 287-88
Proteomics, 259, 262, 263, 435
Pugwash Conferences on Science and World Affairs, 450-56
Pulse crops, 124-26, 174, 497
Punjab, 89, 273, 313, 560
Raman, C.V., 49-50, 608-09
Ramana, 38-39
Reaching the unreached, 374, 501,545
Relay cropping, 100
Research,
accelerated, 200-02
breeding procedures, 119-21
cropping strategies, 126
Resource mobilization, 448
Rice, 86, 276
Beta-carotene rich, 465-66
biotechnology, 599
genetic engineering, 298
genome of, 381
golden, 465
research in China, 295-98
Rich-Poor divide, 231, 371
Rilke, Maria Ranier, 56, 436
Rockefeller Foundation, 500, 596
Rome Declaration on World Food Security, 352-53
Rural development, 183, 185, 186, 227, 246, 277, 293, 312, 313, 353, 395, 440, 477, 483, 494, 538, 555, 563, 588, 601, 604
assets & liabilities, 194-96
education for, 204
science and integrated, 192-209
Rural Township Enterprises Program, China, 253, 581
Rural-Urban relationship, 194
Russia, 67, 68, 91, 188, 598
Sacred groves, 392
Sambasivan, Monkombu Krishna, 37-38
Sanches, Pedro, 457
Sasakawa Peace Foundation, 565, 600
Science and agriculture, 290, 587
Science and society, 241, 368, 398, 535, 567
Science for Basic Human Needs Movement, 386-87
Science and technology, 235, 279, 369, 372, 563, 565, 568, 600
basic human needs, 369-89, 453-55
demonstration programs, 484-85
distinction between, 372
food and water security, and, 376
proprietary, 370, 419-20
public funds, 370
revolution, 257, 380

scientific resources, transfer of, 307
social responsibility of , 206
sustainable development, 226-33
with humanism, 388
Seed production, 109, 166, 174, 367, 427, 597
hybrid, 329, 330, 605
Semi-arid areas, 105, 125, 145, 152, 166, 216
Semi-dwarf varieties, 222, 223, 275-76, 415, 558, 559
Sen, Amartya, 89, 98, 309, 311
Shifting cultivation, 96, 159, 160, 169, 195, 398
Small farm agriculture, 278, 312
Small farm, technologies for, 297
Social and economic progress, 434-36
Social inequity, 235, 372, 387, 404, 453, 561
Socio-Demographic charters, 316-17
Soil, 140-44, 457,
desert soil, 126, 156, 165
Soil conservation, 160-62
research & demonstration centers, 167
Soil fertility, 83, 100, 139, 154, 160, 176, 200, 222, 295, 297, 313, 344, 384, 411, 415
Soil groups, 136, 154-56
Somatic cell culture, 285
Sorghum, 113-14
South Africa, 67, 438
South Asia, 227, 239, 263, 274, 285, 286, 295, 322, 445, 461, 467, 523
South Asian Association for Regional Cooperation (SAARC), 263
South-east Asian Nations (ASEAN), 263
Spaceship Earth, 162, 494
Speth, James Gustave, 562
Sri Lanka, 460, 461, 462, 596
Stability of performance, 114, 126, 416
Standards of living, 146, 150, 179, 253, 373, 494, 581
Subramaniam, C., 29, 45, 392, 557
Sugarcane, 111, 115-16, 120, 152, 180, 198-99, 285, 324
Sunlight, utilization of , 97, 139, 145-46, 147, 148, 150, 152, 153, 161, 163, 193, 279, 390, 558
Sustainable agriculture, 334, 335-36, 401-02, 406, 450, 453, 479, 553-55; 565-66
biovillages and, 335-36
and Evergreen Revolution, 567-68
and food security, 565-66
global environmentally sustainable, 331
low-input, 334
Sustainable and humanistic civilization, 578, 582
Sustainable development, 226-33, 241, 251, 261
Sustainable food security, 240, 257, 315-16, 351-53
Sustainable human development, 502
Sustainable human future, 251-61
Sustainable livelihoods, 254, 330-42, 371, 462
Sustainable nutrition security, 310
Swaminathan, M.S.
Agriculture as applied ecology, 531, 615
"A rare breed of man", 532-33
Aziz, Sartaj,on 27-28
Citations: Award for international cooperation on environment and development, 551-52; Award for outstanding contributions to the integration of women in development, 550-51
AWID honors Swaminathan, 531-32; Borlaug Award, 548; First World Food Prize citation by Norman Borlaug, 299-301, 533-34; Franklin Delano Roosevelt Four Freedoms Medal, 403-04; Indian National Science Academy Silver Jubilee Commemoration Medal, 547-48; Indira Gandhi Award, 234; Magsaysay Award, 99-100; Planet and Humanity Medal, 552; Tyler Prize, 551; Volvo Environment Prize, 539-41

childhood of MSS, 607
concern for children, 616-17
daughters of MSS, 302
Dil, Anwar, on, 29-56
education, of, 545, 558
environmental protection, 616
"Face of food security", 545
father of MSS, 302
global outreach of MSS, 594-99
"God father of millions of Indian
farmers", 543
Green Revolution, Father of, 511;
hero of, 509; pioneer of, 535-36
Honorary doctorate degrees
citations: Asian Institute of
Technology, Thailand, 549-
50; University of
Massachusetts, 552;
University of the
Philippines, 549; University
of Wisconsin, 548
human development, 522
humanitarian causes, 523-24
Indira Gandhi and, 30, 47, 48, 53
institution builder, 591-606,
IARI, at, 591-93
ICAR, at, 593-94
Indian Science Congress,
revitalizing, 594
IRRI, as Director General of,
529-30, 596-98
leader, as, 100, 543-44
Livelihood Box, 506
"Living legend", 512
"Modern Gandhi", 523, 541-43,
Inspired by Gandhi, 534-35, 606
mother of MSS, 39-40, 302
motivation, 543
Planning Commission of India,
at, 594
Pal, B.P., 302, 587-89
person, as, 513, 517, 606-09,
611-12, 613-14
personal philosophy of MSS,
303-10, 606-09
Prefaces, Forewords, and
Introductions to books by
MSS: *Biotechnology in
Agriculture: A Dialogue,
1991*, 553;
*From Stockholm to Rio de
Janeiro The Road to
Sustainable Development,
1991,* 553-55;
*Information Technology: A
Dialogue, 1993,* 555-56;
*Wheat Revolution: A
Dialogue,1993,* 313, 555-60;
*Uncommon Opportunities,
1994,* 560-62;
*Ecotechnology and Rural
Employment,1994,* 562-65;
*Sustainable Agriculture:
Towards Food Security,
1996,* 565-67;
*Sustainable Agriculture:
Towards an Evergreen
Revolution, 1996,* 567-68;
*Agrobiodiversity and
Farmers' Rights, 1996,* 568-
74;
*Gender Dimensions in
Biodiversity Management,
1998,* 574-76;
*A Century of Hope: Towards
an Era of Harmony with
Nature and Freedom from
Hunger, 1999,* 576-79;
*Rio de Janeiro to
Johannesburg, 2002,* 579-
83;
*Groves of Beauty and
Plenty:An Atlas of Major
Flowering Trees in India,*
2003, 584-85;
Books on MSS: *Science and
Agriculture, M.S.
Swaminathan and the
Movements for Self-reliance,*
by S. Ramanujam, et al.,
587-88;
*Man Who Harvests
Sunshine: The Modern
Gandhi: M.S. Swaminathan,*
by Erdélyi, András, 609-10;
*M.S. Swaminathan: One
man's Quest for A Hunger-
Free World, by Gita
Gopalkrishnan,* 610-19;
Scientist and Humanist: M.S.

Swaminathan, by R.D. Iyer, 589-609
Interviews: 489-510
Tributes to MSS: Sartaj Aziz, 27-28, 524; Anwar Dil, 29-56; Norman Borlaug, 302, 303, 511, 513-14; Ronald Reagan, 511; Joshua Lederberg, 511-12; Frank Press, 512; Jimmy Carter, 512; Javier Perez de Cueller, 512; R. Venkataraman, 513; K.R. Narayanan, 513; Narasimha Rao, 513; Abdus Salam, 513; Jacques Diouf, 514; James Gustave Speth, 514; Elizabeth Dowdeswell, 515; Sir Michael Atiyah, 515; Wang Lianzheng, 515; N.S. Randhawa, 516; R.S. Paroda, 516; Manju Sharma, 517; Board of Trustees of The Ford Foundation, 518; Norman Myers, 518-24; Sartaj Aziz, 524; Wakako Hironaka, 524; Ismail Serageldin, 524-25; Queen Noor of Jordan, 525; R. Martin Lees, 525-27
Excerpts from Press Coverage: 529-46
pro-nature, pro-poor, pro-woman development, 492, 523,
president of Pugwash Conferences on Science and World Affairs, 546, 450-56
relatives of MSS, 302
Russell-Einstein Manifesto, on, 451
scientist, as, 99, 511-27, 547-48, 588, 589-90, 609
seed village in Delhi State, 302
Sen, Amartya, 309
student, as, 403
Subramaniam, C., 302
teacher, as, 94,99, 515-16, 607
Time Magazine award, 502, 524, 538-39
transformer of Indian agriculture, 543-44
uncles of MSS, 38, 302
women's empowerment, 617-18
Swaminathan, K., 38
Swaminathan, Mina, 55, 302, 403, 469, 501, 518, 613, 614
Symbiosis (see Synergy)
Symphonic agricultural system, 297, 305
Synergy/Symbiosis, 88, 125, 140, 147, 171-72, 187-88, 197-99
Tagore, Rabindranath, 238, 435
Taiwan, 78, 123, 197, 322, 323
Tamil Nadu, 137, 154, 156. 157, 228, 256, 351, 377, 413, 429-31, 444, 500, 522, 545, 558, 590, 594, 600, 602, 603, 604, 605
Technical literacy/Techniracy, 148, 186-87, 189, 190, 593, 618
Technology, 178-80, 274, 297, 459, 504
access to, disparities in, 370
and equity, 578
for the poor farmer, 188-90
mission approach 237
mismatch between, 491
production technology and post-harvest technology, 163, 197, 202, 213, 233, 247, 249, 274, 315, 345, 350, 384, 478, 481, 483, 490, 491, 495, 507, 532, 596
spatial technology, 508
transferring to the masses, 190-91
wheat, 300
Technological and economic apartheid, 374, 578
Technology Achievement Index, 459
Thailand, 324, 327, 329, 445, 462, 596
Tissue culture, 284
Third world, 275-79, 299, 305, 308, 451, 479, 480, 549, 551, 609, 610
Trade barriers, elimination of, 330-31
Trade-related Intellectual Property Rights (TRIPS), 233,263, 379, 424-25, 427, 471, 473-74
Traditional knowledge and wisdom, 258, 349, 357, 364, 379, 380, 382, 391, 395,424, 427, 470,

473, 478, 491, 501, 520, 542, 572-73, 603, 618
Trees, 87, 88, 139, 143, 157, 165, 166, 169, 176, 202, 285, 292, 366-67, 392, 498, 584-85, 615
Turkey, 72, 409
Twentieth century, balance sheet, 339-40, 369, 402, 407, 576
Twenty-first century, 493, 569, 579
UN Commission on Nutrition, 371, 501
UN Conference on Environment and Development, 226, 298, 311, 315, 338, 390, 470, 526, 554, 555
UN Conference on Human Development, 554
UN Conference on Human Environment, 567
UN Development Force for Food Deficit Regions, 332
UN Environment Programme (UNEP), 251, 367, 391, 579
UN Law of the Sea Convention, 227, 233, 391
UNDP Human Development Reports, 45, 252, 318, 319, 349, 354, 374, 459, 461, 492, 564, 567
UN Millenium Development Goals, 459, 483, 485
United Nations Environmental Protection Agency (UNEP), 251, 256, 364, 367, 368
UN Summit on Sustainable Development, 251, 579
UNESCO, 207, 335, 340, 363, 365, 373, 374, 377, 393
UNICEF, 207, 309, 310, 373, 462, 568, 617
United States Agency for International Development (USAID), 301, 597
Universal Declaration on the Human Genome and Human Rights, 369, 499
Urban slums, 126, 205, 232, 317, 341, 373, 443
US Department of Agriculture, 222, 347, 415
USA, 72, 85, 135, 212, 237, 374, 406, 409, 416, 417, 503, 435, 496, 501, 563, 592, 603
USSR, 87, 89, 107, 121, 220, 327, 598
Under-nourishment (See Malnutrition)
Vasal, S., 465
Vavilov, N.I., 423, 571, 595-96
Vavilov Agricultural Academy, Moscow, 598
Vavilov wheat, 222, 223
Venkataraman, R., 589-91
Vietnam, 520, 595-96
Village Seed Bank, 428, 429, 604
Villegas, Evangelina, 465
Vogel, Orville, 415, 558
War, 234, 560
    Dwight D. Eisenhower on, 234, 387
Ward, Barbara & Rene Dubois, 131
Wasteland development, 242, 247
Water,
    conflicts, 370
    conservation, 333-34
    contaminated, 370
    management of, 100, 595
    requirements, 144-45
    sewerage, 370
    supplies, 370
    table, 237
Weather, Code, Good, 166, 509
Wheat, 29, 30, 36, 41, 44, 45, 47, 55, 133, 472
    breeders, challenges to, 418-20
    Durum, 416
    global demand, 407, 409
    Impact of dwarfing genes on, 72-78
    Miracle or Mummy wheat, 83
    revolution, 556-60
    varieties, 222
Whole Lifecycle Approach, 256, 444, 446, 447, 467, 468, 481
Wilson, Edward O., 52, 436, 441, 572
Women, 95, 185, 203, 204, 205,230, 231, 232, 236, 237, 242, 244, 253, 254, 256, 261, 266, 269, 279, 298, 319, 341, 443, 476, 530
    elimination of all forms of discrimination, against (CEDAW), 435

empowerment of, 210, 330, 375, 479, 482, 523, 567, 614, 618
farmers, 308, 344, 356, 357, 366, 376, 382, 402, 418, 422, 429, 431, 442, 478, 481, 520, 551
feminization of poverty, 495
health, 317, 386, 536, 540
poverty, 375, 439, 501, 577
pro-women, 346, 362, 492, 502, 564, 589, 613
rural, 214, 259, 330, 335, 343, 572, 575
science and technology, 374-75, 503, 532, 540, 551, 555, 604, 605
Women's Biotechnology Park, 601, 602
Women's theater project, 602
World Conference on Women, 374-75
Work projects for students, 184-86
World Bank, 149, 369, 457, 460, 524-25, 554
World Commission on Environment and Development, 296, 304, 554
World food model, 332-33
World Food Prize, 299-301
World Food Programme (WFP), 265, 442, 459, 481-82
World Food Summit, 255, 264, 353, 438
World Health Organization (WHO), 370
World Institute for Development Economics Research (WIDER), 308-09
World Trade Agreement, 214, 241, 242, 367, 504
Impact of WTO Pact, 504-08
World Trade Organization (WTO), 263, 427, 496
Worth of the individual human being, 93
Yield potential, increasing of, 70, 75, 77, 78, 100, 103, 110-11, 116, 121, 133, 140, 171, 172, 196, 197, 216, 222, 297, 300, 347, 361, 409-20
limits to yield production, 115, 117
Yoshida, Shouichi, 280-82
Youth, 90, 92-93, 94, 206, 253, 254, 293, 581, 582, 601, 603
educated, 306
employment, 254, 254, 582
in farming. 211, 224, 293, 346, 478, 489
nutrition, 444
poverty, 162
rural, 207, 345
Zakir Husain, 89, 91
Zero Hunger Program, Brazil, 457, 481